Decision Analysis, Location Models, and Scheduling Problems

Springer
*Berlin
Heidelberg
New York
Hong Kong
London
Milan
Paris
Tokyo*

H. A. Eiselt · C.-L. Sandblom

Decision Analysis, Location Models, and Scheduling Problems

With Contributions by

J. Blazewicz, R. L. Church, A. Drexl,
G. Finke, C. S. ReVelle

With 147 Figures
and 48 Tables

Springer

Prof. H. A. Eiselt
University of New Brunswick
Faculty of Administration
P.O. Box 4400
Fredericton, NB E3B 5A3
Canada

Prof. C.-L. Sandblom
Dalhousie University
Department of Industrial Engineering
P.O. Box 1000
Halifax, B3J 2X4, Nova Scotia
Canada

ISBN 3-540-40338-8 Springer-Verlag Berlin Heidelberg New York

Cataloging-in-Publication Data applied for
A catalog record for this book is available from the Library of Congress.
Bibliographic information published by Die Deutsche Bibliothek
Die Deutsche Bibliothek lists this publication in the Deutsche Nationalbibliografie; detailed bibliographic data is available in the Internet at <http://dnb.ddb.de>.

This work is subject to copyright. All rights are reserved, whether the whole or part of the material is concerned, specifically the rights of translation, reprinting, reuse of illustrations, recitation, broadcasting, reproduction on microfilm or in any other way, and storage in data banks. Duplication of this publication or parts thereof is permitted only under the provisions of the German Copyright Law of September 9, 1965, in its current version, and permission for use must always be obtained from Springer-Verlag. Violations are liable for prosecution under the German Copyright Law.

Springer-Verlag Berlin Heidelberg New York
a member of BertelsmannSpringer Science+Business Media GmbH

http://www.springer.de

© Springer-Verlag Berlin · Heidelberg 2004
Printed in Germany

The use of general descriptive names, registered names, trademarks, etc. in this publication does not imply, even in the absence of a specific statement, that such names are exempt from the relevant protective laws and regulations and therefore free for general use.

Hardcover-Design: Erich Kirchner, Heidelberg

SPIN 10935781   42/3130/DK – 5 4 3 2 1 0 – Printed on acid-free paper

"How much, thought I, has each of these volumes, now thrust aside with such indifference, cost some aching head! how many weary days; how many sleepless nights! How have their authors buried themselves in the solitude of cells and cloisters; shut themselves up from the face of man, and still more blessed face of nature; and devoted themselves to painful research and intense reflection! And all for what? to occupy an inch of dusty shelf—to have the title of their works read now and then in a future age, by some drowsy churchman or casual straggler like myself; and in another age to be lost, even to remembrance. Such is the amount of this boasted immortality. A mere temporary rumour, a local sound; like the tone of that bell which has just tolled among these towers, filling the ear for a moment—lingering transiently in echo—and then passing away like a thing that was not!"

Washington Irving, (1783-1859) "The Mutability of Literature", in *The Sketch Book of Geoffrey Crayon, Gent*, E.P. Dutton & Co., New York, 1906 (first published in 1819).

PREFACE

The purpose of this book is to provide readers with an introduction to the fields of decision making, location analysis, and project and machine scheduling. The combination of these topics is not an accident: decision analysis can be used to investigate decision scenarios in general, location analysis is one of the prime examples of decision making on the strategic level, project scheduling is typically concerned with decision making on the tactical level, and machine scheduling deals with decision making on the operational level. Some of the chapters were originally contributed by different authors, and we have made every attempt to unify the notation, style, and, most importantly, the level of the exposition. Similar to our book on Integer Programming and Network Models (Eiselt and Sandblom, 2000), the emphasis of this volume is on models rather than solution methods. This is particularly important in a book that purports to promote the science of decision making. As such, advanced undergraduate and graduate students, as well as practitioners, will find this volume beneficial.

While different authors prefer different degrees of mathematical sophistication, we have made every possible attempt to unify the approaches, provide clear explanations, and make this volume accessible to as many readers as possible. What is required in terms of prerequisites is collected in the introduction. Here, we provide some basics of algorithms and computational complexity, matrix algebra, graphs and networks, linear and integer optimization, and statistics. This part is not meant as a substitute for a study of these fields, it is merely included here as support methodology.

This is a truly international work: parts of the manuscript were written in Poland, France, Germany, Sweden, the United States, and Canada, most editing was done in New Brunswick and Nova Scotia, Canada, but parts were also completed in Norrköping, Sweden and Cairo, Egypt.

Our thanks go to Springer-Verlag, and particularly Dr. Müller, whose patience and support helped the project finally come to fruition. We would like to express our thanks to Zhijian Yin who provided us with many of the figures for the book, and and #13, a.k.a. Khaled Abou-Zied, who prepared the index and the bibliography, made last minute corrections, and acted as gofer. Many thanks are also due to Lisa Shepard from #1 Copyediting Services for her meticulous work, particularly considering the tight time frame. We are also grateful to colleagues and students who have helped with the book.

H.A. Eiselt
C.-L. Sandblom

CONTENTS

Introduction	1
Notation	1
Support Methodology	3
a. Algorithms and Computational Complexity	3
b. Matrix Algebra	5
c. Graphs and Networks	9
d. Linear and Integer Optimization	11
e. Statistics	13

Part I: Analysis of Decision Making	19
1 Multicriteria Decision Making	23
1.1 Vector Optimization	24
1.2 Basic Ideas of Multicriteria Decision Making	29
1.3 Reference Point Methods	37
1.4 Data Envelopment Analysis	40
1.5 Preference Cones	43
1.6 Multiattribute Value Functions	46
1.7 Outranking Methods	50
1.8 Methods Allowing Inconsistent Estimates	61
2 Games Against Nature	73
2.1 Elements of Games Against Nature	73
2.1.1 Basic Components	73
2.1.2 Lotteries and Certainty Equivalents	76
2.1.3 Visualizations of the Structure of Decision Problems	81
2.2 Rules for Decision Making Under Uncertainty and Risk	87
2.3 Multi-Stage Decisions and the Value of Information	99

3 Game Theory — 111

- 3.1 Features of Game Theory — 112
  - 3.1.1 Elements and Representations of Games — 112
  - 3.1.2 Solution Concepts — 116
- 3.2 Two-Person Zero-Sum Games — 122
- 3.3 Extensions — 133
  - 3.3.1 Bimatrix Games — 133
  - 3.3.2 Multi-Stage Games — 140
  - 3.3.3 n-Person Games — 143

Part II: Location and Layout Decisions — 151

1 Fundamentals of Location and Layout Problems — 153

- 1.1 The Nature of Location Problems — 153
- 1.2 The History of Location Models — 155
- 1.3 The Major Elements of Location Problems — 157
- 1.4 Applications of Location Problems — 164

2 Location Models on Networks — 169

- 2.1 Covering Models — 170
  - 2.1.1 The Location Set Covering Problem — 171
  - 2.1.2 The Maximal Covering Location Problem — 175
- 2.2 Center Problems — 178
  - 2.2.1 1-Center Problems — 179
  - 2.2.2 p-Center Problems — 186
- 2.3 Median Problems — 188
  - 2.3.1 Basic Results and Formulation of the Problem — 189
  - 2.3.2 1-Median Problems — 192
  - 2.3.3 p-Medians Problems — 194
- 2.4 Simple and Capacitated Plant Location Problems — 205
- 2.5 An Application of the Capacitated Facility Location Problem — 208

3 Continuous Location Models — 211

- 3.1 Covering Problems — 212
- 3.2 Single-Facility Minimax Problems — 214
- 3.3 Minisum Problems — 220
  - 3.3.1 Single-Facility Problems — 220
  - 3.3.2 Multi-Facility Problems — 228

4 Other Location Models — 237

4.1 The Location of Undesirable Facilities — 237
4.2 *p*-Dispersion Problems — 243
4.3 Location Models with "Equity" Objectives — 244
4.4 Hub Location Problems — 247
4.5 Competitive Location Problems — 248
4.6 Locating Extensive Facilities and Routing in Irregular Spaces — 252

5 Layout Models — 255

5.1 Facility Layout Planning — 256
5.2 Formulations of the Basic Layout Problem — 260
5.3 Special Cases of the Quadratic Assignment Problem — 266
   5.3.1 Triangulation Problems — 267
   5.3.2 Traveling Salesman Problems — 268
   5.3.3 Matching Problems — 269
5.4 Applications — 270
   5.4.1 Relay Team Running — 270
   5.4.2 Backboard Wiring — 271
   5.4.3 Building Layout Planning — 272
   5.4.5 Keyboard Design — 273
5.5 Solution Methods — 275
   5.5.1 Exact Solution Methods — 275
   5.5.2 Heuristic Solution Methods — 290
   5.5.3 Solving General Facility Layout Problems — 292

Part III: Project Scheduling — 295

1 Unconstrained Time Project Scheduling — 297

1.1 Network Representations — 297
1.2 The Critical Path Method — 302
1.3 Project Acceleration (Crashing) — 309
1.4 Incorporating Uncertainties (*PERT*) — 313

2 Project Scheduling with Resource Constraints — 319

2.1. The Problems and its Formulation — 319
2.2 Exact Solution methods — 326
2.3 Heuristic Methods — 328

Part IV: Machine Scheduling Models — 333

1 Fundamentals of Machine Scheduling — 335

1.1 Introductory Examples — 335
1.2 The Models and Their Components — 340
   1.2.1 Basic Concepts, Notation and Performance Criteria — 340
   1.2.2 Interpretation and Discussion of Assumptions — 345
   1.2.3 A Classification Scheme — 346
1.3 Algorithmic Approaches — 350

2 Single Machine Scheduling — 353

2.1 Minimizing Makespan — 353
2.2 Minimizing Mean Flow Time — 353
   2.2.1 The Shortest Processing Time Algorithm — 354
   2.2.2 The Mean Weighted Flow Time and Other Problems — 356
2.3 Minimizing Objectives Involving Due Dates — 360
   2.3.1 Earliest Due Date Scheduling — 360
   2.3.2 Other Problems — 362

3 Parallel Machine Models — 367

3.1 Minimizing Makespan — 367
   3.1.1 Identical Machines and Tasks of Arbitrary Lengths — 367
   3.1.2 Other Algorithms for Identical Machines — 375
   3.1.3 Algorithms for Uniform and Unrelated Machines — 381
3.2 Minimizing Mean Flow Time — 387
   3.2.1 Identical Machines — 387
   3.2.2 Uniform and Unrelated Machines — 389
3.3 Minimizing Maximal Lateness — 393
   3.3.1 Identical Machines — 393
   3.3.2 Uniform and Unrelated Machines — 396

4 Dedicated Machine and Resource-Constrained Models — 399

4.1 Open Shop Scheduling — 399
4.2 Flow Shop Scheduling — 401
4.3 Job Shop Scheduling — 403
   4.3.1 Basic Ideas — 403
   4.3.2 A Branch and Bound Algorithm — 407
   4.3.3 The Shifting Bottleneck Heuristic — 411
4.4 Resource-Constrained Machine Scheduling — 415

References — 421
Subject Index — 453

INTRODUCTION

This part introduces the reader to some of the support methodology used in this book. We have made every possible attempt to keep the exposition as brief and concise as possible. Readers who are interested in more in-depth coverage are referred to the pertinent literature.

Notation

$\mathbb{N} = \{1, 2, ...\}$: Set of natural numbers
$\mathbb{N}_0 = \{0, 1, 2, ...\}$: Set of natural numbers including zero
$\mathbb{R}$: Set of real numbers
$\mathbb{R}_+$: Set of nonnegative real numbers
$\mathbb{R}^n$: n-dimensional real space
$\in$: Element of
$\subseteq$: Subset
$\subset$: Proper subset
$\cup$: Union of sets
$\cap$: Intersection of sets
$\varnothing$: Empty set
$\rightarrow$: Implies
$\exists$: There exists at least one
$\forall$: For all
$|S|$: Cardinality of the set S
inf: infimum
sup: supremum
$x \in [a, b]$: $a \leq x \leq b$
$x \in [a, b[$: $a \leq x < b$
$x \in]a, b]$: $a < x \leq b$
$x \in]a, b[$: $a < x < b$
$\lceil x \rceil$: Ceiling of x, the smallest integer greater or equal to x
$\lfloor x \rfloor$: Floor of x, the largest integer smaller or equal to x
$|x|$: Absolute value of x
$a := a + b$: Valuation, a is replaced by $a + b$
$\dfrac{\partial f(x)}{\partial x}$: partial derivative of the function $f(x)$ with respect to x

Support Methodology

a. Algorithms and Computational Complexity

Definition a1: The *instance of a problem* is the realization of a problem type with a given size and given parameters. The size of the instance is typically expressed in terms of the number of variables, constraints, machines, customers, decisions, criteria, or similar factors. The magnitude of parameters is not to be used as an indicator of the size of a problem.

Definition a2: An *algorithm* is a set of instructions designed to solve any instance of a given problem. There are two basic classes of algorithms, *exact solution methods* and *heuristic methods* (or *heuristics* for short).

Exact methods often employ enumeration methods on a finite, albeit sometimes very large, set. As far as heuristics are concerned, we distinguish between two distinct classes: construction heuristics which, as the name implies, construct a solution, and improvement heuristics which start with a given solution, typically obtained by a construction heuristic, and attempt to improve it. Heuristics are designed to find (hopefully) good solutions quickly. Sometimes there are proven error bounds that specify how bad a heuristic solution can be. In many cases there is only empirical evidence about the performance of heuristics.

Among heuristic methods, Greedy methods are some of the most popular construction methods, while the concept of pairwise exchange is often used as an improvement method. We will demonstrate these two in the following example. We have chosen to illustrate them on an assignment problem, despite the fact that assignment problems can be solved very efficiently with exact methods.

Example: Consider an assignment problem (see Definition d4 below) with the following cost matrix

$$\mathbf{A} = \begin{bmatrix} 5 & 7 & 4 & 6 & 5 \\ 8 & 4 & 2 & 1 & 9 \\ 5 & 5 & 4 & 7 & 8 \\ 4 & 5 & 3 & 6 & 2 \\ 3 & 7 & 6 & 4 & 8 \end{bmatrix}.$$

A *Greedy algorithm* could then be: choose the least-cost element in each row and make an assignment on it, provided that no element in that column has been assigned yet. If it has, choose the next cheapest element. Ties are broken in favor of the column with the smaller subscript.

In this example, the assignments are made on the elements (1, 3), (2, 4), (3, 1), (4, 5), and (5, 2).

The pairwise exchange method will then examine all pairs (i, j) and (k, ℓ) on which assignments have been made, and attempt to replace them by pairs (i, ℓ) and (k, j) instead. If the new assignment has lower costs, it will replace the previous solution, otherwise another pair of assignments is chosen for examination. The procedure continues until no more pairwise improvements are possible.

Consider, for example, the two pairs (1, 3) and (4, 5), whose costs are $4 + 2 = 6$. The pairwise exchange would attempt to replace with the pairs (1, 5) and (4, 3), whose costs are $5 + 3 = 8$. As the costs of the replacement are higher, the replacement will not be made. On the other hand, consider the pairs of assignments (3, 1) and (5, 2) with costs $5 + 7 = 12$. They could be replaced by assignments of the two pairs (3, 2) and (5, 1) with costs $5 + 3 = 8$. This replacement represents an improvement, so that the new solution is (1, 3), (2, 4), (3, 2), (4, 5), and (5, 1), whose costs are lower by 4 as compared to the previous solution. This process would continue until no further cost reductions by pairwise exchange are possible.

Definition a3: Given an algorithm, the *time complexity function* specifies how many elementary operations the algorithm requires to solve a problem of a given size in the worst case.

Example 1: (Matrix multiplication). Consider the multiplication of two $[n \times n]$-dimensional matrices **A** and **B**. Each element of the resulting matrix **C** is calculated as the sum of n products. Each product takes constant time, so that each element of **C** requires order n, or $O(n)$, calculations. As **C** includes n^2 elements, the multiplication of two matrices is said to take $O(n^3)$ time. Note that in this case, there is no distinction between worst, average and best case: it always takes $O(n^3)$ time to calculate the product of two matrices.

Example 2: (Sorting a linear file). Let a file of length n be given, so that each element of the file is a number. The objective is to sort these numbers in nondecreasing order, and the procedure is to find the smallest element, erase it from the original file and write it onto a new file. Each time the original file is scanned, there are $O(n)$ comparisons. Since each scan determines one element, this process has to be repeated n times, so that the resulting algorithm is of complexity $O(n^2)$. We wish to point out that this procedure is not the most efficient way to sort a file.

Example 3: Consider again an assignment problem with an $[n \times n]$-dimensional cost matrix. One way to solve the problem is by complete enumeration. Since there are a total $n!$ possible solution, complete enumeration is of complexity $O(n!)$

Definition a4: The *complexity of a problem* is defined as the complexity of the best-known algorithm that solves this problem.

Example: While the total enumeration technique of Example 3 above solves the assignment problem, its complexity function $O(n!)$ is inferior to the best known algorithm that is of complexity of less than $O(n^3)$, which is then also the complexity of the assignment problem.

Definition a5: If the complexity function of a problem is bounded by a polynomial function, the problem is said to be in the class **P**. The class of **NP**-hard optimization problems includes problems between which polynomial reductions have been performed, meaning that if a polynomial algorithm exists for one problem in this class, polynomial algorithms exist for all problems in this class. If a problem is proved to be **NP**-hard, it is thought to be highly unlikely that a polynomial algorithm exists for the problem. Hence the class of **NP**-hard problems consists of difficult problems.

Note, however, that problems are not necessarily difficult in practice just because they were proved to be **NP**-hard. The result pertains exclusively to the worst case. Very large instances of some **NP**-hard problems have been solved very efficiently on average.

b. Matrix Algebra

Definition b1: A *matrix* is a two-dimensional array with m rows and n columns. If $m = n$, it is said to be *square*; if $m = 1$, it is called a *row vector*, if $n = 1$, it is a *column vector*, and if $m = n = 1$, it is a *scalar*.

Typically, matrices are denoted by boldface capital letters, vectors are shown as boldface small letters, and scalars are shown as italicized small letters. The i-th row of a matrix $\mathbf{A} = (a_{ij})$ is $\mathbf{a}_{i\bullet}$ and the j-th column of the matrix $\mathbf{A}$ is $\mathbf{a}_{\bullet j}$.

Definition b2: An $[n \times n]$-dimensional matrix $\mathbf{A} = (a_{ij})$ is called an *identiy matrix*, if $a_{ij} = 1$ if $i = j$, and zero otherwise.

An identity matrix is usually denoted by $\mathbf{I}$. The i-th row of an identity matrix is called a *unit vector* $\mathbf{e}_i = [0, 0, \ldots 0, 1, 0, 0, \ldots 0]$ which has the "1" in the i-th position, and zeroes otherwise.

Definition b3: The *sum of* two [m × n]-dimensional *matrices* **A** and **B** is an [m × n]-dimensional matrix **C**, such that $c_{ij} = a_{ij} + b_{ij} \ \forall \ i=1, \ldots, m; j=1, \ldots, n$. The difference of two matrices is defined similarly.

Definition b4: The *product* of an [m × n]-dimensional *matrix* **A** = (a_{ij}) and an [n × p]-dimensional matrix **B** (b_{jk}) is an [m × p]-dimensional matrix **C** = (c_{ik}), such that
$$c_{ik} = \sum_{j=1}^{n} a_{ij} b_{jk} \ \forall \ i,k \ .$$

Definition b5: The *transpose* of an [m × n]-dimensional matrix **A** = (a_{ij}) is a matrix $\mathbf{A}^T = (a_{ij}^T)$, such that $a_{ij}^T = a_{ji} \ \forall \ i,j$. If $\mathbf{A} = \mathbf{A}^T$, the **A** is said to be *symmetric*. The *trace* of an [n × n]-dimensional matrix **A** is defined as
$$tr(\mathbf{A}) = \sum_{j=1}^{n} a_{jj} \ .$$

Definition b6: The *inverse* of an [n × n]-dimensional matrix **A** = (a_{ij}) is a matrix $\mathbf{A}^{-1}$, such that $\mathbf{A}\mathbf{A}^{-1} = \mathbf{A}^{-1}\mathbf{A} = \mathbf{I}$.

Example: Let $\mathbf{A} = \begin{bmatrix} 2 & 0 & 2 \\ 4 & -2 & 2 \\ -1 & 1 & -1 \end{bmatrix}$. Then $\mathbf{A}^{-1} = \begin{bmatrix} 0 & \frac{1}{2} & 1 \\ \frac{1}{2} & 0 & 1 \\ \frac{1}{2} & -\frac{1}{2} & -1 \end{bmatrix}$.

Proposition b7: The following results hold when multiplying, transposing, and inverting matrices **A**, **B**, and **C**:

- $\mathbf{AI} = \mathbf{IA} = \mathbf{A}$
- $\mathbf{A(BC)} = \mathbf{(AB)C}$
- $(\mathbf{A}^T)^T = \mathbf{A}$
- $(\mathbf{AB})^T = \mathbf{B}^T \mathbf{A}^T$
- $(\mathbf{A}^{-1})^{-1} = \mathbf{A}$
- $(\mathbf{AB})^{-1} = \mathbf{B}^{-1} \mathbf{A}^{-1}$
- $(\mathbf{A}^T)^{-1} = (\mathbf{A}^{-1})^T$

Definition b8: The *determinant* of an [n × n]-dimensional *matrix* **A** is defined recursively as $\det(\mathbf{A}) = \sum_{j=1}^{n} (-1)^{i+j} a_{ij} \det(\mathbf{A}_{ij})$, where $\mathbf{A}_{ij}$ denotes the matrix that results from the given matrix **A** by deleting the *i*-th row and the *j*-th column.

This development of the determinant via the *i*-th row is due to Laplace. As a starting condition, the determinant of a [1 × 1]-dimensional matrix **A** is det (**A**) =

a_{11}. Applying Laplace's formula to a $[2 \times 2]$-dimensional matrix $\mathbf{A} = \begin{bmatrix} a_{11} & a_{12} \\ a_{21} & a_{22} \end{bmatrix}$, we obtain det $(\mathbf{A}) = a_{11}a_{22} - a_{12}a_{21}$. The inverse $\mathbf{A}^{-1}$ of a square matrix $\mathbf{A}$ exists if and only if $\det(\mathbf{A}) \neq 0$.

Example: Let again

$$\mathbf{A} = \begin{bmatrix} 2 & 0 & 2 \\ 4 & -2 & 2 \\ -1 & 1 & -1 \end{bmatrix}.$$

Evaluating the determinant with respect to the first row, we obtain

$$\mathbf{A}_{11} = \begin{bmatrix} -2 & 2 \\ 1 & -1 \end{bmatrix}, \mathbf{A}_{12} = \begin{bmatrix} 4 & 2 \\ -1 & -1 \end{bmatrix}, \text{ and } \mathbf{A}_{13} = \begin{bmatrix} 4 & -2 \\ -1 & 1 \end{bmatrix}.$$

so that $\det(\mathbf{A}) = (-1)^{1+1}(2) \det(\mathbf{A}_{11}) + (-1)^{1+2}(0) \det(\mathbf{A}_{12}) + (-1)^{1+3}(2) \det(\mathbf{A}_{13}) = (1)(2)(0) + (-1)(0)(-2) + (1)(2)(2) = 4$. The determinant can be interpreted as the volume of the n-dimensional parallelepiped spanned by the column vectors of the matrix.

Definition b9: An *eigenvalue* or *characteristic value* λ of a square matrix $\mathbf{A}$ is a number, possibly complex, which solves the equation $\det(\mathbf{A} - \lambda \mathbf{I}) = 0$.

One can show that if λ is a real-valued eigenvalue of the square matrix $\mathbf{A}$, then there exists a corresponding vector $\mathbf{x} \neq \mathbf{0}$, such that $\mathbf{A}\mathbf{x} = \lambda \mathbf{x}$. This vector is called an *eigenvector* of $\mathbf{A}$ corresponding to the eigenvalue λ. It may happen that no real-valued eigenvalue exists. Geometrically, an eigenvalue can be interpreted as some form of "scaling" resulting from the linear transformation $\mathbf{x} \to \mathbf{A}\mathbf{x}$. For instance, if [1, 0] and [0, 1] are the eigenvectors for a $[2 \times 2]$-dimensional matrix $\mathbf{A}$, with eigenvalues ½ and 3, respectively, then every transformed vector $\mathbf{x}$ will have its first component halved and second component tripled.

Example: Using again the matrix $\mathbf{A}$ in the example of Definition b8, the eigenvalues of $\mathbf{A}$ can be calculated as 2, –1, and –2, and the corresponding eigenvectors are [1, 1, 0], [2, 2, –3], and [1, 3, –2], respectively.

Procedure b10: The Newton-Raphson method to determine roots of polynomials. *Procedure:* Given the polynomial functions $f(x)$, the Newton-Raphson method for finding a real root of the polynomial, i.e., a real solution to the equation $f(x) = 0$ proceeds as follows. Beginning with an initial estimate x_1, subsequent estimates are calculated iteratively by the relation

$$x_{i+1} := x_i - \frac{f(x_i)}{f'(x_i)}, \quad i=1,2,\ldots$$

Example: Consider the polynomial $y = x^2 - 3$, whose derivative with respect to x is $f'(x) = 2x$. Starting with the initial guess $x_1 = 1.5$ for a root of this polynomial, we obtain $x_2 = 1.5 - \frac{(1.5)^2 - 3}{2(1.5)} = 1.75$. In the subsequent iterations we obtain $x_3 = 1.73214$, $x_4 = 1.73205$, and $x_5 = 1.73205$. Here, the process has terminated. In general, a suitable stop criterion has to be devised.

Procedure b11: An $O(n^3)$ algorithm for the *evaluation of a determinant* is available by *pivoting*. The iterative step is to choose a nonzero element on the main diagonal on the matrix and define it as a pivot. The procedure is initialized by setting $\mathbf{A}^1 = (a_{ij}^1) := \mathbf{A} = (a_{ij})$. With the pivot a_{kk}^k in iteration k we then calculate

$$a_{\ell k}^{k+1} = \begin{cases} 1, & \text{if } \ell = k \\ 0 \ \forall \ \ell \neq k \end{cases},$$

$$a_{kj}^{k+1} = \frac{a_{kj}^k}{a_{kk}^k} \ \forall \ j,$$

$$a_{ij}^{k+1} = a_{ij}^k - \frac{a_{ik}^k a_{kj}^k}{a_{kk}^k} \ \forall \ i \neq k \neq j.$$

Applying this iterative step n times results in an identity matrix $\mathbf{A}^{n+1} = \mathbf{I}$. The determinant is then the product of the pivots, i.e., $\det(\mathbf{A}) = \prod_{k=1}^{n} a_{kk}^k$. Pivoting is also used for tableau transformations in the simplex algorithm for linear programming.

Example: Consider again the matrix

$$\mathbf{A} = \begin{bmatrix} 2 & 0 & 2 \\ 4 & -2 & 2 \\ -1 & 1 & -1 \end{bmatrix}.$$

The pivot steps are shown in the matrices $\mathbf{A}^1$, $\mathbf{A}^2$, $\mathbf{A}^3$, and $\mathbf{A}^4$ with their respective pivots in boldface:

$$\mathbf{A}^1 = \begin{bmatrix} 2 & 0 & 2 \\ 4 & -2 & 2 \\ -1 & 1 & -1 \end{bmatrix}, \mathbf{A}^2 = \begin{bmatrix} 1 & 0 & 1 \\ 0 & -2 & -2 \\ 0 & 1 & 0 \end{bmatrix}, \mathbf{A}^3 = \begin{bmatrix} 1 & 0 & 1 \\ 0 & 1 & 1 \\ 0 & 0 & -1 \end{bmatrix}, \text{and}$$

$$\mathbf{A}^4 = \mathbf{I} = \begin{bmatrix} 1 & 0 & 0 \\ 0 & 1 & 0 \\ 0 & 0 & 1 \end{bmatrix}.$$

We then obtain det (**A**) = (2)(–2)(–1) = 4.

Proposition b12: The following rules apply to determinants of [$n \times n$]-dimensional matrices **A** and **B**:

det (**I**) = 1
det (**AB**) = det (**A**)det (**B**)
det (**A**) = det (**A**^{T})
det (α**A**) = α^n det (**A**)
det (**A**$^{-1}$) = $\dfrac{1}{\det(\mathbf{A})}$.

Procedure b13: Define a *simplex* S in $\mathbb{R}^d$ as a set of (d + 1) points $\mathbf{x}^k$, k=1, 2, ..., d+1, such that there exists no hyperplane H: $\mathbf{ax} = b$ with the property that $\mathbf{ax}^k = b$ $\forall$ k=1, 2, ..., d+1. In other words, a simplex is a minimal independence structure in the sense that all but one of its expreme points are located on a hyperplane. In order to determine the volume of a d-dimensional polyhedron P, it is first necessary to subdivide P into simplices. The *volume of a simplex S* with extreme points $\mathbf{x}^k$, k = 1, 2, ..., d+1 can then be expressed as

$$v(S) = \tfrac{1}{d} \det(\mathbf{A}) = \begin{bmatrix} \mathbf{x}^1 & 1 \\ \mathbf{x}^2 & 1 \\ \vdots & \vdots \\ \mathbf{x}^{d+1} & 1 \end{bmatrix}.$$

c. Graphs and Networks

Definition c1: A *graph* or *network* is a tuple $G = (N, A)$, where $N = \{n_1, n_2, ..., n_n\}$ is a set of *nodes* and $A = \{a_{ij} : (n_i, n_j) \in N\}$ is a set of *arcs*. Arcs are either directed or undirected. Undirected arcs are sometimes called *edges*. A graph with only undirected arcs is called an *undirected graph*, a graph with only directed arcs is a *directed graph* or *digraph*, and a graph with both types of arcs is referred to as a *mixed graph*. A graph is called *bipartite* if its node set can be decomposed into two sets N_1 and N_2 with $N_1 \cup N_2 = N$ and $N_1 \cap N_2 = \emptyset$, such that $a_{ij} \in A$ implies that either $n_i \in N_1$ and $n_j \in N_2$, or $n_i \in N_2$ and $n_j \in N_1$. A graph is *complete bipartite* if there exists at least one arc for each $n_i \in N_1$ and $n_j \in N_2$.

Definition c2: A *path* is an ordered sequence of directed and/or undirected arcs, such that any two successive arcs share one node between them. A *circuit* is a path whose first and last nodes are identical. Sometimes (somewhat incorrectly), a circuit is also called a *cycle*.

Definition c3: A *graph* is *connected*, if there exists at least one path between each pair of nodes. A node that is connected to the remaining tree by only a single edge is called a *pendant node* or *leaf*.

Definition c4: A *forest* is an undirected graph without cycles. A connected forest is called a *tree*. Sometimes (formally incorrectly), the terms forest and tree are also used when the graph is directed.

Definition c5: The set of *neighbors* of a node n_i is defined as $\mathcal{N}(n_i) = \{n_j: a_{ij}$ or $a_{ji} \in A\}$. In directed graphs this set consists of two subsets, $\mathcal{S}(n_i)$ which is the set of *successors* of node n_i and $\mathcal{P}(n_i)$, which is the set of *predecessors* of node n_i. Formally, $\mathcal{S}(n_i) = \{n_j : a_{ij} \in A\}$ and $\mathcal{P}(n_i) = \{n_j: a_{ji} \in A\}$.

Procedure c6: For any given graph, define $\mathbf{C} = (c_{ij})$ as the *distance matrix*, where c_{ij} denotes the length of the arc between n_i and n_j. In order to determine the lengths of the *shortest paths* between all pairs of nodes, define the $[n \times n]$ –dimensional matrix $\mathbf{D}^1 := \mathbf{C}$, and set $k := 1$. The Floyd-Warshall algorithm for the determination of the lengths of the shortest paths can then be described as follows (for details and an example, see, e.g., Eiselt and Sandblom, 2000).

The Floyd – Warshall Algorithm

Step 1: Is $k = n + 1$?
If yes: Stop, $\mathbf{D} = \mathbf{D}^{n+1}$ is the matrix of shortest path distances.
If no: Go to Step 2.

Step 2: Determine $d_{ij}^{k+1} = \min\{d_{ij}^k; d_{ik}^k + d_{kj}^k\}$. Set $k := k + 1$ and go to Step 1.

d. Linear and Integer Optimization

Definition d1: Given an n-dimensional row vector of coefficients $\mathbf{c}$, a column m-vector $\mathbf{b}$ and an $[m \times n]$-dimensional matrix $\mathbf{A}$ as well as the scalar variable z and a column n-vector $\mathbf{x}$ of variables, then the problem

$$\text{P: } \underset{\mathbf{x}}{\text{Max}} \; z = \mathbf{cx} \qquad \text{objective (function)}$$
$$\text{s.t.} \; \mathbf{Ax} \leq \mathbf{b} \qquad \text{(structural) constraints}$$
$$\mathbf{x} \geq \mathbf{0} \qquad \text{nonnegativity constraints,}$$

is called a *linear programming (LP) problem*.

Definition d2: Associated with a linear programming problem P (the so-called *primal problem*) is a *dual problem*, with the m-dimensional row vector $\mathbf{u}$ of variables,

$$\text{P}_D\text{: } \underset{\mathbf{u}}{\text{Min}} \; z_D = \mathbf{ub}$$
$$\text{s.t.} \; \mathbf{uA} \geq \mathbf{c}$$
$$\mathbf{u} \geq \mathbf{0}.$$

Note that while the primal problem has n variables and m structural constraints, the dual problem has m variables and n structural constraints. There is a one-to-one correspondence between each primal variable and dual constraint, and between each primal constraint and dual variable.

Theorem d3: (Weak complementary slackness): Assume that $(\bar{\mathbf{x}}, \bar{\mathbf{u}})$ is a pair of feasible solutions for P and P_D. Then $(\bar{\mathbf{x}}, \bar{\mathbf{u}})$ is a pair of optimal solutions if and only if

$$\bar{\mathbf{u}}(\mathbf{A}\bar{\mathbf{x}} - \mathbf{b}) = 0, \text{ and } (\bar{\mathbf{u}}\mathbf{A} - \mathbf{c})\bar{\mathbf{x}} = 0$$

Definition d4: A special structure of a linear programming problem is the so-called *Transportation Problem*. It is defined on a bipartite graph with $|N_1| = m$, and $|N_2| = n$ with a supply $s_i > 0$ assigned to each node in N_1 and a demand d_j assigned to each node in N_2. Furthermore, each arc a_{ij}, $n_i \in N_1$ and $n_j \in N_2$ has associated with it a unit cost c_{ij}. The objective is to determine a flow pattern from N_1 to N_2 that ships all supplies out of the nodes in N_1, satisfies all demands at the nodes in N_2, and does so while minimizing the total costs of transportation. This problem can be formulated as

$$P: \text{Min } z = \sum_{i=1}^{m}\sum_{j=1}^{n} c_{ij} x_{ij}$$

$$\text{s.t.} \quad \sum_{j=1}^{n} x_{ij} = s_i \; \forall \; i = 1,...,m$$

$$\sum_{i=1}^{m} x_{ij} = d_j \; \forall \; j = 1,...,n$$

$$x_{ij} \geq 0 \quad \forall \; i = 1, ..., m; j = 1, ..., n$$

An *Assignment Problem* is a transportation problem with $m = n$, all supplies and demands equal to one, and whose nonnegative variables are replaced by zero-one variables.

Definition d5: A *mixed-integer linear programming problem* (*MILP*) is a linear programming problem with the additional constraint that at least one, but not all, variables are restricted to be integer. An *all-integer linear programming problem* (*AILP or ILP*) is a linear programming problem all of whose variables are required to be integer.

Given an n-dimensional row vector of objective function coefficients **c**, an m-dimensional column vector of right-hand side values **b**, an $[m \times n]$-dimensional matrix of technological coefficients **A**, a k-dimensional column vector of integer variables **y**, and an $(n-k)$-dimensional column vector of continuous variables **x**, then a general integer linear programming problem can be written as

P_{ILP}: Max $z = \mathbf{c}(\mathbf{y},\mathbf{x})$

s.t. $\mathbf{A}(\mathbf{y}, \mathbf{x}) \leq \mathbf{b}$

$\mathbf{y} \in \mathbb{N}_0^n$

$\mathbf{x} \in \mathbb{R}_+^n$

Definition d6: Given an integer programming problem P_{IP}: Max **cx**, s.t. $\mathbf{Ax} \leq \mathbf{b}$, $x_j \in \mathbb{N}_0 \; \forall \; j \in K$, $x_j \in \mathbb{R}_+ \; \forall \; j \in J$, then the *linear programming relaxation* of P_{IP} is defined as P_{LP}: Max **cx**, s.t. $\mathbf{Ax} \leq \mathbf{b}$, $x_j \in \mathbb{R}_+ \; \forall \; j \in J \cup K$.

Simply speaking, the linear programming relaxation deletes the integrality requirements of all variables.

Lemma d7: Denoting the optimal objective values of the original problem P_{IP} and its linear programming relaxation P_{LP} by $\bar{z}_{IP}$ and $\bar{z}_{LP}$ respectively, we obtain the relation

$$\bar{z}_{LP} \geq \bar{z}_{IP}.$$

Definition d8: Given an integer programming problem P_{IP}: Max **cx**, s.t. **Ax** ≤ **b**, **Dx** ≤ **d**, $x_j \in \mathbb{N}_0 \ \forall\ j \in K$, $x_j \in \mathbb{R}_+ \ \forall\ j \in J$, the *Lagrangean relaxation* with any given so-called *multiplier* vector **u** ≥ **0** is P_{LR}: Max **cx** − **u** (**Ax** − **b**), s.t. **Dx** ≤ **d**, $x_j \in \mathbb{N}_0 \ \forall\ j \in K$, $x_j \in \mathbb{R}_+ \ \forall\ j \in J$. In this relaxation, the constraints **Ax** ≤ **b** are said to have been *dualized*.

It is apparent that Langrangean relaxations have many degrees of freedom. First, it must be decided which constraints **Ax** ≤ **b** to dualize. In some applications, relaxing certain classes of constraints leaves a system **Dx** ≤ **d** that has a special structure which makes it easy to solve, such as a network structure. Alternatively, the remaining system may be decomposable, again aiding the solution process. Another choice to be made is that of the vector **u** of multipliers.

Proposition d9: Between the optimal value of the objective function of the Lagrangean Relaxation and that of the original integer linear programming problem, the following relation holds:

$$\bar{z}_{LR} \geq \bar{z}_{IP} .$$

e. Statistics

Definition e1: The *probability* of an event E_i, which is the random outcome of an experiment, is a number $p(E_i) = p_i \in [0; 1]$, which measures the likelihood that E_i will occur.

If p is near 1, the event is very likely to happen; if p is near 0 the event is unlikely to happen. We can think of $p(E_i)$ as the proportion of times that E_i will occur, if the experiment is repeated a very large number of times. With this view, p is the long-run relative frequency of occurrence of the event E_i.

Theorem e2: If E_i, $i=1, ..., n$ denotes all the events of an experiment, then $\sum_i p_i = 1$.

Example: Consider the experiment of flipping a fair coin twice. Denoting the outcome "head" by H and "tail" by T, the outcome "head on the first flip and tail on the second" is then denoted by HT. The experiment then has four outcomes: HH, HT, TH, and TT. Since the coin is assumed to be fair, all outcomes are equally likely, so that each outcome has a probability of ¼.

Definition e3: A (one-dimensional) *random variable* X is a real-valued function defined on the event space of a given random experiment.

Example: Let X denote the number of heads that come up in an experiment in which a fair coin is flipped twice. If the outcome is HH, then $X = 2$. For the outcomes HT and TH, we have $X = 1$, while for TT, the random variable $X = 0$. In other words, the set $\{X = 2\}$ equals the event set $\{HH\}$. Similarly, $\{X = 1\} = \{HT, TH\}$, and $\{X = 0\} = \{TT\}$. Assuming that the probability of a set equals the sum of probabilities of the events forming the set, we find that $P(X = 2) = P(HH) = ¼$, $P(X = 1) = P(HT) + P(TH) = ¼ + ¼ = ½$, and $P(X = 0) = P(TT) = ¼$.

Generalizing the above example, suppose that the coin is no longer necessarily fair, so that $P(H) = p$. Counting the number X of heads after n flips of the coin, it can be shown that $P(X = j) = \binom{n}{j} p^j q^{n-j}$, where $\binom{n}{j} = \dfrac{n!}{j!(n-j)!}$ and $q = 1-p$, for $j = 0, 1, \ldots, n$.

Definition e4: If the random variable X can take the values a_j, $j \in J$, then the function $P_X(a_j) := p(a_j) := P(X = a_j)$ is called the *(discrete) probability distribution function* of X. The function $F_X(a_j) := F(a_j) := P(X \leq a_j)$ is called the *cumulative probability distribution function* of X.

Example: The number X of heads after n flips of a coin has the probability distribution function $p(j) = \binom{n}{j} p^j q^{n-j}$. We say that X follows a *binomial distribution function* and write $X \in Bin(n, p)$. In the previous example of the number of heads of a fair coin shown after two flips we have $X \in Bin(2, ½)$.

Definition e5: A random variable X is called *continuous* if there exists a function $f_X(x)$, such that $F_X(x) = \int_{-\infty}^{t} f_X(t)dt$. The function $f_X(x)$ is called the *continuous probability density function* of X.

Definition e6: A continuous random variable X with a density function $f(x) = \dfrac{1}{\sigma\sqrt{2\pi}} e^{-\dfrac{(x-\mu)^2}{2\sigma^2}}$ is said to have a *normal* (or *Gaussian*) *distribution* with parameters μ and $\sigma \geq 0$; we write $X \in N(\mu, \sigma)$.

Definition e7: The *expected value* (also called *mean* or *expectation*) $E(X)$ of a discrete random variable X is defined as $E(X) = \sum_j a_j P(X = a_j) = \sum_j a_j p_X(a_j)$. For a continuous random variable X, the expected value $E(X)$ is defined as $E(X) = \int_{-\infty}^{\infty} t f_X(t)dt$.

There are discrete as well as continuous distributions for which the mean does not exist. The mean $E(X)$ is often denoted by μ.

Example: Consider again flipping a fair coin twice and counting the number X of heads. We have already seen that $X \in Bin(2, \frac{1}{2})$. We can then write $E(X) = \sum_j a_j p_X(a_j) = 0\, p_X(0) + 1\, p_X(1) + 2\, p_X(2) = 0\,(\frac{1}{4}) + 1(\frac{1}{2}) + 2(\frac{1}{4}) = 1$. In general, we can show that for a binomially distributed random variable $X \in Bin(n, p)$, the expected value is $E(X) = np$. For the normal distribution, we can show that

$$E(X) = \int_{-\infty}^{\infty} t f_X(t) dt = \int_{-\infty}^{\infty} \frac{t}{\sigma\sqrt{2\pi}} e^{-\frac{(t-\mu)^2}{2\sigma^2}} dt = \mu.$$

Definition e8: The *variance* $V(X)$ (or $Var(X)$ or σ^2) of a discrete random variable X with mean μ is defined as $V(X) = E((X-\mu)^2)$.

There are discrete and continuous distributions for which the variance does not exist.

Example: Consider again a discrete random variable $X \in Bin(n, p)$. It can be shown that $V(X) = np(1-p) = npq$. For a continuous random variable $X \in N(\mu, \sigma)$, we obtain the variance $V(X) = \sigma^2$. In general, it can be shown that $V(X) = E(X^2) - (E(X))^2 = E(X^2) - \mu^2$ and $V(X) \geq 0$.

Theorem e9 (The *addition law of probabilities*): Let A and B each denote sets of events. Then $P(A \cup B) = P(A) + P(B) - P(A \cap B)$.

Example: Let $X \in Bin(3, \frac{1}{2})$, which corresponds to flipping a fair coin three times and counting the number X of heads that come up. Let $A = \{X \leq 1\} = \{HTT, THT, TTH, TTT\}$ and $B = \{X \text{ is odd}\} = \{X = 1 \text{ or } 3\} = \{HHH, HTT, THT, TTH\}$. Then we find $A \cup B = \{X \leq 1 \text{ or } X = 3\} = \{HHH, HTT, THT, TTH, TTT\}$ and $A \cap B = \{X \leq 1 \text{ and } X \text{ odd}\} = \{X = 1\} = \{HTT, THT, TTH\}$. Since the probability of each single event $HHH, HHT, \ldots, TTT$ is $1/8$, it follows that $P(A) = 4/8$, $P(B) = 4/8$, $P(A \cup B) = 5/8$, and $P(A \cap B) = 3/8$. Then $P(A \cup B) = 5/8 = 4/8 + 4/8 - 3/8$ in accordance with the theorem.

Denoting by E the event set which includes all events (i.e., the *universe*), it is clear that all events sets are included in E and from Definition e1, $P(E) = 1$. We now denote by $\overline{A}$ (or CA or $\neg A$) the complement of A, i.e., the set of events that includes all events that are not included in A. In other words, $A \cup \overline{A} = E$ and $A \cap \overline{A} = \emptyset$, and we say that A and $\overline{A}$ are *collectively exhaustive* and *mutually exclusive*. Using the addition law, we obtain $1 = P(E) = P(A \cup \overline{A}) = P(A) + P(\overline{A})$

$- P(A \cap \overline{A}) = P(A) + P(\overline{A})$, so that $P(\overline{A}) = 1 - P(A)$ and we have proved

Lemma e10: For any event set A and its complement $\overline{A}$, we have $P(\overline{A}) = 1 - P(A)$.

Clearly, $0 \leq P(A) \leq 1$ for any set A. Applying the above argument to random variables, it can be shown that for any discrete random variable X and any given numbers $a < b$, $P(a < X \leq b) = F_X(b) - F_X(a)$, and for continuous random variables, $P(a < X \leq b) = P(a \leq X \leq b) = P(a \leq X < b) = P(a < X < b) = F_X(b) - F_X(a) = \int_a^b f_X(t)\, dt$. Furthermore, $P(X = a) = 0$ for any number a and any continuous random variable X.

The addition law can be generalized to three or more event sets. For the case of mutually exclusive event sets A_i, $i = 1, \ldots, m$, i.e., with $A_i \cap A_j = \emptyset \ \forall\ i < j$, we obtain $P\left(\bigcup_{i=1}^{m} A_i\right) = \sum_{i=1}^{m} P(A_i)$.

In order to establish a multiplication law for probabilities, we need to introduce the concept of conditional probability.

Definition e11: Given any event sets A and B with $P(B) \neq 0$, the *conditional probability* $P(A|B)$ *of* A, *given* B is then defined as $P(A|B) = \dfrac{P(A \cap B)}{P(B)}$.

Example: Flip a fair coin three times and count the number X of heads. Let $A := \{X \leq 1\}$ and $B := \{X \text{ odd}\} = \{X = 1 \vee 3\}$. We know that $X \in \mathrm{Bin}(3, \tfrac{1}{2})$ and that $P(A) = \tfrac{1}{2}$, $P(B) = \tfrac{1}{2}$, and $P(A \cap B) = 3/8$. Therefore $P(A|B) = \dfrac{P(A \cap B)}{P(B)} = \dfrac{3/8}{1/2} = \tfrac{3}{4}$.

The conditional probability can be seen as the number of times event A is observed, given that the universe is defined as event B. Here, $B = \{HTT, THT, TTH, HHH\}$, and event A occurs in the first three elements of this set, hence three out of four times.

Theorem e12 (The *multiplication law of probabilities*): For any event sets A and B, the probability $P(A \cap B) = P(A|B)\, P(B)$.

In the theorem, the right-hand side is understood to be zero, if $P(B) = 0$, even though $P(A|B)$ is then not defined. Since $A \cap B = B \cap A$, the right-hand side could also be written as $P(B|A)\, P(A)$. The theorem is a simple rearrangement of the expression for defining $P(A|B)$ or $P(B|A)$.

Definition e13: The event sets A and B are said to be *statistically independent* if $P(A \cap B) = P(A) P(B)$.

By the definition of conditional probability, statistical independence between A and B also implies $P(A|B) = P(A)$ and $P(B|A) = P(B)$. It is now not difficult to prove the Reverend Thomas Bayes's (1702–1761) theorem of "inverse probabilities:"

Theorem e14 (*Bayes's theorem*): Assume that the event sets $A_1, A_2, \ldots, A_m$ are mutually exclusive and collectively exhaustive. Then for any event set B we have

$$P(A_i | B) = \frac{P(B | A_i) P(A_i)}{\sum_{k=1}^{m} P(B | A_k) P(A_k)} \quad \forall \ i = 1, \ldots, m.$$

Suppose now that there are two families of event sets $A_1, A_2, \ldots, A_m$ and $B_1, B_2, \ldots, B_n$, both of which are mutually exclusive and collectively exhaustive. We may then be interested in the conditional probabilities $P(A_i|B_j)$ and $P(B_j|A_i)$ as well as the so-called *joint probabilities* $P(A_i \cap B_j)$ and *marginal probabilities* $P(A_i)$ and $P(B_j)$. The term "joint" probability refers to the fact that both, A_i and B_j have occurred in the event set $A_i \cap B_j$, the expression "marginal" is due to the fact that

$$P(A_i) = \sum_{j=1}^{n} P(A_i \cap B_j) \ \text{ and } \ P(B_j) = \sum_{i=1}^{m} P(A_i \cap B_j), \text{ so that } P(A_i) \text{ and } P(B_j) \text{ are}$$

the row and column sums in the matrix formed by the joint *probabilities* $P(A_i \cap B_j)$, i.e., in the margins of that matrix. Bayes's theorem provides the vehicle for transformations between joint and marginal probabilities. An example of how this is done is presented at the end of Chapter I.2 on games against nature, and will therefore not be restated here. Such transformations of probabilities are usually referred to as *revisions of probabilities*.

Area under the normal curve, i.e., values of $\int_{-\infty}^{x} \frac{1}{\sqrt{2\pi}} e^{-t^2/2} \, dt$

x	0.00	0.01	0.02	0.03	0.04	0.05	0.06	0.07	0.08	0.09
.00	.5000	.5040	.5080	.5120	.5160	.5199	.5239	.5279	.5319	.5359
.10	.5398	.5438	.5478	.5517	.5557	.5596	.5636	.5675	.5714	.5754
.20	.5793	.5832	.5871	.5910	.5948	.5987	.6026	.6064	.6103	.6141
.30	.6179	.6217	.6255	.6293	.6331	.6368	.6406	.6443	.6480	.6517
.40	.6554	.6591	.6628	.6664	.6700	.6736	.6772	.6808	.6844	.6879
.50	.6915	.6950	.6985	.7019	.7054	.7088	.7123	.7157	.7190	.7224
.60	.7258	.7291	.7324	.7357	.7389	.7422	.7454	.7486	.7518	.7549
.70	.7580	.7612	.7642	.7673	.7704	.7734	.7764	.7794	.7823	.7852
.80	.7881	.7910	.7939	.7967	.7996	.8023	.8051	.8079	.8106	.8133
.90	.8159	.8186	.8212	.8238	.8264	.8289	.8315	.8340	.8365	.8389
1.0	.8413	.8438	.8461	.8485	.8508	.8531	.8554	.8577	.8599	.8621
1.1	.8643	.8665	.8686	.8708	.8729	.8749	.8770	.8790	.8810	.8830
1.2	.8849	.8869	.8888	.8907	.8925	.8944	.8962	.8980	.8997	.9015
1.3	.9032	.9049	.9066	.9082	.9099	.9115	.9131	.9147	.9162	.9177
1.4	.9192	.9207	.9222	.9236	.9251	.9265	.9279	.9292	.9306	.9319
1.5	.9332	.9345	.9357	.9370	.9382	.9394	.9406	.9418	.9430	.9441
1.6	.9452	.9463	.9474	.9485	.9495	.9505	.9515	.9525	.9535	.9545
1.7	.9554	.9564	.9573	.9582	.9591	.9599	.9608	.9616	.9625	.9633
1.8	.9641	.9649	.9656	.9664	.9671	.9678	.9686	.9693	.9700	.9706
1.9	.9713	.9719	.9726	.9732	.9738	.9744	.9750	.9756	.9762	.9767
2.0	.9773	.9778	.9783	.9788	.9793	.9798	.9803	.9808	.9812	.9817
2.1	.9821	.9826	.9830	.9834	.9838	.9842	.9846	.9850	.9854	.9857
2.2	.9861	.9865	.9868	.9871	.9875	.9878	.9881	.9884	.9887	.9890
2.3	.9893	.9896	.9898	.9901	.9904	.9906	.9909	.9911	.9913	.9916
2.4	.9918	.9920	.9922	.9925	.9927	.9929	.9931	.9932	.9934	.9936
2.5	.9938	.9940	.9941	.9943	.9945	.9946	.9948	.9949	.9951	.9952
2.6	.9953	.9955	.9956	.9957	.9959	.9960	.9961	.9962	.9963	.9964
2.7	.9965	.9966	.9967	.9968	.9969	.9970	.9971	.9972	.9973	.9974
2.8	.9974	.9975	.9976	.9977	.9977	.9978	.9978	.9980	.9980	.9981
2.9	.9981	.9982	.9983	.9983	.9984	.9984	.9985	.9985	.9986	.9986
3.0	.9987	.9987	.9987	.9988	.9988	.9989	.9989	.9989	.9990	.9990
3.5	.999767									
4.0	.9999683									
4.5	.99999660									
5.0	.999999713									

PART I: ANALYSIS OF DECISION MAKING

Each day, millions of people make thousands of decisions, ranging from small personal decisions ("Where do we eat out tonight?") to major personal decisions ("which vehicle or house are we going to purchase?") and major corporate decisions regarding new product lines, locations of nuclear waste dumps, and the determination of the types of vehicles in a fleet of transport trucks. Each of these decisions has its own structure, context, and idiosyncrasies. In this book, we use the term "decision making" in a specific context. With the exception of the first section of this chapter, we consider a scenario that consists of a finite number of possible decision alternatives that are to be ranked from most desirable to least desirable, most often—but not always—in order to choose the most desirable option. Such problems are sometimes referred to as *selection problems*. In the initial stages of the process, it is important to ensure that as many of the available alternatives are included, as premature restriction to a subset of "reasonable" alternatives may unduly restrict the choices that are available to the decision maker.

The available decisions will then be evaluated on a number of relevant criteria chosen by the decision maker. Most criteria are what may be called "unbounded," i.e., a higher profit or revenue is always preferable to a lower value (or, similarly, lower costs are always preferable to higher costs). This is, however, not always the case. Consider the problem of buying a house. A decision maker usually has an "optimal size" of his dream home in mind; any house that is much smaller or much larger will be less desirable. Such apparent nonlinearities also occur in business contexts. Furthermore, as Simon (1979) pointed out, many decision makers attempt to "satisfice", i.e., are content once they have achieved a certain level of profit or costs. A typical outgrowth of this concept are target values; see, e.g., their use in goal programming. In this part we will deal with target values by choosing appropriate value functions.

Consider now the desires of a single decision maker and how they could be structured. This is frequently done as a hierarchy. On the highest level is a *goal*. The goal specifies the overall desire of the decision maker on a strategic level.

Examples are a decision maker's goal to determine a new location of the company headquarters, the composition of a new fleet of vehicles, or the design of a hospital. Each of these goals can be decomposed into a number of *objectives*. An objective typically expresses features associated with the decision under consideration, e.g., its total cost, the environmental impact, or the dissatisfaction of the labor force with the decision. It is apparent that objectives are located below goals on the tactical level. While some objectives are naturally quantitative (such as profits and costs), others (such as customer satisfaction or environmental damage) are not. Their evaluation presents much more serious problems. In order to allow quantitative analyses to be performed, each objective will be measured and expressed by at least one *criterion*. This is the operational scale. For instance, water pollution may be measured by the presence of PCBs, heavy metals, sulfur, etc., each being a measurable criterion. Some criteria may still be qualitative, e.g., the impact of a nuclear power plant or maximum security prison on a neighboring community. Such criteria may be expressed on a five- or seven-point scale, ranging from very poor, to poor, medium, good, and very good (possibly augmented by the extremes "terrible" and "excellent"). It should be pointed out, though, that many authors use the terms goal, objective, and criterion interchangeably.

So far, we have investigated factors that are, at least in part, under the control of the decision maker. Each decision maker can choose his own goals, objectives, and criteria, and can determine which decision alternatives to consider. A factor beyond the decision maker's control is the outcome of his decision. In some cases, the outcome is certain, as in the case of contracted sales (assuming that the customer does not go bankrupt). In most cases, there is at least some uncertainty about the outcome. Normally, decision makers have some idea about the likelihood of the outcomes. This may be based on similar decisions that were made in the past, or on educated guesses. Sometimes, though, decision makers can specify some possibilities but have no idea about their likelihoods.

In summary, the major three components of selection problems are as follows:

(1) Decisions. We assume that there is a finite number of potential decision alternatives, one of which is to be chosen. Sometimes, it is also possible to choose a combination of the available decisions. We will discuss this possibility in detail when the occasion arises.

(2) Criteria. They provide the yardstick to evaluate the decisions. It is important to note the potential correlation between different criteria.

(3) Outcomes. Each decision, evaluated on any of the given criteria, will result in exactly one of a number of possible outcomes. An outcome is also frequently referred to as an *attribute*.

Chapter 1: Multicriteria Decision Making

Formally, we will consider decisions d_i, $i=1, \ldots, m$; criteria c_k, $k=1, \ldots, q$; and outcomes that are distinguished by their *states of nature* s_j, $j=1, \ldots, n$. Given these three main components, there are a total of three possible pairs, two of which include decisions. The first pair involves decisions and criteria, so that for each chosen pair (d_i, c_k), the outcome is considered certain. This type of deterministic model is referred to as *multicriteria decision problems*, which are discussed in Chapter 1. The second type of problem considers only a single criterion (usually called "payoff"), coupled with a number of possible, but uncertain, outcomes. Here, each pair (d_i, s_j) represents a possible outcome in *decision-making under uncertainty*. Such scenarios are discussed in Chapter 2. Finally, the problems considered in Chapter 3 of this part also involve uncertainty. However, the uncertainty here is not the result of some random force beyond anybody's control, but it derives from the actions of other decision makers, who have their own objectives and decision spaces. Based on these, the decision makers in the problem may choose to compete or cooperate. Such problems are said to belong to the class of *game theory*.

CHAPTER 1 MULTICRITERIA DECISION MAKING

This chapter analyzes a variety of approaches dealing with decision-making in the presence of multiple criteria. Work in the field is based on early studies by Simon (1982), Keeney and Raiffa (1976), and the psychological investigations by Kahneman and Tversky (1979). Zimmermann (1991) points out that *multicriteria decision making*, or *MCDM* for short, can be subdivided into two subsets. They are *multiobjective decision making*, or *MODM*, and *multiattribute decision making*, or *MADM*. The difference is that in multiobjective models, variables tend to be continuous, while multi-attribute decision-making problems are typically discrete problems that choose one out of a given number of options. Many, if not most, authors, however, employ the term multicriteria decision-making in the narrow sense to refer to a multiattribute decision making problem. Given the diversity of the strands of research, it is perhaps not surprising that each approach has its proponents, who work in their own area and hardly ever venture outside. There are, however, some notable exceptions, such as the recent books by Hanne (2001), Belton and Stewart (2002), and Ehrgott and Gandibleux (2002) who attempt to integrate the field.

The first section of this chapter will present the fundamentals of multiobjective decision-making in continuous space. In particular, we will give an account of vector optimization, along with some of its uses and properties. Section 2 outlines the basic ideas of decision-making in discrete space. We then distinguish between problems in which we compare the achievements of decisions with respect to the criteria with the achievements of other decisions under consideration (internal comparisons), and those models that compare the achievements of decisions with external yardsticks (external comparisons). One class of methods that employ external comparisons comprises the reference point methods of Section 3; all others use internal comparisons. The next distinction concerns the consistency of the decision makers' estimates. In the case where the estimates are consistent, we discuss data envelopment analysis in Section 4, preference cones in Section 5, multiattribute value theory in Section 6, and outranking methods in Section 7. The case in which the decision maker's input is possibly inconsistent is discussed in Section 8 of this chapter.

1.1 Vector Optimization

This section considers one way of dealing with continuous problems whose main feature are its multiple objectives. Problems of this nature date back to Charnes and Cooper in the early 1960s; a short historical survey and a fairly comprehensive treatment is provided by Zeleny (1974). In order to formally state the problem, define $\mathbf{x}$ as a column vector of decision variables, let $\mathbf{Ax} \leq \mathbf{b}$ denote a set of linear constraints, and let $\mathbf{c}^k$ denote the row vector of objective function coefficients of the k-th objective, $k = 1, \ldots, q$. The problem can then be written as

$$\begin{aligned} \text{P: Max } z_1 &= \mathbf{c}^1 \mathbf{x} \\ \text{Max } z_2 &= \mathbf{c}^2 \mathbf{x} \\ &\ldots \\ \text{Max } z_q &= \mathbf{c}^q \mathbf{x} \end{aligned}$$

$$\text{s.t. } \mathbf{Ax} \leq \mathbf{b}$$
$$\mathbf{x} \geq \mathbf{0}.$$

Defining the $[q \times n]$- dimensional matrix $\mathbf{C} = \begin{bmatrix} \mathbf{c}^1 \\ \mathbf{c}^2 \\ \vdots \\ \mathbf{c}^q \end{bmatrix}$, the objective functions can also be written as Max $\mathbf{Cx}$. Since the matrix $\mathbf{C}$ consists of vectors, each of which belongs to an objective, the problem P is generally known as a *vector optimization problem*. It is clear that problem P is nothing but a linear programming problem with multiple objectives.

In order to develop a methodology for vector optimization problems, ignore the constraints for the time being. Consider any (feasible) point $\hat{\mathbf{x}}$ and plot the gradients $\mathbf{c}^1$ and $\mathbf{c}^2$ rooted at that point. We can now construct iso-profit hyperplanes $\mathbf{c}^1\mathbf{x} = \mathbf{c}^1\hat{\mathbf{x}}$ and $\mathbf{c}^2\mathbf{x} = \mathbf{c}^2\hat{\mathbf{x}}$. These two hyperplanes subdivide the space into four cones, which share the same common vertex $\hat{\mathbf{x}}$: $C^{++} = \{\mathbf{x}: (\mathbf{c}^1\mathbf{x} \geq \mathbf{c}^1\hat{\mathbf{x}}) \wedge (\mathbf{c}^2\mathbf{x} \geq \mathbf{c}^2\hat{\mathbf{x}})\}$, $C^{+-} = \{\mathbf{x}: (\mathbf{c}^1\mathbf{x} \geq \mathbf{c}^1\hat{\mathbf{x}}) \wedge (\mathbf{c}^2\mathbf{x} \leq \mathbf{c}^2\hat{\mathbf{x}})\}$, $C^{-+} = \{\mathbf{x}: (\mathbf{c}^1\mathbf{x} \leq \mathbf{c}^1\hat{\mathbf{x}}) \wedge (\mathbf{c}^2\mathbf{x} \geq \mathbf{c}^2\hat{\mathbf{x}})\}$, and $C^{--} = \{\mathbf{x}: (\mathbf{c}^1\mathbf{x} \leq \mathbf{c}^1\hat{\mathbf{x}}) \wedge (\mathbf{c}^2\mathbf{x} \leq \mathbf{c}^2\hat{\mathbf{x}})\}$. These four sets are shown in Figure I.1.

Still disregarding feasibility, moving from $\hat{\mathbf{x}}$ into the interior of the cone C^{--} means that, as compared to $\hat{\mathbf{x}}$, the values of both objective functions deteriorate,

Chapter 1: Multicriteria Decision Making 25

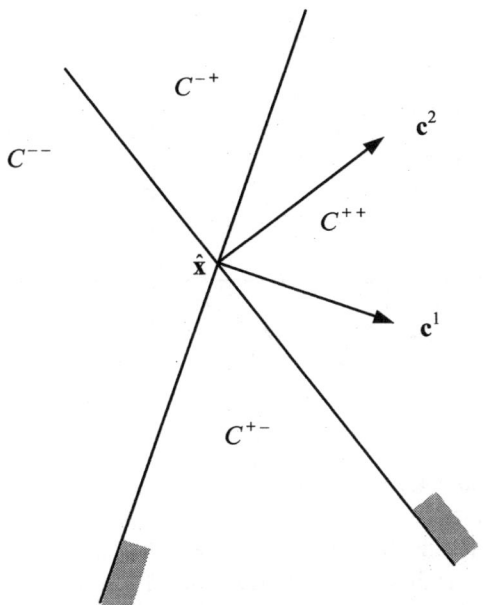

Figure I. 1

so that all solutions in the interior of the cone C^{--} are inferior to $\hat{x}$. Similarly, any solution in the interior of the cone C^{++} is superior to $\hat{x}$ with respect to both objectives, which is why we call $C=^{++}$ an *improvement cone*. The interiors of the sets C^{+-} and C^{-+} do not allow such comparisons with $\hat{x}$: while one objective deteriorates when moving from $\hat{x}$ to some point in either of these cones, the other improves, thus not allowing unambiguous conclusions. The extension of the concept to multiple objectives is straightforward. The improvement cone is then simply defined as the intersection of the half spaces $c^k x \geq c^k \hat{x} \ \forall \ k = 1, ..., q$.

Example 1: Consider the objectives Max $z_1 = x_1 + x_2$, and Max $z_2 = x_1 + 2x_2$, which is a case of little conflict that results in a large improvement cone. This case is shown in Figure I.2a. In the extreme case, all objective functions are identical, i.e., all vectors c^k are identical, so that there is no conflict at all. In this case, the improvement cone at $\hat{x}$ coincides with the half space $c^k x \geq c^k \hat{x} \ \forall \ k = 1, ..., q$: this is standard linear programming. On the other hand, consider the objectives Max $z_1 = 3x_1 + x_2$, and Max $z_2 = -2x_1 - x_2$. As the gradients of the objective functions indicate, there is extensive conflict. This case is characterized by a small improvement cone, as shown in Figure I.2b. Considering again the extreme case, the objectives would be diametrically opposed, resulting in a degenerate improvement cone whose interior is empty.

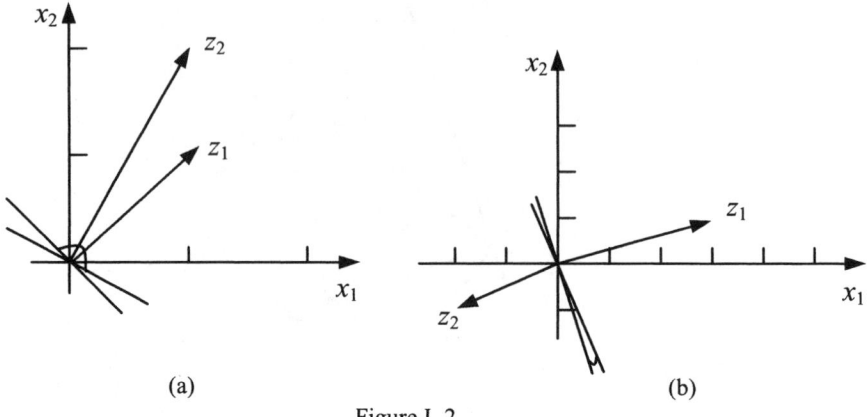

Figure I. 2

Introduce now the feasible set and, for convenience, denote it by $X = \{\mathbf{x}: \mathbf{Ax} \leq \mathbf{b}\}$. By assumption, $\hat{\mathbf{x}}$ is feasible. If the interior of the intersection of the improvement cone and the feasible set includes $\hat{\mathbf{x}}$, then there exists at least one point $\tilde{\mathbf{x}} \neq \hat{\mathbf{x}}$, such that the solution $\tilde{\mathbf{x}}$ is feasible *and* at least one of its objective function values is better than that of $\hat{\mathbf{x}}$. This implies that the solution $\tilde{\mathbf{x}}$ dominates $\hat{\mathbf{x}}$, and $\hat{\mathbf{x}}$ is of no interest to us. This leads to

Definition I.1: A solution $\hat{\mathbf{x}}$ is called *nondominated* (or *noninferior*, or *efficient*, or *Pareto-optimal*), if $C^{++} \cap X = \{\hat{\mathbf{x}}\}$.

The concept of nondominance replaces that of optimality in case of more than one objective. Coupled with the convexity of the feasible set X, we can easily prove

Lemma I.2: All nondominated solutions are located on the boundary of X.

Due to their boundary location, they are also referred to as the *nondominated frontier* or *efficient frontier*.

Example 2: Consider the following problem.

P: Max $z_1 = 2x_1 + x_2$
   Max $z_2 = -3x_1 + 2x_2$

$$\begin{aligned}
\text{s.t.} \quad -x_1 + x_2 &\leq 5 & (1)\\
x_2 &\leq 7 & (2)\\
x_1 + x_2 &\leq 10 & (3)\\
x_1 &\leq 6 & (4)\\
x_1, x_2 &\geq 0.
\end{aligned}$$

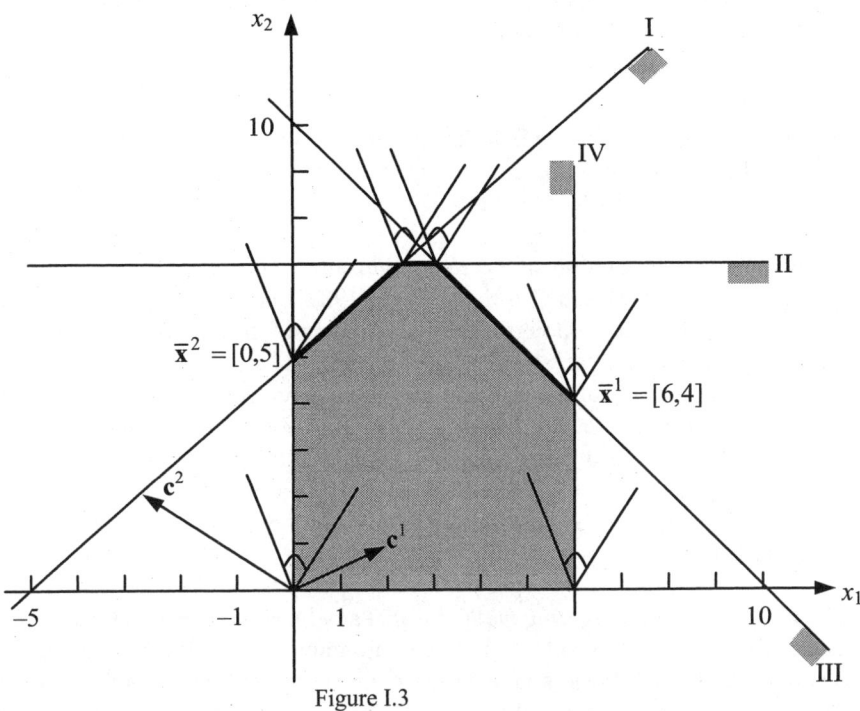

Figure I.3

Figure I.3 shows the feasible set and the gradients of the two objective functions. The improvement cone is also indicated at each of the six extreme points of the shaded feasible region. The nondominated solutions are shown by the bold line. It is apparent that $\bar{\mathbf{x}}^1 = [6, 4]$ is an optimal solution with respect to the first objective, while $\bar{\mathbf{x}}^2 = [0, 5]$ is optimal with respect to the second objective. This demonstrates the relativity of the concept of optimality: The concept applies only to a single objective (which is why prudent analysts will never state that some solution is optimal, but "optimal with respect to a certain objective function"). Moving from $\mathbf{x}^1$ to $\mathbf{x}^2$ along the nondominated frontier leads first from point (6, 4) with objective values (16, −10) to the point (3, 7) with objective values (13, 5), further to the point (2, 7) with objective values (11, 8), and finally to the point (0, 5) with objective values (5, 10). If the decision maker were to prefer the solution (3, 7) over (6, 4), then he would value an increase of 15 in the second objective higher than a decrease of 3 in the first objective, giving an indication of the decision maker's preferences as far as tradeoffs are concerned. The arguments concerning further moves along the efficient frontier are similar. Comparing now a straight-line move from $\mathbf{x}^1$ to $\mathbf{x}^2$, the *average* tradeoff is 20/11, i.e., for each unit of the first objective that we sacrifice, we gain $20/11 \approx 1.8182$ units of the second objective. Whether or not such a move is desirable will depend on whether or not the decision maker accepts this 20: 11 ratio. If one unit of the first objective is worth more than 1.8182 units of the second objective to the decision-maker, then $\bar{x}^1$ is preferable over $\bar{x}^2$. Similarly, if the decision-maker assesses the value of one unit of the first objective as less than 1.8182 units of the second objective, then the decision maker prefers $\bar{x}^2$ over $\bar{x}^1$. We will return to the issue of tradeoffs below. A discussion concerning the problems and pitfalls associated with tradeoffs is provided by Keeney (2002).

Consider now the solution of vector optimization problems. Early attempts go back to Zeleny (1974); see also Eiselt et al. (1987). Zeleny's exact simplex-based method finds all extreme points on the nondominated frontier, but this is also the weakness of the approach—since the number of efficient extreme points is exponential in the size of the problem. An obvious remedy is the approximation of the efficient frontier—quite meaningful, as no decision maker will be able or willing to examine and compare even a small proportion of the efficient set, anyway. Cohon (1978) outlines a number of approaches, two of which, the weighting method and the constraint method, stand out due to their effectiveness and simplicity.

The idea behind the *weighting method* is that any linear convex combination of the given objectives will result in a linear programming problem, one of whose (nondegenerate) solutions is an extreme point on the efficient frontier. Formally, define a row vector of weights $\mathbf{w} \geq \mathbf{0}$, and compute the *composite objective function* Max $z = \mathbf{wCx}$. Replacing the q objectives in the problem P by this composite objective, a standard linear programming problem results, which can be solved by the pertinent methods. An optimal solution is one point on the efficient frontier. The process is repeated for a variety of different weight vectors $\mathbf{w}$.

Example 3: Consider the problem P in Example 2. The weights $\mathbf{w} = [w_1, w_2] = [2, 1]$ result in the composite objective $2z_1 + 1z_2 = 2(2x_1 + x_2) + 1(-3x_1 + 2x_2) = x_1 + 4x_2$, which has the optimal solution $\bar{x} = [3, 7]$. The weight vector $\mathbf{w} = [6, 1]$ generates the composite objective function $6z_1 + 1z_2 = 6(2x_1 + x_2) + 1(-3x_1 + 2x_2) = 9x_1 + 8x_2$, which has the optimal solution $\bar{x} = [6, 4]$. Similarly, the weights $\mathbf{w} = [2, 8]$ generate the composite objective $2z_1 + 8z_2 = 2(2x_1 + x_2) + 8(-3x_1 + 2x_2) = -20x_1 + 18x_2$, which has the optimal solution $\bar{x} = [0, 5]$. As can be seen in Figure I.3, all generated solutions are located on the efficient frontier.

Another possible approach is the *constraint method*. The idea is to choose k, $1 \leq k \leq q$, and transform all but one of the objectives Max $z_v = \mathbf{c}^v\mathbf{x}$ to constraints $\mathbf{c}^v\mathbf{x} \geq \tilde{z}_v$ with some target value $\tilde{z}_v$, $v = 1, 2, \ldots, k-1, k+1, \ldots, q$ and add them to the already existing constraints. The result is a linear programming problem that can easily be solved with any of the pertinent methods. The decision maker will typically transform objectives to constraints in those cases, in which target values $\tilde{z}_v$ are agreed upon, or appear reasonable. Clearly, there are many degrees of freedom to choose k and the target values, thus demanding extensive sensitivity analyses. The optimal solutions to the transformed problem are not necessarily extreme points of the original problem, but they will be part of the nondominated frontier (provided the transformed problem has feasible solutions).

Example 4: Consider again Example 2. Suppose that we have decided to keep the first objective as an objective function, while the second objective will be transformed to a constraint $-3x_1 + 2x_2 \geq \tilde{z}_2$, where the value of $\tilde{z}_2$ is modified parametrically. The optimal solutions $\tilde{x}$ and the associated values of the objective function $\tilde{z}_1$ are shown in Table I.1.

Table I.1

$\tilde{z}_2$	$\tilde{x}$	$\tilde{z}_1$
0	[4, 6]	14
2	[3.6, 6.4]	13.6
5	[3, 7]	13
6	[2.67, 7]	12.33
8	[2, 7]	11
10	[0, 5]	5
> 10	no feasible solution exists	

Again, simple inspection can verify that all optimal solutions obtained in this manner are points on the nondominated frontier.

It is apparent that with searches as demonstrated for the weighting method and the constraint method, a large number of linear programming problems need to be solved. This is not really a drawback; the real difficulty is that these solutions will have to be evaluated and compared with each other by a decision maker. In order to ensure the cooperation of a decision maker, we have to ascertain whether the number of comparisons is limited to a reasonable level. Whatever method we use, we will have to determine a finite—and fairly small—number of solutions that are shortlisted for further consideration. The remainder of this chapter assumes that such a finite number of options exists, among which the decision maker will have to choose.

1.2 Basic Ideas of Multicriteria Decision Making

This section will introduce and discuss some of the components of multiattribute decision making, in the following referred to as *multicriteria decision making*.

Throughout the remainder of this chapter, we assume that the decision maker has m decisions $d_1, d_2, ..., d_m$ that are to be evaluated on q criteria $c_1, c_2, ..., c_q$. The evaluations are collected in an $[m \times q]$-dimensional *evaluation matrix* (or *scoring matrix*) $\mathbf{A} = (a_{ik})$, where a_{ik} denotes the decision maker's evaluation of d_i with respect to the criterion c_k. Typically, the criteria under consideration are a mixed bag: some criteria are quantitative while others are qualitative, some are expressed as utilities, while others are disutilities, and the scales of all criteria are usually different. In order to put the evaluations into some kind of order, we first translate the information in the matrix $\mathbf{A}$ into utilities, that are collected in a matrix $\mathbf{U} = (u_{ik})$. (At this point we wish to point out that we will use the terms "utility" and "value" as synonyms, even though the more recent literature tends to favor the term "utility" for expressions involving risk, while the term "value" is used in all other cases). While this step expresses the value a decision maker associates with different levels of achievements with respect to a given criterion, the second step involves the determination of tradeoffs between criteria. In general, this may involve more or less complicated functions.

Here, we utilize simple weights w_k that are associated with the criteria c_k, $k = 1, ...q$. If, for example, $w_1 = 5$ and $w_2 = 2$, then criterion c_1 is deemed $5/2 = 2.5$ times as important as criterion c_2 by the decision maker, regardless of the achievement level on either criterion. In the next step, we need to combine the individual achievements, considering the tradeoffs between the criteria, in order to compute values of the individual decisions $v(d_i)$. This is achieved by using an *aggregation function* f chosen by the decision maker. The two most frequently used aggregation functions are the weighted sum and the weighted product of the utilities of the achievements, i.e., $v(d_i) = \sum_{k=1}^{q} u_{ik} w_k$, and $v'(d_i) = \prod_{k=1}^{q} u_{ik}^{w_k}$, respectively. While both models have their advantages, and more sophisticated aggregation functions are possible, we will follow a large part of the literature and employ the weighted sum model for most of our illustrations. In the last step, the decision maker will choose a decision by favoring d_i over d_j, if and only if $v(d_i) \geq v(d_j)$. Before summarizing the procedure as an algorithm and illustrating it by means of an example, we will investigate each of the three individual components introduced above.

Consider first the conversion of achievement levels a_{ik} to utilities u_{ik}. In order to do so, we first express all evaluations in terms of utilities rather than disutilities. The second step is to express the qualitative evaluations on a quantitative scale. For that purpose, we typically use a 5- (or 7- or 9-) point *Likert scale* that ranges from "very good" to "good", "medium", and "poor", to "very poor" in equidistant steps. In other words, these qualitative evaluations are reexpressed by utilities of 1, .75, .5, .25 and 0, respectively, or similar in case of larger or smaller steps. For "maximization criteria" c_k, i.e., criteria for which high values are deemed better

Chapter 1: Multicriteria Decision Making

than low values by the decision maker, authors such as Edwards (1971, 1977) have suggested the linear interpolation function $u_{ik} = \dfrac{a_{ik} - \underline{a}_{\bullet k}}{\overline{a}_{\bullet k} - \underline{a}_{\bullet k}}$, where $\overline{a}_{\bullet k} \geq \max\limits_{\ell}\{a_{\ell k}\}$ and $\underline{a}_{\bullet k} \leq \min\limits_{\ell}\{a_{\ell k}\}$. In case of "minimization criteria," utilities are calculated as $u_{ik} = \dfrac{\overline{a}_{\bullet k} - a_{ik}}{\overline{a}_{\bullet k} - \underline{a}_{\bullet k}}$. As a result, all utilities assume values between zero and one. As an example with $m = 4$, consider some decisions whose payoffs are 200, 350, 800, and 600, so that $\underline{a}_{\bullet k} = 200$ and $\overline{a}_{\bullet k} = 800$. The corresponding utilities are then $u_{1k} = 0$, $u_{2k} = .25$, $u_{3k} = 1$, and $u_{4k} = .6667$. The meaning of the utility is the proportion that a given payoff is through the range of values. Here, the range equals 800–200 = 600, and the payoff of, say, 350 is 150 units above the minimum and thus 150/600 = ¼ through the range.

In general, we can distinguish three different types of utility functions as shown in Figure I.4.

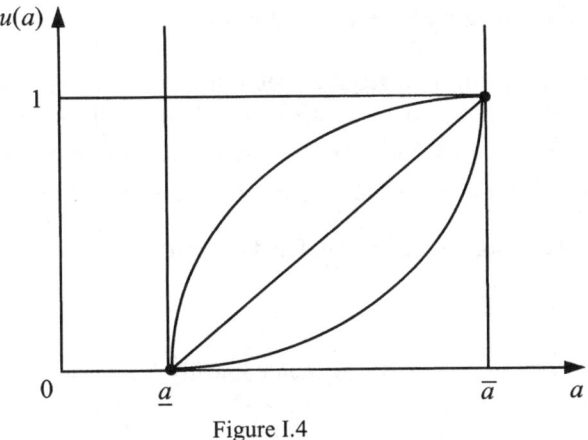

Figure I.4

The linear utility function in Figure I.4 is given by the so-called *Edwards procedure* that was used in the above example. The concave utility function assigns diminishing utility increases to the scores. For example, the first dollar in profits has a higher value to the decision maker than an additional dollar that is gained when the profit has already reached a higher level. Concave utility functions typically infer that some degree of saturation is achieved at higher levels. Concave utility functions are found in many situations. Satisficing behavior of decision makers who use target values have utility functions that stay at the same level, once the prespecified target value has been reached. Clearly, each criterion can have its own utility function. In general, most utility functions are assumed to be

- increasing, i.e., $a_{ik} > a_{\ell k}$ implies that $u(a_{ik}) > u(a_{\ell k})$ for all finite outcomes
- concave, i.e., the increase occurs at a decreasing rate.

Consider now the construction of the utility function for a given criterion. Such a task will necessarily require some input by the decision maker. As a general rule, an experienced analyst will attempt to elicit all necessary information from a decision maker, while keeping the number of questions asked to a minimum. In general, five points in the outcome or value space should give a pretty good idea of the shape of the function. These points could be determined by a version of the bisection method. As an illustration, consider a decision maker's responses to the potential profit in the next quarter. Assume that $10,000 is the low, and $80,000 the high estimate, to which we assign utilities of 0 and 1, respectively. The analyst could then elicit the decision maker's opinion as to what constitutes a utility of 0.5. Assume that some probing yields a value of $55,000. At this point it is already apparent that the decision maker's utility function is concave. Further questions could attempt to specify dollar values for utilities of 0.25 and 0.75, respectively. The ($, u($)) combinations obtained can then be used to determine a utility function for the entire range between $10,000 and $80,000.

Many popular utility functions require even less input from the decision maker. In case of a linear utility function, $u(a) = \alpha + \beta a\, u(a)$, there are only the two parameters α and β that have to be determined. As $u(\underline{a}) = \alpha + \beta \underline{a} = 0$ and $u(\overline{a}) = \alpha + \beta \overline{a} = 1$, these two equations are sufficient, so that no further user input is required. The result is again the function $u(a) = \dfrac{a - \underline{a}}{\overline{a} - \underline{a}}$ introduced in the previous section.

Another possibility is to consider a (monotonically increasing) exponential utility function of the form

$$u(a) = \alpha + \beta e^{-\gamma a}$$

with unknown parameters α, β and γ. Note that setting $\beta < 0$ and $\gamma > 0$ results in a concave utility function, while $\beta > 0$ and $\gamma < 0$ models a convex function. In order to determine these three unknown parameters, three equations are required. The two equations $u(\underline{a}) = \alpha + \beta e^{-\gamma \underline{a}} = 0$ and $u(\overline{a}) = \alpha + \beta e^{-\gamma \overline{a}} = 1$ are known, so that the decision maker must specify only one additional point.

Typically, the analyst would ask the decision maker for a payoff a^*, such that $u(a^*) = \frac{1}{2}$. (While a value $\hat{a}$ with, say, $u(\hat{a}) = .42$ would be equally sufficient to determine the parameters, a decision maker will usually find it considerably more difficult to specify such a value, as opposed to a^*). Solving the three nonlinear equations $u(\underline{a}) = 0$, $u(\overline{a}) = 1$, and $u(a^*) = \frac{1}{2}$ results in

$$2e^{-\gamma a^*} = e^{-\gamma \overline{a}} + e^{-\gamma \underline{a}}$$

$$\alpha = \frac{-e^{-\gamma \underline{a}}}{e^{-\gamma \overline{a}} - e^{-\gamma \underline{a}}}, \text{ as well as}$$

$$\beta = \left[e^{-\gamma \overline{a}} - e^{-\gamma \underline{a}}\right]^{-1}.$$

Example: Let again $\underline{a} = 200$ and $\overline{a} = 800$, and suppose that the decision maker feels that $a^* = 300$, such that $u(300) = \frac{1}{2}$. Solving for γ results in $\gamma = .0067597469$, which, in turn, leads to $\alpha = 1.017626$ and $\beta = -3.933076$, so that the utility function is

$$u(a) = 1.01763 - 3.933076 e^{-.00676a}$$

Other functions can be used as well. One obvious choice is a quadratic equation $u(a) = \alpha a^2 + \beta a + \gamma$. Again, three parameters α, β and γ have to be determined, so that again, in addition to the equations $u(\underline{a}) = 0$ and $u(\overline{a}) = 1$, a third utility estimate is needed, e.g., a score a^* with $u(a^*) = \frac{1}{2}$. Given that, the system of simultaneous linear equations is

$$\alpha \underline{a}^2 + \beta \underline{a} + \gamma = 0$$
$$\alpha a^{*2} + \beta a^* + \gamma = \frac{1}{2}$$
$$\alpha \overline{a}^2 + \beta \overline{a} + \gamma = 1.$$

The previous example with $\underline{a} = 200$, $\overline{a} = 800$, and $a^* = 300$ results in the system

$$40{,}000\alpha + 200\beta + \gamma = 0$$
$$90{,}000\alpha + 300\beta + \gamma = \frac{1}{2}$$
$$640{,}000\alpha + 800\beta + \gamma = 1.$$

The system has the unique solution $\alpha = -.000006667$, $\beta = .008333$, and $\gamma = -1.4$, so that the utility function is $u(a) = -.000006667 a^2 + .008333 a - 1.4$. However, this function has a maximum at $a = 625$ with $u(625) = 1.2042 > 1$, rendering the approach unusable, at least for this example.

The next step involves the determination of weights associated with the criteria. These weights w_k, $k=1, \ldots, q$ indicate the tradeoffs between the criteria as seen by the decision maker. One systematic way to develop such weights is the use of a *value tree.* A value tree is constructed sequentially in the following way. Starting with the decision at hand (the root node of the tree graph, or, more precisely, an arborescence), the main criteria are outlined, each of which is again represented at a node in the tree, which is reached from the root node by a directed arc. Whenever applicable, each of the main criteria is then investigated and subdivided into lower-level subcriteria, which are, again, represented as nodes and reachable from the main criterion by directed arcs, etc. The decision maker is then asked to assign values to each of the main criteria. A weight w_k that is assigned to criterion c_k is associated with the arc leading into the node that represents criterion c_k. The weights are chosen, so that the sum of weights of all arcs that lead out of each node equals one. Once all arcs have weights assigned to them, we can calculate the weights of all criteria that are not further subdivided as the product of the weights along the unique path from the root node to the criterion under consideration.

The idea behind value trees is that with them, the decision maker is only asked to specify weights and relations between criteria at the same level, which is arguably easier than comparing the relative importance of very diverse criteria. It is apparent that only the criteria at the leaves of the value tree are the criteria that are ultimately used in the decision making process. Ideally, these criteria are mutually exclusive and collectively exhaustive (as far as the issues are concerned that are relevant to the decision at hand). This process can be explained by the following

Example: A traveler would like to investigate a variety of potential destinations for his annual vacation. As the main criteria for the trip he has identified "Status," "Enjoyment," and "Relaxation" with which he associates weights of .2, .5, and .3, respectively. While the vacationer does not believe that the criterion "Status" should be further subdivided, he feels that the main sources of enjoyment are "Cultural experience", "Nature," and the "Company of friends." The weights associated with these subcriteria are .4, .4, and .2, respectively. Similarly, the criterion "Relaxation" could be subdivided into "Weather/sun," and "Sport activities" with weights of .7 and .3, respectively. Finally, the planner feels that the most important "Sport activities" are "Tennis" and "Hang gliding" with weights of .8 and .2, respectively. Figure I.5 shows the value tree with the weights next to the arcs specified by the decision maker. The numbers next to the nodes show the weights associated with the subcriteria.

Consider now the third step, the aggregation of utilities. As mentioned above, most analysts use weighted additive or multiplicative models, even though both approaches have significant shortcomings. This can be demonstrated by considering complementary goods, such as a camera and film. By themselves, each has a very limited utility to the user, whereas the combination of the two allows the user to take photographs and enjoy the two products together. This simple illustration demonstrates that utilities cannot simply be determined independently

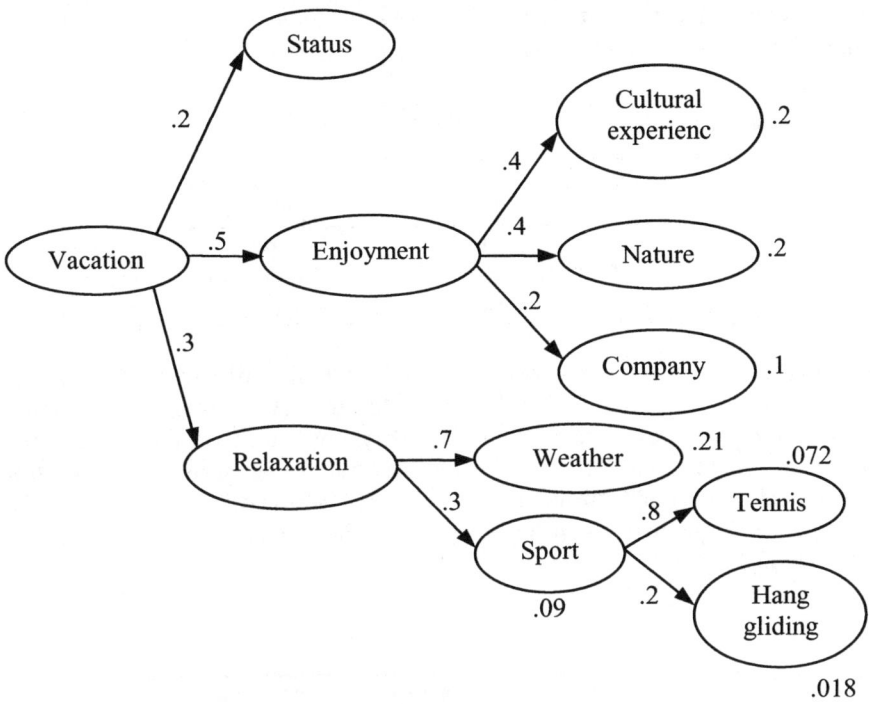

Figure I.5

and then added to each other, or combined in similar ways. However, this is precisely what most models do. Despite the shortcomings of such an approach, we will use it in the remainder of this chapter.

We are now able to summarize the steps above in a prototype of a multicriteria decision making procedure.

A Generic MCDM Procedure

Step 1: Determine utilities u_{ik} for all evaluations a_{ik}; $i = 1, ..., m$; $k = 1, ..., q$.
Step 2: Determine weights w_k for all criteria c_k, $k = 1,..., q$.
Step 3: Compute values $v(d_i) = f(\mathbf{u}_{i\bullet}, \mathbf{w})$ for all $i = 1, ..., m$.
Step 4: Choose decision d_{i*}, such that $v(d_{i*}) = \max_i \{v(d_i)\}$.

In order to illustrate the generic procedure above, we consider the following

Example: A decision maker faces the task of choosing one of three sites for the location of a sanitary landfill. The three decisions are then d_1, d_2, and d_3. The criteria are cost (in millions of dollars), impact on the surrounding communities, and the environmental protection afforded by the solution. The decision maker's evaluations are summarized in Table I.2.

Table I.2

	Cost	Community impact	Environmental protection
d_1	20	high	poor
d_2	80	medium	very good
d_3	40	medium	medium

Note that the first two criteria are of the "minimization" type whereas the last is of the "maximization" type. Using Edwards' linear scaling technique introduced above for c_1, a 5-point scale starting at "very low", then continuing to "low", "medium", "high", and "very high" for c_2, and the 5-point scale that ranges from "very good", to "good", "medium", "poor", and "very poor" for criterion c_3. The scores on both qualitative scales are assumed to have equidistant utilities, and are collected in Table I.3.

Table I.3

	Cost	Community impact	Environmental protection
d_1	1	.25	.25
d_2	0	.5	1
d_3	.6667	.5	.5

If the decision maker estimates the weights of the criteria at $\mathbf{w} = [.4, .25, .35]^T$, the weighted sum model obtains the values $v(d_i) = .55, .475, .5667$, so that $d_3 \succ d_1 \succ d_2$, suggesting that decisions d_3 and d_1 are superior to d_2. However, should the decision maker revise his estimates of the weights to $\mathbf{w} = [.25, .4, .35]^T$, then the values are $v(d_i) = .4375, .55, .5417$, so that $d_2 \succ d_3 \succ d_1$, so that d_2 is now considered the best of the three available choices, albeit at a narrow margin.

Similarly, the weighted product model with the weights $\mathbf{w} = [.4, .25, .35]^T$ results in the values $v(d_i) = .4353, 0, .5610$ and the preference ranking $d_3 \succ d_1 \succ d_2$, and the weights $\mathbf{w} = [.25, .4, .35]^T$ generate the values .3536, 0, and .5373 for the same preference ranking. Note that the weighted product model always produces a value of zero for a decision that has a zero utility in one of its components. This feature is quite desirable in weeding out decisions that perform very poorly on one or more criteria.

1.3 Reference Point Methods

The approaches discussed in this section have in common that they do not compare potential decisions and their consequences with each other. Instead, they define one, or possibly two, reference points that provide yardsticks that are used to measure and evaluate the decisions and their outcomes. To explain the concept, we will revert back to vector optimization problems discussed in Section 1.1. Suppose that the problem under consideration has a nonempty feasible set, and two objectives, Max $z_1 = \mathbf{c}^1\mathbf{x}$ and Max $z_2 = \mathbf{c}^2\mathbf{x}$ that are to be maximized. The optimal solutions with respect to the two objectives are $\bar{\mathbf{x}}^k$, $k = 1, 2$, and their associated objective function values are $\bar{\mathbf{z}}^1 = (\mathbf{c}^1\bar{\mathbf{x}}^1, \mathbf{c}^2\bar{\mathbf{x}}^1)$ and $\bar{\mathbf{z}}^2 = (\mathbf{c}^1\bar{\mathbf{x}}^2, \mathbf{c}^2\bar{\mathbf{x}}^2)$. In the vector optimization problem of Section 1.1, $\bar{\mathbf{x}}^1 = [6, 4]$ and $\bar{\mathbf{x}}^2 = [0, 5]$, so that $\bar{\mathbf{z}}^1 = [16, -10]$ and $\bar{\mathbf{z}}^2 = [5, 10]$.

Given the two objective function values $\bar{\mathbf{z}}^1$ and $\bar{\mathbf{z}}^2$, we can now construct a *reference point* $\hat{\mathbf{z}}$ which, while it typically does not belong to a feasible solution, combines the best features of both optimal solutions. This reference point is also referred to as *ideal point*. For our discussion below, the reference point could be any arbitrary point (and not necessarily $\bar{\mathbf{z}}^k = (\mathbf{c}^1\bar{\mathbf{x}}^k, \mathbf{c}^2\bar{\mathbf{x}}^k)$), as long as a regularity condition is satisfied, which requires that $\bar{\mathbf{z}}^k \leq \hat{\mathbf{z}} \; \forall \; \bar{\mathbf{z}}^k \in Z$, where $Z = \{\mathbf{z} = (z_1, z_2): z_k = \mathbf{c}^k\mathbf{x}; k=1,2; \mathbf{Ax} \leq \mathbf{b}, \mathbf{x} \geq \mathbf{0}\}$ denotes the set of objective values that belong to feasible solutions. In the vector optimization problem of Section 1.1, any ideal point $\hat{\mathbf{z}}$ must satisfy $\hat{\mathbf{z}} \geq [16, 10]$.

Optimization with reference points is by no means new. Zeleny's "de novo programming" is an example of the use of ideal points; see Zeleny (1981, 1986, 1995). The idea is now to find compromise solutions that are "as close as possible" to the ideal point $\hat{\mathbf{z}}$. In order to operationalize the idea, we need a measure of closeness. Rather than introduce and discuss different distance measures at this point, we refer to our discussion of distances in general (and Minkowski metrics in particular) in Section 1.3 of Part II. For given points $\mathbf{v}, \mathbf{w} \in \mathbb{R}^q$, define the ℓ_p norm as $\|\mathbf{v} - \mathbf{w}\|_p = \left[\sum_k |v_k - w_k|^p\right]^{1/p}$. For $p = 1$, we obtain the *Manhattan* distance with $\sum_k |v_k - w_k|$, for $p = 2$, the *Euclidean* distance results with $\sqrt{\sum_k (v_k - w_k)^2}$, while $p \to \infty$ yields the *Chebyshev* distance $\max_k \{|v_k - w_k|\}$.

We will now describe a weighted distance function based on ℓ_p distance functions by defining $\hat{d}_k^p = |\hat{z}_k - z_k|^p$, where $\hat{\mathbf{z}} = (\hat{z}_k)$ is the chosen ideal point, and $z_k = \mathbf{c}^k\mathbf{x}$ is the k-th objective with the feasible set $X = \{\mathbf{Ax} \leq \mathbf{b}, \mathbf{x} \geq \mathbf{0}\}$.

The idea is now to find a solution that is as close to the ideal point as possible. In order to do so, we can write the problem for ℓ_p distances as

$$\text{Min } z(\mathbf{w}, p) = \sum_k w_k \hat{d}_k^p$$

s.t. $\mathbf{x} \in X$,

where $\mathbf{w} = (w_j)$ is a given vector of weights, so that w_j denotes the importance of the deviation of the k-th objective from the ideal point. For $p = 1$, the objective can then be rewritten as $\text{Min } z(\mathbf{w}, 1) = \sum_k w_k |\hat{z}_k - z_k| = \sum_k w_k |\hat{z}_k - \mathbf{c}^k\mathbf{x}|$. As $\hat{z}_k \geq \mathbf{c}^k\mathbf{x}$ by virtue of the regularity condition, the objective can be written as $\sum_k w_k(\hat{z}_k - \mathbf{c}^k\mathbf{x}) = \sum_k w_k \hat{z}_k - \sum_k w_k \mathbf{c}^k\mathbf{x}$, where the first sum is a constant, so that the objective can be written as $\text{Max } z(\mathbf{w},1) = \sum_k w_k \mathbf{c}^k\mathbf{x}$, which is nothing but a linear combination of the objective functions, exactly as it is used in the weighting method.

Following the same logic as the above arguments is *TOPSIS* (*T*echnique for *O*rder *P*reference by *S*imilarity to *I*deal *S*olution), a method suggested by Hwang and Yoon (1981). We define u_{ik} as the utility of decision i with respect to criterion k, and $\bar{\mathbf{u}} = (\bar{u}_{\bullet k})$ with $\bar{u}_{\bullet k}$ as the ideal utility of criterion c_k, where $\bar{u}_{\bullet k} \geq \max_i \{u_{ik}\}$, which is usually, but not necessarily, satisfied as equality. The ideal point is used and referred to as *reference point*. Furthermore, define $\bar{d}_{ik}^p = (\bar{u}_{\bullet k} - u_{ik})^p$, as well as $\mathbf{w} = (w_k)$ as the vector of importance of the criteria. The TOPSIS method can then formally be described as follows.

The TOPSIS Method

Step 1: Determine the distance between the outcome of decision d_i and the ideal point as $\bar{\delta}_i = \left[\sum_k w_k \bar{d}_{ik}^p\right]^{1/p}$.

Chapter 1: Multicriteria Decision Making

Step 2: Solve $\text{Min}_i \bar{\delta}_i$. The decision d_ℓ with $\bar{\delta}_\ell = \min_i \{\bar{\delta}_i\}$ is optimal with respect to the chosen reference point and metric.

The method may be explained by the following

Example: A retail franchise is planning to locate a new store in a rapidly developing town. Management has delineated three important features: (annualized) costs, the quality of the neighborhood, and the future potential of the site. Letting c_1, c_2, and c_3 symbolize the three criteria, the potential locations d_1, d_2, d_3, and d_4 have been evaluated as shown in Table I.4.

Table I.4

	c_1 Cost in \$	c_2 Neighborhood	c_3 Site
d_1	40	medium	good
d_2	50	medium	very good
d_3	30	good	poor
d_4	60	very good	medium

Translating these evaluations into a utility matrix results in

$$\mathbf{U} = (u_{ik}) = \begin{bmatrix} .6667 & .5 & .75 \\ .3333 & .5 & 1 \\ 1 & .75 & .25 \\ 0 & 1 & .5 \end{bmatrix}.$$

Defining the ideal point as the column maxima, we obtain $\bar{\mathbf{u}} = [1, 1, 1]$. For the weights $\mathbf{w} = [.5, .2, .3]^T$, we obtain the distances $\bar{\delta}_i$ for the different distance functions as shown in Table I.5.

Table I.5

	ℓ_1	ℓ_2	ℓ_∞
d_1	.3417	.3526	.5000
d_2	.4333	.5217	.6667
d_3	.2750	.4257	.7500
d_4	.6500	.7583	1.0000

Considering the ideal point, the rankings for the three distance functions are $d_3 \succ d_1 \succ d_2 \succ d_4$ for ℓ_1, $d_1 \succ d_3 \succ d_2 \succ d_4$ for ℓ_2, and $d_1 \succ d_2 \succ d_3 \succ d_4$ for ℓ_∞ distances, clearly favoring decisions d_1 and d_3.

An alternative to the above procedure is to maximize the (weighted) distance from a "least ideal point." While the concept may sound appealing at first, at least to users who aim to avoid the worst case, it has a number of problems associated with it. Most prominently, moving away from the least ideal point is not equivalent to moving towards the ideal point; in fact, we may simultaneously move away from both. Another issue relates to the use of ℓ_∞ distances. Whereas the Chebyshev distance between a decision and the ideal point is expressed as the largest difference with respect to any criterion (i.e., the criterion on which our decision performs worst), in the case of the least ideal point, this distance expresses the largest difference between the least ideal point and our decision, i.e., the criterion on which our decision performs best. This is a fundamentally different underlying philosophy. If both, an ideal point and a least ideal point are given, a decision maker could either use one of them or any appropriate linear combination of them.

Another possibility is the use of target values. Similar to goal programming (see, e.g., Eiselt *et al.* (1987), Ignizio (1976, 1982), Lee (1972)), define the utilities of the target values as u_k^*, $k = 1, \ldots, q$, and let w_k^+ and w_k^- denote the penalty for deviating from the target value. The achievement for each decision on all objectives can then be calculated as a weighted average of the individual over- and underachievements. As an illustration, consider again the same

Example: Suppose that the target values are [45, good, good], or, in terms of utilities, $\mathbf{u}^* = [.5, .75, .75]$, and the two sets of weights are $\mathbf{w}^+ = [.3, .1, 0]^T$, and $\mathbf{w}^- = [.8, .4, .2]^T$. The over- and underachievements of decision d_3 are then [.5, 0, 0] and [0, 0, .5], respectively, so that the weighted average deviation is .25. The values for the decisions d_1, d_2, and d_4 are .15, .2333, and .475, respectively, resulting in a ranking of $d_1 \succ d_2 \succ d_3 \succ d_4$.

1.4. Data Envelopment Analysis

The method described in this section is not a selection tool, but a tool that evaluates one of the given decision alternatives with a hypothetical decision that is generated as a linear convex combination of all possible alternatives.

It is a very useful tool for decision makers in that it provides a simple way to evaluate the relative efficiency of different units. While its roots go back to the early 1950s, data envelopment analyses (*DEA*) were first suggested by Charnes *et al.* (1978), who formulated efficiency as the ratio of weighted output to weighted input. They then applied the well-known transformation to the formulation with a fractional objective to rewrite the problem as a linear programming problem. A good introduction and interesting survey can be found in Belton and Vickers (1993). The setting is similar to that of the other techniques described above, except that minimization criteria are usually referred to as input factors, while maximization criteria are typically thought of as output factors. Furthermore, the different decisions are often thought of as branches or franchises of a company that are to be compared with each other. This is not much more than a semantic change, as any economic analysis either minimizes a decreasing function of the input, or maximizes an increasing function of the output (or a combination thereof). Suppose that the attributes are a_{ik}^+ and a_{ik}^-, where the former denotes the output, while the latter denotes the input of factor k in branch i. Assume that we try to evaluate the efficiency of "branch" or decision d_ℓ.

We can now define w_i as the fraction of decision d_i that is included in the hypothetical composite that decision d_i will be compared with. The composite will then require inputs of $\sum_{i=1}^{m} a_{ik}^- w_i$ $\forall$ k related to input, and produce an output of $\sum_{i=1}^{m} a_{ik}^+ w_i$ $\forall$ k related to output. We can now define the efficiency E_ℓ associated with decision d_ℓ which is determined by the following linear programming problem:

$$P_\ell: \text{Min } E_\ell$$
$$\text{s.t.} \sum_{i=1}^{m} a_{ik}^- w_i \leq a_{\ell k}^- E_\ell \quad \forall \text{ } k \text{ related to inputs,}$$
$$\sum_{i=1}^{m} a_{ik}^+ w_i \geq a_{\ell k}^+ \quad \forall \text{ } k \text{ related to outputs,}$$
$$\sum_{i=1}^{m} w_i = 1$$
$$w_i \geq 0 \quad \forall \text{ } i = 1, \ldots, m.$$

The basic idea of this problem is to find a composite which is as efficient as, or possibly even more efficient than, decision d_ℓ. This is accomplished by ensuring that the composite requires inputs that are no larger than a fraction E_ℓ of the input of d_ℓ, while, at the same time, producing an output that is at least as large as that of d_ℓ. This is to be accomplished while trying to find an efficiency of decision d_ℓ that is a small as possible. If, for instance, the composite could produce at least as much output as d_ℓ with, say, only 80% of the input required by d_ℓ, then decision d_ℓ would be called no more than 80% efficient. If the optimal value $\overline{E}_\ell$ is less than one, then there exists a composite that can produce the same output as decision d_ℓ with less resources; hence decision d_ℓ is not efficient. On the other hand, if the optimal value $\overline{E}_\ell$ equals one, then there exists no composite that can produce the same output as d_ℓ with less resources, and hence d_ℓ can be called efficient. Note that the concept of efficiency employed here is based exclusively on the comparison between existing options. In the extreme case in which all existing options are bad from an objective point of view, we will still conclude that quite a few of the options are "efficient." This is to be understood only in the comparative, and not the absolute, sense. In order to communicate the ideas, consider the following

Example: A city council has to evaluate its four historic buildings. In particular, it operates a lighthouse museum with original Fresnel lens, a downtown townhouse whose main feature is its 19th century collection of paintings, an old factory on the fringes of town that sports avant-garde art exhibitions, and a historic farm near a modern-day suburb. The main criteria used in the evaluation are the operating costs and the loss of revenue based on potential alternative uses of the buildings (the input factors), as well as the number of visitors, the architectural integrity, and the historical significance of the buildings (the output factors). The numerical evaluations are shown in Table I.6.

Table I.6

	Input factors		Output factors		
	Operating costs	Loss of revenue	Number of visitors (in 1,000)	Architectural integrity	Historical significance
Lighthouse	20	100	18	100	90
Townhouse	30	70	25	90	90
Factory	10	60	5	10	70
Farm	5	50	3	50	80

Evaluating the Factory, we can set up the following linear programming problem:

P_3: Min E_3
s.t. $20w_1 + 30w_2 + 10w_3 + 5w_4 \leq 10E_3$
$100w_1 + 70w_2 + 60w_3 + 50w_4 \leq 60E_3$
$18w_1 + 25w_2 + 5w_3 + 3w_4 \geq 5$
$100w_1 + 90w_2 + 10w_3 + 50w_4 \geq 10$
$90w_1 + 90w_2 + 70w_3 + 80w_4 \geq 70$
$w_1 + w_2 + w_3 + w_4 = 1$
$w_1, w_2, w_3, w_4 \geq 0.$

The solution to this problem is $\overline{w}_1 = 0, \overline{w}_2 = .0909, \overline{w}_3 = 0$, and $\overline{w}_4 = .9090$, with $\overline{E}_3 = .8636$. This result indicates that the Factory is only about 86% efficient as compared with a combination of the existing facilities, more specifically, a fictitious combination of 9% of the townhouse and 91% of the farm).

Solving linear programming problems for the other three projects, we learn that $\overline{E}_1, \overline{E}_2$ and $\overline{E}_4 = 1$, respectively, i.e., they are each considered 100% efficient. The reason for this can be explained as follows. Consider the following two scenarios.

(1) Suppose the analysis is for unit ℓ, and assume that there exists an output factor r, such that unit ℓ has a higher output than any other unit with respect to that factor, i.e., $a_{\ell r}^+$ is the unique maximum in the r-th column. Given the second set of constraints that require that the composite has an output at least as large as that of unit ℓ, this can only be achieved by setting $w_\ell = E_\ell = 1$. As a result, the ℓ-th unit will be considered efficient, even though its output on one factor may be just marginally higher than that of other units, and its input may be arbitrarily large.

(2) A similar argument can be made for the input factors. Suppose again that we analyze unit ℓ, and assume that there exists an input factor s, that is lower than any of that of the other input units, i.e. $a_{\ell s}^-$ is the unique column minimum in column s. If we had $w_\ell < 1$, then the left-hand side of the s-th input constraint would be larger than $a_{\ell s}^-$, so that $E_\ell > 1$ is required. However, this solution cannot be optimal as $w_\ell = E_\ell = 1$ and all other variables equal zero is feasible and has a lower value of the objective function. This indicates that the basic model of data envelopment analysis as presented here requires some refinements.

1.5 Preference Cones

This section investigates an input by the decision maker that consists of direct preference statements of the type $d_i \succ d_\ell$. Given m decision alternatives, it would be sufficient if the decision maker were to make $(m-1)$ such statements. Always deleting the dominated alternative, such a process would eliminate all but one decision, which would solve the problem. However, preference statements such as this can only be derived through in-depth comparison of the two alternatives. Typically, decision makers do not have the time to answer questions that require a major involvement in the problem itself. In order to minimize the involvement of decision makers beyond the bare minimum, we can try to extract as much information from each preference statement as possible.

In order to do so, suppose that the decision maker has expressed that $d_i \succ d_\ell$. Assuming that the weighted sum model applies, $v(d_i) = \sum_k u_{ik} w_k$ with unknown weights w_k and similarly for $v(d_\ell)$, so that we can write $v(d_i) \geq v(d_\ell)$, or, equivalently, $\sum_k (u_{ik} - u_{\ell k})w_k \geq 0$ for all weights w_k that satisfy $\sum_k w_k = 1$ and $w_k \geq 0 \ \forall \ k$. For convenience, any set of weights that satisfies these two conditions will be referred to as feasible weights.

Consider now any decision d_v with $v \neq i, \ell$. The purpose is to determine whether or not $d_i \succ d_v$ as well. If this were so, then $\sum_k (u_{ik} - u_{vk})w_k \geq 0$ for all feasible weights. Conversely, we cannot conclude that $d_i \succ d_v$, if there exists at least one set of feasible weights for which $\sum_k (u_{ik} - u_{vk})w_k < 0$ or, alternatively, $\sum_k (u_{vk} - u_{ik})w_k > 0$. Hence, if the problem

$$\text{P: Max } z = \sum_k (u_{vk} - u_{ik})w_k \geq 0$$

$$\text{s.t.} \quad \sum_k (u_{ik} - u_{\ell k})w_k \geq 0$$

$$\sum_k w_k = 1$$

$$w_k \geq 0 \ \forall \ k$$

has an optimal solution $\bar{z} > 0$, then the decision d_v is not dominated by the decision d_i. However, if $\bar{z} \leq 0$, then $d_i \succ d_v$ and the decision d_v can be deleted. This testing can be repeated with all existing decisions, and whenever new input from the decision maker becomes available. (The "iterated dominance" procedures employed in the theory of games (Section 3.2 of this part) and certain location algorithms (Chapter 3 of Part II) are somewhat reminiscent of this iterative process). In order to illustrate the procedure, consider the following

Table I.7

	Costs (in million $)	Benefit to motorists	Inconvenience to residents
d_1	125	3	8
d_2	150	4	7
d_3	210	10	0
d_4	180	6	5
d_5	175	6	6

Chapter 1: Multicriteria Decision Making

Example: A decision maker has delineated three main concerns when routing a new stretch of highway. The main criteria are cost, the benefit to motorists (in terms of reduced driving time), and the inconvenience and increased pollution to residents. Table I.7 provides details about the five possible decisions and their achievements on the three criteria.

Using Edwards's linear scale, we can translate these achievements into the utility matrix

$$\mathbf{U} = \begin{bmatrix} 1 & 0 & 0 \\ .7059 & .1429 & .125 \\ 0 & 1 & 1 \\ .3529 & .4286 & .375 \\ .4118 & .4286 & .25 \end{bmatrix}.$$

Sometimes it is possible to reduce the number of decisions by using the concept of dominance. A decision d_i is called dominated, if it performs worse on all criteria than any of the other decisions, i.e., if there exists a decision d_ℓ, such that $u_{ik} \leq u_{\ell k}$ $\forall\ k$. The concept will be formally introduced in Definition I.3 in Chapter 2 of this part. It is apparent that there are no dominances in this example. Assume now that the decision maker has indicated that, in his estimation, $d_1 \succ d_5$. This statement results in the constraint

$$.5882w_1 - .4286w_2 - .25w_3 \geq 0. \tag{5}$$

Testing now whether or not $d_1 \succ d_2$, the objective is

$$\text{Max } z = -.2941w_1 + .1429w_2 + .125w_3,$$

subject to constraint (5), $\sum_k w_k = 1$, and $w_k \geq 0\ \forall\ k$. The optimal solution is $\bar{z} = 0$, so that we can delete decision d_2.

Testing whether or not $d_1 \succ d_3$, we obtain the objective

$$\text{Max } z = -w_1 + w_2 + w_3,$$

subject to the same constraints as before, resulting in $\bar{z} > 0$, so that no dominance can be detected.

Finally, we wish to assert if $d_1 \succ d_4$, which is done by considering the objective function

Max $z = -.6471w_1 + .4286w_2 + .375w_3$,

subject to the same constraints as above. The result is $\bar{z} > 0$, so again there is no dominance.

This leaves decisions d_1, d_3, and d_4, and the decision maker must provide further information in order to make additional deletions. The above process demonstrates that even a single preference statement may be used to eliminate a number of other decision alternatives—or none at all. Determining preference statements that, if made, allow the deletion of a large number of other decisions, is an interesting problem in itself.

1.6 Multiattribute Value Functions

Multiattribute Value Functions were pioneered by Keeney and Raiffa (1976). In their simplest version, the functions address Steps 1 and 2 of the generic procedure described in Section 1.2. They attempt to do so by transforming user input into utility functions. As demonstrated below, this is possible with minimal user input.

One aspect of multiattribute value functions deals with the determination of the weights w_k of the criteria c_k, $k = 1, ..., q$. To facilitate the discussion, consider criterion c_1 as the "base criterion," typically a benchmark criterion such as cost or profit. In order to determine the tradeoffs between the criteria, we will examine all $(q - 1)$ pairs (c_1, c_k), $k = 2, ..., q$. Denote again by $\underline{a}_{\bullet k}$ and $\bar{a}_{\bullet k}$ the lowest and highest score achieved on criterion c_k, respectively, let $u(a)$ again express the utility of score a, and let $A \succ B$ ($B \succ A$) indicate a decision maker's preference of an option A over a competitor B (preference of B over A). Here, we assume that the utilities have been normalized, so that the lowest (highest) utility on any criterion equals zero (one). The method then asks the decision maker to compare the hypothetical pair of outcomes $[u(\bar{a}_{\bullet 1}), u(\underline{a}_{\bullet k})]$ with the hypothetical pair $[u(\underline{a}_{\bullet 1}), u(\bar{a}_{\bullet k})]$, i.e., a combination of outcomes that consists of the highest score on the benchmark criterion and the lowest possible score on criterion c_k with a combination of the lowest possible score on the benchmark criterion and the highest score of criterion c_k. Three cases can occur.

Case 1: $[u(\underline{a}_{\bullet 1}), u(\bar{a}_{\bullet k})] = [0, 1] \succ [1, 0] = [u(\bar{a}_{\bullet 1}), u(\underline{a}_{\bullet k})]$ This situation is shown in Figure I.6.

The decision maker is then asked to supply a score $\hat{a}_{\bullet k}$ of criterion c_k so that he feels indifferent between the pairs $[u(\underline{a}_{\bullet 1}), u(\bar{a}_{\bullet k})]$ and $[u(\bar{a}_{\bullet 1}), u(\hat{a}_{\bullet k})]$, i.e., $[0, 1] \sim [1, u(\hat{a}_{\bullet k})]$. We now know that $w_1 u(\underline{a}_{\bullet 1}) + w_k u(\bar{a}_{\bullet k}) = w_1 u(\bar{a}_{\bullet 1}) +$

$w_k u(\hat{a}_{\bullet k})$, or, as $u(\underline{a}_{\bullet 1}) = 0$ and $u(\overline{a}_{\bullet k}) = u(\overline{a}_{\bullet 1}) = 1$,

$$w_k = w_1 + w_k u(\hat{a}_{\bullet k}). \tag{6}$$

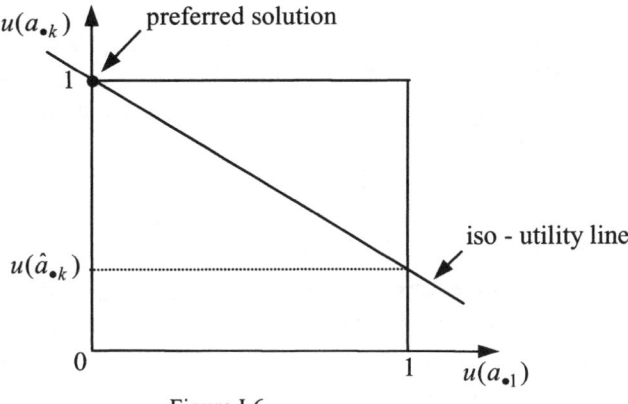

Figure I.6

Case 2: $[u(\underline{a}_{\bullet 1}), u(\overline{a}_{\bullet k})] = [0, 1] \prec [1, 0] = [u(\overline{a}_{\bullet 1}), u(\underline{a}_{\bullet k})]$. This scenario is shown in Figure I.7.

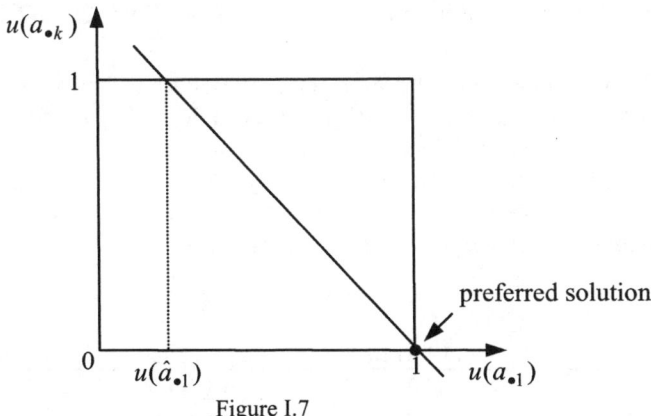

Figure I.7

The decision maker is then asked to identify a score $\hat{a}_{\bullet 1}$ of the base criterion, so that he is indifferent between the pairs $[u(\overline{a}_{\bullet 1}), u(\underline{a}_{\bullet k})]$ and $[u(\hat{a}_{\bullet 1}), u(\overline{a}_{\bullet k})]$ i.e., $[1, 0] \sim [u(\hat{a}_{\bullet 1}), 1]$.

We can then write $w_1\ u(\overline{a}_{\bullet 1}) + w_k u(\underline{a}_{\bullet k}) = w_1 u(\hat{a}_{\bullet 1}) + w_k u(\overline{a}_{\bullet k})$ which, with $u(\underline{a}_{\bullet k}) = 0$ and $u(\overline{a}_{\bullet 1}) = u(\overline{a}_{\bullet k}) = 1$, reduces to

$$w_1 = w_1 u(\hat{a}_{\bullet 1}) + w_k. \qquad (7)$$

Case 3: $[u(\underline{a}_{\bullet 1}), u(\overline{a}_{\bullet k})] = [0, 1] \sim [1, 0] = [u(\overline{a}_{\bullet 1}), u(\underline{a}_{\bullet k})]$. This type of input implies that

$$w_1 = w_k. \qquad (8)$$

Each of the $(q-1)$ pairs (c_1, c_k), $k = 2, ..., q$ results in one linear equation, (6), (7), or (8) for a total of $(q-1)$ linearly independent equations. Adding the equation

$$\sum_{k=1}^{q} w_k = 1,$$

we have q equations in q unknowns $w_1, ..., w_q$. Let the resulting solution be a set of weights $\overline{w}_k, k = 1, ..., q$. We can then determine the values of the decisions as

$$v(d_i) = \sum_{k=1}^{q} \overline{w}_k u(a_{ik}) \qquad \forall \ i = 1, ..., m.$$

The decisions can now be ranked with respect to their values, providing the decision maker with a useful tool to create a short list and make a final decision. As usual, sensitivity analysis, e.g., with a number of alternative choices of $\hat{a}_{\bullet 1}$ and $\hat{a}_{\bullet k}$, can provide additional insight with respect to the robustness of the ranking.

Example: Consider the problem of locating a polluting facility at one of four available sites. Each site allows a different technology, impacting upon the cost structure and pollution emission. The three criteria deemed relevant are c_1: profit, c_2: the impact of the facility on the community, and c_3: the emission of pollutants in tons per day. The four potential sites are d_1, d_2, d_3, and d_4, and the scores of the decision alternatives on the three criteria are shown in Table I.8.

Table I.8

	c_1: Profit	c_2: Impact	c_3: Emissions
d_1	500	very little	40
d_2	200	medium	30
d_3	800	significant	60
d_4	350	little	50

In order to transform these scores into utilities, it is first required to bring them into an appropriate form. While c_1 is a quantitative, maximizing criterion, c_2 and c_3 are not. In order to transform the quantitative scores in c_2 to utilities, it is felt that a score of "very little" would represent a utility of 1, while $u(\text{significant}) = 0$ is considered appropriate, with $u(\text{little}) = .75$ and $u(\text{medium}) = .5$ being the

Chapter 1: Multicriteria Decision Making 49

intermediate values. Finally, the emission criterion c_3 is of the minimization type, so that $u(30) = 1$ and $u(60) = 0$. Intermediate values for c_3 are generated by a linear utility function, whereas the decision maker has decided to use an exponential utility function for the profit criterion. The decision maker states that, in his opinion, $u(300) = ½$, resulting again in the concave utility function $u(\text{profit}) = 1.01763 - 3.933076 \, e^{-0.00676(\text{profit})}$ (see the example in Section 1.2). The resulting utilities for the four decisions and the three criteria are then shown in the matrix

$$\mathbf{U} = \begin{bmatrix} .8837 & 1 & .6667 \\ 0 & .5 & 1 \\ 1 & 0 & 0 \\ .6485 & .75 & .3333 \end{bmatrix}.$$

The next task is to specify the tradeoffs between the criteria. Here we have chosen profit c_1 as a base criterion. First, consider the pair (c_1, c_2). Comparing (200, very little) with utility $[0, 1]$ with (800, significant), whose utility is $[1, 0]$, the decision maker turns out to prefer the latter alternative, i.e., $[0, 1] \prec [1, 0]$ and Case 2 applies. Upon further investigation, the decision maker feels that (400, very little) has the same utility as (800, significant), so that $\hat{a}_{\bullet 1} = 400$. As $u(\hat{a}_{\bullet 1}) = u(400) = .7544$, this results in the relation

$$w_1 = 0.7544 w_1 + w_2.$$

The next step deals with the relation between the first and third criterion, i.e., the pair (c_1, c_3). Comparing (200, 30) with (800, 60), the decision maker prefers the former alternative, i.e., $[0, 1] \succ [1, 0]$ and Case 1 applies. Upon further investigation, the decision maker opines that (800, 50) is comparable to (200, 30), i.e., $\hat{a}_{\bullet 3} = 50$ and $u(\hat{a}_{\bullet 3}) = u(50) = .3333$. This leads to the relation

$$w_3 = w_1 + .3333 w_3.$$

The resulting system of simultaneous linear equations is then

$$\begin{aligned} .2456 w_1 - w_2 &= 0 \\ -w_1 \qquad\quad + .6667 w_3 &= 0 \\ w_1 \quad + w_2 + w_3 &= 1, \end{aligned}$$

which has the solution $\overline{\mathbf{w}} = [\overline{w}_1, \overline{w}_2, \overline{w}_3]^T = [.3642, .0895, .5463]^T$. Given these weights and the utility matrix $\mathbf{U}$, we can now calculate the values of the decisions as

$$v(d_1) = .7756, \; v(d_2) = .5911, \; v(d_3) = .3642, \text{ and } v(d_4) = .4854.$$

This result provides a clear ranking $d_1 \succ d_2 \succ d_4 \succ d_3$.

Suppose now that the decision maker does not feel confident in his statement that $(c_1, c_3) = (800, 50)$ and $(200, 30)$ have a very similar utility. Suppose he wishes to revise that to $(800, 40)$ and $(200, 30)$. With $\hat{a}_{\bullet 3} = 40$ and $u(\hat{a}_{\bullet 3}) = u(40) = .6667$, the relation between the base criterion c_1 and c_3 is then $w_3 = w_1 + .6667w_3$ so that $-w_1 + .3333w_3 = 0$ now replaces the second equation of the above system. The solution of this modified problem is $\overline{w} = [.2355, .0580, .7065]^T$, resulting in $v(d_1) = .7371$, $v(d_2) = .7355$, $v(d_3) = .2355$, and $v(d_4) = .4317$, and with it the same order of decision alternatives, except that d_1 and d_2 are now almost tied for best. This result stresses the need for sensitivity analyses.

1.7 Outranking Methods

The major tasks of multiattribute utility theory in the previous section was to avoid asking decision makers directly to specify utilities of outcomes and weights of criteria, but derive them based on information that decision makers are more likely able to specify. Outranking methods, as pioneered by Benayoun et al. (1966), Roy (1971, 1975, 1978), and Brans and Vincke (1985) take a different route. These methods take utilities of scores and weights of criteria as given; based on that information, compare decisions with each other, deriving (a weak version of) dominance between them. The final result of all of these approaches is a structure, usually represented by a graph, that indicates which decision is preferred to, or "outranks," another. This outranking may or may not be complete.

Roy entitles his method *ELECTRE* (Elimination et (and) Choice Translating Algorithms), and it has been presented over the years in a variety of versions, viz., *ELECTRE I, II, III, IV,* and *TRI*. Here we will introduce the basic ideas put forward in the first two versions. Let a utility matrix **U** and a weight vector **w** be specified by the decision maker. The first step is then to set up two $[m \times m]-$ dimensional matrices $\mathbf{C} = (c_{i\ell})$ and $\mathbf{D} = (d_{i\ell})$, called *concordance* and *discordance matrix*, respectively. An element $c_{i\ell}$ will indicate the strength of the preference of decision d_i over decision d_ℓ while an element $d_{i\ell}$ in the discordance matrix will express how much worse decision d_i is considered as compared to d_ℓ. Both concordances and discordances are calculated based on the scores of the decisions on *all q* criteria.

In particular, the concordance $c_{i\ell}$ between two decisions d_i and d_ℓ is defined as

$$c_{i\ell} = \frac{\sum_{k: u_{ik} \geq u_{\ell k}} w_k}{\sum_{k=1}^{q} w_k}.$$

In other words, $c_{i\ell}$ measures the proportion of weights of those criteria for which decision d_i has a higher utility than decision d_ℓ. Often, the sum of weights add to one, so that the denominator equals one.

Similar to the concordance matrix $\mathbf{C}$, a $[m \times m]$-dimensional discordance matrix $\mathbf{D}$ can be set up. Each element of the discordance matrix compares two decisions with each other. In particular the discordance of decision d_i as compared to decision d_ℓ is defined as

$$d_{i\ell} = \frac{\max_{k:u_{\ell k} > u_{ik}} \{w_k(u_{\ell k} - u_{ik})\}}{\max_{k} \{\max_{\mu,\nu} \{w_k |u_{\mu k} - u_{\nu k}|\}\}}.$$

The denominator is the same for all elements in the discordance matrix; it ensures that discordances are normalized between zero and one. The numerator is the largest weighted difference between the utilities of two decisions d_i and d_ℓ, given that d_ℓ is strictly preferred over d_i. Again, a large value of $d_{i\ell}$ shows that decision d_i is felt to be strongly inferior to decision d_ℓ.

As the discordance as defined here only applies to criteria that are measured on a cardinal scale, an alternative discordance can be defined as

$$d_{i\ell} = \begin{cases} 1, & \text{if } u_{\ell k} - u_{ik} > t_k \text{ for any } k \\ 0 & \text{otherwise} \end{cases},$$

where t_k is a threshold value defined by the decision maker. Here, a value $d_{i\ell} = 1$ indicates that decision d_i cannot outrank decision d_ℓ.

Concordance and discordance between a decision and itself, i.e., the diagonal elements c_{ii} and d_{ii}, are undefined.

Defining now $\underline{c}$ as the minimal required concordance and $\overline{d}$ as the maximal allowable discordance, we can then say that decision d_i outranks decision d_ℓ (in symbols: $d_i \blacktriangleright d_\ell$), if and only if

$$c_{i\ell} \geq \underline{c} \text{ and } d_{i\ell} \leq \overline{d}.$$

While the parameters $\underline{c}$ and $\overline{d}$ are typically user-defined, some authors use the average concordance $\underline{c} = \frac{1}{m(m-1)} \sum_i \sum_{k \neq i} c_{ik}$ and the average discordance

$\bar{d} = \frac{1}{m(m-1)} \sum_{i} \sum_{k \neq i} d_{ik}$. Here, we assume that $\underline{c}$ and $\bar{d}$ are set by the decision maker.

The resulting outranking relations can be displayed in a graph $G = (N, A)$ with nodes n_i, $i = 1, ..., m$ (each representing a decision d_i) and directed arcs $a_{i\ell}$, each representing an outranking relation. Formally, $a_{i\ell} \in A$ if and only if $d_i \blacktriangleright d_\ell$. We can then define the *kernel* of the graph as the node set $K \subseteq N$, such that $\{a_{i\ell} : n_i, n_\ell \in K\} = \emptyset$ and $N \setminus K = \{n_\ell : \exists a_{i\ell} \text{ with } n_i \in K\}$, i.e., no decision in the kernel outranks another decision in the kernel, and each decision not in the kernel is outranked by at least one decision in the kernel. The kernel then consists of the preferred decisions. In case the graph G includes one or more circuits, each of these circuits would have to be broken up and the resulting kernel must be determined. The union of all such kernels will then represent the relevant decisions.

Example: A department store considers a variety of options involving an expansion and complete renovation. Management has narrowed down the available options to four, d_1, d_2, d_3, and d_4. These options are to be compared with each other on the basis of short and long-term objectives. The only short-term criterion is c_1: profit, while the long-term criteria are c_2: customer satisfaction (creating loyalty and potential future sales) and c_3: employee satisfaction (for good labor relations). Management considers short and long-term objectives equally important and believes that customer satisfaction is slightly more important than employee satisfaction, resulting in weights $w_1 = .5$, $w_2 = .3$, and $w_3 = .2$. The utilities of the four decision alternatives on the three criteria are expressed in the utility matrix

$$\mathbf{U} = \begin{bmatrix} 1.0 & .3 & 0 \\ 0 & .4 & .6 \\ .8 & 1.0 & .5 \\ .5 & 0 & 1.0 \end{bmatrix}.$$

Given the weights $\mathbf{w} = [.5, .3, .2]^T$, the concordance and discordance matrices are then

$$\mathbf{C} = \begin{bmatrix} - & .5 & .5 & .8 \\ .5 & - & .2 & .3 \\ .5 & .8 & - & .8 \\ .2 & .7 & .2 & - \end{bmatrix}, \quad \mathbf{D} = \begin{bmatrix} - & .24 & .42 & .40 \\ 1.0 & - & .80 & .50 \\ .20 & .04 & - & .20 \\ .50 & .24 & .60 & - \end{bmatrix}.$$

If the decision maker specifies values $\underline{c} = .5$ and $\overline{d} = .3$, then the graph in Figure I.8a results, whereas for $\underline{c} = .6$ and $\overline{d} = .6$, the graph is shown in Figure I.8b.

Figure I.8

The graph in Figure I.8a is acyclic and $\{n_3\}$ is the unique kernel. On the other hand, the graph in Figure I.8b has a circuit and $\{1\}$ and $\{3\}$ are both kernels. It appears reasonable to shortlist the decisions d_1 and d_3. However, extensive sensitivity analyses will typically follow such calculations. Such sensitivities will not only examine changes for various values of $\underline{c}$ and $\overline{d}$, but also changes in the utilities and weights.

While outranking methods do not generally produce rankings, it appears that d_3 is the most preferred alternative followed by d_1, d_4, and finally d_2. This also reflects the ranking obtained by the usual approach that bases rankings on the weighted sum of utilities.

Arbitrarily consider two other graphs with circuits. These graphs are shown in Figures I.9a and I.9b. The graph in Figure I.9a has no kernel, while the graph in Figure I.9b has a unique kernel $\{2, 4\}$.

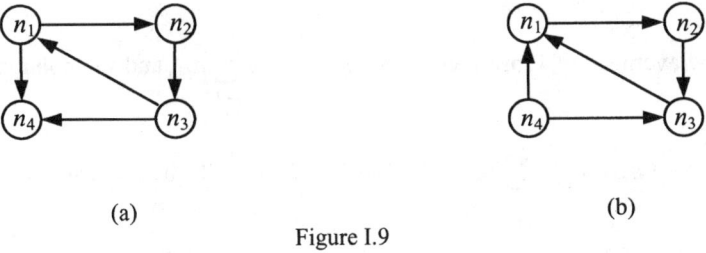

Figure I.9

ELECTRE II is an extension of *ELECTRE I* that produces a ranking of the given decisions, i.e., a partial order. It is more elaborate in that rather than a single outranking relation between each pair of decision alternatives d_i and d_ℓ, it uses strong and weak outranking relations. Also, in order to avoid the frequent undesirable occurrence of a decision d_i outranking some decision d_ℓ and vice versa, for an outranking of d_i over d_ℓ to exist, it requires not only that the concordance (discordance) of d_i over d_ℓ exceed (fall short of) a given bound, but also that the concordance of d_i over d_ℓ exceeds that of d_ℓ over d_i. A similar approach, a so-called *compromise ranking algorithm* (*VIKOR*), was developed by Opricovic (1998). A good description can be found in Tzeng *et al.* (2002).

A different approach to outranking was proposed by Brans and Vincke (1985), following earlier work by Brans in the early 1980's. Like *ELECTRE*, Brans's *PROMETHEE* method comes in various versions. Also like *ELECTRE*, *PROMETHEE* (*P*reference *R*anking *O*rganization *Met*hod for *E*nrichment *E*valuations) directly compares decision alternatives with each other. However, rather than working with a single pair of concordance and discordance matrices, *PROMETHEE* compares pairs of decisions on each criterion individually by means of a preference matrix, and then aggregates the individual preference scores to an overall preference index. This overall index is then used to produce outranking relations. Whereas *PROMETHEE I* produces a partial order of the given decision alternatives, thus leaving room for decisions to be considered unrelated, *PROMETHEE II* will produce a complete order or ranking of the alternatives. Both methods work with the utilities u_{ik}.

For each criterion c_k, $k = 1, \ldots, q$, the method determines an $[m \times m]$-dimensional *individual preference matrix* $\mathbf{P}^k = (p_{i\ell}^k)$, where $p_{i\ell}^k$ expresses the degree of preference of the decision d_i over d_ℓ with respect to criterion c_k. Given user-defined weights w_k, $k = 1, \ldots, q$, with $\sum_{k=1}^{q} w_k = 1$, the utility-like expressions $p_{i\ell}^k$ are then aggregated to preferences $p_{i\ell}$, which are collected in the overall preference matrix $\mathbf{P} = \sum_{k=1}^{q} w_k \mathbf{P}^k$, where the diagonal elements p_{ii} are undefined.

The row averages of $\mathbf{P}$ are then $\mathbf{p}^+ = (p_i^+) = \frac{1}{m-1} \sum_{\substack{\ell=1 \\ \ell \neq i}}^{m} p_{i\ell}$, and the column averages are $\mathbf{p}^- = (p_\ell^-) = \frac{1}{m-1} \sum_{\substack{i=1 \\ i \neq \ell}}^{m} p_{i\ell}$. *PROMETHEE I* will then construct outranking relations in which

$$d_i \blacktriangleright d_\ell, \text{ if and only if } p_i^+ \geq p_\ell^+ \text{ and } p_i^- \leq p_\ell^-$$

(with one of the inequalities strict). The outranking relations can then be visualized by constructing the graph $G = (N, A)$, where the node $n_i \in N$ represents decision d_i,

$i = 1, ..., m$, and $a_{i\ell} \in A$ if $d_i \blacktriangleright d_\ell$, $i, \ell = 1,..., m, i \neq \ell$. In general, the outranking relations generated by *PROMETHEE I* will not result in a complete order of the decision alternatives. *PROMETHEE II*, on the other hand, produces a complete order by calculating $p_v = p_v^+ - p_v^- \; \forall \, v = 1, ... \, m$, which order is then the basis of the ranking.

In our discussion above we introduced preference functions that transform scores u_{ik} and $u_{\ell k}$ "somehow" into a preference $p_{i\ell}^k$. Brans and Vincke have proposed a variety of functions that could be used for that purpose. Our terminology differs somewhat from that used by the authors of the original contributions. In order to formalize matters, define $\Delta_{i\ell}^k = u_{ik} - u_{\ell k}$ as the difference of the utilities of decision d_i and d_ℓ with respect to criterion c_k, and let Δ, $\underline{\Delta}$, and $\overline{\Delta}$ be user-defined parameters. The following four transformations are then among the multitude of functions that translate differences of scores $\Delta_{i\ell}^k$ into preferences $p_{i\ell}^k$.

(1) Binary preferences $p_{i\ell}^k = \begin{cases} 0, & \text{if } \Delta_{i\ell}^k \leq \Delta \\ 1 & \text{otherwise} \end{cases}$

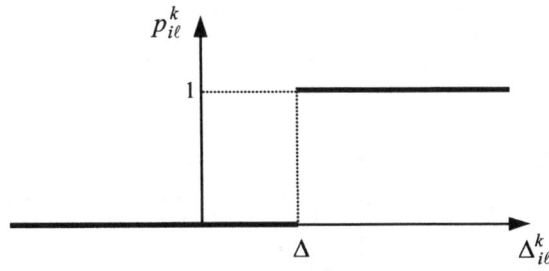

Figure I.10

The parameter Δ can assume positive or negative values or zero. In the special case of $\Delta = 0$, $\Delta_{i\ell}^k > 0$ is equivalent to $u_{ik} > u_{\ell k}$, which is reflected by the preference index $p_{i\ell}^k = 1$. To the left of the origin, $u_{ik} \leq u_{\ell k}$, and $p_{i\ell}^k = 0$, indicating that decision d_i is not preferred to decision d_ℓ.

(2) Stepwise preferences $p_{i\ell}^k = \begin{cases} 0, & \text{if } \Delta_{i\ell}^k \leq \underline{\Delta} \\ \frac{1}{2}, & \text{if } \Delta_{i\ell}^k \in\,]\underline{\Delta}; \overline{\Delta}\,[\\ 1, & \text{if } \Delta_{i\ell}^k \geq \overline{\Delta} \end{cases}$

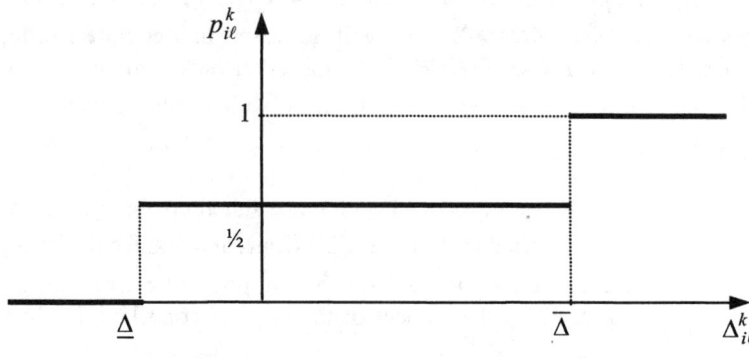

Figure I.11

This preference function assigns a value of ½ to $p_{i\ell}^k$, if the utility is within a prespecified range, given by $\underline{\Delta}$ and $\overline{\Delta}$. Outside these limits the function is again binary.

(3) Piecewise linear preferences $p_{i\ell}^k = \begin{cases} 0, & \text{if } \Delta_{i\ell}^k \leq \underline{\Delta} \\ (\Delta_{i\ell}^k - \underline{\Delta})/(\overline{\Delta} - \underline{\Delta}), & \text{if } \Delta_{i\ell}^k \in]\underline{\Delta}; \overline{\Delta}[\\ 1, & \text{if } \Delta_{i\ell}^k \geq \overline{\Delta} \end{cases}$

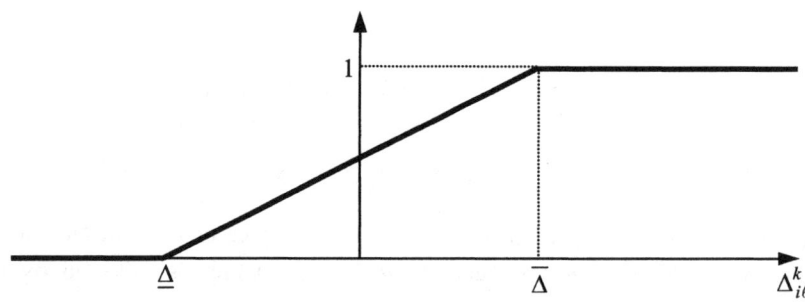

Figure I.12

Similar to the stepwise preferences in (2), piecewise linear preferences assign values of 0 and 1, if the scores of the two decisions d_i and d_ℓ are not within a certain range of each other. Inside this range the preference function increases linearly.

Chapter 1: Multicriteria Decision Making

(4) Exponential preferences $p_{i\ell}^k = \begin{cases} 1 - \alpha e^{-\beta \Delta_{i\ell}^k}, & \text{if } \Delta_{i\ell}^k \geq \dfrac{1}{\beta} \ln \alpha \\ 0, & \text{otherwise} \end{cases}$.

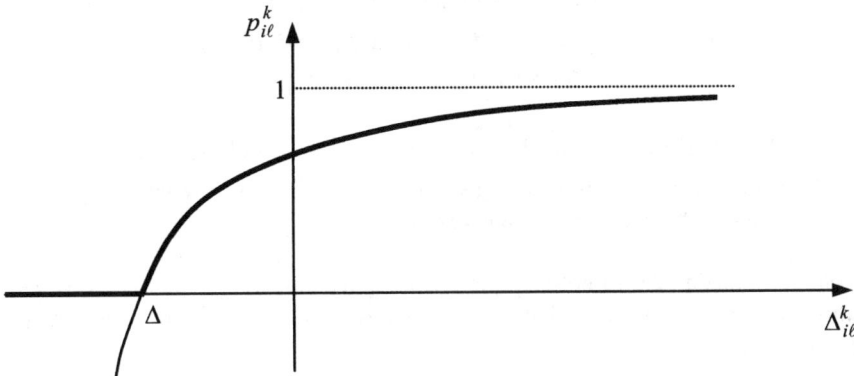

Figure I.13

At $\Delta = \frac{1}{\beta} \ln \alpha$, the preference function is zero, and to the right of Δ, it increases in a concave fashion, and approaches a value of 1 as $\Delta_{i\ell}^k$ increases.

Clearly, many other preference functions can be thought of. Their choice will depend on the tradeoffs between the criteria in the way the decision maker evaluates them. To reiterate, the decision maker may use any preference function $p_{i\ell}^k = p^k(\Delta_{i\ell}^k)$ for each of the criteria, and there is no need to choose the same function $p^k(\Delta_{i\ell}^k)$ for all criteria. We now are able to formally describe the method.

The *PROMETHEE* Method

Step 1: For each criterion c_k choose a preference function $p^k(\Delta_{i\ell}^k)$. For each pair of decisions (d_i, d_ℓ), calculate $\Delta_{i\ell}^k = u_{ik} - u_{\ell k}$ and the preference $p_{i\ell}^k$ according to the chosen function $p^k(\bullet)$.

Step 2: (Aggregate the preferences). Determine the aggregated preference matrix $\mathbf{P} = (p_{i\ell})$ with $p_{i\ell} = \sum_{k=1}^{q} p_{i\ell}^k w_k \quad \forall i, \ell = 1, ..., m; \ i \neq \ell$.

Step 3: Determine $p_i^+ = \frac{1}{m-1}\sum_{\substack{\ell=1\\\ell\neq i}}^{m} p_{i\ell}$ and $p_\ell^- = \frac{1}{m-1}\sum_{\substack{i=1\\i\neq \ell}}^{m} p_{i\ell}$.

PROMETHEE I: Construct a graph $G = (N, A)$, with n_i representing $d_i \ \forall \ i$, and $a_{i\ell} \in A$, if $p_i^+ \geq p_\ell^+$ and $p_i^- \leq p_\ell^-$, with one of the inequalities strict. The graph will present a partial order.

PROMETHEE II: Calculate $p_v = p_v^+ - p_v^- \ \forall \ v$ and order the decision with respect to nonincreasing values of p_v. This is a complete order of the decision alternatives.

Example: Consider the same example used for the illustration of the *ELECTRE* method (and ignore the fact that the scores have already been transformed into utilities) with utility matrix

$$\mathbf{U} = \begin{bmatrix} 1.0 & .3 & 0 \\ 0 & .4 & .6 \\ .8 & 1.0 & .5 \\ .5 & 0 & 1.0 \end{bmatrix}$$

and criteria weights $\mathbf{w} = [.5, .3, .2]^T$. It has been decided that preferences in the first criterion follow the exponential function $p_{i\ell}^1 = 1.6161 - e^{-.48\Delta_{i\ell}^1}$ with $\Delta = -1$, preferences in the second criterion should be modeled as a stepwise function with $\underline{\Delta} = -.1$ and $\overline{\Delta} = .2$, and preferences in the third criterion can be represented by a piecewise linear function with $\underline{\Delta} = -.3$ and $\overline{\Delta} = .2$. This results in the three preference matrices

$$\mathbf{P}^1 = \begin{bmatrix} - & .9973 & .7076 & .8295 \\ 0 & - & .1480 & .3449 \\ .5153 & .9350 & - & .7502 \\ .3449 & .8295 & .4612 & - \end{bmatrix}, \ \mathbf{P}^2 = \begin{bmatrix} - & 0 & 0 & 1 \\ .5 & - & 0 & 1 \\ 1 & 1 & - & 1 \\ 0 & 0 & 0 & - \end{bmatrix}, \text{ and}$$

$$\mathbf{P}^3 = \begin{bmatrix} - & 0 & 0 & 0 \\ 1 & - & .8 & 0 \\ 1 & .4 & - & 0 \\ 1 & 1 & 1 & - \end{bmatrix}.$$

The overall preference matrix **P** with row averages $\mathbf{p}^+$ and column averages $\mathbf{p}^-$ appended is then

$$\mathbf{P} = \begin{bmatrix} - & .4987 & .3538 & .7148 \\ .3500 & - & .2340 & .4724 \\ .7576 & .8475 & - & .6751 \\ .3724 & .6148 & .4306 & - \end{bmatrix} \qquad \mathbf{p}^+ = \begin{bmatrix} .5224 \\ .3521 \\ .7601 \\ .4726 \end{bmatrix}$$

$$\mathbf{p}^- = \begin{bmatrix} .4933 & .6537 & .3395 & .6208 \end{bmatrix}.$$

The resulting graph G with *PROMETHEE I*'s outranking relations is shown in Figure I.14.

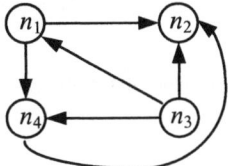

Figure I.14

The relations suggest the superiority of d_3 and the inferiority of d_2 with d_1 and d_4 in between, where d_1 appears to be preferred to d_4. The same complete outranking relation $d_3 \blacktriangleright d_1 \blacktriangleright d_4 \blacktriangleright d_2$ is generated by *PROMETHEE II*, where $\mathbf{p} = \mathbf{p}^+ - \mathbf{p}^- = [.0291, -.3016, .4206, -.1482]^T$.

In addition to the outranking relations provided by the *PROMETHEE* methods, it is also possible to generate a two-dimensional display that may provide further insight into the problem. In order to do so, set up the $[m \times q]$-dimensional matrix

$$\mathbf{G} = (g_{ik}), \text{ with } g_{ik} = \frac{1}{m-1} \left[\sum_{\substack{\ell=1 \\ \ell \neq i}}^{m} p_{i\ell}^k - \sum_{\substack{\ell=1 \\ \ell \neq i}}^{m} p_{\ell i}^k \right] \forall\, i, k,$$

i.e., g_{ik} denotes the average advantage the decision d_i has over the other decisions with respect to criterion c_k. We can then compute the $[q \times q]$-dimensional matrix $\mathbf{H} = \mathbf{G}^T \mathbf{G}$. Note that $\frac{1}{m}\mathbf{H}$ is the covariance matrix of the criteria.

Mareschal and Brans (1988) then propose the use of principal components analysis to find a projection from the q-dimensional criterion space onto a 2-dimensional plane. Their analysis (for details, we refer readers to their paper) has

λ_1 and λ_2 as the largest and second-largest eigenvalues of the matrix **G** with **u** and **v** being the corresponding eigenvectors. It is now possible to construct the *GAIA plane* (*G*eometric *A*nalysis for *I*nteractive *A*id) as follows:
- the criterion c_k is displayed by the vector (u_k, v_k)
- the decision alternative d_i is represented by the point $(\mathbf{g}_{i\bullet}\mathbf{u}^T, \mathbf{g}_{i\bullet}\mathbf{v}^T)$
- the overall direction of the objective is

$$\pi = \left(\left[\sum_k w_k^2\right]^{-\frac{1}{2}} \sum_k w_k u_k ; \left[\sum_k w_k^2\right]^{-\frac{1}{2}} \sum_k w_k v_k \right).$$

If the vector π is short, then there is much conflict between the criteria, and the best solutions are decision alternatives whose points are located near the origin. It is also worth pointing out that the proportion of the available information contained in the *GAIA* plane is $\delta = (\lambda_1 + \lambda_2)/ \sum_{k=1}^{q} \lambda_k$.

The representation of the criteria and decision alternatives can then be evaluated with the following in mind:

- criteria with similar directions in the *GAIA* plane indicate similar preferences
- if there exists a criterion on which the scores of the decision alternatives differ significantly, its vector will be long
- the distances between the points representing decision alternatives indicate how similar the decisions score on the criteria
- a point near the direction of the vector of a criterion will indicate a high score of the decision alternative on that criterion.

Example: Continue with the example used to illustrate the *PROMETHEE* method. Given the preference matrices $\mathbf{P}^k$, we can calculate

$$\mathbf{G} = \begin{bmatrix} .5581 & -.1667 & -1.0000 \\ -.7563 & .1667 & .1333 \\ .2946 & 1.0000 & -.1333 \\ -.0963 & -1.0000 & 1.0000 \end{bmatrix}.$$

Then

$$\mathbf{H} = \mathbf{G}^T\mathbf{G} = \begin{bmatrix} .97953 & .17179 & -.79448 \\ .17179 & 2.05558 & -.94438 \\ -.79448 & -.94438 & 2.03554 \end{bmatrix}.$$

The three eigenvalues of the matrix **H** are .47844, 1.38497, and 3.20724, so that

$\lambda_1 = 3.20724$ and $\lambda_2 = 1.38497$ with $\mathbf{u} = [-.30334, -.63086, .71415]$ and $\mathbf{v} = [-.52515, .73605, .42715]$, with $\delta = .9056$, i.e., close to 91% of the variability is explained in the GAIA plane.

The directions of the three criteria are $[-.3033, -.5252]$ for $\mathbf{c}_1$, $[-.6309, .7360]$ for $\mathbf{c}_2$, and $[.7141, .4271]$ for $\mathbf{c}_3$, while the points associated with the four decision alternatives are $[-.7783, -.8429]$ for d_1, $[.2194, .5768]$ for d_2, $[-.8154, .5244]$ for d_3, and $[1.3742, -.2583]$ for d_4. The overall direction of the criteria is $\pi = [-.3214, .0708]$. Figure I.15 shows the directions of the criteria and the points of the decision alternatives in the *GAIA* plane.

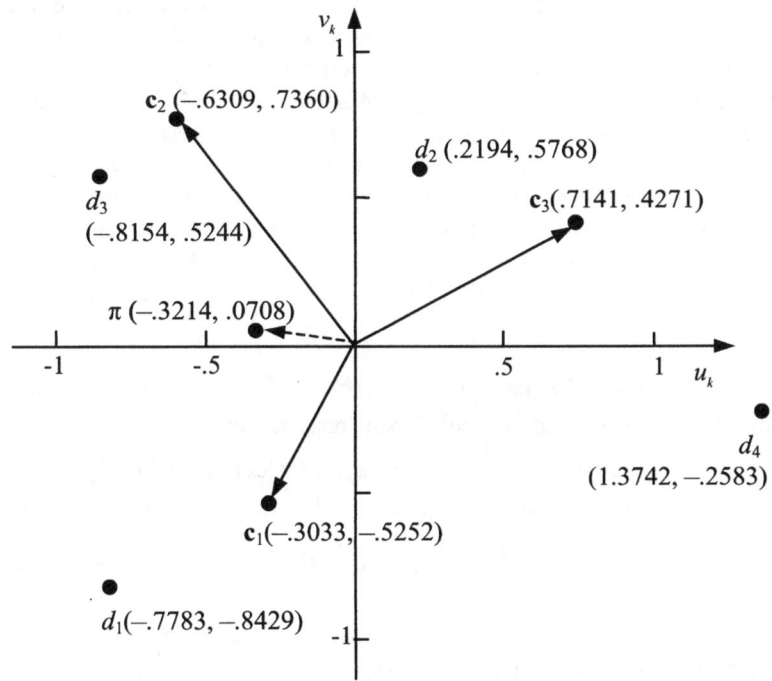

Figure I. 15

The graph clearly indicates that there is significant disagreement between the criteria. With the decisions pretty much spread out across the plane, there is little agreement between decisions and criteria.

1.8 Methods Allowing Inconsistent Estimates

So far, we have assumed that decision makers are able to correctly, reliably, and consistently specify utilities u_{ik} and weights w_k for all decisions and criteria. While this assumption may be justified in the case of hard data such as money, size, length, or any other quantitative data, it will be problematic in case of approximate, uncertain, or qualitative data. All methods discussed in this section have in common that utilities and weights are no longer known exactly.

For the purpose of our discussion it is useful to realize that both utilities and weights indicate payoffs. For instance, if for a given criterion c_k, the utilities of decisions d_1 and d_2 are $u_{1k} = 2$ and $u_{2k} = 5$, then the second decision is considered 2½ times as valuable as the first, with respect to the criterion c_k. If utilities and weights are not known exactly, a standard method is to ask the decision maker to make pairwise companions between decisions and criteria, respectively. The resulting data will then determine the tradeoffs and, along with them, the utilities and weights. While in the case of m decisions it would be sufficient to determine $(m-1)$ tradeoffs, it is useful to introduce some redundancy by requiring the decision maker to directly specify tradeoffs that were already implied. For instance, if the decision maker has specified that on some criterion c_k, the decision d_1 is three times as good as the d_2 which, in turn, is twice as good as decision d_3, then it is implied that d_1, is six times as valuable as d_1 (where all tradeoffs relate exclusively to the criterion c_k). The decision maker may, however, indicate that d_1 is only five times as good as d_1, thus creating an inconsistency. The degree of inconsistency in the tradeoffs will give an indication of the reliability of the estimates.

The concept of pairwise comparisons goes back to Fechner (1860). Formally, for each criterion c_k we set up an $[m \times m]$-dimensional matrix $\mathbf{C}^k = (c_{i\ell}^k)$ for $k = 1, ..., q$, where $c_{i\ell}^k$ indicates that the decision maker considers decision d_i exactly $c_{i\ell}^k$ times as important as decision d_ℓ with respect to criterion c_k. While we assume *pairwise consistency*, i.e., $c_{i\ell}^k = \dfrac{1}{c_{\ell i}^k}$, *general consistency*, defined by the triangular equation $c_{i\ell}^k = c_{i\nu}^k c_{\nu\ell}^k \ \forall \ i, \ell, \nu, k$, is not required. Since $c_{ii}^k = 1 \ \forall \ i, k$, general consistency implies pairwise consistency, following from the condition $1 = c_{ii}^k = c_{i\ell}^k c_{\ell i}^k$. Similarly, tradeoffs between the criteria are shown in the $[q \times q]$-dimensional matrix $\mathbf{C} = (c_{k\mu})$, where $c_{k\mu}$ indicates that the decision maker considers the criterion c_k to be $c_{k\mu}$ times as important as criterion c_μ. For qualitative data, Saaty (1980) suggests a value of $c_{i\ell}^k = 1$, if decisions d_i and d_ℓ are considered *equal* with respect to criterion c_k, $c_{i\ell}^k = 3$, if d_i is *preferred* over d_ℓ, $c_{i\ell}^k = 5$ if d_i is *strongly preferred* over d_ℓ, $c_{i\ell}^k = 7$ if d_i is *very strongly preferred* over d_ℓ, and $c_{i\ell}^k = 9$, if d_i is *extremely preferred* over d_ℓ, and intermediate values filling the gaps. As obvious as this scale may be, it is not without its critics. They point out that, for instance, while the sequence of weights 1, 2, 3, ..., 9 is uniformly distributed, its reciprocal values $1/1, 1/2, 1/3, ..., 1/9$ are skewed to the right.

Chapter 1: Multicriteria Decision Making

There are also a number of paradoxes that have been pointed out in this context; see, e.g., Belton and Gear (1983), and, for the contentious issue of *rank reversal* (which is said to happen if the addition of a decision alternative causes a change in the ranking of some of the decision alternatives), Barzilai and Golany (1994). Lootsma (1988, 1990) has devised exponential scales as an alternative.

As an example of the concept of consistency, consider four decisions and some criterion c_k, such that

$$\mathbf{C}^k = \begin{bmatrix} 1 & 2 & 1/3 & 5 \\ 1/2 & 1 & 1/4 & 2 \\ 3 & 4 & 1 & 10 \\ 1/5 & 1/2 & 1/10 & 1 \end{bmatrix}.$$

This matrix includes four sets of estimates regarding the tradeoffs between the decisions with respect to the criterion c_k, each represented by one row of the matrix $\mathbf{C}^k$. Again, the estimates in the first row show that the decision maker considers d_1 twice as important as d_2, $1/3$ as important as d_3, and five times as important as d_4. Although pairwise consistency holds, we notice general inconsistencies between the estimates, e.g., regarding the tradeoff between d_1 and d_4. While the first row sets a relation of $c_{14}^k / c_{11}^k = 5:1$ between d_1 and d_4, the second row indicates that the relation is $c_{24}^k / c_{21}^k = 2 : 1/2 = 4:1$. Similarly, the third row shows a ratio of $10 : 3 = 3 \, 1/3 : 1$, and row four displays a ratio of $1 : 1/5$ or $5 : 1$. (Similarly, $c_{13} c_{32} = 1/3 \, (4) \neq 2 = c_{12}$). On the other hand, the matrix

$$\mathbf{C}^k = \begin{bmatrix} 1 & 1/2 & 3 \\ 2 & 1 & 6 \\ 1/3 & 1/6 & 1 \end{bmatrix}$$

satisfies the condition of general consistency.

Estimates concerning weights of criteria (rather than utilities of decisions) are handled equivalently, starting with the matrix $\mathbf{C}$ rather than $\mathbf{C}^k$. Clearly, if the evaluations by the decision maker are highly inconsistent, it is advisable to find a consensus before proceeding any further.

The purpose of all methods described in this section is twofold. The first objective is to determine a set of utilities and weights that properly reflect the decision maker's evaluations. These utilities and weights can then be used in the usual way to determine values of decisions $v(d_i) = \sum_{k=1}^{q} u_{ik} w_k$, thus providing a ranking of the available alternatives. The second objective is to provide a measure that indicates how consistent the decision maker's estimates are. Such a measure will be associated with each of the matrices $\mathbf{C}^k$, $k = 1, ..., q$ as well as $\mathbf{C}$.

The first method for determining values of decisions is the *normalization technique*. It is a very simple procedure that allows quick calculations. We will first describe the method formally, and then explain and illustrate it.

The Normalization Technique

Step 1: Calculate the normalized matrices $\tilde{\mathbf{C}}^k = (\tilde{c}_{i\ell}^k) \ \forall \ k = 1, ..., q$ and $\tilde{\mathbf{C}} = (\tilde{c}_{k\mu})$ by dividing each element in $\mathbf{C}^k$ and $\mathbf{C}$ by its column sum, i.e.,

$$\tilde{c}_{i\ell}^k = c_{i\ell}^k \Big/ \sum_{v=1}^{m} c_{v\ell}^k \ \forall \ i, \ell, k, \text{ and}$$

$$\tilde{c}_{k\mu} = c_{k\mu} \Big/ \sum_{v=1}^{q} c_{v\mu}.$$

Step 2: Calculate the column vectors of the utility matrix $\mathbf{U} = \mathbf{u}_{\bullet k} = (u_{ik}) \ \forall \ k$ and the weight vector $\mathbf{w} = (w_k)$, as the row averages of $\mathbf{C}^k$ and $\mathbf{C}$, respectively. Formally,

$$u_{ik} = \frac{1}{m} \sum_{\ell=1}^{m} \tilde{c}_{i\ell}^k, \text{ and}$$

$$w_k = \frac{1}{q} \sum_{\mu=1}^{q} \tilde{c}_{k\mu}.$$

Step 3: The values associated with the decisions are then

$$v(d_i) = \sum_{k=1}^{q} u_{ik} w_k.$$

In summary, the algorithm proceeds as follows. Step 1 divides each matrix element by its column sum, and Step 2 calculates the average value in each row of the matrix. We will illustrate this procedure by means of the following

Example: The International Olympic Committee has narrowed down the choice of the location of the 2016 Summer games to three choices, *viz.*, Kabul, Lima, and

Luanda. The Committee has four main criteria for its selection: the revenue generated for the Committee, the facilities planned (and likely to exist) for the events and the accommodation of the participants, the impact on the community, city, and country, and the political impact made by the choice. The Committee has agreed on the following estimates concerning the three decisions and four criteria:

$$\mathbf{C}^1 = \begin{bmatrix} 1 & 6 & 2 \\ \frac{1}{6} & 1 & \frac{1}{2} \\ \frac{1}{2} & 2 & 1 \end{bmatrix}, \; \mathbf{C}^2 = \begin{bmatrix} 1 & 2 & 3 \\ \frac{1}{2} & 1 & 1 \\ \frac{1}{3} & 1 & 1 \end{bmatrix}, \; \mathbf{C}^3 = \begin{bmatrix} 1 & \frac{1}{5} & \frac{1}{2} \\ 5 & 1 & \frac{1}{3} \\ 2 & 3 & 1 \end{bmatrix},$$

$$\mathbf{C}^4 = \begin{bmatrix} 1 & \frac{1}{2} & \frac{1}{4} \\ 2 & 1 & \frac{2}{3} \\ 4 & 1\frac{1}{2} & 1 \end{bmatrix}, \text{ and } \mathbf{C} = \begin{bmatrix} 1 & 3 & 9 & \frac{1}{2} \\ \frac{1}{3} & 1 & 2 & \frac{1}{4} \\ \frac{1}{9} & \frac{1}{2} & 1 & \frac{1}{9} \\ 2 & 4 & 9 & 1 \end{bmatrix}.$$

Notice the overriding importance the Committee associates with the political aspect c_4, followed by financial considerations c_1 and, at a much lower level, the expected conditions for the participants c_2 and the impact on the community c_3. Given the high utility of decision d_3 on criterion c_4, even a cursory glance will have d_3 as a strong contender. Formally, the normalized columns are collected in the matrices

$$\tilde{\mathbf{C}}^1 = \begin{bmatrix} .6 & .6667 & .5714 \\ .1 & .1111 & .1429 \\ .3 & .2222 & .2857 \end{bmatrix}, \; \tilde{\mathbf{C}}^2 = \begin{bmatrix} .5455 & .5 & .6 \\ .2727 & .25 & .2 \\ .1818 & .25 & .2 \end{bmatrix}, \; \tilde{\mathbf{C}}^3 = \begin{bmatrix} .125 & .0476 & .2727 \\ .625 & .2381 & .1818 \\ .25 & .7143 & .5455 \end{bmatrix},$$

$$\tilde{\mathbf{C}}^4 = \begin{bmatrix} .1429 & .1667 & .1304 \\ .2857 & .3333 & .3478 \\ .5714 & .5 & .5217 \end{bmatrix} \text{ and } \tilde{\mathbf{C}} = \begin{bmatrix} .2903 & .3529 & .4286 & .2687 \\ .0968 & .1176 & .0952 & .1343 \\ .0323 & .0588 & .0476 & .0597 \\ .5806 & .4706 & .4286 & .5373 \end{bmatrix}.$$

Notice that the tradeoffs have not been changed by this operation. For instance, consider the estimates that compare the given decisions with decision 1, based on criterion three, i.e., the first column of the matrix $\mathbf{C}^3$. It specifies that decision d_2 is 5 times more valuable than decision d_1, a ratio unchanged in $\tilde{\mathbf{C}}^3$, where $\tilde{c}^3_{21}/\tilde{c}^3_{11} = .625/.125 = 5$. Computing the row averages of the matrices $\tilde{\mathbf{C}}^k$, $k = 1, ..., 4$ results in the columns $\mathbf{u}_{\bullet k}$ which, in turn, form the matrix

$$U = \begin{bmatrix} .6127 & .5485 & .1484 & .1467 \\ .1180 & .2409 & .3483 & .3223 \\ .2693 & .2106 & .5033 & .5310 \end{bmatrix},$$

while the row averages of $\tilde{C}$ result in $w = [.3351, .1110, .0496, .5043]^T$. We can now calculate the values associated with the decisions as

$$v(d) = Uw = [.3475, .2461, .4064]^T.$$

As hypothesized earlier, the strong weight of the political criterion with $w_4 = .5043$ and the high performance of d_3 on that criterion makes it an obvious choice.

The reliability of results such as those obtained above depends, among other factors, on the consistency of the decision maker's evaluations. There are various ways to measure consistency. The first employs a measure from elementary statistics. It can be motivated as follows. Consider a matrix C (the arguments for the matrices C^k are similar) that includes evaluations, which are consistent. For instance, let

$$C = \begin{bmatrix} 1 & 2 & 6 \\ \frac{1}{2} & 1 & 3 \\ \frac{1}{6} & \frac{1}{3} & 1 \end{bmatrix}.$$

Step 1 of the normalization technique will then compute

$$\tilde{C} = \begin{bmatrix} .6 & .6 & .6 \\ .3 & .3 & .3 \\ .1 & .1 & .1 \end{bmatrix}.$$

In other words, if the evaluations are consistent, then all elements in a row of $\tilde{C}$ will be identical, as $\tilde{c}_{i\ell} = \dfrac{c_{i\ell}}{\sum_r c_{r\ell}} = \dfrac{c_{iv} c_{v\ell}}{\sum_r c_{rv} c_{v\ell}} = \dfrac{c_{iv}}{\sum_r c_{rv}} = \tilde{c}_{iv}$. This suggests the coefficient of row-wise variation as an appropriate measure of consistency. To formalize, let $\mu_i^k = \dfrac{1}{m}\sum_{\ell=1}^{m} \tilde{c}_{i\ell}^k$ denote the mean of row i in the matrix $\tilde{C}^k$. The coefficient of variation in the i-th row is then $\sigma_{\mu_i^k}^k = \dfrac{1}{\mu_i^k}\sqrt{\dfrac{1}{m}\sum_{\ell=1}^{m}\left(\tilde{c}_{i\ell}^k - \mu_i^k\right)^2}$ and the average coefficient is $\sigma_\mu^k = \frac{1}{m}\sum_{i=1}^{m}\sigma_{\mu_i}^k$. Similarly, let μ_k denote the mean of the k-th row of $\tilde{C}$, then the coefficient of variation is $\sigma_{\mu_k} = \frac{1}{\mu_k}\sqrt{\frac{1}{q}\sum_{\ell=1}^{q}(\tilde{c}_{k\ell}-\mu_k)^2}$ and the average coefficient is $\sigma_\mu = \frac{1}{q}\sum_{k=1}^{q}\sigma_{\mu_k}$. The average coefficient of variation measures the average deviation of the decision maker's evaluation from the mean. As a rule of thumb, such deviations should typically not exceed .10 – .15. In the above example $\mu_1^1 = .6127$, so that $\sigma_{\mu_1}^1 = .0652$. Similarly, we obtain $\mu_2^1 = .118$ and $\sigma_{\mu_2}^1 = .1541$, and $\mu_3^1 = .2693$ and $\sigma_{\mu_3}^1 = .1256$, so that $\sigma_\mu^1 = .1150$. The average coefficients of variation of the remaining matrices $\tilde{C}^2$, $\tilde{C}^3$, $\tilde{C}^4$, and $\tilde{C}$ are .1125, .5253, .0804, and .1676, respectively. What stands out is the very high degree of inconsistency in matrix $\tilde{C}^3$. The message to the analyst is to return to the decision maker and double-check his preferences with respect to the impact of his choice on the community.

Another possibility to measure the consistency of the pairwise comparisons is Saaty's *consistency index*. We first calculate eigenvalue-like parameters associated with the matrices C^k, $k = 1, ..., q$, and C and denote them by λ_{max}^k and λ_{max}, respectively. Their values can be defined as

$$\lambda_{max}^k = \frac{1}{m}\sum_{i=1}^{m}\frac{1}{u_{ik}}\sum_{\ell=1}^{m}c_{i\ell}^k u_{\ell k} \quad \forall \ C^k, k = 1, ..., q, \text{ and}$$

$$\lambda_{max} = \frac{1}{q}\sum_{k=1}^{q}\frac{1}{w_k}\sum_{\ell=1}^{q}c_{k\ell}w_\ell \quad \text{for the matrix } C.$$

The consistency index of the matrices C^k is $CI_k = \frac{\lambda_{max}^k - m}{m-1} \ \forall \ k$, while the consistency index for the matrix C is $CI = \frac{\lambda_{max} - q}{q-1}$. The *consistency ratio* is then $CR^k = CI^k/RI$ and $CR = CI/RI$, respectively, where the empirically calculated *random index RI* depends on the size of the matrix and is taken from the Table I.9; see also Saaty (1994, 1996) and Belton and Stewart (2002).

Table I.9

Matrix size	3	4	5	6	7	8	9	10
RI	.52	.89	1.11	1.25	1.35	1.40	1.45	1.49

Similar to the approach with the coefficient of variation, any consistency ratio in excess of 0.1 should be viewed with suspicion and double-checked. In the above example, $\lambda_{max} = 4.0521$ and $CR = .01737/.89 = .0195 < .1$. Similarly, $\lambda^1_{max} = 3.0183$ and $CR^1 = .00915/.52 = .0176 < .1$, $\lambda^2_{max} = 3.0183$ and $CR^2 = .00915/.52 = .0176 < .1$, $\lambda^3_{max} = 3.4810$ and $CR^3 = .2405/.52 = .4625 > .1$, and $\lambda^4_{max} = 3.0093$ and $CR^4 = .0046/.52 = .0089 < .1$. Again, the estimates relating to the third criterion are way off and have to be checked with the decision maker.

An automatic method to reconcile the decision maker's evaluations was recently suggested by Eiselt (2001). The idea is to modify the weights and/or utilities so as to deviate as little as possible from the decision maker's original estimates while the modified weights and/or utilities are consistent.

An approach similar to the normalization technique above was suggested by Saaty (1980). It is called the *analytic hierarchy process* or *AHP*. It has found many applications; for pertinent references see, e.g., Olsen (1995). The basic procedure can be described as follows.

The Analytic Hierarchy Process

Step 1: Calculate the largest eigenvalues λ^k_{max}, $k = 1, ..., q$ and λ_{max} for the matrices $\mathbf{C}^k$ and $\mathbf{C}$, respectively, by solving the characteristic equations $\det|\mathbf{C}^k - \mathbf{I}\lambda| = 0$ and $\det|\mathbf{C} - \mathbf{I}\lambda| = 0$, respectively, where $\mathbf{I}$ denotes an identity matrix of appropriate size.

Step 2: Given the eigenvalues λ^k_{max} and λ_{max} determine the corresponding normalized eigenvectors $\mathbf{u}_{\bullet k}$ and $\mathbf{w}$ by solving the system of simultaneous linear equations $(\mathbf{C}^k - \mathbf{I}\lambda^k_{max})\mathbf{u}_{\bullet k} = \mathbf{0}$, augmented by $\mathbf{eu}_{\bullet k} = 1 \; \forall \; k$ with $\mathbf{e} = [1, 1, ..., 1]$, and $(\mathbf{C} - \mathbf{I}\lambda_{max})\mathbf{w} = \mathbf{0}$, augmented by $\mathbf{ew} = 1$, respectively.

Step 3: Compute the values associated with the decisions as $v(d) = \mathbf{Uw} = [\mathbf{u}_{\bullet 1}, \mathbf{u}_{\bullet 2}, ..., \mathbf{u}_{\bullet q}]\mathbf{w}$.

Example: Consider again the selection problem of the Olympic Committee. The characteristic equation of the matrix $\mathbf{C}^1$ is

$$\det \begin{vmatrix} 1-\lambda & 6 & 2 \\ \frac{1}{6} & 1-\lambda & \frac{1}{2} \\ \frac{1}{2} & 2 & 1-\lambda \end{vmatrix} = 0,$$

or $(1-\lambda)^3 + \frac{6}{4} + \frac{4}{6} - (1-\lambda) - (1-\lambda) - (1-\lambda) = 0$, or simply $-\lambda^3 + 3\lambda^2 + \frac{1}{6} = 0$, which has the single real root $\lambda^1_{max} = 3.0183$. The system of simultaneous linear equations is then

$$\begin{bmatrix} (1-3.0183) & 6 & 2 \\ \frac{1}{6} & (1-3.0183) & \frac{1}{2} \\ \frac{1}{2} & 2 & (1-3.0183) \end{bmatrix} \begin{bmatrix} u_{11} \\ u_{21} \\ u_{31} \end{bmatrix} = 0$$

and $u_{11} + u_{21} + u_{31} = 1$. One of the first three equations is linearly dependent, and the remaining system has the solution $u_{\bullet 1} = [.6144, .1172, .2684]^T$. Repeating the procedure for the matrices $\mathbf{C}^2$, $\mathbf{C}^3$, $\mathbf{C}^4$, and $\mathbf{C}$ results in the eigenvalues $\lambda^2_{max} = 3.0183$, $\lambda^3_{max} = 3.4683$, $\lambda^4_{max} = 3.0092$, and $\lambda_{max} = 4.0518$, as well as the eigenvectors

$$\mathbf{U} = (\mathbf{u}_{\bullet 1}, \mathbf{u}_{\bullet 2}, \mathbf{u}_{\bullet 3}, \mathbf{u}_{\bullet 4}) = \begin{bmatrix} .6144 & .5499 & .1339 & .1463 \\ .1172 & .2402 & .3420 & .3220 \\ .2684 & .2098 & .5241 & .5317 \end{bmatrix},$$

and $\mathbf{w} = [.3346, .1099, .0487, .5068]^T$. The result are the values of the decisions

$$\mathbf{v}(d) = \mathbf{Uw} = \begin{bmatrix} .3467 \\ .2455 \\ .4079 \end{bmatrix}.$$

Notice the similarity between the results of the analytic hierarchy process and the normalization procedure. As a matter of fact, their results differ by less than 0.4%. The consistency can be checked in a way similar to that discussed above. Here, λ_{max} and λ^k_{max} are the largest eigenvalues calculated in the analytic hierarchy matrix. The remaining procedure is identical. In our numerical example, $\lambda_{max} = 4.0518$ and $CR = .0194$, $\lambda^1_{max} = 3.0183$ and $CR^1 = .0176$, $\lambda^2_{max} = 3.0183$ and $CR^2 = .0176$, $\lambda^3_{max} = 3.4683$ and $CR^3 = .4503$, and $\lambda^4_{max} = 3.0092$ and $CR^4 = .0088$.

These results are almost identical to those obtained by the normalization technique. An up-to-date survey of the *AHP* is provided by Foreman and Gass (2001), and new research results are included in Golden and Wasil (2003).

A number of authors have criticized the assumptions and properties of the analytic hierarchy method. Barzilai et al. (1987) have suggested the use of the *Geometric Mean Method* instead. Given matrices $\mathbf{C}^k$, $k = 1, ..., q$ and $\mathbf{C}$, it can be described as follows.

The Geometric Mean Method

Step 1: Calculate the utility of decision d_i with respect to criterion c_k as

$$u_{ik} = \left[\prod_{\ell=1}^{m} c_{i\ell}^{k} \right]^{1/m},$$

i.e., as the geometric mean of the rows of the matrices $\mathbf{C}^k$, $k = 1, ..., q$.

Step 2: Calculate the weights of the criteria as the normalized geometric means of the rows of matrix $\mathbf{C}$, i.e.,

$$w_k = \left[\prod_{v=1}^{q} c_{iv} \right]^{1/q} \bigg/ \sum_{\mu=1}^{q} \left[\prod_{v=1}^{q} c_{\mu v} \right]^{1/q}.$$

Step 3: The values associated with the decisions are then computed as

$$v(d_i) = \left[\prod_{k=1}^{q} u_{ik}^{w_k} \right] \bigg/ \sum_{\mu=1}^{q} \left[\prod_{v=1}^{q} u_{\mu v}^{w_v} \right].$$

In our numerical example, the geometric means of the rows of $\mathbf{C}^1$ are 2.2894, .4368, and 1, respectively. Repeating the calculations for all matrices $\mathbf{C}^k$, we obtain

$$\mathbf{U} = \begin{bmatrix} 2.2894 & 1.8171 & .4642 & .5 \\ .4368 & .7937 & 1.1856 & 1.1006 \\ 1 & .6934 & 1.8171 & 1.8171 \end{bmatrix}.$$

The geometric means of the rows of the matrix $\mathbf{C}$ are 1.9168, .6389, .2803, and 2.9130, respectively. Normalizing those results in the vector of weights $\mathbf{w} =$

[.3334, .1111, .0488, .5067]^T. The non-normalized values of the decisions (the numerators of the expression in Step 3) are then

$$\begin{bmatrix} (2.2894)^{.3334}(1.8171)^{.1111}(.4642)^{.0488}(.5000)^{.5067} \\ (.4368)^{.3334}(.7937)^{.1111}(1.1856)^{.0488}(1.1006)^{.5067} \\ (1.0000)^{.3334}(.6934)^{.1111}(1.8171)^{.0488}(1.8171)^{.5067} \end{bmatrix} = \begin{bmatrix} .9549 \\ .7828 \\ 1.3379 \end{bmatrix},$$

which, normalized, produces the values

$$\mathbf{v}(d) = [.3105, .2545, .4350]^T,$$

which are, again, quite similar to those calculated with the other two methods. Also, the consistency of the decisions maker's evaluations can be checked as discussed above.

CHAPTER 2   GAMES AGAINST NATURE

This chapter investigates situations in which one decision maker is pitted against an opponent whose actions do not consider the payoffs that result from the combination of strategies chosen by the two players. For simplicity, the second player will be called "nature." Clearly, the choice of strategy chosen by the "intelligent" or "rational" player (represented by us) against such an opponent will depend on the information we have about nature's choice of strategies. The first section of this chapter introduces the basic elements of games against nature, including their components, certainty equivalents, and visualization. The second section provides a number of rules that we may wish to follow in the absence of much detailed information, and the third section investigates the value of certain types of information.

2.1   Elements of Games Against Nature

This section first defines games against nature and looks at possible extensions of the concept. It then describes a type of lottery in order to introduce the concept of certainty equivalent. Finally, two types of visual tools are presented that inform decision makers about the main features of the problem and the sensitivity of its parameters.

2.1.1   Basic Components

Games against nature are a special variety of two-person games, involving two players, each of whom has a finite number of decision alternatives. In addition, a payoff function is associated with each pair of decisions by the two players, indicating the loss or gain of the first player. To formalize matters, let the two players be denoted by I and II, let the decision alternatives of I be d_i, $i = 1, ..., m$, while the strategies of II are s_j, $j = 1, ..., n$. The payoff associated with the pair (d_i, s_j) is then a_{ij}, which player I receives from player II, and all payoffs are collected in the $[m \times n]$-dimensional matrix $\mathbf{A}$. There is a fundamental difference between the two players: whereas *I* is assumed to be a rational player whose objective is to maximize his own gain, player *II* chooses his strategies randomly. Based on this assumption, *II* is often referred to as "nature". Typically, d_i are called *decisions* while s_j are referred to as *states of nature*. In general, a game against nature may have any number of stages that are played sequentially. For instance, a rational decision maker, player *I*, may make a strategic decision regarding the expansion or relocation of a firm. This decision is followed by some customer demand for the firm's products. The magnitude of the demand is, of course, beyond the control of player *I*. It is possible to model the demand as a state of nature, as the forces that determine the magnitude of the demand do not choose the demand in order to maximize their gain, but follow some unknown process. Based on the demand, player *I* then has to decide whether to hire new personnel, lay off employees, increase or cut advertising, declare bankruptcy, or make a similar choice. Such a choice is then followed by another state of nature. Notice that if, to continue with this example, player *I*'s last decision were followed by a stockholders' meeting, then this part is not properly modeled by nature, as stockholders do not follow a random process with their decisions, but a rational objective, albeit not necessarily the same as player *I*. This chapter will consider only single-stage games against nature.

All games against nature are selection problems, i.e., the task is to choose exactly one among the given alternatives. This should be kept in mind, as many situations allow mixing strategies, so that modeling the problem as a standard game against nature unnecessarily restricts the available options. As an example, consider an investor who can invest $100 in stocks or bonds. Let nature's strategies be restricted to a booming and a declining economy, and the returns to the investor after one year are $120 and $70 for the stocks after a good or bad year, respectively, while bonds yield $106 either way. Rather than choosing stocks *or* bonds, the investor may choose to diversify his investment, thus capitalizing on the potential of a growing economy, while avoiding the sharp decline of his portfolio by holding stocks. A fifty-fifty split between stocks and bonds would, for instance, yield $113 and $88 respectively, more opportunity than bonds (but less than stocks), while less risk than stocks (but more than bonds). As in all modeling, the nature of the problem will determine if a model such as games against nature is applicable in a specific situation.

In order to motivate our discussion, consider the following

Example: A farmer makes plans regarding the use of his land in the next year. His options are to plant wheat, corn, potatoes, or to breed cattle. His choice is largely determined by the future weather which is, of course, beyond his control, following a random process instead. As such, a game against nature may be the appropriate modeling tool. The farmer has determined that the relevant weather patterns can be classified as very wet, wet, normal, dry, and very dry. These are

Chapter 2: Games Against Nature

the states of nature in the model. The payoffs are typically dependent on the yield and are transformed into dollars as a common unit, whereas in developing countries using nutritional value of the yield may be more appropriate. In this example the payoff matrix is

$$\mathbf{A} = \begin{bmatrix} 3 & 5 & 7 & 9 & 1 \\ 6 & 5 & 4 & 2 & 8 \\ 4 & 4 & 5 & 7 & 3 \\ 6 & 4 & 3 & 2 & 7 \end{bmatrix}.$$

Note that decision d_1 has a much wider range than decision d_3 making the former a more risky and the latter a more conservative choice. Also note that for any state of nature, player I's payoff from decision d_2 is never worse than the corresponding payoff from d_4. This indicates that player I is never worse off using d_2 as compared to d_4. In general, we can formulate

Definition I.3: Decision d_i is said to *dominate* decision d_k, if and only if $a_{ij} \geq a_{kj}$ $\forall$ j and $a_{ij} > a_{kj}$ for at least one j. We will write such dominance as $d_i \succ d_k$.

Dominated rows can be deleted from the problem, as the corresponding decisions will never be chosen by a rational player. In this example, $d_2 \succ d_4$ hence player I's decision d_4 can be deleted. Note that nature cannot have any dominated strategies—the concept does not apply.

An important feature of games against nature is the type of information available to player I regarding nature's choice of strategy. There are two extremes, *certainty* and *uncertainty*. In the case of certainty, player I knows exactly what state of nature will occur, while in the case of uncertainty, player I had no information whatsoever concerning what state of nature will occur. The case of certainty is easily dealt with. If the future state of nature is known in advance, all other columns of the payoff matrix are irrelevant and can be deleted. All player I has to do is choose the most beneficial, i.e., highest payoff, in the applicable column and choose the pertinent decision. In the above example, if the farmer knew that the coming year would be s_3 or "medium", the known payoffs for his decisions are $\mathbf{a}_{\bullet 3}$ = $[7, 4, 5, 3]^T$, so that he will choose d_1, grow wheat, and receive a payoff of 7.

The other extreme is uncertainty. Here, player I does not even know how likely it is that any of the states of nature occurs. This type of situation occurs frequently in the advent of new technologies, when little, if anything, is known about their performance. Typical examples include the performance of the heat shields required for the space shuttle during reentry into the earth's atmosphere, and customer reaction to internet banking when it was first introduced. In all problems involving uncertainty, no past data are available that might give some guidance for the future. The next section will provide a number of rules for decision making under uncertainty. All of these rules have in common that they generate an *anticipated payoff* vector **a** whose largest element will then determine player I's optimal decision. The methods differ by the way they generate vector **a**.

There is, of course, a wide area between certainty and uncertainty. Somewhere in-between is an area where decision making is said to be under *risk*. In that case, player I does not know which state of nature will occur, but he does know the probability used by nature to choose his strategies. Between uncertainty and risk is the field of *partial information*, where player I does not know nature's probability distribution, but some of its parameters such as mean, median, mode, or variance.

2.1.2. Lotteries and Certainty Equivalents

So far, we have been optimizing a function of payoffs. However, a certain payoff may be valued differently by different people. Furthermore, decision makers' perceptions of wealth do not really follow the standard rules of computation used so far. As an example, a wealthy individual will consider an additional dollar less valuable than a poor individual. Often, economists work with decreasing marginal utilities of money, i.e., each additional dollar does increase an individual's utility, but to smaller and smaller degrees. This is the reason why we will replace the payoffs used so far in our discussion by *utilities*. Once these utilities are properly defined, we can use them in any of the methods described above; all we have to do is replace the word "payoff" by "utility".

One reason for someone's utility deviating from the actual payoff in a situation is the inclusion of risk. While most people will try to avoid risk, some—the gambler-type—are actually seeking it. In order to get a mathematical grip on this, consider the following lottery. A decision maker is offered two choices: either play a game of chance, or accept a certain amount in cash. The game consists of the following two outcomes: there is a chance of p to win a prespecified amount of money, and there is a chance of $(1 - p)$ to win nothing. Here, we are trying to find the cash amount that a decision maker will consider equivalent to the opportunity of playing the game. As an example, consider the following game. There is a 0.5 chance to win $100 and a 0.5 chance to win zero. Clearly, the expected monetary value of playing the game is 50. The dollar amount that any specific individual will consider equivalent to the opportunity of playing this game will be called the decision maker's *certainty equivalent*, or *CE* for short. For instance, someone who has an equivalent of $40 in our game, is apparently prepared to accept $10 less than the expected value of the game (the so-called *risk premium RP*) to avoid the risk associated with playing the game. Such an individual is referred to as *risk-averse* shown by his $RP > 0$. On the other hand, someone who has a certainty equivalent to $70 is a risk-seeking player who considers the risk associated with the gamble worth $20 to him, i.e., his $RP < 0$. A risk neutral player has a risk

Chapter 2: Games Against Nature   77

premium $RP = 0$. Repeating the game in our example for different values of p and determining a decision maker's certainty equivalents for each probability, we can profile a decision maker by plotting his certainty equivalents against the probabilities p. The prototypical types of risk behavior are shown in Figure I.16.

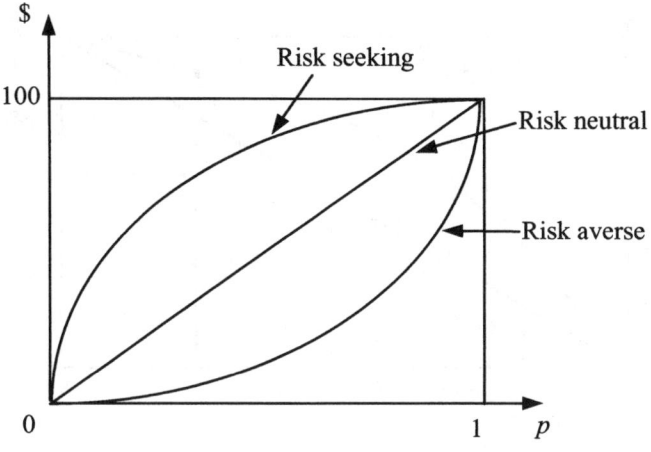

Figure I.16

We now describe a method that determines a decision maker's utility function in practice. We assume that all relevant payoffs $\$_1, \$_2, ..., \$_t$ are ordered in nondecreasing order, i.e., $\$_1 \geq \$_2 \geq ... \geq \$_t$, where ties are broken arbitrarily. We arbitrarily set $u(\$_1) = 1$ and $u(\$_t) = 0$ (other values may be chosen if so desired). We then apply the following:

Iterative step:
> For any $\$_k \in [\$_i; \$_j]$ with $u(\$_i)$ and $u(\$_j)$ known, find the probability p, so that the decision maker is indifferent between $\$_k$ with certainty and the lottery: $\$_i$ with probability p and $\$_j$ with probability $(1-p)$. Define $u(\$_k) = p\, u(\$_i) + (1-p)\, u(\$_j)$.

This step is repeated until utilities for all relevant payoffs have been determined.

As an example, consider the following situation. The relevant payoffs (in $100,000) are $\$_1 = 8 \geq \$_2 = 6 \geq \$_3 = 2 \geq \$_4 = 1$. Without a loss of generality, let $u(\$_1) = u(8) = 1$ and $u(\$_4) = u(1) = 0$. Determine the utility of $\$_2$ by the following gamble: receive 6 with certainty or 8 with probability p and 1 with probability $(1-p)$. Suppose the decision maker decides that $p = 0.8$. Then $u(\$_2) = u(6) = (0.8)(1) + (0.2)(0) = 0.8$. The next step is to determine the utility for $\$_3$. The decision maker is presented with the choice of 2 with certainty or 6 with probability p and 1 with probability $(1 - p)$. Assume the decision maker feels the two choices are equivalent for $p = 0.4$. Then $u(\$_3) = u(2) = (0.4)(0.8) + (0.6)(0) = 0.32$. Plotting the $-values against the utilities $u(\$)$ and connecting the points for better visibility we obtain Figure I.17.

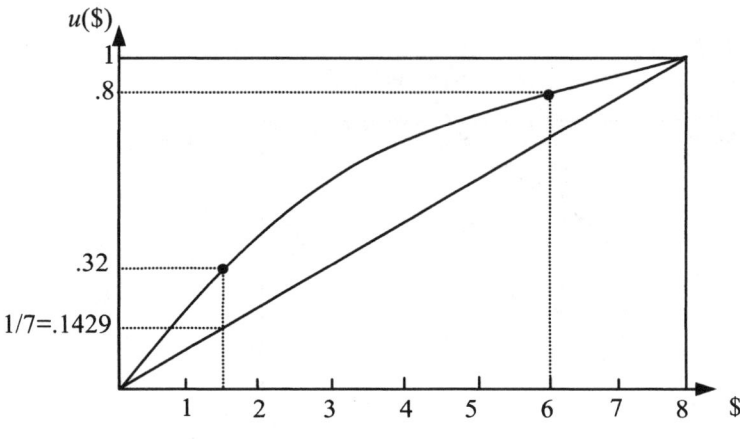

Figure I.17

The straight line in Figure I.17, $u(\$) = (1/7)(\$ -1)$, represents risk neutrality whereas the curve that incorporates the points determined in the above procedure represents our decision maker's certainty equivalents; we will call it the CE-line for short. Its shape can be used to explain the behavior of a decision maker. Here, for any given $-value, the CE-line is higher than the risk neutrality line, i.e., the decision maker's utility for the certainty equivalent is higher than for the gamble, indicating risk aversion. In general, if a utility function in a $(\$, u(\$))$ diagram is concave, then the decision maker is risk averse, if it is convex, the decision maker is risk seeking. Many decision makers do not, however, readily fall into one of the simple categories discussed above. For instance, if we were trying to determine a decision maker's certainty equivalent in the above game for, say, $10, we may find that $CE = 6$, i.e., risk-seeking behavior with a $1, i.e., 10% premium paid for participation in the game. This type of behavior is exhibited week after week by the millions of people playing the lotteries. On the other hand, if the same individual were to ask to participate in the same game for, say, $100,000, we will invariably find that $CE < 50,000$. Typically, students in decision analysis classes came up with CE-values between $25,000 and $40,000—and that is "in theory", i.e., without someone actually offering the money! This behavior is modeled by a typical S-shaped utility function, convex, i.e., risk seeking, for small amounts, and concave, i.e. risk averse, for larger amounts. A measure of risk behavior is $r(\$) = - u''(\$)/u'(\$)$. If $r(\$) > 0$, the decision maker is risk averse, for $r(\$) < 0$ he is risk seeking, and for $r(\$) = 0$, he is risk neutral.

Example: A decision maker has $5,000,000 to invest and two possibilities to do

Chapter 2: Games Against Nature

so. The first possibility is to buy stocks and the second to buy bonds. There are three states of nature: a booming economy, a slowly growing economy, and a declining economy. The returns are shown in the following matrix.

$$\begin{bmatrix} 20\% & 5\% & 2\% \\ 10\% & 10\% & 10\% \end{bmatrix}$$

Given the amount to invest, the payoff matrix (in millions of dollars) is then

$$\begin{bmatrix} 1.0 & .25 & .1 \\ .5 & .5 & .5 \end{bmatrix}$$

Suppose that the decision maker's utility function has been established as

$$u(\$) = \begin{cases} 5\$^2, & \text{for } \$ \in [0;.2] \\ -(5/6)\$^2 + 2\$ - (1/6), & \text{for } \$ \in [.2; 1]. \end{cases}$$

This S–shaped utility function is convex between 0 and .2, and concave between .2 and 1. We can now convert the payoffs to utilities, resulting in the utility matrix

$$U(\$) = \begin{bmatrix} 1.0 & .28125 & .05 \\ .625 & .625 & .625 \end{bmatrix}$$

Suppose the probabilities for the states of nature have been estimated as [0.5; 0.4; 0.1]. Then the expected utilities are 0.6175 and 0.625, respectively, so that the decision maker will choose the investment in bonds, i.e. the less risky alternative. Suppose now that the investment is done through a broker, who charges a flat rate of $100,000 for any investment with a positive return. The new payoffs are

$$\begin{bmatrix} .9 & .15 & 0 \\ .4 & .4 & .4 \end{bmatrix},$$

and the utility matrix is

$$U(\$) = \begin{bmatrix} .9583 & .1125 & 0 \\ .5 & .5 & .5 \end{bmatrix}$$

with expected utilities of 0.52415 and 0.5 respectively, so that the decision maker will now choose the first investment alternative. The reason for the apparent change of mind on the decision maker's part is found in his risk behavior. First observe that a risk-neutral decision maker would have considered expected payoffs of .61 and .5 before the fee, and .51 and .4 after the fee was levied (all in millions of dollars), so that he would have opted for stocks in both cases.

In order to explain the decision maker's behavior, consider the two parts of the utility function. The first part of the utility function (for low $-values) is $u(\$) = 5\$^2$. Hence $u'(\$) = 10\$$ and $u''(\$) = 10$, so that r($) = $-10/10\$ < 0$ for all $\$ > 0$, so that the decision maker is risk seeking in that part. For $\$ \in [.2; 1]$, $u(\$) = -(5/6)\$^2 + 2\$ - (1/6)$, so that $u'(\$) = -(5/3)\$ + 2$ and $u''(\$) = -5/3$, resulting in a risk coefficient $r(\$) = 5/3/(-(5/3)\$ + 2)$. As $\$ \leq 1$, the denominator of the fraction is positive, so that $r(\$) > 0$ and the decision maker is risk averse in this interval. Now the reason for the "change of mind" becomes apparent. Before the brokerage fee was considered, 90% of the states of nature fell into the decision maker's risk aversion range, so that he opted for the safer investment in bonds. Once the brokerage fees were included, only 50% of the states of nature fell into the decision maker's risk aversion range, which resulted in his choice of stocks over bonds.

A famous example to motivate the inclusion of risk in a decision maker's utility function is the *St. Petersburg paradox*. It describes a simple game: a fair coin is flipped until "heads" come up for the first time. Suppose this is the case at the *n*-th trial. The payoff is then $\$2^n$. A risk neutral player would determine that the probability of "heads" coming up at the *n*-th trial is $1/2^n$, the expected monetary value of the game is

$$\sum_{n=1}^{\infty} \frac{1}{2^n} 2^n = 1 + 1 + 1 + \ldots \to \infty.$$

With an expected payoff approaching infinity, a risk neutral player would be prepared to pay any finite sum to participate in the game as this cost would still be less than the expected payoff. However, few individuals will offer more than, say, $10 to participate in the game. One reason for this seemingly irrational behavior is as follows. Most individuals consider events whose probability of occurrence is below some threshold as so unlikely that they think of them as "impossible," i.e., as having a probability of zero. Assuming that someone uses a threshold of one tenth of a percent, i.e., about $1/2^{10}$, then the expected payoff of the truncated game is $10. This line of reasoning omits the very high payoffs that are associated with very low probabilities. Another reason is that most people do not maximize expected payoffs, but expected utilities. And as most people assign about the same utility to, say, ten million dollars, hundred million dollars, or any sum even larger, they will not be enticed to pay any significant amount to play the game.

In general, utilities should be used only for the purpose of comparison of strategies and their ranking, as their absolute values have no meaning. Also note that for a given a utility function $u(\$)$, the linear transformation $\hat{u}(\$) = a + b\, u(\$)$ with parameters $b > 0$ and $a \in \mathbb{R}$ results in the same ranking of decision alternatives.
Many models have attempted to incorporate payoff and risk in one function. One

popular example is the standard portfolio selection model; see, e.g., Eiselt *et al.*(1987). There we have expected return and risk of a portfolio to consider and two standard ways to deal with the problem are to either maximize return, subject to an upper bound on the acceptable risk, or minimize risk with a lower bound on the return. A similar problem is commonplace in the insurance industry. Take a decision maker with annual after-tax income of $W = \$50,000$ who considers taking out insurance on his vehicle. If he does not, then there may be damage $D = \$10,000$ with a probability of $p = 0.01$. Let us assume our decision maker has a logarithmic utility function, i.e, $u(\$) = \ln \$$. A quick check reveals that $u'(\$) = 1/\$$ and $u''(\$) = -1/\$^2$, so that $r(\$) = 1/\$ > 0$, meaning the decision maker is risk averse. If no accident happens, the individual is left with $50,000 for which he has a utility of $u(50,000) = \ln 50,000 = 10.8198$, and if an accident happens, there are only $40,000 left with a utility of $u(40,000) = \ln 40,000 = 10.5966$. Then the expected utility if there is no insurance is

$$E(u(\$)) = p\,u(W-D) + (1-p)\,u(W).$$

In our example, $E(u(\$)) = 10.8176$. The $-value that leads to this utility (i.e., $ for which $\ln \$ = 10.8176$) is the certainty equivalent $\$ = e^{-10.8176} = 49,891.20$. Then the *insurance premium IP* is defined as $W - \$ = 50,000 - 49,891.20 = 108.80$ which is the amount the individual should be prepared to pay for the vehicle insurance.

2.1.3 Visualizations of the Structure of Decision Problems

This section will explore a number of different tools that are useful for the visualization of decision scenarios. In particular, we will concentrate on three types of diagrams: influence diagrams, decision trees, and tornado diagrams. Crudely speaking, influence diagrams are mostly used for planning on the macro (or strategic or tactical) level, decision trees are most frequently applied to planning on the micro (or operational) level, while tornado diagrams are very useful in displaying results in sensitivity analysis. A comparison of graphical techniques for problems in decision analysis is provided by Bielza and Shenoy (1999).

The structure of *influence diagrams* is similar to that of general networks in that they comprise elements, called nodes, and binary relations, called (directed) arcs. The main difference between influence diagrams and networks is that influence diagrams distinguish between three different types of nodes: *decision nodes* that are symbolized by rectangular nodes, *chance nodes* that are represented by circles, and *consequence nodes* that are shown by rectangles with rounded corners. As the name suggests, the decision maker will have a choice at the decision node, while nature makes her choice at a chance node. The consequence nodes will list the outcomes that result from a combination of decisions and chance events. Typically, the arcs have one of two meanings: arcs that lead into chance or consequence nodes indicate that the chance or consequence is a result of the predecessor node, while an arc leading into a decision node n_i indicates that the decision n_i follows after its predecessor nodes have occurred.

The most primitive structure of an influence diagram has just three nodes, one of each type. Here, the choice node and the chance node have arcs leading into a consequence node that results from the decision maker's choice and the event "chosen" by nature. In order to explain the concept, consider the following

Example: A Faculty member is in charge of organizing a scientific conference. The site has been determined, but the exact dates have still to be decided upon. The main objectives are the status of participants (i.e., having a few "stars" attend will improve the image of the conference), and making a profit (or, at the very least, not having to sustain a loss). Whether or not people can or will attend the conference will depend on the plans they have already made, and on the alternatives that present themselves at the different dates. One possibility is to obtain additional information in terms of a survey, sent to a selection of potential conference participants. An influence diagram that shows the relation between the components of the decision making process is shown in Figure I.18.

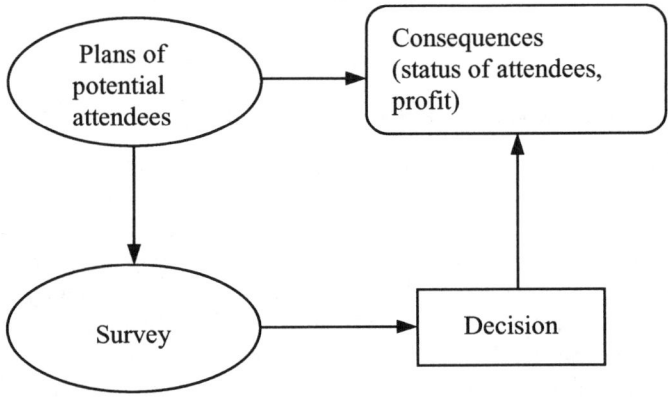

Figure I.18

It is apparent that the structure of Figure I.18 is very simple, but it clearly shows the dependence of the survey results on the actual plans of potential attendees and the alternatives they have, the dependence of the decision on the survey, and the influence of the plans by potential attendees and the decision on the number and type of people who will attend the conference, as well as the revenue generated through registration fees.

While such a diagram is very useful in the planning on the macro level, it is not once the planning has entered the much more detailed micro stage. At this point the use of a *decision tree* may be useful. Similar to an influence diagram, a decision tree includes rectangular nodes for decisions made by the decision maker, while circular nodes again symbolize events controlled by nature. The final consequences of the decisions and the chance events are shown as triangles at leaves of the decision tree. There are no consequence nodes in decision trees. Arcs between nodes show the sequence of events. Each arc leading out of a decision node is associated with a decision, while each arc that leads out of a chance node relates to a random event. Clearly, if the number of possible decisions or events is even moderately large, decision trees will get very large. In order to illustrate this, consider again the previous

Example: Suppose that there are only three possible decisions regarding the conference dates: June, July, and August. As far as the uncertain events are concerned, there would—realistically speaking—be an almost infinite number of combinations of the type and number of attendees of the conference and the profit made. In order to limit the size of the tree, we consider only two possible outcomes of a decision, many and few attendees, respectively. The decision tree then starts with the planner's decision whether or not to request a forecast. If no forecast is requested, the planner then has to decide which date to choose for the conference. Once this is done, nature will choose one of her strategies "many" or "few" (attendees). On the other hand, if a forecast was commissioned, there will be one outcome for each of the proposed dates. Based on the outcome, the planner can then choose one of the three possible dates, after which nature will "choose" one of her strategies "many" or "few." The decision tree for this problem is shown in Figure I.19.

The final graphical tool discussed in this section are so-called *tornado diagrams*. The idea is that in virtually all decision problems, sensitivity analyses are an essential part of the decision-making process. As usual, the idea is to avoid the inclusion of uncertainties in the mathematical problem in order to keep the complexity of the model within reason. In principle, it would be possible to perform sensitivity analyses on as many different parameters as desired, even at the same time. However, performing sensitivity analyses on more than a single parameter at a time will necessarily lead to statements that link the changing parameters to each other, making the information thus found hard to comprehend in the best case, and useless in the worst. One type of sensitivity analysis allows the decision maker to specify the parameters of the problem not as single numbers, but as intervals with lower and upper bounds. The analysis will then determine what the effects on the solution are, when the parameter is set to the extremes. While this is an easy task when the parameters are independent of each other, additional assumptions are needed when this is not the case. This latter case will be addressed in the next section. Clearly, some parameters will have much larger

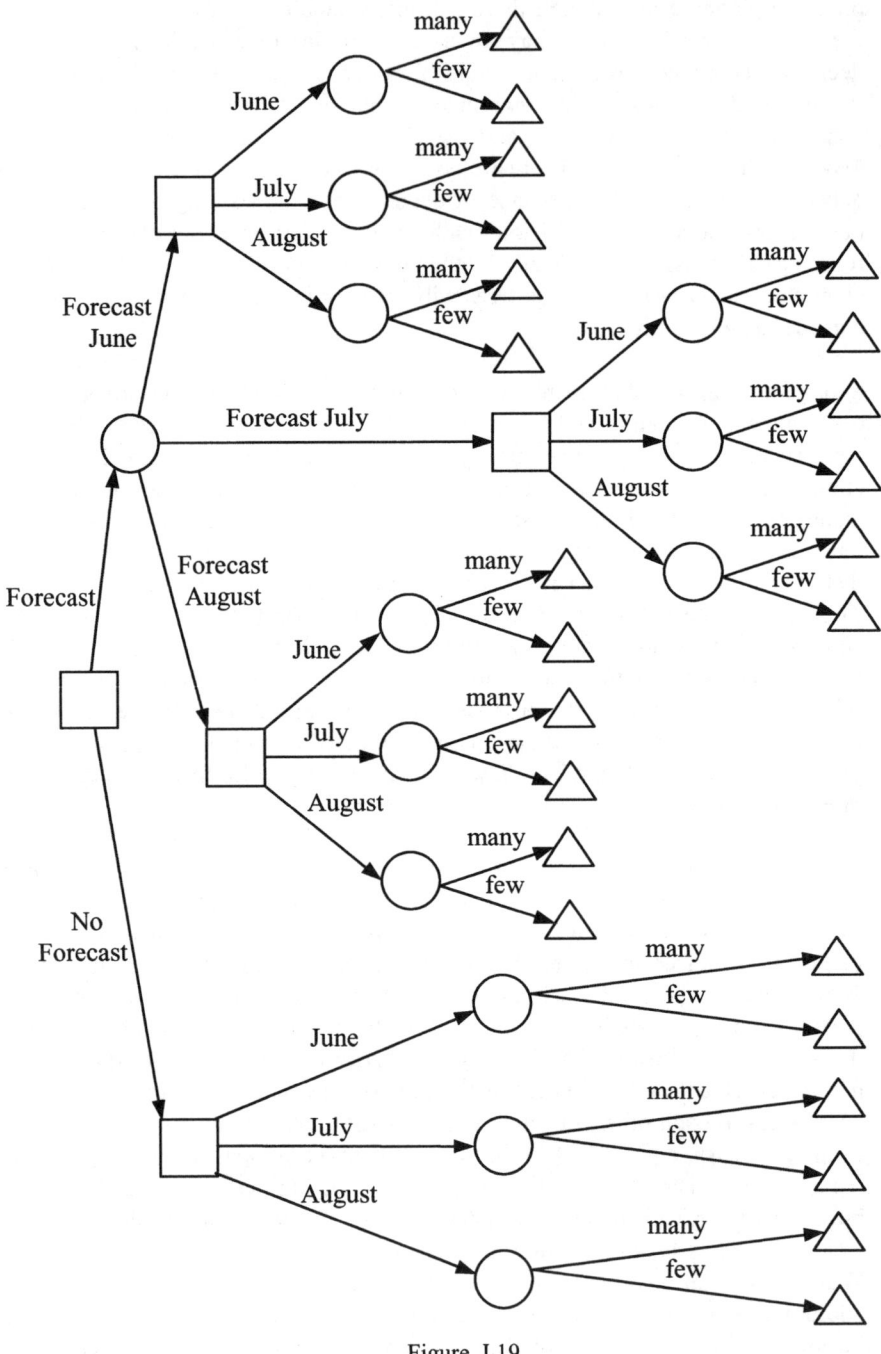

Figure I.19

effects on the value of the objective function than others. Such knowledge is invaluable to decision makers, and tornado diagrams provide a visualization of the sensitivity of the objective with respect to the different parameters. This concept is best explained by a simple

Example: The city council of a medium-sized town ponders the future of their concert hall. They have the option to either organize a number of concerts, operas, and special performances in the hall, or rent it out to a company that organizes conventions. However, this is not an either-or decision. It will be possible to rent out the hall for a part of the year, and organize performances during the remainder of the year. So far, it has been contemplated to rent out the hall between 20% and 50% of the year, with most councilors leaning toward 30% (so that 70% of the time performances are organized, thus retaining the basic mission of the building). The biweekly revenues that are derived from the shows range wildly between $1,000 and $25,000, with $15,000 being assumed to be a realistic figure. Furthermore, there are additional revenues that derive from radio and television rights.

On the other hand, there are fixed costs for basic cleaning, lighting, maintenance, etc., plus additional costs if shows are put on. Whenever the hall is rented out, the renter is held responsible for all of these costs, which include lighting, security, damages, and cleaning. The pertinent figures (all in $1,000) are summarized in Table I.10.

Table I.10

Item	Cost		
	Anticipated	minimum	maximum
Fixed costs	6	6	6
Show lighting	.8	.6	.9
Security	5	1	6
Damages	2	0	7
Cleaning	.2	.1	.3
Revenues	15	2	25
Radio/TV rights	3	0	8
Rental income	2	1.8	2.3
Planned proportion of rental	.3	.2	.5

Presently, there is a dispute between the city and the convention organizers regarding the fees, and the organizers have threatened to move to another venue.

The city planners' first move is to set up their own profit function, which turns out to be

𝒫 = (revenue from shows)+(revenue from rental)−(fixed cost)−(cost for shows)
  = [(revenue from shows)+(radio/TV rights)](1 − planned proportion of rental)
  + (rental income)(planned proportion of rental) − (fixed cost)
  − (costs for lighting + security + damages + cleaning)(1−planned proportion of rental).

Table I.11 provides the profit for the lowest and the highest values of each of the individual parameters, along with the width of the interval, given that all other parameters are at their anticipated volume.

Table I.11

Item	Profit at min	Profit at max	Width of interval
Show lighting	1.74	1.53	.21
Security	4.4	.9	3.5
Damages	3.0	−1.9	4.9
Cleaning	1.67	1.53	.14
Revenues	−7.5	8.6	16.1
Radio/TV rights	−.5	5.1	5.6
Rental income	1.54	1.69	.15
Planned proportion of rental	2.4	0	2.4

These results are now arranged in a tornado diagram by first arranging them in increasing order with respect to the width of the interval. A two-dimensional diagram is then constructed with the profit plotted on the abscissa, while each bar

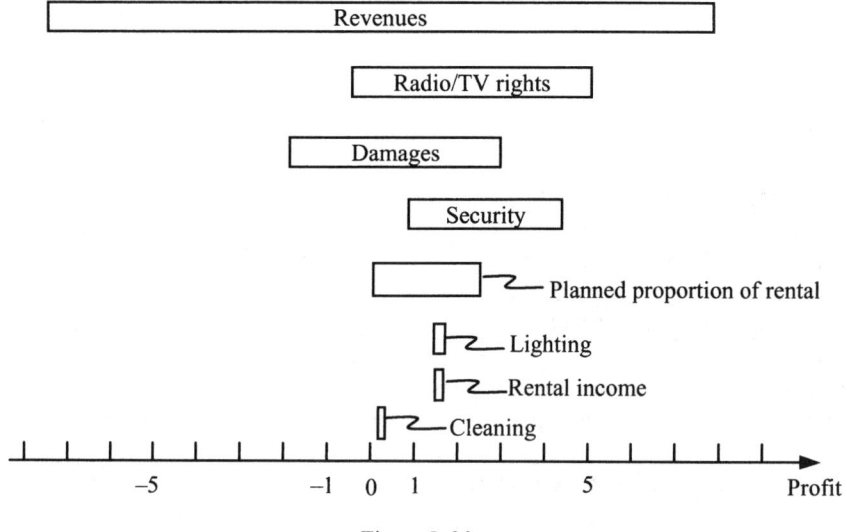

Figure I. 20

Chapter 2: Games Against Nature

at the ordinate represents one of the factors in the profit function, with the factor with the largest width of its interval plotted on top, and the one with the smallest width at the bottom. The resulting graph has the shape of a tornado, giving the type of graph its name. The plot for our example is shown in Figure I.20.

The diagram clearly indicates that the disputed revenues from the rental of the hall are quite insignificant. On the other hand, the revenues from the shows, the proportion of rental, and the radio and television rights have a significant effect on the profitability of the venture. This emphasizes the need for the organizers to very carefully screen the potential shows for their profitability and set up a promising program.

2.2 Rules for Decision Making Under Uncertainty and Risk

This section provides a number of rules for choosing a decision against nature. The first rules are designed for decision making under uncertainty, while the last are applicable under risk. One feature all these rules have in common is the simplicity of their assumptions. All rules will be explained by using a numerical example with three decisions and five states of nature in which the payoffs, to be maximized, are given in the matrix

$$\mathbf{A} = \begin{bmatrix} 3 & 5 & 7 & 9 & 1 \\ 6 & 5 & 4 & 2 & 8 \\ 4 & 4 & 5 & 7 & 3 \end{bmatrix}.$$

Rule 1 (Wald criterion, pessimist's rule, maximin rule): The anticipation in this rule is that nature displays a behavior adversarial to player I who, consequently, assumes that whatever decision he makes, nature will counter it by choosing a state of nature that minimizes player I's payoff. Formally, the anticipated payoff $\mathbf{a}^{(1)}$ consists of the row minima, of which player I chooses the maximum, i.e., he will choose $d_i = \arg\max_\ell \{\min_j \{a_{\ell j}\}\}$. In our example, $\mathbf{a}^{(1)} = [1, 2, 3]^T$, whose maximum is "3", so that the player will choose decision d_3 and anticipate a payoff of 3.

A major shortcoming of Wald's rule is that it only considers the worst outcomes and ignores all others. Consider, for example, a situation in which the payoff matrix is

$$\mathbf{A} = \begin{bmatrix} 100 & 100 & 100 & 100 & 49 \\ 50 & 50 & 50 & 50 & 50 \end{bmatrix}.$$

According to Wald's rule, a pessimistic player will anticipate payoffs of 49 and 50 respectively, and consequently choose the second strategy. This exclusive focus on the worst case appears rather limiting.

Also note that with this rule, player I could gain by mixing strategies if the original problem were to allow it. For instance, in a problem with two decisions, two states of nature, and payoffs of 10 and 2, and 4 and 12, respectively, player I could mix his strategies, say, sixty-forty and thus obtain another decision with payoffs of $(.6)(10, 2) + (.4)(0, 12) = (7.6, 6)$. The row minima of the original strategies are 2 and 4, respectively, while the row minimum of the new mixed decision is 6, making it the preferred choice.

Rule 2 (The optimist's rule, maximax rule): This rule is similar to Wald's rule above, except that it focuses on the best case, rather than the worst. Formally, the anticipated payoffs are $a_i^{(2)} = \max_j \{a_{ij}\} \; \forall \; i$ among which player I chooses the maximum of the row maxima. In the above example, $\mathbf{a}^{(2)} = [9, 8, 7]^T$, so that an optimistic planner will choose the first strategy and anticipate a payoff of 9. As the maximax rule chooses a decision on the basis of the largest payoff in the matrix, and a mixed strategy has payoffs that are linear convex combinations of existing payoffs, mixed strategies cannot find better solutions. Again the exclusive focus on an extreme outcome severely limits the value of this criterion.

Rule 3 (Hurwicz criterion): This rule is an improvement over the extreme positions of the previous two criteria as it combines worst and best cases. More specifically, the anticipated payoff is $\mathbf{a}^{(3)} = \lambda \mathbf{a}^{(1)} + (1 - \lambda)\mathbf{a}^{(2)}$ with $\lambda \in [0; 1]$. In other words, the anticipated outcome is a linear convex combination of worst and best case with λ as *coefficient of pessimism*. For $\lambda = 1$, there is total pessimism and, as expected, the rule coincides with Wald's criterion. Similarly, if $\lambda = 0$, the planner exhibits no pessimism and Hurwicz's rule coincides with the optimist's rule. Clearly, for any given decision maker it will be extremely difficult to pinpoint an individual's exact coefficient of pessimism. However, exact knowledge of λ may not be necessary. Letting λ remain unspecified, the anticipated payoffs of the three *strategies* in the above example are $a_1^{(3)}(\lambda) = 9 - 8\lambda$, $a_2^{(3)}(\lambda) = 8 - 6\lambda$, and $a_3^{(3)}(\lambda) = 7 - 4\lambda$. Plotting $\mathbf{a}^{(3)}$ against λ for the three strategies results in the three linear functions shown in Figure I.21.

In this example, all three lines happen to intersect at $\lambda = \frac{1}{2}$. Note that the relevant function is the *upper envelope*, as for each given value of λ, the decision maker will choose the strategy that has the highest anticipated payoff. The upper envelope, shown as the broken line in Figure I.21, is characterized by $\mathbf{a}^{(1)}$ for $\lambda \leq \frac{1}{2}$ and $\mathbf{a}^{(3)}$ for $\lambda \geq \frac{1}{2}$. This leads to the following decision rule: as long as the decision maker's parameter of pessimism λ is within the range $[0; \frac{1}{2}[$, choose decision d_1, if $\lambda > \frac{1}{2}$ choose d_3, while for $\lambda = \frac{1}{2}$, the decision maker can either choose decision d_1 or d_3. This approach does not require knowledge of exact values of λ. However, with many available strategies individual ranges may be very small, thus again requiring fairly precise estimates of the decision maker's estimation of λ. As Hurwicz's rule is partly based on a maximin criterion, its outcome may be improved by the use of mixed strategies.

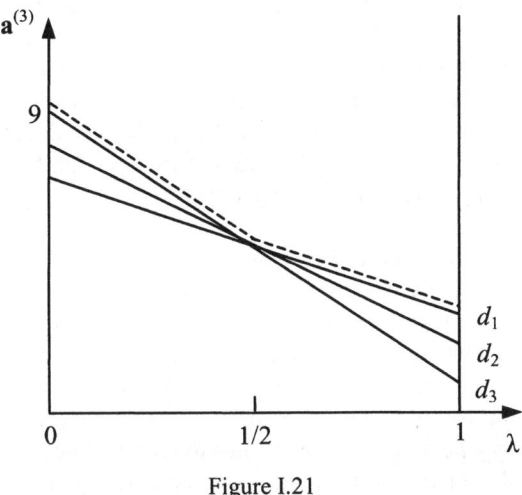

Figure I.21

Rule 4 (*Savage-Niehans rule, regret criterion*): In order to apply this rule, we first define a *regret matrix* (also referred to as *opportunity loss matrix*) $\mathbf{R} = (r_{ij})$ by subtracting each element of the payoff matrix $\mathbf{A}$ from its column maximum, i.e., $r_{ij} = \max_{k} \{a_{kj}\} - a_{ij} \ \forall \ i, j$. In other words, for each state of nature, the decision maker compares the payoff of each strategy with the best possible payoff given that state of nature. Note that the regret matrix consists of disutilities (as opposed to the original payoff matrix that consists of utilities). Applying the Wald criterion on the regret matrix requires then the use of a minimax (rather than a maximin) rule. In our numerical example, the regret matrix is

$$R = \begin{bmatrix} 3 & 0 & 0 & 0 & 7 \\ 0 & 0 & 3 & 7 & 0 \\ 2 & 1 & 2 & 2 & 5 \end{bmatrix},$$

so that the anticipated regrets are $\mathbf{r} = [7, 7, 5]^T$, and the decision maker chooses the third strategy and anticipates a regret of 5.

Again, mixing strategies may, if permissible, lead to improvements of the solution. In our example, consider a 50:50 mix of d_1 and d_2 whose payoffs are then [4.5, 5, 5.5, 5.5, 4.5]. The regrets of this new decision are then [1.5, 0, 1.5, 3.5, 3.5], so that the maximal regret of this mixed decision is 3.5, which is lower than that of the best individual decision d_3.

All rules described above have in common that they assume no knowledge on the part of the decision maker about the frequency or likelihood of nature's choice of strategies. Suppose now that the rational decision maker has acquired information concerning the probability distribution used by nature. Such knowledge could be based on past observations, expert forecasts, or similar information. Formally, the decision making is now under risk and player I is assumed to have knowledge about the probability distribution $\mathbf{p} = (p_j)$, where p_j denotes the probability that the state of nature s_j will occur. This additional information can then be incorporated in the decision making process in a variety of ways.

Rule 5 (Bayes expected value criterion): Bayes's criterion is probably best introduced by what is known as the *Newspaper Boy Problem* in the literature. It is a typical game against nature in which a newspaper boy must decide how many newspapers to buy, whereas the demand for the newspaper is uncertain and modeled as states of nature. The boy's past experience will provide the probabilities for the different levels of demand. The payoff a_{ij} then equals the revenue of the newspapers that are sold minus the costs of buying them. The actual sales are the smaller of the newspapers purchased according to d_i and the sales according to s_j, while the purchasing costs depend entirely on d_i. A similar, and more substantial, application occurs in the airline industry, where the decisions relate to the different types of aircraft (with their different capacities) and the states of nature model again levels of demand, i.e., the number of seats that could be sold on some route. The objective is then to choose a type of aircraft for a specific route so as to maximize the profit of the assignment.

Formally, the anticipated outcome of a decision with Bayes's criterion is its statistical expectation, i.e., $a_i^{(5)} = \sum_j a_{ij} p_j \ \forall \ i$. The *expectation* $a_i^{(5)}$ is also frequently called the *expected (monetary) value* of the decision d_i, $EV(d_i)$. This criterion is in essence a weighted maxsum rule. In our example, let the vector of probabilities be $\mathbf{p} = (p_j) = [.1, .2, .1, .4, .2]$. We then obtain the expected monetary values $EV(d_i)$ as

$$a_1^{(5)} = (3)(.1) + (5)(.2) + (7)(.1) + (9)(.4) + (1)(.2) = 5.8,$$
$$a_2^{(5)} = (6)(.1) + (5)(.2) + (4)(.1) + (2)(.4) + (8)(.2) = 4.4, \text{ and}$$
$$a_3^{(5)} = (4)(.1) + (4)(.2) + (5)(.1) + (7)(.4) + (3)(.2) = 5.1,$$

so that the decision maker's optimal choice is decision d_1 with an expected payoff of 5.8. It is apparent that mixed strategies cannot improve the expected payoff. The reason is that the expected payoff of a mixed strategy is a linear convex combination of the expected payoff of the existing decisions, which obviously cannot exceed the highest expected payoff of any of the existing decisions.

It should also be mentioned that Bayes's rule also allows the possibility to couple the regret approach with Bayesian expectation. In particular, such a decision rule will calculate the regret matrix and then choose the decision that minimizes the expected (rather than maximal) regret. This criterion is frequently referred to as the minimization of the *expected opportunity loss* (*EOL*) with value *EOL** at optimum. Formally, $EOL(d_i) = \sum_j r_{ij} p_j \; \forall \; i$, and in our example $EOL(d_1) = 3(.1) + 0(.2) + 0(.1) + 0(.4) + 7(.2) = 1.7$ is the minimal expected opportunity loss, so that decision d_1 is optional with this criterion.

It seems obvious that whenever some information is available, the decision maker should take this information into consideration. This is, however, not necessarily the case. Consider again the farmer who must choose a crop, and assume that the payoff specified in the payoff matrix is all the farmer will get with nothing else to fall back upon. Furthermore, assume that a payoff of "2" or lower means starvation. In such a case, even if probabilities for the different states of nature are available, the farmer will ignore them and play a minimax strategy and choose his third decision as this is the only decision that assures that starvation can be avoided. It is, however possible to suitably modify the payoff so that Bayes's rule can still be applied. In our example, define utilities u_{ij} for all payoffs a_{ij}, so that $u_{ij} = a_{ij}$, if $a_{ij} > 2$ and $u_{ij} = -M$ otherwise, where $M \gg 0$. The utility matrix is then

$$\mathbf{U} = \begin{bmatrix} 3 & 5 & 7 & 9 & -M \\ 6 & 5 & 4 & -M & 8 \\ 4 & 4 & 5 & 7 & 3 \end{bmatrix}$$

Applying Bayes's rule with the above probabilities result in the expected utilities [5.6–.2M, 3.6–.4M, 5.1] so that the farmer will opt for the third strategy.

We can also apply sensitivity analyses to the probabilities of the events by using tornado diagrams. However, as the probabilities are not independent of each other, additional assumptions are required. Consider, for instance, the possibility of the probability of one event increasing by some small increment ε. As the sum of probabilities of these mutually exclusive and collectively exhaustive events must equal one, the probability of at least one other event must decrease. Assume now that an increase of the probability of one event by some ε is offset by a simultaneous decrease of the probabilities of the other events by equal amounts, here $\frac{1}{4}\varepsilon$. As an example, suppose that p_4 increases by ε, then p_1, p_2, p_3, and p_5 all decrease by $\frac{1}{4}\varepsilon$ each. Now $p_1 = .1 - \frac{1}{4}\varepsilon$, $p_2 = .2 - \frac{1}{4}\varepsilon$, $p_3 = .1 - \frac{1}{4}\varepsilon$, $p_4 = .4+\varepsilon$, and $p_5 = .2 - \frac{1}{4}\varepsilon$. Given the modified probabilities, the expected values are then $EV(d_1, \varepsilon) = 5.8 + 5\varepsilon$, $EV(d_2, \varepsilon) = 4.4 - 3.75\varepsilon$, and $EV(d_3, \varepsilon) = 5.1 + 3\varepsilon$. As a decision maker is interested in the highest expected payoff, the optimal payoff will be a point on the upper envelope of the three functions. Assume now that $p_4 \in [.2, .5]$, so that the range of ε, as $p_4 = .4$, is $\varepsilon \in [-.2, .1]$. In the (ε, EV) diagram shown in Figure I.22, this upper envelope is then given by a line segment connecting $(-.2, 5.15)$ and $(-.16, 5)$ (this segment belongs to the decision d_2), and another segment connecting $(-.16, 5)$ and $(.1, 6.3)$; this segment belongs to the decision d_1.

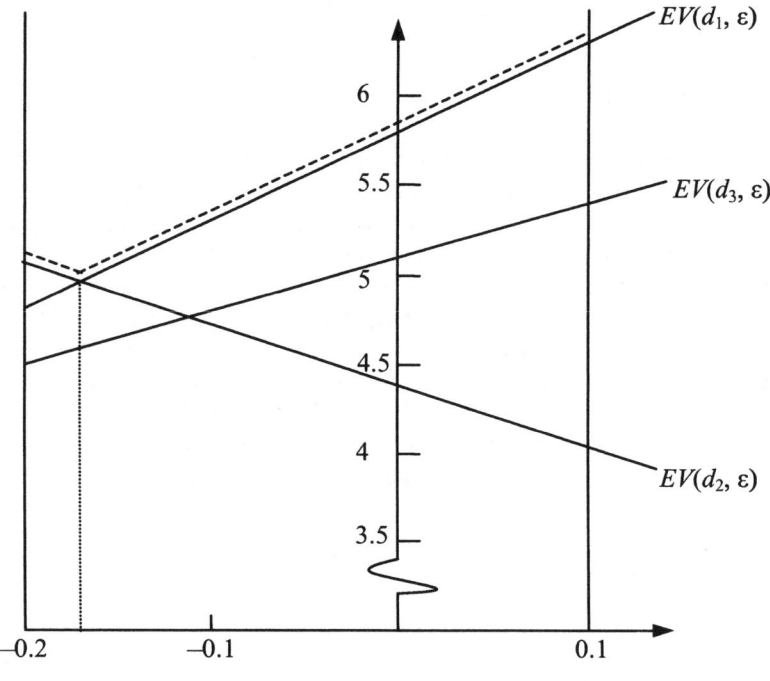

Figure I. 22

Consequently, for any $\varepsilon \in [-.2, .1]$, the optimal expected profit is in $[5, 6.3]$. The intervals for changes of the other probabilities can be calculated similarly. Note that some of these intervals may not have any break points. In particular, for $p_1 \in [0, .2]$, the profit is between 5.55 and 6.05, for $p_2 \in [.1, .5]$, the profit is 5.8 regardless of the changes, for $p_3 \in [.1, .2]$, the profit ranges from 5.8 to 6.05, for $p_4 \in [.2, .5]$, the profit is between 5 and 6.3, and for $p_5 \in [0, .4]$, the profit is between 5 and 6.8. This results in the tornado diagram shown in Figure I.23.

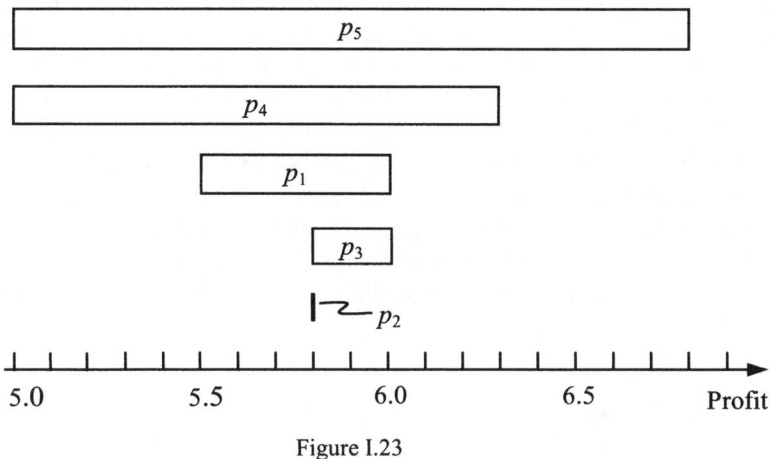

Figure I.23

Rule 6 (*Laplace rule*): The roots of this decision rule go back thousands of years. According to Manser (1988) the principle was "associated with the French philosopher Jean Buridan (c. 1295–1356), although it was first found in the philosophy of Aristotle." The story goes that Buridan put two identical bales of hay at identical distances from an ass's head. Now the animal had no reason to reject one bale of hay for the other, couldn't decide, and consequently starved to death. While the story appears somewhat contrived, the underlying *principle of insufficient reason* is less so. In simple words: if a decision maker does not know the probabilities of the individual states of nature, there is no reason to assume they are different from each other; hence $p_j = 1/n \ \forall \ j = 1, ..., n$. As such, Laplace's rule is on the boundary between uncertainty and risk: it assumes that nothing is known about the probabilities (i.e., uncertainty), but then, based on the principle of insufficient reason, assumes they are all equal (i.e., risk). In our example the anticipated payoffs are $\mathbf{a}^{(6)} = [5, 5, 4.6]$, so that the decision maker will choose the first or second strategy. Ties are either broken arbitrarily or secondary objectives are used. As Laplace's rule is a special case of Bayes's rule with equal probabilities, mixed strategies cannot improve the solution.

Rule 7 (*Hodges-Lehmann rule*): Similar to Hurwicz's rule that employs a linear convex combination of worst and best possible outcomes, Hodges and Lehmann define a linear convex combination of the worst possible anticipated outcome (Wald rule) and the expected (i.e., Bayes) outcome. Again, $\lambda \in [0; 1]$ denotes the parameter of pessimism. In other words, λ is the weight associated with the pessimistic outcome, whereas $(1 - \lambda)$ is the weight of the expected outcome. Formally, $\mathbf{a}^{(7)} = \lambda \mathbf{a}^{(1)} + (1 - \lambda)\mathbf{a}^{(5)}$. In our example, the pessimistic outcome is $[1, 2, 3]^T$ and the expected outcome is $[5.8, 4.4, 5.1]^T$, so that we obtain $[5.8 - 4.8\lambda, 4.4 - 2.4\lambda, 5.1 - 2.1\lambda]^T$. Just as in the Hurwicz rule, we can plot the anticipated outcome against the λ values. The corresponding graph is shown in Figure I.24.

Again, for any specified value of λ, the decision maker will choose the decision with the highest anticipated payoff, hence the relevant function is the upper envelope of the anticipated payoff functions, shown in Figure I.21 is the broken line. Note that d_2 is nowhere part of the upper envelope, hence it will never be used even though it is *not* dominated. Intersecting the anticipated payoff functions of the first and third strategies, we obtain $5.8 - 4.8\lambda = 5.1 - 2.1\lambda$ or $\lambda = 7/27 \cong 0.2593$. Thus the decision rule: for $\lambda \leq 7/27$ choose d_1, for $\lambda \geq 7/27$ choose d_3. It is also clear that for $\lambda = 1$, the Hodges-Lehmann rule coincides with the Wald criterion, whereas for $\lambda = 0$, it simplifies to Bayes's rule. If permitted, mixed strategies may improve the solution, as they can in Wald's rule, which is part of this rule.

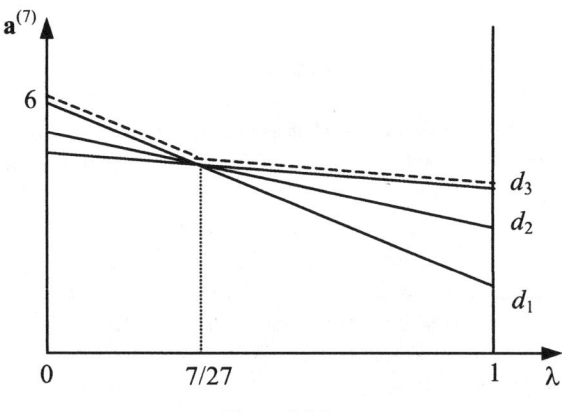

Figure I.24

Rule 8 (*Maximize the probability of exceeding a target value*): The application of this rule requires the decision maker to exogenously set a target value T and then compute the probabilities that the expected payoff meets or exceeds T. The task is then to add the probabilities of all states of nature that result in a payoff greater or equal to the specified target value for each decision. Using the probabilities in Bayes's rule, Table I.12 displays the probabilities of exceeding some given (integer) target values.

Table I.12

Target value	1	2	3	4	5	6	7	8	9	10
d_1	1.0	.8	.8	.7	.7	.5	.5	.4	.4	0
d_2	1.0	1.0	.6	.6	.5	.3	.2	.2	0	0
d_3	1.0	1.0	1.0	.8	.5	.4	.4	0	0	0

The information provided in Table I.12 can be visualized in Figure I.25.

Chapter 2: Games Against Nature 95

Again, the decision maker will make a decision on the basis of the upper envelope of the functions. Figure I.25 shows that d_3 is best for target values less or equal to 4, and d_1 is the best decision for target values exceeding 4. Note that d_2 is not part of the upper envelope and thus will never be chosen. This criterion is reminiscent of the well-known concept of (first-order) stochastic dominance.

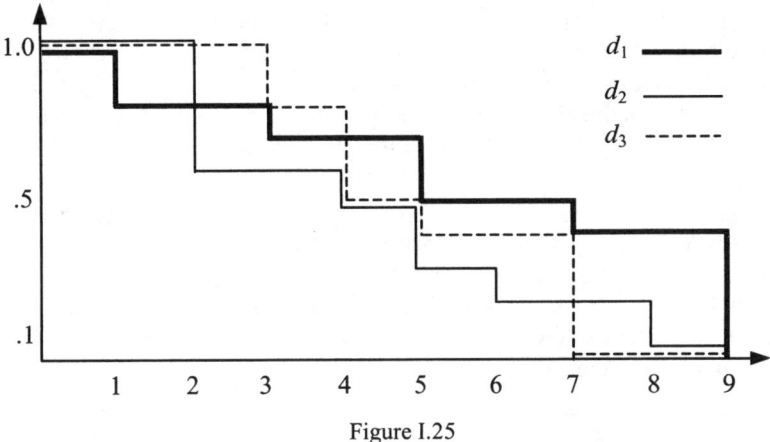

Figure I.25

A similar criterion was suggested by Starr and Zeleny (1977). Rather than calculate the probability of meeting or exceeding a given target value, they first compute the probability of making the best possible decision given that the state of nature was known in advance. The decision with the highest such probability is then chosen. Consider again the example introduced at the beginning of this section and the probabilities used in Bayes's rule. They were

$$\mathbf{A} = \begin{bmatrix} 3 & 5^* & 7^* & 9^* & 1 \\ 6^* & 5^* & 4 & 2 & 8^* \\ 4 & 4 & 5 & 7 & 3 \\ 6^* & 4 & 3 & 2 & 7 \end{bmatrix}$$

$$\mathbf{p} = [.1 \quad .2 \quad .1 \quad .4 \quad .2]$$

The starred elements are the highest payoffs of any decision, given that the state of nature is known. As an example, if it were known that s_3 were to occur, then the best decision is d_1 as it results in a payoff of 7. Once optimal strategies for all states of nature have been determined, we calculate the probability for each decision to be best. In our example, decision d_1 is best when the states of nature s_2, s_3, and s_4 occur, whose probabilities are .2, .1, and .4, respectively, so that d_1 yields the highest possible result $.2 + .1 + .4 = .7$ or 70% of the time. The probabilities that d_2, d_3, and d_4 is best are .5, 0, and .1, respectively, so that decision d_1 would be chosen. (The sum of these probabilities exceeds 1 because of ties, i.e., multiple column maxima). It is interesting to note that decision d_3 never comes out best, even though its expected payoff (as calculated in Rule 5) is the second highest. Furthermore, its probability of being the best decision is even lower than that of the dominated decision d_4. However, one can easily show that dominated strategies will never be chosen with the Starr-Zeleny rule. An obvious criticism of this rule is, of course, its exclusive focus on "winning". A similar rule could be set up with focus on losing, subsequently minimizing the probability to do so. In our example, the probabilities of losing are .3, .4, .2, and .7, so that d_3 is the decision that has the smallest probability to be the worst. Similar to Hurwicz's rule, this rule could be combined with the Zeleny-Starr rule in a linear combination.

Rule 9 (*Starr's domain criterion*): Similar to Laplace's rule, Starr (1962) devised a technique for decision making problems under uncertainty that uses probabilistic tools normally reserved for decision making under risk. As the technique is fairly complex if more than a few states of nature are involved, we introduce another, smaller example with the payoff matrix

$$\mathbf{A} = \begin{bmatrix} 7 & 3 & 4 \\ 2 & 6 & 3 \\ 1 & 1 & 9 \\ 5 & 2 & 3 \end{bmatrix}.$$

The probabilities of the states of nature are unknown, and we denote them by p_1, p_2, and p_3. The expected monetary values are then

$EV(d_1) = 7p_1 + 3p_2 + 4p_3,$
$EV(d_2) = 2p_1 + 6p_2 + 3p_3,$
$EV(d_3) = 1p_1 + 1p_2 + 9p_3,$ and
$EV(d_4) = 5p_1 + 2p_2 + 3p_3.$

As they are probabilities, $p_j \geq 0$, $j = 1, 2, 3$, and $p_1 + p_2 + p_3 = 1$. The latter equation allows us to substitute one of the probabilities by a function of the others. Arbitrarily choosing p_3, we can write

$EV(d_1) = 4 + 3p_1 - p_2,$
$EV(d_2) = 3 - p_1 + 3p_2,$
$EV(d_3) = 9 - 8p_1 - 8p_2,$ and
$EV(d_4) = 3 + 2p_1 - p_2.$

The main idea is now as follows. We will subdivide the space of all possible probability distributions into domains D_i, $i = 1, ..., m$, so that D_i is the set of all probability distributions for which the i-th decision is best. In general, the i-th decision is preferred over the k-th decision, formally written as $d_i \succ d_k$ if and only if $EV(d_i) \geq EV(d_k)$. The domain D_i is then formally defined as $D_i = \{\mathbf{p}: EV(d_i) \geq EV(d_k) \; \forall \; k \neq i\}$. As each domain is defined as the intersection of half spaces, it denotes a polytope. Each domain D_i is a polyhedron, and the union of all domains is the unit simplex $\{\mathbf{p}: \mathbf{p} \geq 0, \sum_j p_j = 1\}$.

In our example, the decision maker's first strategy is best if $EV(d_1) \geq EV(d_2)$, $EV(d_1) \geq EV(d_3)$, and $EV(d_1) \geq EV(d_4)$. Applying the above expressions to the respective expected values, we obtain the inequalities $-4p_1 + 4p_2 \leq 1$, $11p_1 + 7p_2 \geq 5$, and $-p_1 \leq 1$. Now, D_1 is the set of all probability distributions p_j that satisfy these inequalities. Similarly, the decision maker's second strategy is best if its expected monetary value exceeds those of all other decisions, i.e., $D_2 = \{\mathbf{p}: -4p_1 + 4p_2 \geq 1, 7p_1 + 11p_2 \geq 6, -3p_1 + 4p_2 \geq 0\}$. Decision d_3 is superior within $D_3 = \{\mathbf{p}: 11p_1 + 7p_2 \leq 5, 7p_1 + 11p_2 \leq 6, 10p_1 + 7p_2 \leq 6\}$. For the fourth strategy, $EV(d_4) \geq EV(d_1)$ implies that $p_1 \leq -1$ which contradicts $p_j \geq 0$ required by probabilities. It can easily be verified that d_1 dominates d_4. All feasible probability distributions in a (p_1, p_2) diagram are then found within the triangle with vertices $(0, 0)$, $(1, 0)$, and $(0, 1)$. A tessellation of this triangle into domains D_i, $i = 1, 2, 3$ is shown in Figure I.26.

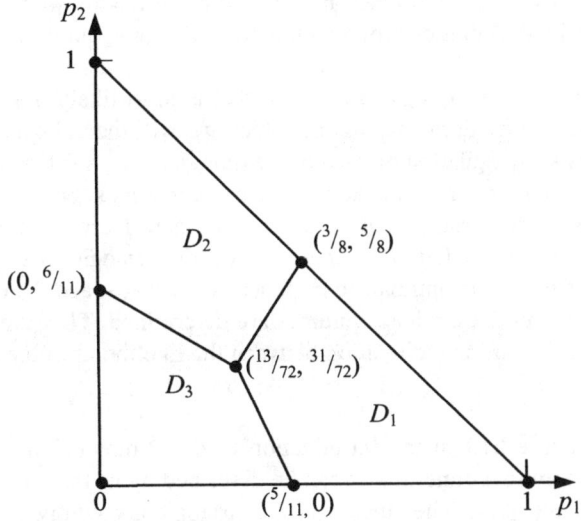

Figure I.26

Let us review the ideas that lead to Figure I.26. With each decision d_i we can associate an expected monetary value which, in the (p_1, p_2, EV) space, can be represented by a plane $EV(d_i)$. As the decision maker is only interested in the best decision, the upper envelope of all these planes, i.e., $EV^* = \max_i \{EV(d_i)\}$, is relevant. Projecting the faces of the polytope of EV^* onto the (p_1, p_2) plane, the boundaries of the projected faces are the tessellation of Figure I.26.

Even if the decision maker is aware of the actual probability distribution **p** and hence could simply apply Bayes's criterion, domains can be quite useful. The location of **p** in the graph will determine how stable the solution is. If **p** is close to the border of its domain, say D_i, then a small change could cause d_i to become nonoptimal. On the other hand, if **p** is located squarely within D_i the solution is stable and small or even moderate changes of the probability distribution **p** do not change the optimality of d_i.

Assume now that the actual probability distribution **p** is unknown. If nothing is known, then we may assume—similar, but *not* equivalent, to Laplace's assumption of equal likelihood of all states of nature—that all probability distributions are equally likely. Given that, all points in the triangle in Figure I.26 are equally likely to occur. Consequently, the decision with the largest domain is the one most likely to be best among the available decisions. Note again the exclusive focus on "winning:" The rule is not concerned about how much better a strategy is as compared to others. For instance, the strategy most likely to be best may only be marginally better than the second best and may be considerably worse than others, for those probability distributions for which it does not come out best.

In order to determine the strategy or decision that is most likely best, we need to determine the size of the domains, among which we will then choose the largest. This first requires a triangulation of each of the domains, i.e., a subdivision of each polyhedron into triangles. Then, the area of each triangle must be determined. The sum of all areas of the triangles of a domain is then a crude indicator of its likelihood. Setting it in relation to the area of the triangle will provide the likelihood that a decision is optimal. In d–dimensional real space $\mathbb{R}^d$, each domain is subdivided into simplices, whose volumes are determined. The sum of all these volumes will then determine the relation between the likelihood of the decisions to be optimal.

Apply now Procedure b13 in the Introduction to our 2-dimensional problem. A triangle (a simplex in two dimensions) can be described by its three extreme points (x_1, y_1), (x_2, y_2), and (x_3, y_3). Then the area of the triangle is then given by

$$\text{Area(triangle)} = \tfrac{1}{2}\det \begin{bmatrix} x_1 & y_1 & 1 \\ x_2 & y_2 & 1 \\ x_3 & y_3 & 1 \end{bmatrix}$$

In our example, the domain D_1 can be decomposed into two triangles with vertices (13/72, 31/72), (1, 0), (5/11, 0), and (13/72, 31/72), (1, 0), (3/8, 5/8). Their respective sizes are 93/792 and 70/576 for a combined area if $v(D_1) \cong 0.2389$. Similarly, we obtain $v(D_2) \cong 0.1471$ and $v(D_3) \cong 0.1139$. Clearly, the sum of areas of the three domains equals the area of the triangle, which is $v(T) = 0.5$. Now, the probability of strategy d_i being best is $v(D_i)/v(T)$ $\forall$ i. In our example, these probabilities are 47.78%, 29.42%, and 22.78%. In other words, there is almost a 50 percent probability that the first strategy is best and the domain criterion prescribes that it be chosen.

Extensions of this concept include the use of different likelihoods for the probability distributions, or the original assumption of equal likelihood of all probabilities being restricted to a subset of the triangle. For a further discussion, see Eiselt and Langley (1990). Another interesting aspect is to determine the average height of the expected value functions of all decisions in the (p_1, p_2, EV) space. This average height represents the expected payoff given that all probability distributions are equally likely and it coincides with the Laplace criterion which gives it a different, and more general, interpretation.

The most serious drawback of Starr's domain criterion is the "curse of dimensionality". The number of dimensions of the problem depends on n, the number of states of nature, so that the domains associated with each of the decisions are then $(n–1)$-dimensional polyhedral, which have to be triangulated into simplices whose volumes have to be determined. This seriously limits the applicability of the method to decision scenarios with a small number of states of nature.

2.3 Multi-Stage Decisions and the Value of Information

We begin this section with a game against nature that has more than one stage. Multistage games are best motivated by means of a small

Example : When hiring new employees, decision makers typically first choose a handful of candidates for a short list. Shortlisted individuals are then interviewed and then compared with each other, resulting in a ranking. Then, the highest-ranking candidate is made an offer; if it is accepted, the process terminates, otherwise the candidate with the next-highest ranking is chosen, and so forth. This example will demonstrate that such a procedure is not necessarily optimal.

Suppose that three candidates A, B, and C have been shortlisted. Furthermore, suppose that the hiring department feels that, based on time constraints, it will be able to make an offer to its second choice if the first offer is rejected, but there will not be enough to make three sequential offers in the case of two rejections. Table

I.13 shows the weights the decision maker associates with the three candidates, as well as the probabilities of an offer being made to the candidate in round 1 (early) and in round 2 (late), respectively.

Table I.13

Candidate	Weight	Probability of acceptance in	
		Round 1	Round 2
A	5	.1	.05
B	4	.2	.15
C	2	.5	.4

The decision problem can then be described by the decision tree in Figure I.27, where the upper (lower) arcs out of an event node indicates that the candidate accepts (rejects) the offer. The arcs out of decision nodes show which candidate receives an offer, the number in the terminal nodes display the payoff to the organization, and the numbers next to the arcs that lead out of event nodes are the

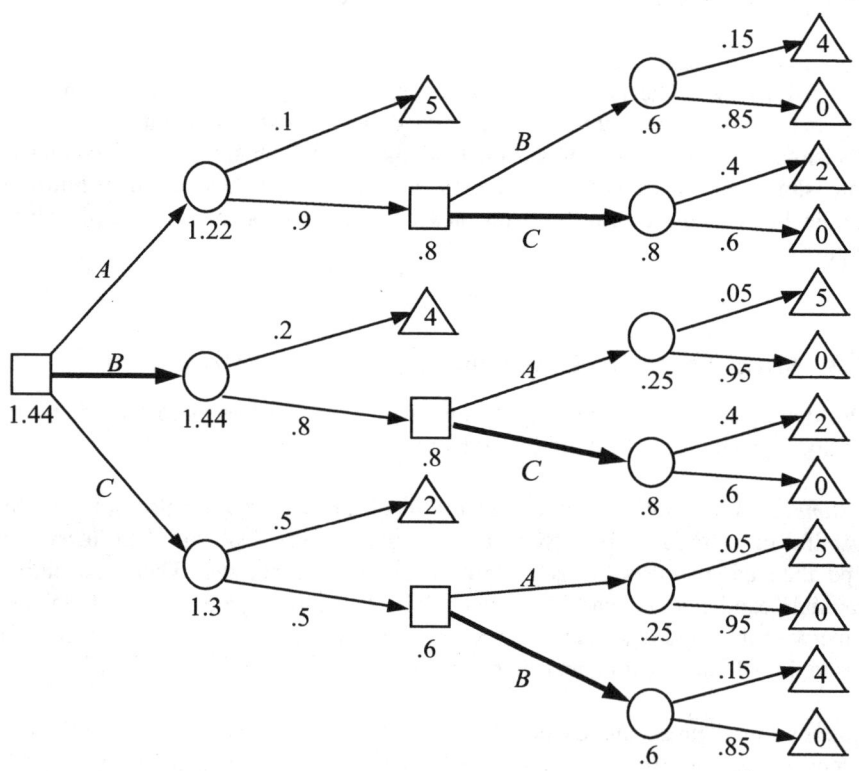

Figure I.27

probabilities that a candidate accepts or rejects an offer. The backward recursion proceeds in exactly the same way it does in games against nature. The numbers next to the nodes show the expected payoffs to the decision maker's organization, and the bold arcs indicate optimal decisions.

The optimal strategy is to make an offer to candidate B and, should this offer be rejected, make an offer to candidate C. At first glance, it is surprising and counterintuitive that the highest ranking candidate will not receive an offer at all. The reason for this is that the probability that candidate A will accept the offer is very small, and in the second round, typically a couple of months later, the probabilities that any of the shortlisted candidates will accept are lower. This problem would be exacerbated if the situation in which no candidate accepts the offer has not a payoff of "0", but, say, a value of -2. Such a negative payoff could be justified by the position being left unoccupied, and consequently, some customers are left without service, causing them to search for service elsewhere. In such a case it would be optimal to make the first offer to candidate C. Situations such as this are typically found in smaller University departments that are not very attractive to many better job applicants, due to the lack of programs and graduate students.

In the remainder of this section we will explore the possibilities of using the concept of multistage games against nature in order to determine the value of different types of information in games against nature. The information we are concerned with is anything *beyond* the knowledge of the probability distribution **p**. In one extreme, there is *no* (additional) *information*, in which case we may use Bayes's rule. In the other extreme we have *perfect information*. In that case, we know beforehand precisely which state nature will choose. However, nature still plays her strategies following the probability distribution **p**. We may think of a game with perfect information as a sequential game in which nature chooses her strategy first, following the probability distribution **p**; this choice is then announced, so that for player I the game is then under certainty. Perfect information is certainly an extreme concept as such information is highly unlikely to exist, but the value of perfect information can provide an upper bound on the value of any information. Between the two extremes of no information and perfect information is the large area of *sample information* (also referred to as *imperfect information*). Based on the specifics of that information, we will also be able to compute the value of sample information. In order to formalize the discussion, some definitions are needed. Let EV^* denote the highest expected payoff without information, i.e.,

$$EV^* = \max_i \{EV(d_i)\} = \max_i \left\{ \sum_j a_{ij} p_j \right\},$$

so that $\arg\max_i\{EV(d_i)\}$ is the optimal Bayes strategy. On the other hand, the expected payoff with perfect information *EPPI* is obtained by choosing the column maxima so that $EPPI = \sum_j p_j \max_i \{a_{ij}\}$. The difference between these two payoffs is then the *expected value of perfect information EVPI*, i.e., $EVPI = EPPI - EV^*$.

Example: As an example illustration of the concept, consider again the example introduced at the beginning of this chapter. For convenience, we display again the payoff matrix for the three nondominated decision alternatives and with the five states of nature. It is

$$\mathbf{A} = \begin{bmatrix} 3 & 5^* & 7^* & 9^* & 1 \\ 6^* & 5^* & 4 & 2 & 8^* \\ 4 & 4 & 5 & 7 & 3 \end{bmatrix}$$

and the probability distribution is $\mathbf{p} = [.1, .2, .1, .4, .2]$. As a result, the expected monetary values are $EV(d_i) = 5.8, 4.4$, and 5.1 with $EV^* = 5.8$ and the first strategy is the optimal Bayes strategy. For the computation of the expected payoff with perfect information, the best decisions for each state of nature are indicated in the payoff matrix by stars. As a result, the $EPPI = 6(.1) + 5(.2) + 7(.1) + 9(.4) + 8(.2) = 7.5$, so that $EVPI = 7.5 - 5.8 = 1.7$.

There is an interesting relation between the expected value of perfect information and the expected opportunity loss. Recall that

$$EVPI = EPPI - EV^* =$$

$$= \sum_j p_j \max_k \{a_{kj}\} - \max_i \left\{ \sum_j a_{ij} p_j \right\} = \min_i \left\{ \sum_j \left\{ \max_k \{a_{kj}\} - a_{ij} \right\} p_j \right\} =$$

$$= \min_i \left\{ \sum_j r_{ij} p_j \right\} = \min_i EOL(d_i) = EOL^*,$$

so that the *EVPI* can also be obtained by setting up the regret matrix, computing expected regrets for all strategies, and then selecting the strategy with the smallest expected regret. Consequently, the decisions that maximize the expected payoff also minimize the expected regret; furthermore, $EV(d_i) + EOL(d_i) = EPPI \; \forall \; i$.

In addition, while the concept of perfect information is almost always discussed in the Bayesian context, it is in no way limited to it. Similarly to the procedure above, we could compute the value of perfect information in the case of pessimistic

Chapter 2: Games Against Nature

behavior of the decision maker. In our example, a decision maker without information would choose d_3 with an anticipated payoff of 3 and with information, the worst case (i.e. the minimum among all starred elements in **A**) is 5, so that the value of perfect information to such a decision maker is then $5 - 3 = 2$.

Consider now the possibility of information that is less than perfect. The general idea is to employ a forecasting service, such as a marketing research firm or a similar institute. The result of this survey is one of a number of indicators, I_1, I_2, ..., I_q. These indicators may or may not coincide with the states of nature. However, there should be some correlation between the indicators and the states of nature, otherwise the use of indicators is meaningless. For instance, a decision maker who is interested in the development of the prime rate may observe the supply of money as an indicator instead. The process that leads to the calculation of the expected value of sample information then takes the following sequential form. First, the decision maker will have to decide whether or not to acquire (and pay for) the forecasting service. If not, player I must make do with the limited information available, the probability distribution **p**, and use Bayes's rule to find an optimal decision. Should he, however, decide to acquire the sample information, the next step involves the actual outcome, i.e., the indicators I_k that result from the survey. Once the outcome is known, player I must choose a decision d_i, and finally, nature will choose her strategy s_j.

The calculation of the value of imperfect information is best accomplished by means of a decision tree. The general structure of a decision tree for this purpose is shown in Figure I.28.

Starting from the left at what is known as the *root of the tree,* the decision maker first decides whether or not to obtain additional information by conducting or commissioning a survey. If this is not desired (the lower branch), the decision maker must then choose one of his decision alternatives after which nature chooses her strategy. This lower part of the decision tree is identical to that of the usual Bayesian analysis performed earlier. Note that even if the decision maker and nature are assumed to decide upon their respective strategies simultaneously, it is equivalent to having the decision maker choose first and nature choose second. The reason is that nature's choice is made at random regardless of the decision maker's choice. This equivalence of a sequential game and a simultaneous game cannot be made if the second player is also rational.

Assume now that it has been decided to obtain additional information. The result of the survey will be an indication I_k, $k = 1$, ..., q. Clearly, the indicators are beyond the control of the decision maker, but depend on the probabilities of the states of nature *and* the past record of the institute. Assuming that one of the indicators has been presented to the decision maker, he will then make one of his decisions d_i, $i = 1$, ..., m. Once that is done, nature has that last word in choosing one of her strategies s_j, $j = 1, ..., n$. At that point, we can associate a payoff with each of the pendent vertices of the tree. It is worth noting that the situation at any given node of the tree is determined by all decisions and occurrences on the unique path from the root of the tree to the node under consideration.

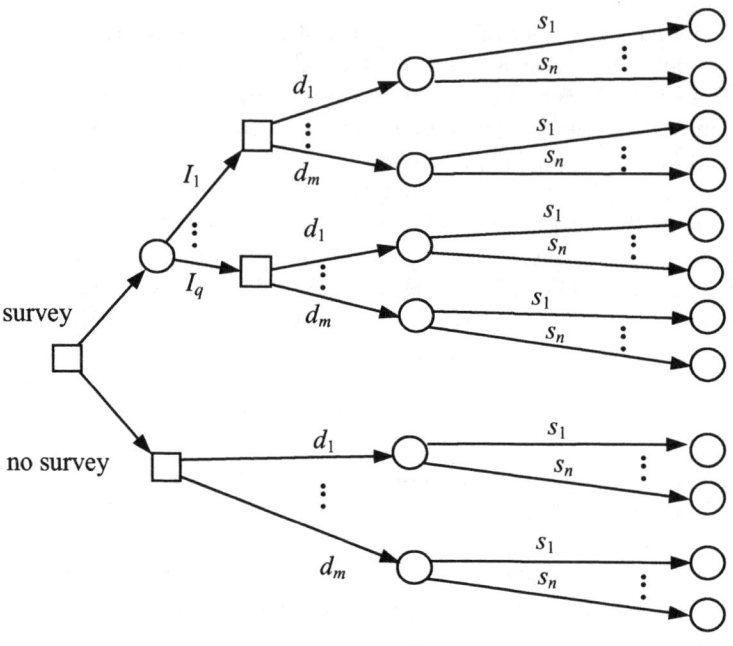

Figure I.28

There are three points in the decision tree at which nature makes decisions. As they follow some probability distribution, the appropriate probabilities have to be determined. In the lower part of the tree where no additional information is sought, nature uses the *prior probabilities* p_j or, as we will call them from now on to avoid confusion, $P(s_j)$. Consider now the upper part of the tree. If a survey is commissioned, nature will choose indicators according to *indicator probabilities* $P(I_k)$. Finally, after player I has made his decisions, nature will decide again on a state of nature. The probabilities associated with these choices of nature are, however, not the prior probabilities $P(s_j)$. The reason is that nature's decisions follow the occurrence of indicators and indicators and states of nature are not independent: if they were, indicators would be meaningless. Nature's decisions chronologically follow the indicators, so that their occurrence has the occurrence of an indicator as a condition. Consequently, we will use *posterior probabilities* $P(s_j | I_k)$. These probabilities give an indication of the strength of the linkage between the indicators and the states of nature. They can either be given as required, as a simple table that reports how many times I_k and s_j were observed together, or in some other form. Frequently, the prior probabilities $P(s_j)$ and the

Chapter 2: Games Against Nature

"inverse" conditional probabilities $P(I_k | s_j)$ are known. The simple transformations provided in Theorem e14 in the Introduction of this book can then be used to calculate the appropriate probabilities.

Having assigned probabilities to all of nature's decisions and the appropriate payoffs to all leaves (terminal nodes) in the tree we can employ a simple backward recursion to determine a sequence of optimal decisions as well as the value of sample information. Let the labels of the leaves of the tree be the payoff. Let n_ℓ be a labeled node and define $\mathscr{P}(n_\ell)$ as the set of all predecessor nodes of n_ℓ. Let N_d denote the set of decision nodes. We can then formally describe the backward recursion method for decision trees as follows.

Backward Recursion in Decision Trees

Step 1: Does the tree include any unlabeled nodes?

If yes: Go to Step 2.
If no: Stop, the label of the root of the tree is the expected payoff with sample information *EPSI*.

Step 2: Choose any labeled node n_ℓ with an unlabeled predecessor, and let $n_v = \mathscr{P}(n_\ell)$. Is $n_v \in N_d$?
If yes: Go to Step 3.
If no: Go to Step 4.

Step 3: Label n_v with the maximum label of all of its successors. Go to Step 1.

Step 4: Label n_v with the expected value of all of its successors. Go to Step 1.

Example: A decision maker has three choices to respond to economic changes: no change ($= d_1$), expand ($= d_2$), or relocate ($= d_3$) his business. The states of nature relate to the economic conditions: a downturn ($= s_1$), a stable economy ($= s_2$), and a minor economic upturn ($= s_3$). The corresponding payoff matrix is given as

$$\mathbf{A} = \begin{bmatrix} 3 & 4 & 7 \\ -1 & 2 & 12 \\ 4 & 5 & 6 \end{bmatrix}.$$

The estimated probabilities of the states of nature are $\mathbf{p} = [0.3; 0.5; 0.2]$, so that the expected monetary values of the three strategies are 4.3, 3.1, and 4.9, respectively. As a result, the decision maker will choose strategy d_3 and $EV^* = EV(d_3) = 4.9$. Simple computations reveal that the expected payoff with perfect information is $EPPI = 6.1$, so that the expected value of perfect information is $EVPI = 6.1 - 4.9 = 1.2$. Consequently, no information, however accurate, is worth more than 1.2.

Suppose now that a survey institute offers its services. Their recommendation is summarized in one of two indicators, I_1: a negative outlook, and I_2: a positive outlook on the economy. The frequencies that were observed in the past are summarized in Table I.14.

Table I.14

	s_1	s_2	s_3	
I_1	36	50	12	98
I_2	24	50	28	102
	60	100	40	200

In other words, there are a total of 200 observations, among which 36 times the indicator was I_1, and the state of nature turned out to be s_1, 50 times the indicator was I_1, and s_2 came up, and so forth. The last column shows that in total, I_1 occurred 98 times, while I_2 came up 102 times. The last row includes the prior probabilities $P(s_j)$, i.e., nature played her strategies s_1, s_2 and s_3, 60, 100, and 40 times, respectively. Based on this table it is easy to calculate the required probabilities: $P(s_j)$ are already known, $P(I_k)$ are $98/200 = .49$ and $102/200 = .51$, respectively, and $P(s_j | I_k)$ is the observed frequency of the pair (I_k, s_j) divided by the frequency of I_k. For instance, $P(s_3 | I_2) = 28/102 = .2745$.

Frequently, the same information is provided in the form of the conditional probabilities $P(I_k | s_j)$. We can then involve the fact that $P(I_k) = \sum_j P(I_k | s_j) P(s_j)$ for the indicator probabilities, and Bayes's rule $P(s_j | I_k) = \dfrac{P(I_k | s_j) P(s_j)}{P(I_k)}$ to compute the posterior probabilities. This can be done in tables, one for each indicator I_k. In our example, we have

$P(I_k | s_j)$:

	s_1	s_2	s_3
I_1	.6	.5	.3
I_2	.4	.5	.7

For I_1, we then obtain the information in Table I.15.

Chapter 2: Games Against Nature

Table I.15

s_j	$P(s_j)$	$P(I_1\|s_j)$	$P(I_1\|s_j)P(s_j)$	$P(s_j\|I_1)$
s_1	0.3	0.6	0.18	0.3673
s_2	0.5	0.5	0.25	0.5102
s_3	0.2	0.3	0.06	0.1224
			$P(I_1) = 0.49$	

Similarly, for I_2 we obtain the information in Table I.16.

Table I.16

s_j	$P(s_j)$	$P(I_2\|s_j)$	$P(I_2\|s_j)P(s_j)$	$P(s_j\|I_2)$
s_1	0.3	0.4	0.12	0.2353
s_2	0.5	0.5	0.25	0.4902
s_3	0.2	0.7	0.14	0.2745
			$P(I_2) = 0.51$	

In each of the above two tables, the first two columns are given in the above problem description. The third column computes the products of the first two columns and, by definition, sums up the probabilities for the indicator probability. Finally, the entries in the third column are each divided by the indicator probability by virtue of Bayes's rule, resulting in the posterior probabilities shown in the fourth column.

Given the indicator and posterior probabilities computed above, we have now all ingredients necessary for the decision tree. The tree in its entirety is shown in Figure I.29 (for now, ignore the labels shown next to the nodes except for the payoffs at the pendent vertices).

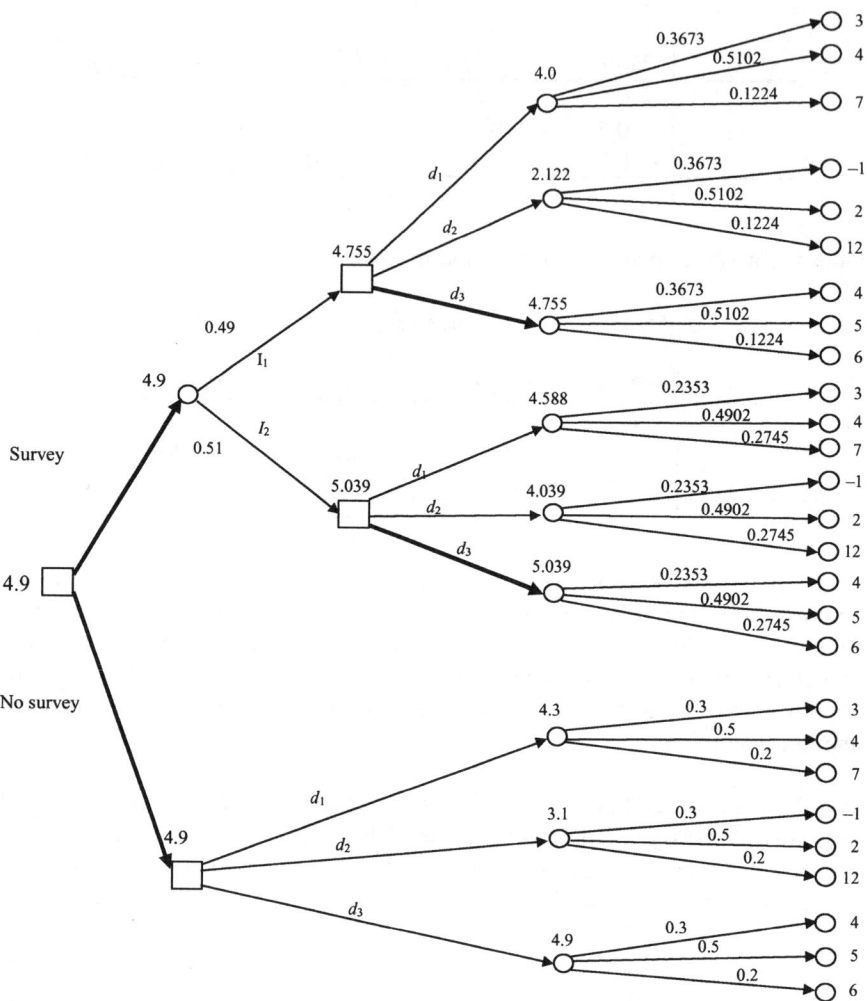

Figure I.29

As prescribed in the algorithm, the backward recursion starts at the leaves of the tree. Going back one level, we will determine new labels of event nodes, so that expected values have to be calculated. For instance, at the top of the tree we compute 3(.3673) + 4(.5102) + 7(.1224) = 4 which is the label of the predecessor. All labels on that level are calculated in that fashion. Going back another level leads us into decision nodes, so that a maximization criterion is applied. At the top of the tree, max{4.0, 2.122, 4.755} = 4.755 and similar for the other decision node. The decisions that belong to the new labels (here d_3 in both cases) are optimal and are shown in bold arcs. Another level back leads into an event node,

Chapter 2: Games Against Nature

whose expected payoff is 4.755(.49) + 5.039(.57) = 4.9. Finally, moving into the root of the tree (a decision node) we maximize again and obtain the value of *EPSI* = 4.9. Similar to the case of perfect information, we can calculate the *expected value of sample information* as $EVSI = EPSI - EV^*$. In this example, $EVSI = 4.9 - 4.9 = 0$, i.e., the sample information is worth nothing. In general, the *EVSI* provides an upper bound for the amount of money that should be spent to purchase sample information of this type. The reason for the apparent worthlessness of the sample lies in the weakness of the link between the indicators and the states of nature as seen in the $P(I_k|s_j)$ matrix. This leads to a decision tree in which the optimal decision is d_3, regardless of what indication comes up. A relative indicator of the value of sample information, define the *efficiency* $E = EVSI/EVPI$. Clearly, $E \in [0, 1]$. As E approaches one, the value of sample information approaches perfection; as it nears zero, it approaches worthlessness. The efficiency in our example is $E = 0$.

Consider now another survey institute that offers its services. Suppose that its past record is described by the following matrix of conditional probabilities.

$$P(I_i|s_j) = \begin{bmatrix} .0 & .5 & 1.0 \\ 1.0 & .5 & 0 \end{bmatrix}$$

Note that the advice of this institute appears much worse than the previous one: whereas the previous institute was ambivalent in its indicators, this one is plain wrong most of the time. For instance, whenever an economic downturn occurred (s_1), they *always* made a positive prediction (I_2) and, similarly, whenever a minor economic upturn occurred (s_3), they *always* predicted a negative outcome (I_1). Computing indicator probabilities we obtain $P(I_1) = 0.45$ and $P(I_2) = 0.55$, so that

$P(s_j|I_1) = [0.0000, 0.5556, 0.4444]$, and

$P(s_j|I_2) = [0.5454, 0.4545, 0.0000]$.

Applying again the recursive procedure in the decision tree, we find that the expected payoff with sample information for this institute is 5.35, so that $EVSI = 0.45$ and the efficiency is $E = 0.45/1.2 = 0.375$. In other words, even though the institute is surely wrong a good part of the time, its value is higher than that of the previous institute. The reason for this is in fact that the type of advice doesn't really matter, it is the consequence of the action that results from the advice. To make things simple, assume an institute forecasting stock prices has the worst record of all—it is always wrong. Actually, that advice is just as good as perfect information; all the decision maker has to do is the opposite of what the institute suggests! The same principle applies in our example; the only reason for the relatively low efficiency is the institute's indecision in the (with $p_2 = .5$ rather likely) second state of nature.

CHAPTER 3   GAME THEORY

Game theory is a field that studies situations which involve multiple players who make independent decisions, so that each combination of decisions results in an outcome with which payoffs for each player are associated.

This—quite general—definition allows for applications from a large variety of disciplines and scenarios. For instance, the determination of prices by gas retailers in an oligopolistic market structure is such a game, as each player has a single variable, his price, as a decision variable, and the customer demand and his profit will be determined not only by his own price, but also that of other competitors. Similar situations are found in many other business, societal, and political situations.

The origins of game theory date back some 200 years and to the economic inquiry of scientists such as Cournot, Bertrand, Edgeworth, and Hotelling. The mathematical branch of game theory originated in the work of John von Neumann in the 1920s, reputedly done in part in the back seat of taxicabs. The first publication of results in the field is the seminal work published in "The Theory of Games and Economic Behavior" by von Neumann and Morgenstern (1944). Later contributions include those by Nobel laureates Nash (1950), Selten (1965), and Harsanyi (1964), as well as many other authors. An early classic is the text by Luce and Raiffa (1957), later milestones are those by Owen (1982) and Fudenberg and Tirole (1993).

Given the limited space available, this chapter can only provide a glimpse of some of the possibilities that game theory allows. The first section will introduce the elements of game theory, the second investigates a class of simple and well-solved problems called two-person games, and the third section extends two-person games to a more general scenario with $n > 2$ players.

3.1 Features of Game Theory

This section will first discuss the elements of game theory and their representations. The second part then surveys the main solution concepts that are commonly applied to the games under consideration.

3.1.1 Elements and Representations of Games

All games include a number of basic ingredients. The first is the number of decision makers, usually called *players*. The number of players is denoted by $n \geq 2$. Typically, the larger the number of players, the more difficult is the analysis of the game. Games with two players have a special role in game theory, and they have been much analyzed. Each player has at his disposal a number of *decisions*, also called *strategies*. These decisions are the variables, and there can be a finite or an infinite number of them. The number of players and their decisions are treated in game-theoretic models as given parameters, so that the preselection of appropriate decisions is a task to be performed prior to the game by the individual players. We emphasize this, as often players restrict the number of their decisions on the basis of preconceived notions that some decisions "can surely not be optimal." For a pertinent discussion, readers are referred to Keeney (1992). In general, it is a good idea to first include as many decisions as possible and later delete them in case it is determined that other choices are better.

To each combination of decisions by all players of the game we assign a set of *payoffs*. The payoffs indicate to the players what they will get for this combination of decisions. These payoffs may or may not be measured in the same units: for example, in the retail industry, one firm may measure the outcome in terms of profit that it makes, while another upstart more interested in market penetration will measure its outcome in terms of sales or market share. Whichever way the payoffs are measured, it is important to realize that none of the players can determine the outcome by himself—it is the interplay of all participants in the game that will result in the final outcome.

If all players' payoffs are measured in the same units, it is possible to aggregate the sum of the payoffs. If the sum of payoffs of all players equals zero for all combinations of decisions, we are talking about a *zero-sum game*. A slight generalization of zero-sum games is the concept of *constant-sum games* in which the sum of payoffs of all players equals some constant for all combinations of decisions. It is easy to demonstrate that each constant-sum game can be transformed into an equivalent zero-sum game. It is also possible to reduce any n-person nonzero-sum game to an equivalent $(n+1)$-person zero-sum game by creating a dummy player who picks up the difference between the sum of payoffs of the first n players and zero.

Whether or not the payoffs of all players are measured in the same units, it is possible that the players will engage in *competition* or in *cooperation*. Games with three or more players are typically played with partial cooperation in the sense that some players cooperate in a block of players which will compete with the other players. Such blocks are called *coalitions*. One important question is which coalitions are likely to form, what's in it for individual players, how much power each player will have, and how the spoils of the combined payoff are going to be distributed among the members of a coalition.

An important issue concerns the information each player has at any point in time. We distinguish between *complete* and *incomplete information* on one hand, and *perfect* and *imperfect information* on the other. Information is considered incomplete if at least one of the players is not fully aware of the rules of the game. Information is imperfect if at least one of the players does not know what move one of the other players made during an earlier stage of the game. This is typically the case in card games, when a player does not know which cards another player discarded during a previous move.

Another issue is whether the players' moves are *simultaneous* or *sequential*. Whereas simultaneous moves do not allow any player knowledge about his opponents' choices, sequential moves do, provided there is perfect information. A two-person game with simultaneous moves is equivalent to a two-person game with sequential moves, given that the second player does not know what move the first player made. For instance, this is the case in games of nature discussed in the previous chapter with nature being the first player in a simultaneous game. Games with sequential moves can, of course, involve any number of moves by the players. Games with multiple moves are discussed at the end of this section.

At this point, it is useful to discuss two distinct representations of games. The first are *games in normal* (or *strategic*) *form*. Each such game is described by its players, the strategies available to each of its players, and the payoffs for all possible combinations of strategies. It is crucial that each strategy provide a decision rule for each possible situation, and a player will make the same decision each time he has the same information. Games in normal form are typically (although not necessarily) associated with matrix games, in which each row (column) is associated with one strategy of a player I (player II). (In cases of games with more than two players, multidimensional matrices have to be used). In two-person games, player I is often referred to as the row player, while player II is called a column player. Each matrix element then expresses the payoffs to all of the players. The case of two players, who, upon using their i-th and j-th strategy, respectively, receive payoffs of a_{ij} and b_{ij}, respectively, is referred to as a *bimatrix game*, as the payoffs are typically shown in a matrix **A**, whose element in row i and column j includes the pair of payoffs (a_{ij}, b_{ij}). Bimatrix games are discussed in some detail in the next section. Games in normal form are best for the description of games in which players choose their moves simultaneously. As an illustration, consider the following

Example: Two players, *I* and *II*, bid on a property. Player *I* considers bidding either $8,000 or $12,000 on the lot, while player *II* considers making a bid of either $7,000, $9,000, or $13,000. The highest bid will win. Player *I*'s utility is $10,000 divided by the amount bid, given that he gets the lot, and it is zero if he does not. Similarly, player *II* has a utility of $12,000 divided by the amount bid if he gets the lot, and zero otherwise. The utilities for all combinations of bids can then be summarized in the matrix

$$\mathbf{A} = \begin{bmatrix} (1.25,0) & (0,1.33) & (0,0.92) \\ (0.83,0) & (0.83,0) & (0,0.92) \end{bmatrix}.$$

Another representation of games are *games in extensive form*. Games in extensive form are described by the players, the strategies available to each player when it is his turn to choose, a player's knowledge when it is his turn to choose, what he knows about the strategies chosen in the past, and the payoffs for all combinations of strategies chosen by the players of the game. Games in extensive form are most often associated with *game trees*, in which, starting with a root node, each player's strategies are represented by arcs leading out of a node that represents the point in time at which it is a player's time to choose. Games in extensive form are more or less straightforward extensions of decision trees used in games against nature. Games are best represented in extensive form when they involve situations in which players act sequentially. This can best be shown in the same

Example: Suppose that player *I* bids first, followed by player *II*, who, at the time of his bidding, knows about player *I*'s bid. The game in extensive form can then be represented by the game tree in Figure I.30a, where each player's bids are indicated next to the arcs, and the payoffs (a_{ij}, b_{ij}) are indicated at the leaves of the tree.

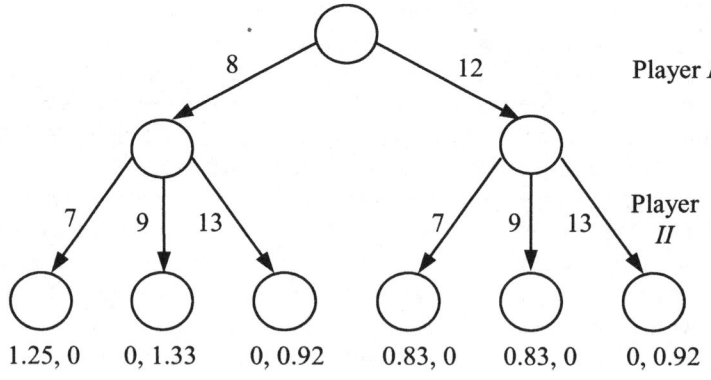

Figure I.30a

Similarly, if player *II* were to bid first, followed by player *I*, who, at the time of his bid, knows about player *II*'s bid, the situation in Figure I.30b would result.

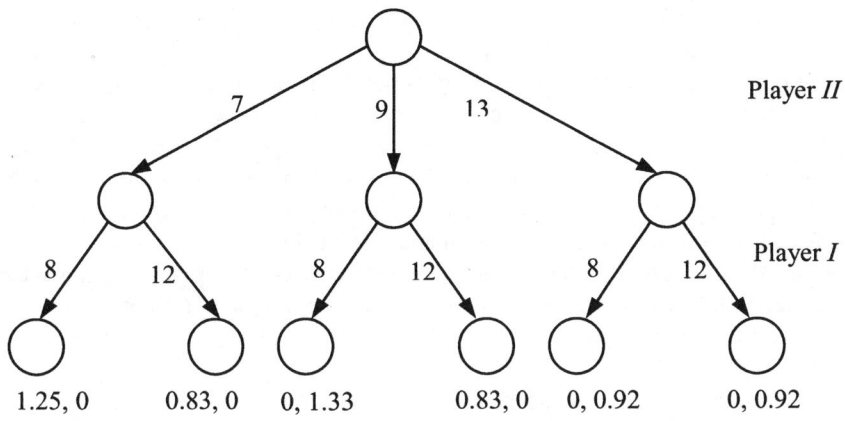

Figure I.30b

Finally, suppose now that both players bid simultaneously. This scenario is shown in Figure I.30c, where the set that includes the two nodes on the second level is called the *information set*. It indicates that when player *II* moves, he has no information about which node in the information set will actually occur. Here, we have arbitrarily chosen an adaptation of Figure I.30a; alternatively, we could have used Figure I.30b and included the three nodes at the bottom of the first level in the information set.

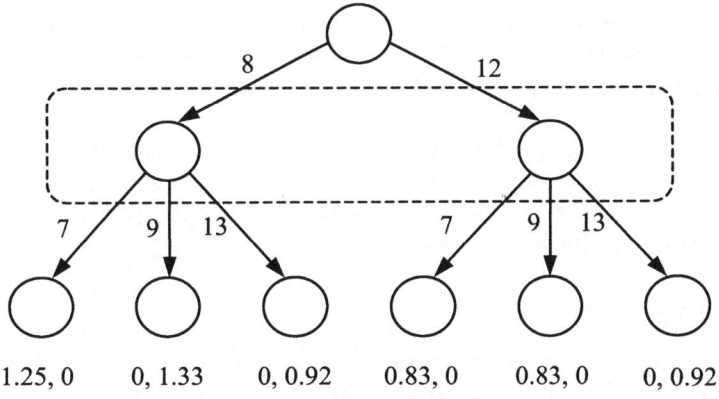

Figure I.30c

The use of these concepts with respect to solution criteria is shown in the next section.

3.1.2 Solution Concepts

This section employs the tools introduced in the previous section to discuss two solution concepts that have been suggested in the literature. Before doing so, we would like to point out that, similar to our discussion of games against nature, it is also possible to apply the concept of dominance. As opposed to games against nature, the concept of dominance in two-person games applies to both players. In other words, each player can delete strategies if they are never better than any other of their strategies. It is also possible to repeatedly apply this concept, shuttling between players *I* and *II*. This process is called *iterated dominance*, which sometimes can not only reduce the size of the game, but actually solve it. However, while the concept of dominance is difficult to dismiss, it is not at all certain that iterated dominance properly reflects actual human decision making. We defer the discussion of dominance to the sections on two-person zero-sum games and bimatrix games.

The first major solution concept to be discussed here was first described by the economist von Stackelberg (1943). It refers to a game in which players make their moves sequentially with foresight. One of the players takes the role of the *leader*, while the other is the *follower*. The leader will make his move first, and the follower will make the move only after the leader has announced and/or made his move. In general, the follower's problem is much easier to solve than that of the leader. The reason is that the follower always faces a problem under certainty, in that the leader's intentions and objectives do not have to be guessed as he has already made his move. In contrast, the leader will have to make assumptions regarding the follower's reaction to each of his possible decisions. This function is called the *reaction function*. Based on the reaction function, (which can be determined from the reaction graph to be discussed below), the leader will then make his choice. The result will, of course, depend on which of the two players is the leader and who assumes the role of the follower.

Without loss of generality, suppose that player *I* is the leader (the row player) and *II* is the follower (the column player). The leader will first study the follower's reaction. In particular, the follower will choose his strategy $j(i)$ given that the leader chooses his i-th strategy), such that

$$b_{i,j(i)} = \max_j \{b_{ij}\} \quad \forall\, i. \tag{1}$$

If two row maxima are identical with respect to the payoffs b_{ij}, then it may be reasonable to assume that the follower chooses among the tied strategies $j(i)$ the one that has the largest payoff to player *I*. Given the follower's choice, the leader then chooses the strategy that maximizes his own payoff, i.e., the row $\bar{i}$, such that

$$a_{\bar{i}\bar{j}} = a_{\bar{i},j(\bar{i})} = \max_i \{a_{i,j(i)}\} \tag{2}$$

Chapter 3: Game Theory 117

The case in which player II is the leader and I is the follower is similar; the von Stackelberg solution in that case is referred to as ($s_{\bar{i}}^I, s_{\bar{j}}^{II}$).

Formally, we can write

Definition I.4: Suppose that player I is the leader and player II is the follower. A pair of strategies ($s_{\bar{i}}^I, s_{\bar{j}}^{II}$) is then called a *von Stackelberg solution*, if conditions (1) and (2) are satisfied.

As an example, consider the following

Example: Consider the payoff matrix

$$\mathbf{A} = \begin{bmatrix} (-2,3) & (2,0) & (0,-1) \\ (4,-2) & (1,-1) & (-3,2) \\ (0,-2) & (2,5) & (0,-3) \end{bmatrix}.$$

Let player I be the leader, while player II is the follower. The follower's choices are then as follows: for $i = 1$, $j(1) = 1$, for $i = 2$, $j(2) = 3$, and for $i = 3$, $j(3) = 2$. The leader will then choose $a_{\bar{i}\bar{j}}$ = max $\{a_{11}, a_{23}, a_{32}\}$ = max $\{-2, -3, 2\} = 2$, so that $\bar{i} = 3$ and $\bar{j} = j(3) = 2$, i.e., ($s_{\bar{i}}^I, s_{\bar{j}}^{II}$) = ($s_3^I, s_2^{II}$). Similarly, we can determine the solution in the case that player I is the follower and player II is the leader. The follower's reaction in this case results in $i(1) = 2$ for $j=1$, $i(2) = 3$ for $j = 2$ (a tiebreaker was needed here), and finally in $i(3) = 1$ for $j = 3$ (again, this situation requires a tiebreaker). The leader will then choose the payoff $b_{\bar{i}\bar{j}}$ = max $\{b_{21}, b_{32}, b_{13}\}$ = max $\{-2, 5, -1\} = 5$. In general, the two von Stackelberg solutions "first I, then II" and "first II, then I" are different.

As mentioned above, a more visually appealing display of sequential games uses game trees. The game tree for the above example for "first I, then II" is shown in Figure I.31.

In a game tree, the von Stackelberg solutions can be determined by a simple modification of the *backward recursion* introduced in games against nature. Starting at a leaf of the tree, the backward recursion uses the "max" criterion in each step, i.e., going back into a node at which player I makes a decision, we use the maximum of player I's payoffs a_{ij}, while going back into a node, at which player II decides, we employ the maximum of player II's payoffs b_{ij}. The bold arcs in Figure I.31 indicate the optimized decisions of the two players at all stages of the game. Note that backward recursion is not possible in games in which an information set includes more than one node.

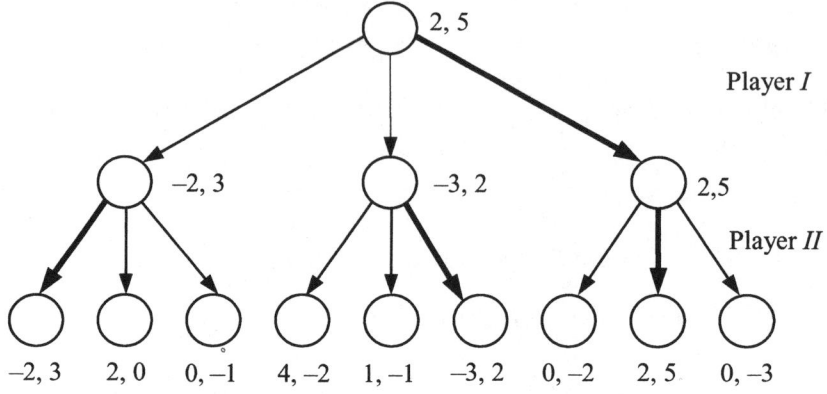

Figure I.31

One issue related to von Stackelberg's leader-follower concept is whether or not a player has an advantage being a leader, or if he would be better off being a follower. Which of the two has an advantage will, of course, depend on the specific situation. For instance, in the above example player *I* is indifferent, whereas player *II* might prefer to be a follower: If player *I* acts as leader, his payoff is 2; if he acts as follower, his payoff is also 2. Similarly, player *II* receives a payoff of 0 or 5 as leader (it depends on player *I*'s choice for which there is a tie), and a payoff of 5 as follower.

There are, however, some problems associated with the concept. Suppose that the leader has an advantage (something also known in the marketing literature as "first mover advantage"), then there is no reason why each player would not want to be a leader—provided he has the ability to fill that role. Similarly (but worse), if both players have an advantage to assume the role of a follower, then, at least without any additional assumptions, the game will never be played.

The second, and arguably most famous, game theory concept also expresses a notion of stability. Nash (1950) described a situation, a *Nash equilibrium*, in which none of the players has an incentive to break out of the arrangement. To formalize our discussion, define a_{ij} and b_{ij} as player *I* and *II*'s payoffs given that they use their *i*–th and *j*–th strategy, respectively. We can then write

Definition I.5: A *Nash equilibrium* is a pair of strategies (s_{i*}^I, s_{j*}^{II}), such that

$$a_{i*j*} \geq a_{ij*} \; \forall \; i, \text{ and}$$
$$b_{i*j*} \geq b_{i*j} \; \forall \; j.$$

Chapter 3: Game Theory

In other words, a Nash equilibrium refers to an element of the game which is the column maximum with respect to the payoffs a_{ij} (indicating that player I has no desire to move out of the present arrangement), and a row maximum with respect to b_{ij}, so that player II has no incentive to change his strategy. In the above example, (s_2^I, s_3^{II}) is a unique Nash equilibrium.

In order to obtain some insight, it is useful to examine the reactions of a player to a strategy of his opponent. Suppose that there are two players, whose payoffs are a_{ij} and b_{ij}, if player I chooses his i-th strategy, while player II opts for his j-th strategy. The result of this combination is a payoff of a_{ij} to player I and b_{ij} to player II. Note that by changing his strategy, player I can force the payoff to any payoff in column j, while player II can force the payoff to any element in row i.

Let now $a_{i(j),j}$ denote a column maximum in column j and $b_{i,j(i)}$ a row maximum in row i. As player I's objective is to maximize his payoffs, he will change from strategy s_i^I to strategy $s_{i(j)}^I$. This move can be captured in the corresponding *reaction graph* $G = (N, A)$, in which a node n_{ij} is assigned to each combination of strategies in which player I plays his i-th, and player II plays his j-th strategy. The reaction graph will then have arcs $(n_{ij}, n_{i(j),j})$ for all i and j with $i \neq i(j)$. If the column maximum is not unique, arcs to all column maxima exist. The argument for player II is equivalent. Player II can force any payoff in row i. In order to maximize his payoff, he will choose strategy $s_{j(i)}^{II}$, so that the reaction graph includes arcs $(n_{ij}, n_{i,j(i)})$ for all i and j with $j \neq j(i)$.

To illustrate the concept of reaction graphs, consider the following

Example: Let the payoff matrix $\mathbf{A} = (a_{ij}, b_{ij})$ be

$$\mathbf{A} = \begin{bmatrix} (-2,3) & (2,0) & (0,-1) \\ (4,-2) & (1,-1) & (-3,2) \\ (0,-2) & (2,5) & (0,-3) \end{bmatrix}.$$

The corresponding reaction graph is shown in Figure I.32.

A Nash equilibrium, provided it exists, describes a (locally) stable situation. In the reaction graph associated with the game, a Nash equilibrium is characterized by a node with zero outdegree. In the above example, the node n_{32} fulfills this condition and is a (unique) Nash equilibrium.

Two open questions related to the concept of Nash equilibria are whether or not a Nash equilibrium actually exists for any given game, and, if it does, whether it can be reached from an arbitrary starting point, given that both players follow a set of rational rules. Neither question can be answered in general; existence and reachability of equilibria depend on the specific game. In general, a Nash equilibrium, provided it exists, may not be reachable by a process in which players sequentially optimize their payoff, assuming that their opponent will not react (i.e., without foresight); see, e.g., Bhadury and Eiselt (1998).

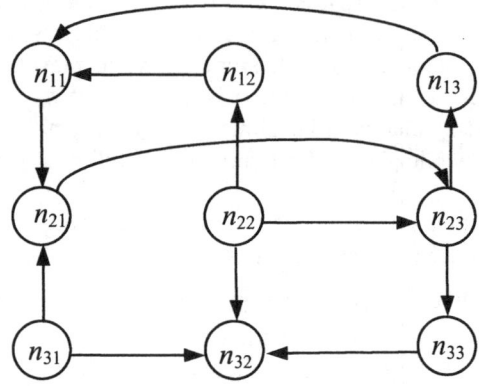

Figure I.32

Another problem relates to cases, in which multiple Nash equilibria exist. Which of them will actually be realized? Selten (1965) devised a refinement of Nash's equilibrium concept that is referred to *subgame perfection*. It is designed to remove those Nash equilibria that are based on non-credible threats. For instance, the statement "If I don't get a piece of cake right now, I am going to kill myself" is non-credible. In order to eliminate such equilibria in a game tree, we have to start at any node that is an information set in itself and examine all subsequent nodes. The concept will be illustrated by an example that loosely follows Baird *et al.*'s (1994) debtor-lender example.

Example: Two companies compete in the fashion market; company *I* is the large market leader, and *II* is a smaller upstart. As the leader, *I* has a choice of either a retro look or a hi-tech look. If *I* chooses the high-tech look, company *II* will introduce their own line of clothes in postmodern look. However, if *I* introduces a retro look, *II* will have to react by choosing between their own "more retro" look and the postmodern look that they had originally planned for. The latter case does not cause *I* to react again, whereas *II*'s former strategy will prompt *I* to either do nothing about *II*'s reaction, or refit their line of clothes to high-tech. This sequence of choices and the appropriate payoffs are shown in Figure I. 33.

The corresponding version of the matrix game is shown in Table I. 17, where each player's strategy is a complete description of his action and reaction. For instance,

Chapter 3: Game Theory

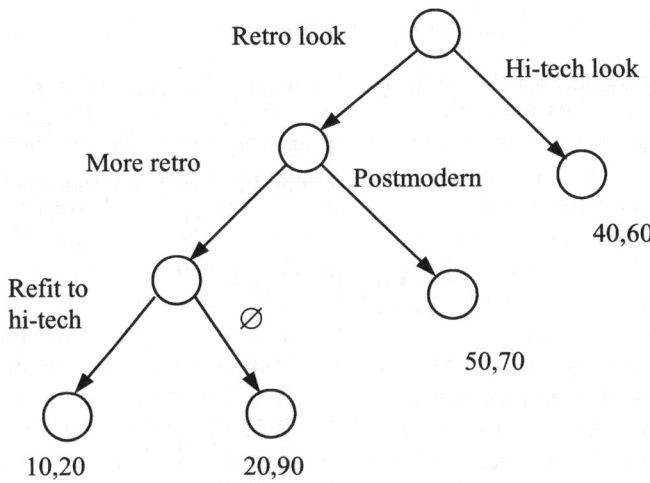

Figure I. 33

Player I's strategy "Retro, *, refit" indicates I's first choice of the retro look, the star shows that player I has no jurisdiction about the second step, and finally I will choose to refit.

Table I. 17

		Player II	
		More retro	Postmodern
	Retro, * refit	10, 20	50, 70
Player I	Retro, *, $\varnothing$	20, 90	50, 70
	High-tech	40, 60	40, 60

Player I will choose the column maxima with respect to the first payoffs, while player II will choose the row maxima with respect to the second payoff value. The game has two Nash equilibria: the strategies (s_3^I, s_1^{II}), i.e., "High-tech, more retro" with payoffs 40 and 60 to the two players, and the combination (s_1^I, s_2^{II}), i.e., "retro, postmodern, refit" with payoffs 50 and 70, respectively.

We will now demonstrate that one of these equilibria will not be realized. In the game tree, consider the node at the second level from the bottom, where player I has to make the choice between no reaction and refit to high-tech. At this point, player I will never choose to refit; while this strategy hurts his opponent quite badly, it also hurts himself. Hence, threatening to refit is not credible. This eliminates the "retro, postmodern, refit" equilibrium from consideration.

3.2 Two-Person Zero-Sum Games

The class of games discussed in this chapter is one of the best known and most easily solved games available. Not surprisingly, the ease with which it can be solved is based on its simplicity, which, in turn, results from a number of very strong assumptions. As the name of the game suggests, there are two players, each of whom has a finite number of strategies. Each player chooses one of these strategies and the result of this simultaneous pair of choices is a payoff. The "zero-sum" part of the name of the game indicates that the sum of the payoffs to the two players equals zero. In other words, the payoff of one player equals the negative payoff of the other, or, more simply, one player's gain is the other player's loss. One of the consequences of the zero-sum assumption is that cooperation between the two players is meaningless, as, regardless of the strategies they choose, their combined payoffs are always zero. This leads immediately to the observation that the players in this game are completely antagonistic. This makes it very difficult to find compromise solutions. Similar to zero-sum games are *constant-sum games*, in which each pair of strategies results in a combined payoff of the same constant. Each constant-sum game can easily be reduced to an equivalent zero-sum game.

One possibility to reduce the size of two person zero-sum games or even solve them altogether is to check the payoff matrix $\mathbf{A} = (a_{ij})$ for dominances. Here the combination of strategies s_i^I and s_j^{II} results in a payoff of a_{ij} to player I and $-a_{ij}$ to player II. Our discussion follows that of the concept of dominance in games against nature; here, however, each of the two rational players will delete dominated strategies.

Rule 1: Row k is dominated by row i, if $a_{kj} \leq a_{ij} \ \forall \ j$ (with strict inequality for at least one j in case of strict dominance).

In contrast to games against nature, dominances for columns can also be meaningful here. In particular, we can state

Rule 2: Column ℓ is dominated by column j, if $a_{i\ell} \geq a_{ij} \ \forall \ i$ (again, with strict inequality for at least one i in case of strict dominance).

Whenever a row or column is determined to be dominated, it can be deleted, as neither player will choose a strategy that is never better, and at least in one outcome worse, than another strategy. The above two rules can then be applied repeatedly in any sequence in order to delete rows and columns. Rows or columns that are deleted based on this process are said to be deleted based on *iterated dominance*. It is important to point out that players who use iterated dominance must have a great deal of foresight, so it is questionable if players, even if they are rational, could be reasonably assumed to employ the concept in practice. A payoff

Chapter 3: Game Theory

matrix that includes neither dominated rows nor dominated columns is called *irreducible*. In order to illustrate the concept, consider the following

Example 1: Consider the payoff matrix

$$\mathbf{A} = \begin{bmatrix} 3 & -4 & 2 & 1 & 2 \\ 0 & 5 & 4 & 2 & 3 \\ 6 & 4 & 2 & 0 & -1 \\ 3 & 2 & 2 & 2 & 3 \end{bmatrix}.$$

Rows 1 and 2 as well as rows 1 and 3 are not comparable. However, row 1 is dominated by row 4, so that row 1 can be deleted. This leaves the reduced payoff matrix

$$\mathbf{A}^1 = \begin{bmatrix} 0 & 5 & 4 & 2 & 3 \\ 6 & 4 & 2 & 0 & -1 \\ 3 & 2 & 2 & 2 & 3 \end{bmatrix}.$$

Here column 3 dominates column 2 (note that column 4 also dominates column 2) and column 4 dominates column 3, so that the second and third columns can be deleted, leaving us with

$$\mathbf{A}^2 = \begin{bmatrix} 0 & 2 & 3 \\ 6 & 0 & -1 \\ 3 & 2 & 3 \end{bmatrix}.$$

At this point, there are no further column dominances. However, row 3 now dominates row 1, so that row 1 can be deleted, leaving

$$\mathbf{A}^3 = \begin{bmatrix} 6 & 0 & -1 \\ 3 & 2 & 3 \end{bmatrix}.$$

Now column 2 dominates column 1, so that column 1 can be deleted, leaving

$$\mathbf{A}^4 = \begin{bmatrix} 0 & -1 \\ 2 & 3 \end{bmatrix}.$$

Here, row 2 dominates row 1, leaving $\mathbf{A}^5 = [2, 3]$, where column 1 dominates column 2, leaving the single element $a_{44} = 2$. This solves the game, i.e., it is optimal for player I to choose s_4^I and for player II to play s_4^{II}, resulting in a gain of 2 for player I and a loss of 2 for player II.

It is, however, not always possible to reduce the game to one strategy for each player which then solves the problem. As a counterexample, consider

Example 2: Let the payoff matrix be

$$\mathbf{A} = \begin{bmatrix} 3 & 7 & -4 & 2 \\ 4 & -5 & -2 & -1 \\ 2 & 5 & 0 & -1 \end{bmatrix}.$$

There are no row dominances, but the first column is dominated by column 3 and 4. Deleting it results in

$$\mathbf{A}^1 = \begin{bmatrix} 7 & -4 & 2 \\ -5 & -2 & -1 \\ 5 & 0 & -1 \end{bmatrix}.$$

Now, the second row is dominated by row 3. Deleting row 2 results in

$$\mathbf{A}^2 = \begin{bmatrix} 7 & -4 & 2 \\ 5 & 0 & -1 \end{bmatrix},$$

where column 1 is dominated by both other columns. Its deletion results in

$$\mathbf{A}^3 = \begin{bmatrix} -4 & 2 \\ 0 & -1 \end{bmatrix},$$

which is irreducible.

Regardless what combination of strategies was selected, one of the players will have an incentive to move out of the present arrangement by changing his strategy.

Formally, a stable element must be

- a column maximum
- a row maximum.

Chapter 3: Game Theory

Suppose now that a_{ij} is a stable element, so that the row minimum in row i is $r_i = a_{ij}$ and the column maximum in column j is $c_j = a_{ij}$. Consider now any row $k \neq i$. Its minimum is r_k and by definition, $r_k \leq a_{kj}$ which, as a_{ij} is the maximum in column j, satisfies $a_{kj} \leq a_{ij}$, so that we obtain $r_k \leq a_{kj} \leq a_{ij} = r_i$, i.e., $r_i = \max_{\nu}\{r_\nu\}$. Similarly, consider some column ℓ whose maximum is c_ℓ, so that $c_\ell \geq a_{i\ell}$. As, by definition, a_{ij} is the minimum in row i, we have $a_{i\ell} \geq a_{ij}$, so that $c_\ell \geq a_{i\ell} \geq a_{ij} = c_j$, i.e., $c_j = \min_{\mu}\{c_\mu\}$. This leads to

Theorem I.6: A two-person zero-sum game has a stable solution called a *saddle point* a_{ij}, if $\max_{\nu} \min_{\mu} \{a_{\nu\mu}\} = a_{ij} = \min_{\mu} \max_{\nu} \{a_{\nu\mu}\}$. The value a_{ij} is then the *value of the game*.

A saddle point in a matrix game is stable in the sense that it corresponds to a pair of equilibrium strategies.

Example: As an illustration of the concept, consider again Example 1 from the beginning of this chapter. The payoff matrix was

$$\mathbf{A} = \begin{bmatrix} 3 & -4 & 2 & 1 & 2 \\ 0 & 5 & 4 & 2 & 3 \\ 6 & 4 & 2 & 0 & -1 \\ 3 & 2 & 2 & 2^* & 3 \end{bmatrix} \begin{matrix} \min \\ -4 \\ 0 \\ -1 \\ 2 \leftarrow \max \end{matrix}$$

$$\max: \quad 6 \quad 5 \quad 4 \quad \underset{\uparrow \min}{2} \quad 3$$

It is apparent that $a_{44} = 2$ is a saddle point of the game and hence a stable solution. Consider now the payoff matrix of Example 2. It was

$$\mathbf{A} = \begin{bmatrix} 3 & 7 & -4 & 2 \\ 4 & -5 & -2 & -1 \\ 2 & 5 & 0 & -1 \end{bmatrix} \begin{matrix} \min \\ -4 \\ -5 \leftarrow \max \\ -1 \end{matrix}$$

$$\max \quad 4 \quad 7 \quad \underset{\uparrow \min}{0} \quad 2$$

In this case, a saddle point does not exist and hence, no combination of strategies of the two players is stable, requiring additional considerations.

Suppose now that the players in Example 1 have chosen strategies s_2^I and s_1^{II}, respectively. Player II can then do no better, but player I will improve his payoff by using s_3^{II}, i.e., moving to n_{31}. Player II will react by changing his strategy to s_5^{II}, i.e., moving to n_{35}. At this point player I will move to either n_{25} or n_{45}, for which there is a tie. If he chooses n_{45} and player II responds by choosing n_{44}, one of his optimal strategies at this point, we end up at the stable solution (s_4^I, s_4^{II}). This coincides with the result in Example 1, which was obtained by dominances. On the other hand, if player I chooses n_{25} instead of n_{45}, player II will move to n_{21}, and we are back where we started. Note that stable solutions do not always exist. A typical example is $\mathbf{A}^3$ in Example 2. We have

$$\mathbf{A}^3 = \begin{bmatrix} -4 & 2 \\ 0 & -1 \end{bmatrix},$$

whose reaction graph is shown in Figure I.34.

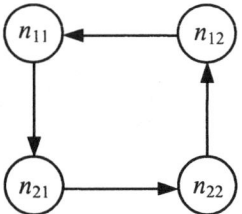

Figure I.34

A very simple, yet interesting, application of two-person zero-sum games was provided by Haywood (1954). It concerns a decision situation faced by Allied and Japanese forces in 1943 in the Battle of the Bismarck Sea near New Guinea. The Japanese forces were planning a convoy that had to bypass the Island of New Britain either to the North or the South. The Allied Forces under General Kenney were aware of the plan in general, but did not know if the Japanese convoy would take the northerly or the southerly route. The options for the two commanders were simple: the Japanese could send the convoy either on the northerly or the southerly route, while Allied Forces could search for them either in the North or the South. The criterion for the decisions was the number of days that the convoy would be discovered and bombed by the Allied Forces. The objectives of the two commanders were completely antagonistic: the Allied Forces wanted to maximize the number of days of bombing, while the Japanese wanted to minimize that number.

Formally let player *I* denote the Allied forces, while player *II* represents the Japanese decision maker. The possible decisions are s_1^I and s_2^I for an Allied decision to search the northerly and southerly route, respectively, while s_1^{II} and s_2^{II} are player *II*'s decisions to route the convoy on the northerly and southerly route, respectively. Player *I*'s payoff matrix is then

$$\mathbf{A} = \begin{bmatrix} 2 & 2 \\ 1 & 3 \end{bmatrix}.$$

It is apparent that the game has a saddle point at (s_1^I, s_1^{II}). The value of the game is $a_{11} = 2$, i.e., there will be two days of bombing of the convoy. Interestingly enough, at the time of the decision, the tools of game theory were not available to the two commanders, yet each of them made a decision that, after the fact, was shown to be optimal. This fact alone would be of mere historic interest. An important lesson can be learned from this example, though. Whereas the decisions of both commanders were probably optimal, the result was a crushing defeat of the Japanese in the Battle of the Bismarck Sea. The lesson is that even optimal decisions can lead to poor outcomes. This situation occurs when a decision maker must choose among options none of which produces a desirable outcome.

The games discussed above are easy to solve when a saddle point exists. Suppose now that a game does not have a saddle point, so that no stable solution exists. In that case we can define a *mixed extension*, in which decision makers do not choose one of their pure strategies each, but specify probabilities with which they intend to play their respective strategies. Formally, player *I* plays his strategies with probabilities $\mathbf{p} = [p_1, p_2,..., p_m]^T$, while player *II* uses probabilities $\mathbf{q} = [q_1, q_2, ..., q_n]^T$. The expected payoff to player *I* is then $\sum_{i=1}^{m} \sum_{j=1}^{n} a_{ij} p_i q_j = \mathbf{p}^T \mathbf{A} \mathbf{q}$.

In order to illustrate the concepts involved in two-person zero-sum games without a saddle point, we will illustrate our discussion with an example in which each of the two players has only two strategies. That allows a graphical representation of the problem as seen from each of the players.

Example: Consider a game with the payoff matrix

$$\mathbf{A} = \begin{bmatrix} 1 & -3 \\ 0 & 2 \end{bmatrix}$$

which has no saddle point. Assuming that player II plays his strategies with probabilities q_1 and q_2, respectively, player I's expected payoffs are $EV(s_1^I) = 1q_1 - 3q_2$ and $EV(s_2^I) = 0q_1 + 2q_2$. As $q_1 + q_2 = 1$, these expected values can be rewritten as $EV(s_1^I) = 4q_1 - 3$ and $EV(s_2^I) = -2q_1 + 2$. These functions are shown in Figure I.35.

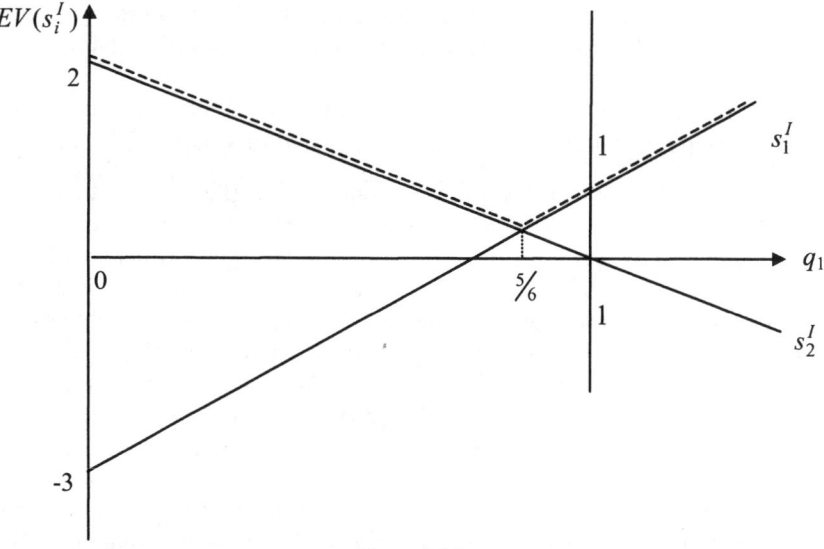

Figure I.35

Whatever probability distribution **q** player II uses, player I will react by choosing his best response, i.e., the strategy that maximizes his expected payoff. In Figure I.35, this reaction is the upper envelope of the functions $EV(s_i^I)$; here shown as a broken line. Since the players in two-person zero-sum games are totally antagonistic (as one player's gain is the other player's loss), player II will chose a probability distribution **q** that minimizes this payoff. The result is the lowest point on the upper envelope. In terms of our example, this is achieved as $q_1 = 5/6$ (and $q_2 = 1/6$) with $EV(s_1^I) = EV(s_2^I) = 1/3$.

The analysis of the game from the point of view of the second player is similar. In our example, the situation is shown in Figure I.36.

The expected payoffs of player II with his two strategies are $EV(s_1^{II}) = -p_1$ and $EV(s_2^{II}) = 5p_1 - 2$ (note that II's payoff is $-a_{ij}$). Equivalently, player II's loss is $-EV(s_1^{II})$ and $-EV(s_2^{II})$. For any probability distribution **p** chosen by player I, player II will react by minimizing his loss, resulting in his choice of the lower envelope in Figure I.33. Player I will, in turn, choose a probability distribution **p** that makes this loss as high as possible, thus realizing the maximal point on the lower envelope. Here, we obtain $p_1 = \frac{1}{3}$ (and hence $p_2 = \frac{2}{3}$), so that $EV(s_1^{II}) = EV(s_2^{II}) = -\frac{1}{3}$.

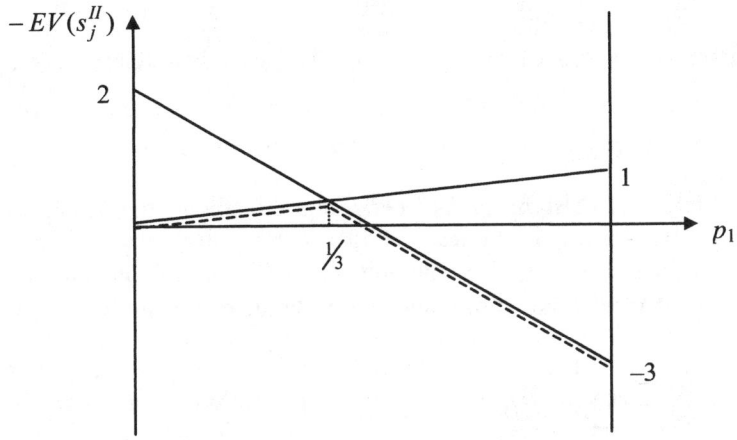

Figure I.36

We can now calculate the probabilities $p_{ij} = p_i q_j$ for each combination of strategies (s_i^I, s_j^{II}); these probabilities are collected in a matrix **P**. In our example,

$$\mathbf{P} = \begin{bmatrix} 5/18 & 1/18 \\ 10/18 & 2/18 \end{bmatrix}.$$

The value of the game is then $v = \mathbf{p}^T \mathbf{A} \mathbf{q}$.

We are now able to generalize and write the two players' problems as mathematical programming problems for any number of strategies. Consider first the planning of player II, who will adapt his choice to player I's behavior. Player I has payoff of a_{ij} and he will choose the maximum among the expected payoffs of his strategies, i.e., he will determine $\arg\max_i \{EV(s_i^I)\}$. Given that, player II will choose a probability distribution **q** that minimizes this function. Formally, player II's problem is to determine probabilities $q_1, ..., q_n$ by solving the minimax problem

$$P_{II} : \underset{\mathbf{q}}{\text{Min }} v = \underset{i}{\text{max}} \{EV(s_i^I)\} = \underset{i}{\text{max}} \left\{ \sum_j a_{ij} q_j \right\} = \underset{i}{\text{max}} \{\mathbf{a}_{i\bullet}\mathbf{q}\}$$

$$\text{s.t. } \sum_{j=1}^{n} q_j = 1$$

$$q_j \geq 0 \; \forall j.$$

This minimax problem can be written as the equivalent linear programming problem

$$P'_{II}: \quad \text{Min } v$$

$$\text{s.t. } v - \sum_j a_{ij} q_j \geq 0 \; \forall i$$

$$q_j \geq 0 \; \forall j.$$

Suppose now that the value of the game $v > 0$. Dividing all constraints by v results in

$$v/v - \sum_j a_{ij} q_j / v \geq 0 \; \forall i$$

$$\sum_j q_j / v = 1/v$$

$$q_j / v \geq 0 \; \forall j.$$

Defining new variables $q'_j = q_j / v \; \forall j$ allows us to write

$$\sum_j a_{ij} q'_j \leq 1 \; \forall i$$

$$\sum_j q'_j = \frac{1}{v}$$

$$q'_j \geq 0 \; \forall j.$$

Consider now the objective function. As $\text{Min } v = \text{Max } \frac{1}{v} = \sum_j q'_j$ as expressed in the above constraint, the problem can now be written as

$$P_{II}^* : \text{Max} \frac{1}{v} = \sum_j q'_j$$

$$\text{s.t. } \sum_j a_{ij} q'_j \leq 1 \; \forall i$$

$$q'_j \geq 0 \; \forall j.$$

Chapter 3: Game Theory

The planning of player I is similar. He will first consider his opponent's behavior. Player II's payoffs are $-a_{ij}$ and his expected payoffs are $EV(s_j^{II}) = \sum_i -a_{ij} p_i \ \forall \ j$, or, equivalently, his expected loss is $-EV(s_j^{II})$. Clearly, player II will attempt to minimize this loss, and player I will choose a probability distribution $\mathbf{p}$ that maximizes the loss. Consequently, player I's problem can be written as

$$P_I: \quad \underset{\mathbf{p}}{\text{Max}} \ v = \underset{j}{\min} \ \{-EV(s_j^{II})\} = \underset{j}{\min} \left\{ \sum_i a_{ij} p_i \right\} = \underset{j}{\min} \ \{\mathbf{p}^T \mathbf{a}_{\bullet j}\}$$

$$\text{s.t.} \ \sum_i p_i = 1$$

$$p_i \geq 0 \ \forall i$$

Applying the same transformations as in the discussion of player II and using new variables $p'_j = p_i / v \ \forall \ i$ we obtain the equivalent problem

$$P_I^*: \quad \text{Min} \ \frac{1}{v} = \sum p'_i$$

$$\text{s.t.} \ \sum_i a_{ij} p'_i \geq 1 \ \forall j$$

$$p'_i \geq 0 \ \forall \ i.$$

Comparing the problems P_I^* and P_{II}^*, it is apparent that they are dual to each other. In order to satisfy the assumption that the value of the game $v > 0$, it is sufficient to add a constant to all elements of the payoff matrix, so that all resulting payoffs are strictly positive. Any game with only positive payoffs has a positive value. This allows us to state

Theorem I. 7: In a two-person zero-sum game with payoff matrix $\mathbf{A}$ there exist probability vectors $\mathbf{p}$ and $\mathbf{q}$ such that

$$\underset{\mathbf{q}}{\text{Min}} \ \underset{i}{\max} \ \{\mathbf{a}_{i \bullet} \mathbf{q}\} = \underset{\mathbf{p}}{\text{Max}} \ \underset{j}{\min} \ \{\mathbf{p}^T \mathbf{a}_{\bullet j}\}$$

Example 1: Consider a two-person zero-sum-game with the payoff matrix

$$\mathbf{A} = \begin{bmatrix} -3 & 3 & 4 & 2 \\ 4 & -4 & 0 & 8 \\ 6 & 2 & -3 & 3 \end{bmatrix}.$$

Simple inspection reveals that $\max_i \{\min_j \{a_{ij}\}\} = -3 < 3 = \min_j \{\max_i \{a_{ij}\}\}$, so that the game in *pure strategies* does not have a saddle point. Arbitrarily adding a constant "8" to all individual payoffs results in the modified payoff matrix

$$\mathbf{A'} = \begin{bmatrix} 5 & 11 & 12 & 10 \\ 12 & 4 & 8 & 16 \\ 14 & 10 & 5 & 11 \end{bmatrix}.$$

Solving either of the problems P_I^* or P_{II}^*, we obtain the optimal solutions $\mathbf{\bar{p}'}$ = $[.0538, .0250, .0308]^T$, $\mathbf{\bar{q}'}$ = $[.0433, .0125, .0538, 0]^T$, and the objective value $1/v = .1096$. As $p_i = v p'_i \ \forall \ i$ and $q_j = v q'_j \ \forall \ j$, we obtain the probability vectors $\mathbf{\bar{p}}$ = $[.4909, .2281, .2810]^T$ and $\mathbf{\bar{q}}$ = $[.3951, .1141, .4909, 0]^T$. The value of the game is calculated by subtracting the constant "8" from $\bar{v} = 9.1241$, so that it is 1.1241. Note that player *II*'s fourth strategy is never played even though it is not dominated.

Definition I.8: A two-person game is called *fair*, if its value of the game is $\bar{v} = 0$.

Clearly, there are very few fair games in real life, other than maybe a friendly poker game at home with friends: lotteries typically only return 50% on the dollar, and in just about all cases, overhead and other costs of running the game have to be deducted before payments can be made.

Another well-known two-person zero-sum game belongs to the class of so-called *Colonel Blotto games*. There are many versions of Blotto games, one of which is provided in

Example 2: Colonel Blotto has three armies at his disposal with which to attack and hopefully conquer a walled city. The city has two gates and is defended by four armies. Each of the opposing commanders has to allocate armies to the city gates in order to attack or defend them, respectively; attacks cannot occur at any place other than the gates. In the case that the attacker has strictly more armies at either gate, Blotto wins and the city is captured; in all other cases the defender has won. Defining (x_1, x_2) as Blotto's allocation of armies to the city gates (and similar for the defenders), Blotto's strategy (2, 1) against (4, 0) results in a successful defense of the first gate, while one of Blotto's armies is able to walk through the undefended second gate. Note the asymmetry of the rule: while Blotto wins the battle if he wins at either gate, the defenders must win at both gates in order to win the battle. Hence, the attacker will concentrate his armies while the defender will have to spread them out. Clearly, this requires the defender to have more armies than Blotto to have a fair game. The payoff matrix is then $\mathbf{P} = (p_{ij})$, where $p_{ij} = 1$, if Blotto wins, and -1 otherwise. The payoff matrix is then shown in Table I.18.

Table I.18

		Defender				
		(4, 0)	(3, 1)	(2, 2)	(1, 3)	(0, 4)
	(3, 0)	−1	−1	1	1	1
Blotto:	(2, 1)	1	−1	−1	1	1
	(1, 2)	1	1	−1	−1	1
	(0, 3)	1	1	1	−1	−1

It is easily determined that the Blotto game does not have a saddle point. Solving the mixed extension results in multiple optima, one of which is $\bar{\mathbf{p}} = [½, 0, ½, 0]$ and $\bar{\mathbf{q}} = [0, ½, 0, ½, 0]$ with a value of $\bar{v} = 0$. In other words, Blotto will use each of his strategies $(x_1, x_2) = (3, 0)$ and $(1, 2)$ with a probability of ½, while the defender will choose his strategies $(3, 1)$ and $(1, 3)$ with a probability of ½ each, resulting in a fair game with a 50% chance to win for each of the players. The problem is, of course, that the game will be played only once and cannot be repeated.

3.3 Extensions

This section discusses three important extensions of the concepts introduced in the previous sections. First we drop the "fixed pie" assumption that the payoffs are zero-sum, and allow the payoffs of the two players to be independent. this allows the discussion of a number of practically relevant scenarios. The second aspect is to do away with the restriction that the players choose their strategies once, and then the outcome of the game is revealed. In a sequential game, we allow players to repeatedly choose strategies, as is often the case in political negotiations or when buying a house. Finally, we relax the assumption that there are only two players in the game. This opens up a whole new range of possibilities, e.g., the forming of coalitions, and the resulting sharing of the payoffs.

3. 3. 1   Bimatrix Games

Similar to two-person-zero-sum games discussed in the previous section, bimatrix games involve two players, each of whom has a finite number of strategies at his disposal. As introduced in Section 3.1, we assume that the combination of strategies (s_i^I, s_j^{II}) results in the payoffs (a_{ij}, b_{ij}) for the two players. The concept of dominance applies in modified form again. In bimatrix games, row k is dominated by row i, if $a_{kj} \leq a_{ij} \; \forall \; j$, and column ℓ is dominated by column j, if $b_{i\ell} \leq b_{ij} \; \forall \; i$.

In contrast to zero-sum games, one player's payoff is generally not related to his opponent's payoff in bimatrix games. That reduces the level of competition and opens up the possibilities of cooperation, a concept that is meaningless in zero sum games (or, for that matter, constant-sum games), where the sum of payoffs is zero (constant) for any pair of strategies chosen by the players.

Example 1 (Prisoner's Dilemma): Two individuals have been arraigned and indicated for a felony. Each of the two prisoners has two options at his disposal, to confess or deny the charge. A deal has been made, so that the confessor will receive preferential treatment. If his confederate chooses to deny the charge, he will be made chief witness and obtain a lesser sentence, and if the other prisoner confesses as well, their confessions will be taken as a sign of remorse, resulting in a shorter sentence for both. In the case that both prisoners deny the charge and stick to it throughout the trial, the state may not have sufficient evidence to convict them for the felony and may have to settle for the conviction for a misdemeanor. Normally, the elements of the payoff matrix are given in terms of years in prison. For reasons of uniformity with the other examples, we have transformed these years into utilities by subtracting the prison sentence in each case from "10". The payoffs are then shown in Table I.19.

Table I.19

		Player *II*:	
		Confess	Deny
Player *I*:	Confess	5, 5	10, 0
	Deny	0, 10	8, 8

It is apparent that in this symmetric game, the strategy "deny" is dominated by "confess" for both players, leaving the pair ("confess", "confess") as the only matrix element. This element must then also be a Nash equilibrium. This assertion is confirmed in the reaction graph shown in Figure I.37, where c denotes "confess" and d symbolizes "deny".

The existence of a unique Nash equilibrium appears to settle the case—both prisoners should confess. However, suppose now that while incarcerated, they meet and discuss their case. They will soon discover that there is tremendous benefit in cooperating. In particular, if both of them agree to deny the charges, their individual utility increases from 5 to 8 (or, in terms of their sentences, the time they will have to do in prison decreases from 5 to 2 years). Obviously, this outcome is preferred by both prisoners. Suppose now that the two agree to cooperate. Back in his cell, each of the prisoners will look at his options again and discover that he can gain even more by double-crossing his partner in crime by turning state's evidence and confess. Putting together the two double-crossing courses of action leads back to the Nash equilibrium. In other words, breaking out of this equilibrium is beneficial to both players, but does not lead to a stable situation.

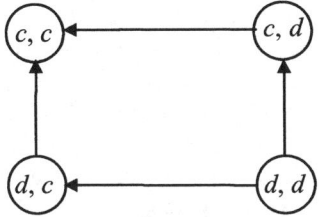

Figure I.37

Consider now the von Stackelberg solution of the game. First, let player *I* be the leader and *II* the follower. If the leader announces that he will confess, the follower will also confess, resulting in a payoff of 5 to player *I*. If the leader decides to deny the charges, then the follower will again confess, resulting in a payoff of 0 to player *I*. These results provide the reaction function to player *I*, who will optimize his own action by confessing. The result is that both prisoners confess, the same as the Nash equilibrium. Due to the symmetry of the payoff matrix, the von Stackelberg solution with player *I* as follower and player *II* as leader also has both prisoners confess.

Example 2 (The *Game of Chicken* and The *Battle of the Sexes*): The idea of the Game of Chicken, supposedly "played" in California, is an adaptation of a "game" shown in the James Dean movie "Rebel Without a Cause." The two players sit in their cars, located a significant distance apart and directed towards each other. At a signal, the two cars drive toward each other at a high speed. As they approach each other, each driver has two choices: either speed up and continue driving toward his opponent, or swerve and avoid collision. If both drivers continue to speed toward each other, their cars will crash and both will die. If one driver speeds and the other swerves, the speeder is considered a "hero", while his opponent loses face and is considered a wimp. Finally, if both drivers swerve, both will be frowned upon. A utility table of the Game of Chicken is shown in Table I.20.

It is apparent that there are no dominated strategies in this game. The unique reaction graph is shown in Figure I.38, where *sp* and *sw* symbolize "speed" and "swerve," respectively.

Table I.20

		Player *II*	
		Speed	Swerve
Player *I*	Speed	(0, 0)	(10, 2)
	Swerve	(2, 10)	(3, 3)

Figure I.38 shows that (speed, swerve) and (swerve, speed) are both Nash equilibria. As far as the von Stackelberg solution (first player *I*, then player *II*) is concerned, player *II*'s reaction function answers player *I*'s strategy "speed" by "swerve", and *I*'s decision "swerve" by "speed". Player *I*'s associated payoffs are 10, and 2, respectively, resulting in the solution ("speed", "swerve"). Similarly, the von Stackelberg solution for the "first player *II*, then player *I*" is the pair of strategies ("swerve", "speed"). It can be observed that the leader in the sequential game has the advantage.

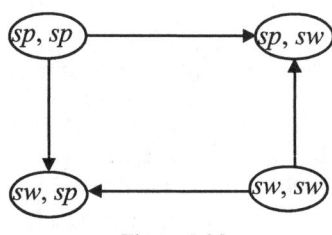

Figure I.38

The game, somewhat pretentiously called "The Battle of the Sexes" involved a couple whose strategies include going to a football match or the ballet. The husband prefers football, his wife favors the ballet. Both, however, prefer going together to an event. The payoffs of this game are shown in Table I.21.

Table I.21

		Player *II*	
		Football	Ballet
Player *I*	Football	(4, 1)	(0, 0)
	Ballet	(0, 0)	(1, 4)

Again, the game does not include dominated strategies. The unique reaction graph of the game, where "f" symbolizes football and "b" stands for ballet is shown in Figure I.39.

Chapter 3: Game Theory   137

As in the Game of Chicken, there are two Nash equilibria. The von Stackelberg solution of the "first player *I*, and then player *II*" is (football, football), while the sequential game "first player *II*, then player *I*" has the solution (ballet, ballet). Again, the leader has an advantage over the follower.

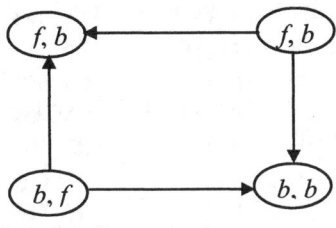

Figure I.39

Example (The *Tax Game*, the *Arms Race*): The Tax Game involves an industrialist, whose strategies are to invest or not to invest in a certain country. His opponent is the chief administrator of the country who can decide to tax the investor.

The investor's best scenario is, of course, to invest and not be taxed, while the worst case to him would be to invest and be taxed. If he does not invest, he prefers the state not to tax. Even though it does not matter to him financially and may even cause some regret, the state's decision may be taken as an encouraging sign for future investments. For the state, the best possible scenario is that the investor invests and it decides to tax, while the combination "do not invest" and "tax" is considered worst. Among the two scenarios that do not involve taxation, the state prefers money to be invested, as that will result in positive spin-offs, such as increased employment resulting, in turn, in increased tax income and reduced unemployment costs. The payoffs are shown in Table I.22.

Table I.22

		State	
		Tax	Do not tax
Investor:	Invest	(0, 10)	(10, 2)
	Do not invest	(2, 0)	(3, 1)

The tax game does not include any dominances. In the unique reaction graph of the game invest (do not invest) and tax (do not tax) are abbreviated as *i*(*ni*) and *t*(*nt*), respectively. The graph is shown in Figure I.40.

It is apparent that the game does not have a Nash equilibrium. Consider now the von Stackelberg solutions. In the "first player *I*, then player *II*" game, the state's reaction to the investor's decision to invest (not to invest) is to tax (not to tax), for payoffs of 0 and 3 to the investor, respectively. Consequently, the investor chooses not to invest, and the state will not tax with payoffs of 3 and 1 for the two players. If the state were the leader of the game, the investor's reaction to the state's decision to tax (not to tax) is not to invest (to invest), so that the payoffs to the state are 0 and 2, respectively, resulting in the state's decision not to tax and the investor investing with payoffs of 10 and 2 for the two players. In reality, it is likely that the investor will not make a decision until the state has announced whether or not it will tax, so that the state will have to assume the leader's role, resulting in the aforementioned (no tax, invest) solution.

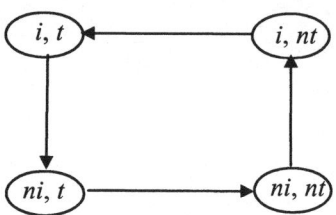

Figure I.40

The "Arms Race" or "Disarmament Problem" has a similar structure. The scenario includes two countries I and II, whose decisions are to arm or disarm, and to attack or not to attack, respectively. Country I's top choice is to disarm while country II does not attack, while in the worst case, it disarms and country II attacks. Given that country I decides to arm, it prefers if its opponent does not attack. Consider now country II. Its best case involves country I to disarm while it attacks, and its worst case has it attack country I which has decided to arm. Given that country II does not attack, it prefers the case in which country I arms, as this ties up valuable resources of country I and allows country II to score propaganda points by being able to paint its opponent as a warmonger. The payoff table for the Arms Race is shown in Table I.23.

Table I.23

		Country II	
		Attack	Do not attack
Country I:	Arm	(3, 0)	(5, 4)
	Do not arm	(0, 10)	(10, 2)

The game does not include any dominances. Its reaction graph is similar to that shown in Figure I.40. Clearly, there is no Nash equilibrium. The von Stackelberg solution to the "country I first, then country II" is "arm" and "do not attack", while the solution of the "country II first, then country I" is to disarm and not to attack.

Chapter 3: Game Theory

We will conclude this section with the illustration of a bimatrix game that involves supply chains. It is a sequential game that lends itself to the von Stackelberg solution. The exposition follows the tutorial paper by van der Veen and Venugopal (2000). Consider a chain that consists of a manufacturer, a retailer, and the customers. All units of a product will flow from the manufacturer first to the retailer, and then on to the customers. We can reasonably expect the manufacturer to be the leader, and the retailer the follower in this game. The decision variables of the manufacturers and the retailer are the price they charge for the product. Let p denote the retail price paid by the customers, and denote by w the wholesale price charged by the manufacturer. On the demand side, we have the customer demand which is assumed to be price-sensitive, so that the demand can be written as $d(p) = a - bp$, where $a, b > 0$ are given parameters. On the supply side, there is the manufacturer's cost function which, for a given technology, is $C(q) = cq + f$ with $c > 0$ and $f \geq 0$ denoting variable and fixed costs, respectively. The link between supply side and demand side is provided by the relation $d(p) = q$. The profit functions of the manufacturer and the retailer are then

$$\mathcal{P}_M = (w - c)q - f, \text{ and}$$
$$\mathcal{P}_R = (p - w)q,$$

so that the total profit in the entire chain is

$$\mathcal{P}_T = \mathcal{P}_M + \mathcal{P}_R = (p - c)q - f.$$

Notice that the total cost function does not include the wholesale price, as it is charged by a unit in the system (the manufacturer) and paid by another unit in the system (the retailer). We can now distinguish between two cases.

Case 1 (cooperation): In this case, the manufacturers and the retailer cooperate so as to maximize their combined profit, which they will divide later. This involves maximizing the profit function $\mathcal{P}_T$. Setting the derivative with respect to p equal to zero, we obtain the price $p^* = \dfrac{a + bc}{2b}$, which results in quantities $q^* = \tfrac{1}{2}(a - bc)$ and a profit of $\mathcal{P}_T^* = \dfrac{(a - bc)^2}{4b} - f$.

Case 2 (competition): In the usual recursive procedure, we first determine the reaction function by optimizing the follower's profit function $\mathcal{P}_R$. This function can be optimized by differentiating with respect to p and equating the result to zero, leading to $\bar{p} = (a + bw)/2b$ and, using this price, to a demand of $\bar{q} = \tfrac{1}{2}(a - bw)$.

The leader now uses this quantity $\bar{q}$ in his profit function, which is then $\mathcal{P}_M = (w-c)\bar{q} - f$ which is optimized by differentiation with respect to the decision variable w. Setting the expression equal to zero results in the wholesale price $\bar{w} = (a+bc)/2b$, and subsequently a retail price $\bar{p} = (3a+bc)/4b$, a quantity $\bar{q} = (a-bc)/4$, as well as profits $\overline{\mathcal{P}_M} = \dfrac{(a-bc)^2}{8b}$, $\overline{\mathcal{P}_R} = \dfrac{(a-bc)^2}{16b}$ and a total profit of $\overline{\mathcal{P}_T} = \dfrac{3}{16b}(a-bc)^2$.

Comparing the two solutions, it is now easy to show that $\mathcal{P}_T^* > \overline{\mathcal{P}_T}$ (ignoring the fixed costs, $\mathcal{P}_T^*$ is one third larger than $\overline{\mathcal{P}_T}$), $q^* = 2\bar{q}$, and, as long as, $a > bc$ (which is guaranteed by the nonnegativity of the quantities in both scenarios), $p^* < \bar{p}$. In other words, both firms, the manufacturer and the retailer, increase their combined profit by cooperating, while consumers also profit in the form of lower prices and larger quantities.

3.3.2 Multi-Stage Games

An extension of the sequential games introduced above includes *multi-stage games*, in which each player has more than a single move. Typical examples of such games are chess, checkers, poker, and others. Frequently, while the numbers of moves are finite, their number is astronomically large. As a simple example, consider the game of chess. In his first move, the white player has 20 possible moves (advance of each of his 8 pawns by one or two steps, or move each of his two knights in two different ways), and the same applies to the black player. Thus, after each player made one move, there are already 400 possible configurations. After that, the analysis gets more tedious, as the pieces start interfering with each other. There are many varieties of multi-stage games: those with two or more players, games with sequential and those with simultaneous moves, and games in which some players are rational, while other player's moves may be partially due to chance, just to name a few. Below, we present an example of a multi-stage game. In purely sequential games, such as that below, a straightforward extension of the backward recursion shown above can be used to solve the game.

Example: Consider a very simple version of the process of labor negotiation. Before the negotiations commence, the two players, *Management* (*M*) and *Labor* (*L*) have announced their base positions. *Management* offers a 1% wage hike and a 1-year job security for everyone, while *Labor* demands 3% higher wages and a 3-year job security. *Management* has calculated that each percent rise in labor costs increases its cost by $1,000,000, while the increase of job security by 1 year adds $800,000 to the costs. It is generally agreed upon that a strike and a lockout adds

$500,000 to the subsequent settlement, which is assumed to be one third of the difference between the most recent offer and demand of any disputed issue away from the current offer (in case of a lockout), or away from the current demands (in case of a strike). If the case goes to arbitration, the arbitration is likely to choose the midpoint between the most recent demand and offer. Both sides are aware of this. *Labor* considers a 1% wage hike roughly equivalent to a 1-year increase in job security. In case of a tie, *Labor* will choose higher job security over higher wages, and both sides will try to avoid a strike and a lockout.

Suppose now that the sequence of offers and demands is a shown in Figure I.41, where the letters *M* and *L* at a decision node indicates the player who moves next. Furthermore, numbers next to the arcs of the game tree denote the wage hike and years of job security in the offer or demand symbolized by the arc. The numbers in the terminal nodes denote the cost to management above the hike, and the percent wage increase and years of job security below it.

In order to explain some of the calculations, consider the case of a lockout in Figure I.41. *Management*'s most recent offer is (1, 2), and *Labor*'s most recent demand is (3,2). Apparently, the second factor is undisputed, so they will agree on 2-year job security. As far as the wage hike is concerned, offer and demand differ by two percentage points. They will settle ⅓(2) = ⅔ **percentage points away from the current offer, i.e., they will agree to 1⅔ = 1.67%. The costs of the settlement in this case are** 1.67 + 2 + .5 = 4.17, i.e., the cost s of the wage hike, the job security, and the costs of the lockout. The other costs are calculated similarly.

The recursive procedure in this example will start at the terminal nodes and work backwards. Going back into a node in which *Management* is the decision maker, the "min" criterion will be applied to the costs; going back into a node at which *Labor* decides, the sum of wage hike and years of job security (based on labor's perception of a 1:1 tradeoff between 1% wage increase and a 1-year increase in job security) is maximized.

The optimal decisions are shown in Figure I.41 as bold arcs. The result shows, given the options provided in this model, that it is optimal for *Management* to make a conciliatory, i.e., high, offer at the beginning. *Labor* will then maximize its benefit by not accepting this offer but demand even more. Their new demands are then accepted by *Management*. Clearly, the settlement of a 2.5% wage hike and a 3-year job security for all is very close to *Labor*'s original demands and, with costs of $4,900,000, very pricey. The only reason for *Management* making the first conciliatory offer is the possibility of a strike if it offers only a 1% wage increase and 2 years of job security; this possibility is even costlier than the high settlement. *Management*'s best option would probably be to devise a strategy that is less costly than the one shown to be optimal, yet that offers more than its hardline strategy which results in a strike. Given that, the solution of the problem is not an

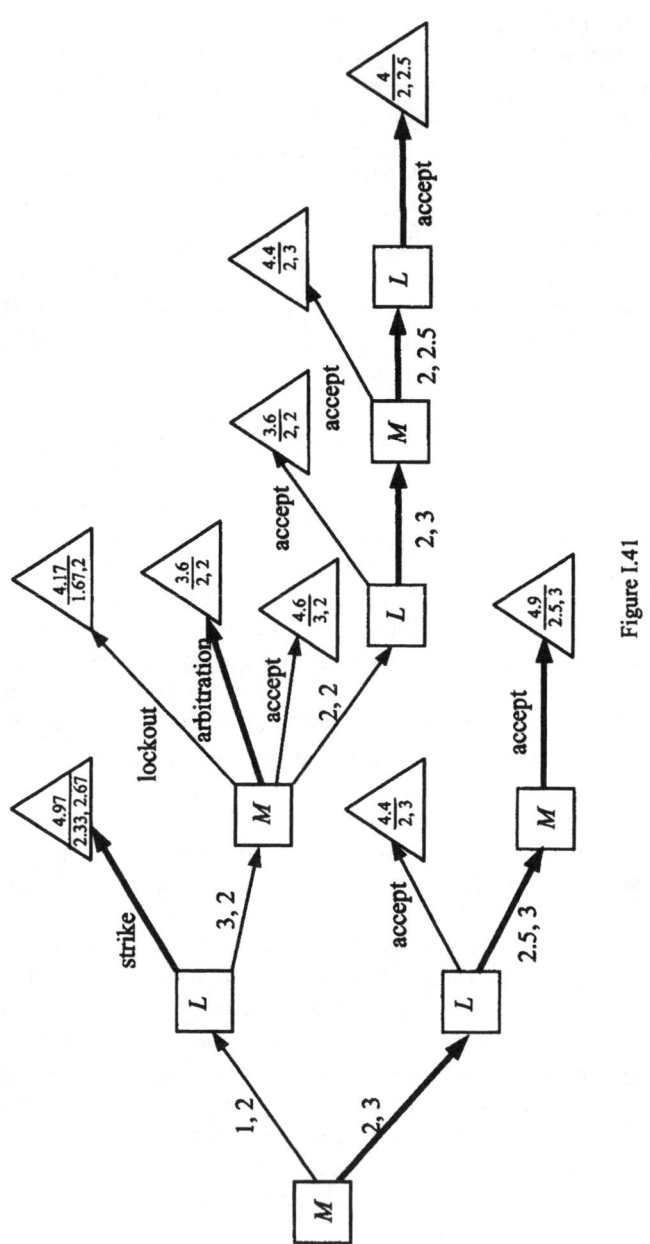

Figure I.41

optimized strategy, but an indication that a new strategy should be devised (together with a general specification of its properties).

3.3.3   *n*-Person Games

In the previous sections on game theory, we have restricted ourselves to just two players who compete with each other. This section generalizes the discussion by including $n > 2$ players. This is much more than a simple extension: it opens up possibilities that do not exist for two players, at least not beyond a very rudimentary form. Most prominently, with more than two players, it becomes possible for some players to cooperate by forming coalitions. Typically, we distinguish competitive *n*-person games and cooperative *n*-person games. The best-known solution concepts for competitive games are Nash equilibria and von Stackelberg solution, both of which were already discussed in the context of two-person games. Their extension to more than two players is straightforward, with the obvious increase in the degree of difficulty of computing them.

In this section, we will concentrate on cooperative solutions to *n*-person games. The fundamental question is, of course which coalitions will eventually form. This question can only be answered by determining how the spoils of a coalition will eventually be distributed, as this indicates to each member of a potential coalition what the incentive is for him to join the coalition. Our arguments below follow loosely the work by Morris (1994). We will say that a coalition S receives a certain *payoff* which it then distributes to its members as a *payout*.

In order to formalize our discussion, suppose that there are *n players* 1, 2, ..., *n*, which are collected in the set *N*. A *coalition* is a (typically nonempty) set $S \subseteq N$, and the *countercoalition* of a coalition S is a coalition $S^c = S \setminus N$. Finally, the set *N* is called the *grand coalition*.

Given a coalition S and its countercoalition S^c in a game, the game between them is reduced to a bimatrix game with the players S and S^c and payoff matrix [**A**, **B**], where **A** includes the payoffs to coalition S, while **B** includes the payoffs to the countercoalition S^c. We can then define

Definition I.9: The *characteristic function* v(S) assigns the value of the two-person-zero-sum-game with players S and S^c and payoff matrix **A** to the coalition S. Similarly, v(S^c) is the minimax solution of the two-person-zero-sum-game with players S^c and S and payoff matrix **B**.

The characteristic function defines a lower bound on the payoff that a coalition S can achieve. It should be noted that this is a very conservative estimate, as the countercoalition S^c will attempt to maximize its own payoff rather than minimize the payoff of the coalition S. The question is now how the payoff that is obtained by a coalition S will be distributed among its members. Define x_j as the amount that player $j \in S$ will receive. The following definition puts some very loose conditions on this payoff.

Definition I.10: An *imputation* $\mathbf{x} = (x_1, x_2, \ldots, x_n)$ is a distribution of the payouts in a coalition S that satisfies the following two conditions:

- $x_j \geq v(\{j\}) \; \forall \, j \in S$ ("individual rationality")
- $\sum_{j=1}^{n} x_j = v(N)$ ("collective rationality").

The requirement of individual rationality states that each member of a coalition must receive at least as much from the coalition as he would if he were to refuse to join the coalition and act as a single player. The collective rationality requires the distribution of the entire amount that the grand coalition would obtain. It is possible to prove that each imputation satisfies the principle of superadditivity, i.e., for any two coalitions S and T with $S \cap T = \emptyset$, $v(S \cup T) \geq v(S) + v(T)$. In other words, joining forces never hurts and may be beneficial.

The conditions for an imputation are so loose that they are not useful as a solution concept. In order to tighten the conditions somewhat, we can define

Definition I.11: An imputation is said to be in the *core*, if and only if

$$\sum_{j \in S} x_j \geq v(S) \; \forall \; S \subseteq N.$$

The requirements for an imputation to be in the core require that the payouts to the members of a coalition do not fall short of the payoff that the coalition receives. While the requirements for an imputation to be in the core put additional constraints upon the payouts, the concept is not very useful as a solution concept. The reason is that the constraints may be so tight that the core of a game is actually empty. On the other hand, they may be loose enough, so that there is a very large number of solutions of the game. A variety of solution concepts can be devised. As early as 1944, von Neumann and Morgenstern suggested *stable sets* for the solution of cooperative games, a concept based on dominances. Here, we introduce a later suggestion by Shapley (1953).

Definition I.12: The *Shapley value* of a player i in an n-person game is defined as

$$\sigma_i = \sum_{S: \{i\} \in S} \frac{(s-1)!(n-s)!}{n!} \Delta(i, S),$$

where $s = |S|$ and $\Delta(i, S) = v(S) - v(S \setminus \{i\})$.

Chapter 3: Game Theory

In the Shapley value as defined above, the term $\Delta(i, S)$ denotes the value that player i contributes to the coalition S. The Shapley value can then be derived by a set of axioms, or, as we will do here, explained by the process of forming a coalition.

Suppose that the coalition forms over time with one member joining after another. More specifically, suppose that the n players are queried one after another whether or not they want to join the coalition. Clearly, given that there are n players, there will be a total of $n!$ sequences of players. Assuming that all of these sequences are equally likely to occur, the probability of any one to actually occur is $1/n!$. All those sequences in which player i is the last one to join the coalition S, have $(s-1)$ players join before him, and $(n-s)$ join the coalition after him. Consequently, there are $(s-1)!(n-s)!$ sequences that have player i in position s. Given that, $(s-1)!(n-s)!/n!$ is the probability of any one of these sequences occurring.

Multiplying this probability by player i's marginal contribution to the coalition and summing up over all coalitions in which player i is the last player to join will result in the expected contribution player i makes to the coalition. This is the Shapley value. Before illustrating the concept by means of an example, we would like to mention the existence of other solution concepts such as the bargaining set, the kernel, and the nucleolus. The discussion is beyond the scope of this book, and readers are referred to the pertinent literature.

Example: There are three major firms that supply the market with their product. Their products are indistinguishable from each other, so that customers cannot distinguish between them. Each of the firms uses its own technology, and thus has its own cost function. Firm 1 faces costs of $100 + 4q_1$, where q_1 denotes the quantity the company makes. Similarly, firms 2 and 3 have cost functions $200 + 3q_2$, and $300 + 2q_3$, respectively. (We could state that, based on the cost functions, firm 3 is the most technologically advanced firm, followed by firm 2 and finally firm 1). The total quantity the market is supplied with is then $q = q_1 + q_2 + q_3$. The price p that can be obtained for a unit of the product is governed by the price-

Table I.24

q_1, q_2, q_3	$p = 15 - 0.05q$	Profits of firms 1, 2, 3	Sum of profits
30, 20, 10	12	140, –20, –200	–80
30, 20, 100	7.5	5, –110, 250	145
30, 80, 10	9	50, 280, –230	100
30, 80, 100	4.5	–85, –80, –50	–215
50, 20, 10	11	250, –40, –210	0
50, 20, 100	6.5	25, –130, 150	45
50, 80, 10	8	100, 200, –240	60
50, 80, 100	3.5	–125, –160, –150	–435

quantity relation $p = 15 - 0.05q$. The three firms must now make decisions regarding their respective outputs q_j, $j = 1, 2, 3$. In particular, firm 1 can make 30 or 50 units, firm 2 can manufacture 20 or 80 units, and firm 3 can produce 10 or 100 units. The prices and profits of all combinations of these decisions are summarized in Table I.24.

Consider now the two-"person" bimatrix game induced by the coalition of players 1 and 2 against player 3. The payoffs with the combinations of q_1 and q_2 for the coalition (row players) and q_3 (the lone column player) are shown in Table I.25.

Table I.25

	10	100
30, 20	120, −200	−105, 250
30, 80	330, −230	−165, −50
50, 20	210, −210	−105, 150
50, 80	300, −240	−285, −150

Consider first the coalition. Their payoffs are

$$\begin{bmatrix} 120 & -105 \\ 330 & -165 \\ 210 & -105 \\ 300 & -285 \end{bmatrix}.$$

Assuming that player 3 tries to minimize the payoff to the coalition, the first column is dominated and can be deleted. The coalition then chooses the best of the remaining strategies which is either the first or the third row, both resulting in $v(\{1, 2\}) = -105$. On the other hand, firm 3 has the payoffs

$$\begin{bmatrix} -200 & 250 \\ -230 & -50 \\ -210 & 150 \\ -240 & -150 \end{bmatrix}.$$

Its second strategy (column) dominates the first, so that column 1 can be deleted. The coalition is then assumed to minimize its opponent's gain, resulting in their choice of strategy 4, so that $v(\{3\}) = -150$. Similarly, we obtain $v(\{1, 3\}) = -135$ and $v(\{2\}) = -130$, and $v(\{2, 3\}) = 20$ and $v(\{1\}) = -85$. This leads immediately to the imputations that must satisfy

$$x_1 + x_2 + x_3 = 145,$$
$$x_1 \geq -85,$$
$$x_2 \geq -130, \text{ and}$$
$$x_3 \geq -150.$$

Clearly, there is an infinite number of imputations that satisfy these very loose constraints. In order to determine the core, we add the following inequalities:

$$x_1 + x_2 \geq -105,$$
$$x_1 + x_3 \geq -135, \text{ and}$$
$$x_2 + x_3 \geq 20.$$

Substituting $x_3 = 145 - x_1 - x_2$, the system of simultaneous linear inequalities is $x_1 \geq -85$, $x_2 \geq -130$, $x_1 + x_2 \leq 295$, $x_1 + x_2 \geq -105$, $x_2 \leq 280$, and $x_1 \leq 125$. The resulting set of solutions in the core is shown in Figure I.42 in (x_1, x_2) space.

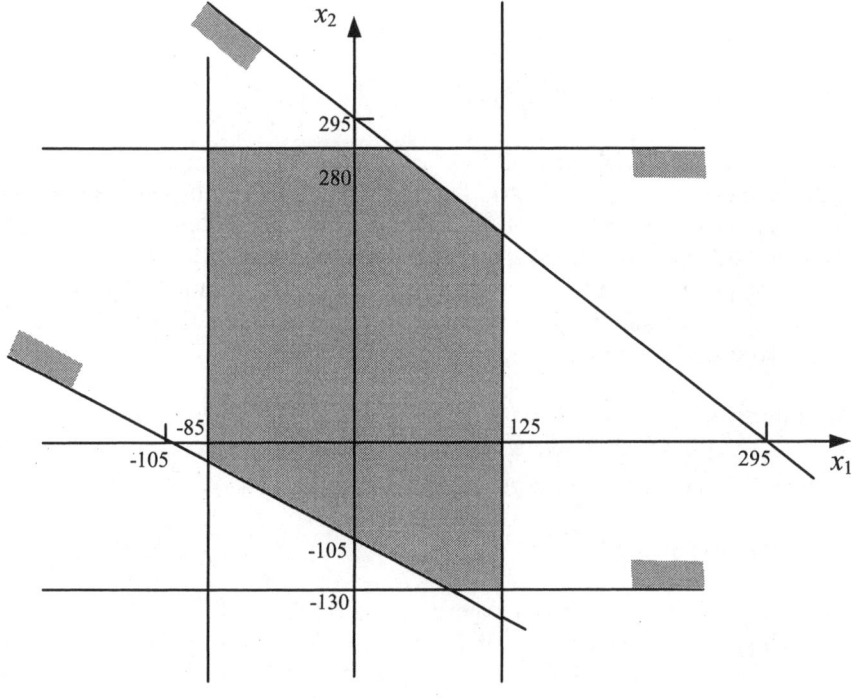

Figure I.42

It is apparent that the core includes an infinite number of solutions, thus this is no real help with finding a solution to the problem.

In order to determine the Shapley value, we first determine the marginal contribution of a player to a coalition. We obtain

$\Delta(1; \{1\}) = -85 - 0 = -85$, $\Delta(1; \{1, 2\}) = -105 + 130 = 25$, $\Delta(1; \{1, 3\}) = -135 + 150 = 15$, $\Delta(1; \{1, 2, 3\}) = 145 - 20 = 125$ for player 1,

$\Delta(2; \{2\}) = 130 - 0 = -130$, $\Delta(2; \{1, 2\}) = -105 + 85 = -20$, $\Delta(2; \{2, 3\}) = 20 + 150 = 170$, $\Delta(2; \{1, 2, 3\}) = 145 + 135 = 280$ for player 2, and

$\Delta(3; \{3\}) = -150 - 0 = -150$, $\Delta(3; \{1, 3\}) = -135 + 85 = -50$, $\Delta(3; \{2, 3\}) = 20 + 130 = 150$, $\Delta(3; \{1, 2, 3\}) = 145 + 105 = 250$ for player 3.

Given these marginal contributions, we can calculate the Shapley values as

$$\sigma_1 = \frac{2}{6}(-85) + \frac{1}{6}(25) + \frac{1}{6}(15) + \frac{2}{6}(125) = 20,$$

$$\sigma_2 = \frac{2}{6}(-130) + \frac{1}{6}(-20) + \frac{1}{6}(170) + \frac{2}{6}(280) = 75, \text{ and}$$

$$\sigma_3 = \frac{2}{6}(-150) + \frac{1}{6}(-50) + \frac{1}{6}(150) + \frac{2}{6}(250) = 50,$$

which is not only an imputation, but is also in the core.

Consider now a special type of *n*-person games that is very important in the organization in democratic societies, as well as in a variety of business contexts, e.g., stockholders' meetings. The games in question are *voting games*. The main difference between the games discussed above and voting games is that in the latter, each player has only two strategies, yes or no, or, quantitatively speaking, 0 or 1. Similarly, a coalition either wins the vote or it does not. Games of this nature can be defined as follows.

Definition I.13: A *weighted majority game* is a simple *n*-person voting game $[q; w_1, w_2, ..., w_n]$ in which q denotes a *quota* required to win the game, and w_i is the weight of player i. Then the imputation of a coalition S is $v(S) = 1$, if $\sum_{i \in S} w_i \geq q$, and 0 otherwise.

A simple example of a weighted majority game is [3; 1, 1, 1, 1, 1]. In this 5-player game, all players have the same weight and a simple majority will win the game. In the 4-player game [12; 5, 4, 6, 3], any coalition S with 3 or more members will win the game, i.e., $v(S) = 1 \ \forall \ |S| = 3$, while any coalition with 2 or less members will lose. Some extreme cases exist. A *dummy player* is a player without influence, i.e., any winning coalition can win with or without him. As an example, consider the game [10; 7, 6, 2] that requires a two-thirds majority of all votes. The only winning coalitions in this game are $S_1 = \{1, 2\}$ and the grand coalition $S_2 = N = \{1, 2, 3\}$. As $\{3\} = S_2 \setminus S_1$, player 3 is a dummy. At the other extreme is a dictator of a game. A player i is a dictator, if $v(i) \geq q$, so that the dictator above can win the game, regardless of the decisions of the other players. The first player in the game [7; 7, 6, 2] is a typical example of a dictator. One step below a dictator is a player with *veto power*. Such a player is no dictator, but no coalition can win without him. An example is the 5-person game [10; 4, 2, 3, 2, 2,] where the first player has veto power as even the coalition $N \setminus \{1\}$ does not have enough weight to win the game.

One important question relates to the power of a player. In order to calculate the power, we can again resort to the Shapley value, which, for voting games, reduces to

$$\sigma_i = \sum_{\substack{S: \{i\} \in S \\ v(S)=1 \\ v(S \setminus \{i\})=0}} \frac{(s-1)!(n-s)!}{n!}.$$

As an example, consider the weighted majority game [10; 6, 4, 5]. The winning coalitions are $\{1, 2\}$, $\{1, 3\}$, and $\{1, 2, 3\}$. For player 1, none of these coalitions can win without him, so that $\sigma_1 = \frac{(2-1)!(3-2)!}{3!} + \frac{(2-1)!(3-2)!}{3!} + \frac{(3-1)!(3-3)!}{3!}$ $= \frac{1}{6} + \frac{1}{6} + \frac{2}{6} = \frac{4}{6}$. Player 2 is involved in the winning coalitions $\{1, 2\}$ and $\{1, 2, 3\}$. The first of these coalitions cannot win without him, whereas the second will. Hence, $\sigma_2 = \frac{(2-1)!(3-2)!}{3!} = \frac{1}{6}$. Similarly, player 3 is involved in the winning coalitions $\{1, 3\}$ and $\{1, 2, 3\}$, the first of which cannot win without him, while the second can. Consequently, $\sigma_3 = \frac{(2-1)!(3-2)!}{3!} = \frac{1}{6}$. These power indexes of .6667, .1667, and .1667 are in contrast to the relative weights of the players of $\frac{6}{15}, \frac{4}{15}, \frac{5}{15}$ or .4, .2667, and .3333.

Another power index was developed by Banzhaf (1965). The idea behind Banzhaf's power index is that of a *marginal* or *swing vote*. Simply speaking, a player is a marginal voter in a coalition, if his vote—one way or the other—changes the outcome of the vote. The sequence in which the players vote is irrelevant. Considering all possible 2^n outcomes, define by m_i the number of times that player i's vote is marginal. The *Banzhaf power index* is then defined as

$$\beta_i = \frac{m_i}{\sum_k m_k} \quad \forall i.$$

As an illustration, consider again the weighted majority game [10; 6, 4, 5]. Table I.26 enumerates all possible outcomes, the resulting decision, and whether or not player 1, 2, and 3 casts a swing vote, respectively.

Table I.26

Case	Votes	Decision	Does player i cast a swing vote?		
			$i = 1$	$i = 2$	$i = 3$
(1)	YYY	Y	Yes	No	No
(2)	YYN	Y	Yes	Yes	No
(3)	YNY	Y	Yes	No	Yes
(4)	NYY	N	Yes	No	No
(5)	YNN	N	No	Yes	Yes
(6)	NYN	N	Yes	No	No
(7)	NNY	N	Yes	No	No
(8)	NNN	N	No	No	No

As an example, consider case (6), where Players 1 and 3 vote "No", while player 2 votes "yes", so that the outcome has 4 votes for "Yes" and 11 for "No", and the outcome is "No". If player 1 would change his vote, the outcome would be YYN, which is case (2) and the decision changes from "No" to "Yes", i.e, in case (6), player 1 casts a swing vote. The other cases are dealt with similarly. As a result, in $m_i = 6$ cases player 1 casts a swing vote, whereas players 2 and 3 cast swing votes in $m_2 = 2$ and $m_3 = 2$ cases each. The resulting Banzhaf power indices are $\beta_1 = .6$, and $\beta_2 = \beta_3 = .2$. An interesting application of the Banzhaf index is found in http://www.esi2.us.es/~mbilbao/niza.htm, where the size and composition of the European Commission is analyzed based on the Treaty of Nice.

PART II: LOCATION AND LAYOUT DECISIONS[1,2]

Location and, to a lesser extent, layout problems have been studied by scientists for many years. While in the early years location problems were of interest mostly to geographers, during the middle of the 20th century researchers and practitioners from other fields became interested in the field as well. Location problems can crudely be described as models in which a number of facilities is to be located in the presence of customers, so as to meet some objective. Obvious applications of the problem occur when facilities such as warehouses, shopping malls, parking garages, or hotels are to be located. Although these few examples are quite different from each other, they share some common features. Most prominently, all of these applications deal with facilities that are expensive to set up, are difficult to remove, and require planning that involves multiple stakeholders. Clearly, this means that the usual algorithmic concerns no longer apply: there is no need to worry about a few minutes (or hours or even weeks) of computing time, when the task is to locate a multimillion dollar facility whose costs could easily double if some of the parties affected by the facility resort to aggressive means, e.g., neighbors of the facility who sue the operator or owner of the facility, call for boycotts, or make life difficult and expensive in many other ways. This is not to say that algorithmic efficiency is not important: enumeration techniques that work with astronomically finite sets are not feasible to employ as they would not finish within a human lifetime. The point is that the emphasis in location models is different from those in other areas, and solution methodologies should reflect this focus if they attempt to be relevant. The large variety of objectives that have been used in several location contexts witnesses this need. The incorporation of multiple decision makers in location models is, however, still lacking. A general theory of compromise solutions in this context would provide a huge boost to the application and the applicability of location models.

---

[1] Chapters 1-4 of this part were coauthored by Professors R.L. Church, University of California, Santa Barbara, CA, U.S.A., and C.S. ReVelle, The Johns Hopkins University, Baltimore, MD, U.S.A.

[2] Chapter 5 of this part was coauthored by Professor G. Finke, University Joseph Fourier, Grenoble, France.

CHAPTER 1 FUNDAMENTALS OF LOCATION AND LAYOUT PROBLEMS

This chapter introduces the basic ideas of location models. We first provide a glimpse into the space of location problems, one of its major components. The second section gives a short account of the long history of location models, starting with its foundations in geometry, to its geographical heyday, the economic inquiry, and finally into the realm of optimization. The third section outlines the major elements that distinguish location models. Understanding these features is mandatory for an appreciation of the different results that are available for the different models, some that are just subtle modifications of others. Finally, the fourth section surveys some of the applications of location models. Here, we have attempted to choose examples that demonstrate the breadth and depth of the study of location models and their uses.

1.1 The Nature of Location Problems

Location problems come in many guises. In this section, we will attempt to convey to the reader some of the basic ideas that are common to most location models. Most location problems can be defined as follows: given a space, a distance function that is defined between any two points in that space, a number of existing "customers" that are located in the space under consideration and who have a certain "demand" for a product, the task is now to locate one or more "facilities" in that space that will satisfy some or all of the customers' demand. As a primitive example, suppose the space is a county, the distance between any two points are the mileages (or, alternatively, the driving time) between the two points, the customers are individual households with certain demands for groceries, and the task is to locate a supermarket that will serve the customers in the county. It is already apparent from a definition as general as this that a multitude of location problems can be thought of by varying the individual factors introduced above.

The definition and example provided above are normative: given a specific scenario, the solution to the location model under consideration will recommend a solution to the decision maker. However, there is another side to location problems. Assume that the customers and the facilities are already located in a particular space. Now, given an address, the task of a *query problem* is to find a customer at that address. This task must often be accomplished in the shortest possible time. While the problem appears simple at first glance, this may not be the case. Consider some examples. In the simplest case, a letter was mailed to an individual, and the query problem attempts to find the addressee. Given that the streets in the county have names, the houses have numbers, and the individuals have their names displayed at the postal boxes on or in front of the house, the problem is not difficult to solve. A similar problem is faced by ambulance drivers in the case of an emergency. Provided the call clearly specifies the location of the patient, this is also not too difficult, in areas where there is a uniform renumbering of the houses along rural routes. It is obvious, though, that in this application, time is of the essence. The problem becomes considerably more difficult if the victim of an accident or a crime does not exactly know his own position and can only, say, specify that he is standing, say, "in front of a WalMart," (of which there may multiple stores in town) or, worse, "on Bennett Street, opposite number 1028," which may actually be Barnett Street number 1620. Similarly, if an infant calls 911 but cannot be clearly understood or does not exactly know where he is right now, things become tricky. To demonstrate the size of real problems in this category, the city of Los Angeles keeps over 14,000 city blocks in their dispatch file records.

Problems that also fall into this general category arise in the quest of military units trying to determine enemy positions given incomplete and unreliable information, divers trying to locate shipwrecks and the golden doubloons they carried, or anthropologists, who attempt to locate ancient settlements on the basis of archaeological finds. The global positioning systems (*GPS*) that are built into some cars attempt to solve some of these problems by giving drivers their exact position in case of an emergency.

Consider now the scale of location problems. There are generally three potential levels of analysis to a locational decision: the macro, meso, and micro levels. These three levels of analysis are based on map scale. Map scale is usually reported as a ratio 1: x, so that 1 distance unit on the map represents x distance units in the field. For example, on a topographical map of scale 1:63,360, 1 inch on the map equals 63,360 inches or 1 mile in the field. The higher the ratio (such as 1:25,000 or 1: 3,000 as is often used for the planning of subdivisions), the greater detail and the closer the map is to the real world. Conversely, the smaller the scale, the larger the area represented and the more generalized the map representation. If we are involved in locating a facility in a large region, the map scale is small and the area is large. We will call this a macro-level location problem. If the area is much smaller, such as a town or a county, then the map scale can be considerably larger than in the macro problem, coupled with an increase in attendant details. We will call this level the meso-level of analysis. Finally, if the area is small, e.g., a building site, the map scale can be very large, such as 1:100, and the level of analysis is what we will term a micro-level analysis. The type of problem will determine the scale of the problem. For example, in a macro problem, we may ask in which three cities we should operate major distribution warehouses that supply the nation. In contrast, in a meso-level analysis, we may ask the question: "Now that we have decided that Pittsburgh is one of the regional locations, where in Pittsburgh should we locate this warehouse?" or, maybe, "which piece of land should be purchased for the new warehouse?" And finally, once we have determined which parcel of land is to be used for the warehouse, our micro-level location question may be: "how should we orient the building on this parcel of land, where should parking be placed, and how should the warehouse floor be laid out? Should this be a multi-level facility?" For the first two levels of analysis, the location question does not involve the exact footprint of the facility that allows us to represent the facility as a point on a map. Problems on the micro level are often termed *layout problems*. In order to formalize matters, define *location problems* as problems in which the facility to be located is small in relation to the space under consideration. On the other hand, in layout problems, the shape of the facility cannot be ignored by representing it as a point on a map: it has to be considered explicitly. Given this additional feature, it is apparent that layout problems are typically more difficult than location problems.

1.2 The History of Location Models

Perhaps the earliest known location model is the problem of locating a point on a Cartesian plane that minimizes the sum of distances to a predefined set of points. This problem can be attributed to Torricelli who worked on it as early as 1648, followed later by Fermat, Steiner, and Weber. An interesting historical account is provided by Wesolowsky (1993). Torricelli, Steiner and Fermat approached the problem as a mathematical puzzle. Weber (1909) raised Torricelli's problem from a mathematical puzzle to a problem in economic efficiency. The problem that bears his name includes a firm that wishes to locate a factory in order to minimize the costs associated with the transport of the raw materials to the factory and the transport costs of the finished product to market. Representing the costs as weights at the customer locations and using straight-line distances, Weber's problem can be shown to reduce to the problem of finding a point that minimizes the sum of weighted distances between the customer locations and the facility location. In addition to solving the problem, Weber demonstrated that with certain combinations of weights the factory location coincides with either the source of the raw material or the market, thereby enabling a distinction between market-oriented activities and raw-material oriented activities. For example, soft drink bottling is market-oriented, whereas coal gasification is raw-material oriented.

The earliest model that locates economic activities in the plane was developed by von Thünen in his 1826 treatise, "Der isolirte Staat" (the isolated state). The author developed a spatial economic theory of the allocation of farm land around an isolated farm town. The main argument is that farming activities would best be organized based on the product value and cost of transport to the town center. Since not all land can be planted with the highest valued crop, he reasoned that land should be allocated in concentric circles about the farm town based on total demand and value of the crop and cost of transport. Thus, optimal layout of land has been of interest for almost 200 years!

One of the next great milestones in location research was that of Christaller (1933), who is known as the originator of central place theory. Starting with a Cartesian plane with an embedded triangular lattice, perfect knowledge on the part of consumers and perfect competition on the part of marketers, Christaller reasoned that certain patterns of economic activities are efficient across the landscape. His stability criterion is that there exists no location position in which enough excess profits are generated, so that another entrepreneur can effectively enter the market. In order to find a stable solution, the assumption of homogeneity on the part of the land is made. Christaller's original area of study was Southern Germany. Nowadays, such patterns have been identified in places like Iowa and regions of China. Central place models have since been formulated as optimization models; see, e.g., Kuby (1989).

In the late 1920s, the economist Hotelling (1929) developed models for spatial competition that includes price competition. His work forms the basis for much of the spatial/market competition research taking place today. The inclusion of price competition turns out to make the problem much more difficult than the corresponding pure location problems would have been.

In the late 1930s, Weiszfeld (1937) published an algorithm for the Fermat-Torricelli-Weber problem. Miehle (1958), Kuhn and Kuenne (1962), and Cooper (1963) independently rediscovered Weisfeld's algorithm. The main attractiveness of Weiszfeld's method is that it generally converges very quickly towards an optimal solution, even though Drezner et al. (2002) and others have identified examples in which convergence is painstakingly slow. A highly entertaining account of Weiszfeld's discovery can be found in Vazsonyi (2002).

Weiszfeld, Cooper, and Kuhn and Kuenne gave location problems a new twist. Even though the computational equipment, even in the 1960s, was still in its infancy, their work focused on the computation of optimal locations. Cooper's (1963) contribution and, even more prominently, Hakimi's (1964, 1965) seminal work is not only a milestone, but it also laid the foundation of location theory as we know it today. Based on his contributions, Hakimi is considered by many as the father of modern location theory. To this day, his network models provide the backbone of an operations researcher's toolkit. As in other areas of operations research, the advances in computational equipment have fostered the advances made in the field.

Since the 1970s, research in location science has increased substantially. Network location model developments by Revelle and Swain (1970), Maranzana (1964), Erlenkotter (1976) and Teitz and Bart (1968) and planar continuous space model developments by Wesolowsky (1973), Hurter and Martinich (1989), and Francis and White (1974) helped popularize the field. Domschke and Drexl's (1985) bibliography already lists in excess of 1,500 references, and activities in the field have by no means abated.

However, to this day many location planners have resisted using one of the many models available in the literature. One of the reasons appears to be the high positioning of the location decision on the strategic – tactical – operational ladder, and its multiple facets regarding diverse stakeholders, different objectives, and a multitude of factors that influence the decision. On the other hand, almost all location models described in the literature use the distance between a facility and it customers as the only criterion for locating the facility. This is in stark contrast to many real-life scenarios, in which factors such as favorable labor conditions, proximity to markets or suppliers, existing infrastructure, and tax structures as often quoted by planners as equally or, more often, more important than transportation costs; see, e.g., Schmenner (1982). The difficulty of quantifying of many of these factors has prevented most algorithm designers from incorporating them in their methods. Some exceptions are the recent use of decision analysis techniques for location problems; see, e.g., Korpela and Tuominen (1996) for the selection of the site of a warehouse, or Larichev and Olson (2001) for the location of solid waste facilities and nuclear waste repositories.

1.3 The Major Elements of Location Problems

This section will discuss the major elements of location models. A variety of authors have attempted to classify location models with respect to their main components. Some of these attempts are found in the contributions by Brandeau and Chiu (1989); Daskin (1995), Mesa and Boffey (1996); Hamacher and Nickel (1996). None, however, has achieved widespread acceptance similar to Kendall's notation in queuing. Here, we resist the temptation to offer another classification scheme and simply list and discuss the main components.

(1) The first component of every location problem is the *space* customers and facilities are, and will be, located in, and the *distances* between facilities and/or customers. We distinguish between *continuous location models*, in which customers and facilities are located in some subset of the d-dimensional real space $\mathbb{R}^d$. In contrast, in *discrete location models* customers and facilities are positioned at an (often finite) number of points. A third, and very popular, alternative are *network location models*. Here, customers are either located at the nodes (which is the case in almost all contributions on network location), or along the edges or

arcs of the network. If the demand is not originally located at the nodes of the network, different demand points are frequently aggregated and then displayed as a node in the network. Aggregation is frequently done on the basis of city blocks or census tracts. For aggregation errors, see Francis *et al.* (2002). Similarly, the facilities may either be located only at the nodes, or also along the edges.

The space is probably the prominent descriptor of location models; a number of standard books on the subject have used the space as an identifier of the special focus of their work. As examples, see Love *et* al. (1988) for continuous models and Daskin (1995) for network location models. This distinction is also very useful with respect to the tools that are used for the solution of the problem. While discrete and network location models tend to require tools from integer programming, continuous location models typically require algorithms from nonlinear programming.

Consider first continuous location models. In addition to the space, we also need a measure for the distances between pairs of points. A large body of literature exists on the subject. Most popular distance functions are metrics and gauges; see, e.g., Plastria (1995). The most popular distance functions for planar problems are various versions of Minkowski's ℓ_p metric that defines d_{ij} as the distance between points (x_i, y_i) and (x_j, y_j) as $d_{ij} = [(x_i - x_j)^p + (y_i - y_j)^p]^{\frac{1}{p}}$. For $p = 1$, the Minkowski metric reduces to *rectilinear, rectangular, Manhattan*, or simply ℓ_1 distances $d_{ij}^1 = [|x_i - x_j| + |y_i - y_j|]$. For $p = 2$, we obtain the *Euclidean* (or *straight-line*, or *as the crow* flies) distance $d_{ij}^2 = \sqrt{(x_i - x_j)^2 + (y_i - y_j)^2}$, which is the most popular distance function in continuous spaces. Finally, for $p \to \infty$, the Minkowski metric reduces to what is known as the *Chebyshev* (or *max norm*, or ℓ_∞) metric $d_{ij}^\infty = \max\{|x_i - x_j|; |y_i - y_j|\}$.

A common objection to the Chebyshev distance is that there does not appear to be any way to physically move from one point to another within that distance. However, consider the following scenario. A robotic arm is to be moved from a point P_i to a point P_j, and we are interested in determining the amount of time that it takes to make that move. In the first case, assume that movements are only possible parallel to the axes, one movement at a time, at identical speeds. For simplicity, assume that the speed equals 1. Given that technology, the amount it takes to move the arm from P_i to P_j is proportional to the ℓ_1 distance. Suppose now that the technology is basically the same, but now we are able to switch on horizontal *and* vertical movements at the same time. This means that if the two points do not have the same x or y coordinate, we will first move at a 45° angle (at a speed of $\sqrt{2}$), and, once we are at the same x or y coordinate as the point we want to reach, we move parallel to the appropriate axis. The total amount of time it

Chapter 1: Fundamentals of Location and Layout Problems

takes to reach that point is proportional to the Chebyshev distance. Finally, if we are not only able to make horizontal and vertical movements at the same time, but can also choose different speeds for that purpose, the time required for the move is proportional to the Euclidean distance as we can adjust the speeds to allow for a straight-line movement.

Starting in the 1970s, Love and several co-researchers have performed research on matching ℓ_p distances to observed road (or rail or air) distances between points in the plane. The idea is this. Suppose that a decision maker is interested in storing the distances between a large number of points. For any given n points, there are $O(n^2)$ distances. If the value of n is very large, then this number may be out of reach for many storage media. On the other hand, if there were a way of closely approximating the true road distance by one that is computed on the basis of the coordinates of the two points, then it would be sufficient to store only the coordinates of the n points, requiring only $O(n)$ storage.

In their studies, Love and Morris (1979) have found that the k-p-s distance was the most accurate; according to it, the distance between (a_i, b_i) and (x, y) is defined as

$$d_i = k(|x - a_{i1}|^p + |y - a_{i2}|^p)^{1/s},$$

where the parameters k, p, and s were determined by fitting k-p-s distances to observed distances in the field. The authors found that in most cases the value of s was "not markedly different from that of p." Since actual routes are not likely to follow purely straight line or rectilinear paths, Brimberg and Love (1992) explored a weighted combination of Euclidean and rectilinear norms where the weight k is strictly positive. This measure is called a weighted one-two norm. This norm was found to be a more accurate distance measure where distances tend to have some directional bias. Thus, in a practical sense, the ℓ_p norm and the k-p-s norm are representative measures which can be used in practice. In many applications, either the Euclidean or Manhattan metric is assumed by the modelers without much discussion. The error involved in making such assumptions has not been subject to much testing.

In continuous location models, the two-dimensional space is by far the most popular space. However, some inherently complex models such as location optimization under competition typically make do with single-dimensional spaces, i.e., location on a line segment. On the other hand, the location of satellites for military and civilian purposes will necessarily take place in three-dimensional space. Some nontraditional applications of location models use higher-dimensional space; they are described below. Another extension deals with location problems on the sphere. These are typically global problems that take the curvature of the earth into consideration.

The case of distances in discrete location problems is considerably simpler: they can be defined whichever way the decision maker pleases. The disadvantage is that for any set of n locations of customers and facilities, all $O(n^2)$ distances will have to be stored.

In network models, common practice is to assume that the movements between customers and facilities take place along the shortest path between the two points. For n points (typically, the nodes of the network), this will require $O(n^3)$ by using one of the appropriate methods, such as the Floyd-Warshall method; see, e.g., Procedure c6 in the Introduction, or, for more detail, Eiselt and Sandblom (2000). It is, however, conceivable to incorporate in the location model congestion in the transportation network and determine individual or system optima according to Wardrop's principles; see again Eiselt and Sandblom (2000).

(2) *The number p of facilities* that are to be located is the second important component that determines the type and the complexity of a location model. Most models include an exogenously specified number of facilities, whereas a few, most competitive location models, choose the number of facilities endogenously. It is of little surprise that in problems with exogenously given numbers of facilities, single-facility location models are not only considerably easier to solve than multi-facility location problems, they frequently provide the extent of what can be solved in polynomial time. Sometimes budget constraints implicitly limit the number (and type) of facilities to be opened. In most applications, the number of facilities is fairly small, so that planners get away with solving the optimization problems independently for a sequence of "reasonable" values of p, and then choose the solution that is most appropriate for the problem at hand.

(3) The *magnitude of demand* of the customers is another important feature of location models. We usually distinguish between *elastic* and *inelastic demand*. A customer's demand is called elastic if the magnitude of the demand will react to other factors, most prominently prices. Most demand, particularly demand for discretionary goods, is elastic. Consider, for instance, the demand for a particular type of vehicle. As long as customers perceive the price and the quality of the vehicle as adequate, good, or even outstanding, they will purchase the vehicle. However, as its price increases (or the perceived quality decreases), customers will consider alternative products, and eventually buy other products. As a result, the demand for the product will decrease. On the other hand, there are essential goods whose demand is independent of its price or even, within reasonable limits, its quality. However, there are very few essential goods: the typical traditional examples are bread (which, however, may be substituted by other foods), and butter (which may be substituted by margarine). Better examples for essential goods are housing and transportation (or, a close relative, the price of gasoline), the latter only as long as the transportation is essential.

(4) Another aspect is the *allocation of demand* to the facilities. Suppose that in a location model with multiple facilities the demand of all customers are known, and the facilities have been located. The next question is which facility (or facilities) each customer's demand is satisfied from. In *allocation problems*, a decision maker at the facility decides from which branch a customer will receive the delivery. Typical examples are warehouses from which furniture is delivered to the customers. In allocation problems, customers have no jurisdiction over (and, generally, do not care) which facility they are served from. Location scientists usually distinguish between models with *fixed interaction* (i.e., those in which it is known in advance which facility a customer is served from), and those with *variable interaction* (i.e., those in which the location of the facility determines which facility will deliver to which customer).

On the other hand, there are *customer choice models*. In these models, customers have jurisdiction over the choice of the facility their demand is satisfied from. The most prominent examples occur in the retail industry, where the customers decide which store they will patronize. Location models that include customer choice frequently include *attraction functions* that express the degree to which a customer is attracted to a facility. Such attraction functions often include the price charged at the facility for the product in question, and an attractiveness parameter that is a composite, which expresses the floor pace of the facility (a proxy expression for the selection at the store), the friendliness of the staff, and similar factors. Huriot and Thisse (2000) refer to allocation and customer choice as *shipping* and *shopping models*.

(5) An essential feature of each model is the level of certainty that is associated with its parameters. As usual, we distinguish between *deterministic models* whose parameters are all assumed to be known with certainty, and *probabilistic* models, in which at least one type of parameter is not known with certainty, but only by some probability distribution. (As introduced in decision analysis in Part 1 of this book, locating one or more facilities in such a case falls under the general heading of planning under risk). It is obvious that, other things being equal, deterministic models are considerably easier to solve than their probabilistic counterparts.

In most probabilistic models, the uncertain parameter is the demand. Sometimes, the customer-facility distances are also considered to be uncertain. Again, following the discussion in decision analysis, there are different ways of dealing with uncertainty. Minimizing the loss in the worst case, or minimizing the regret, are just two examples of these possibilities.

(6) Most location models are *static*, in the sense that they provide one set of demands. However, most facilities are located with a long-term perspective in mind, which would normally prescribe a *dynamic location problem*. Typically, the uncertainty of future demand and the complexity of the resulting model are stated as the reason for dealing with static models.

(7) Similar to the static *vs.* dynamic issue above is the distinction between *single commodity* and *multicommodity models*. While most models involve multiple commodities, virtually all location models ignore this fact and assume a homogeneous good. Again, the main reason for this assumption is the complexity of the model that would result from heterogeneous goods.

(8) Whereas most of the previous components were related to the customers, focus now on the facilities. The facilities can either provide *single level* or *hierarchical service*. A good illustration of hierarchical service is given by Narula (1984) in the context of health care services. Suppose that there are three levels of service: a doctor's (or nurse's) office, a small clinic, and a hospital. The doctor's office provides only basic service, a clinic on the next-higher level can provide all services that a doctor can, but it also can deliver some additional care. Similarly, hospitals on the highest level can provide all the care that a clinic can, but they also provide additional services. Clearly, doctors are distributed more densely across the state than clinics which, in turn, are more frequent that hospitals. The question is now to (a) locate the different facilities, and (b) design a referral plan, so as to maximize the service that can be provided within a given budget.

(9) One of the most important—and most neglected—features of any location model is the *objective* pursued by the decision maker(s). Traditionally, most location models have been assumed to be of one of two types: they would either minimize the sum of weighted distances (which are used as a proxy expression for transportation costs, an objective typically thought of as related to the private sector), or minimize the maximum (weighted) distance between any customer and its closest facility (something that applies to the location of emergency facilities, making this type related to the public sector). In the mid-1970s, Church and Garfinkel were the first to consider the fact that not all facilities are "desirable," in the sense that desirable facilities are best located as close as possible to the customers. Their (ob-) noxious counterpart, however, will be located as far away from the customers as possible. The early literature includes some references to obnoxious = unpleasant facilities, while other papers deal with noxious = toxic facilities. The models associated with such "undesirable" facilities as they are commonly called today make no distinction between unpleasant and toxic facilities: they will attempt to locate such facilities as far away from the general public as is possible. However, even the concept of "undesirable" facilities is not very illuminating. As an example, consider an undesirable facility such as a sanitary landfill. Minimizing the cost of transporting the garbage to the facility will locate the landfill as close to the customers as possible, just as a desirable facility, whereas the general public would probably attempt to push away the very same facility from their own locations. In other words, it is not so much the type of facility that determines where it will be located, but the type of objective that the planners employ in the location process. Eiselt and Laporte (1995) have introduced the terms "push" and "pull" for different classes of objectives, depending on whether the customers attract or deter the facility.

Chapter 1: Fundamentals of Location and Layout Problems

In addition to the pull and push objectives introduced so far, another type of objective has emerged in the not too distant past. These are the so-called "equity objectives". Using the dictionary definition of equity as "fairness", the difficulty of quantification of the expression becomes immediately apparent. To make matters worse, almost all authors who discuss "equity" objectives, actually use the criterion of equality instead, thus attempting to determine locations so as to make the facility-customer distances as equal as possible. This is the reason why Eiselt and Laporte (1995) have coined the term "balancing objectives" for this type of objective. Another type of balancing objective occurs when one firm owns a number of franchises, and the objective is to position them so as to ensure that their sales are about equal. A good introduction to the subject is found in Marsh and Schilling (1994).

Another type of objective is frequently found in the location of facilities in the public sector, where the predominant concern is accessibility for all potential customers. These are the so-called *covering objectives* which attempt to maximize the number of customers that can be reached by the facilities.

Finally, many realistic problems have multiple objectives, Schilling *et al.* (1980) is one of the first references that explicitly deals with multiple objectives and the tradeoffs between them in the context of location problems.

(10) The location model under consideration may be either *competitive*, or it may assume that competition does not exist. Competitive location models were first described by the economist Hotelling (1929) in his seminal paper. Clearly, most situations are competitive, and yet only a small portion of location models are. One reason for the lack of a competitive component in location models is that competition will make even simple location models very difficult, and moderately difficult ones will become completely intractable. Furthermore, an explicit consideration of competition will require important additional information, such as assumptions concerning an opponent's reaction to each possible action on the part of a decision maker. Such assumptions will be very difficult to make. As the annotated bibliography by Eiselt *et al.* (1993) shows, many competitive location models use game theory as a tool for analysis.

It is apparent that this short list of components is by no means exhaustive. For instance, many existing location models assume that the facilities are not constrained by any capacities. Such an assumption is made for the sake of simplifying the computations, while most practical models are, of course, *capacitated*.

Another aspect deals with preexisting conditions. Many location models pretend that the facilities that they locate exist in a vacuum. Frequently, however, similar or possibly competing facilities do already exist, a factor that must be taken into consideration. Such *conditional location models* were first introduced by Drezner

(1982) and are, to this day, mostly found in competitive location models. This is unnecessarily limiting as there are interesting scenarios in which one firm wants to locate additional branches so as to not cannibalize (too much) its own existing facilities. Such models are, for instance, important in the location of automobile dealerships.

Another class of problems that is not considered very often consists of what may be called *unlocation problems*. As the name suggests, these problems attempt to remove or close facilities in the most efficient manner. For instance, it may be desired to close health care clinics in a manner that decreases the service level as little as possible. Formally, unlocation problems can easily be reduced to standard location problems, which is probably the reason for their neglect by almost all researchers. In general, the idea to solve unlocation problems is as follows. Suppose that at present, p facilities are located, and the objective is to optimally remove q facilities. The procedure then is to find optimal locations for the remaining $(p-q)$ facilities, considering only the locations of the p existing facilities as potential locations. The resulting problems are usually small and not difficult to solve.

Another location problem is very prominent in the airline industry. We are referring to *hub location problems* that model the predominant hub-and-spoke system in the industry. The idea is to locate hubs so as to provide acceptable customer service, given that all customers whose present location as well as their destination are not at a hub will have to take a commuter plane to a hub first, and then either fly to another hub, and from there on with another commuter plane to their final destination (or, if their destination is served by the same hub as their origin, they will only take two commuter planes to and from the hub).

There are many other possible features of location models that space allows us here only to mention. Harwitz *et al.* (1998, 2000) consider one aspect in the behavior of customers. In particular, they assume that customers engage in *price search* before they make their purchase. This means that customers will make a certain number of inquiries (either in person, on the internet, or by phone) before deciding which store to patronize. Another aspect that some authors have looked at is *multipurpose shopping*. Both aspects discussed here would make standard location problems virtually intractable. As Beckmann (1972) pointed out, "As everyone knows in location theory one is forced to work with simple assumptions in order to get any results at all".

1.4 Applications of Location Problems

This section will explore some of the many applications that have been described in the literature. Here, we select some typical examples. A good survey of applications throughout the 1980s is provided by Eiselt (1992).

Jacobs et al. (1996) employ a straightforward location method and algorithm for their problem of locating a blood collection facility in the mid-Atlantic States of the United States. The authors use an integer programming problem, based on a small number of alternatives, to minimize the sum of distances as an expression of proximity.

Many authors employ location models when siting emergency facilities, such that no potential "customer" (such as homeowner, when locating fire stations) is further away from any of these stations than a given distance standard, as much as that is possible. Fujiwara et al. (1987) locate ambulances in Bangkok, Thailand. Their objective is to have as many customers as possible within reach of an ambulance. Current and O'Kelly (1992) locate two types of warning sirens in a mid-Western city of the United States. One of their models minimizes the number of citizens that are not within reach of a siren, given a budget constraint. A related problem is solved by Ehrgott (2001) whose model locates emergency helicopters in South Tyrol, Northern Italy, so that the longest distance between any community and its nearest emergency helicopter is as short as possible.

A number of contributions are based on geometric approaches. One such paper is by Wirasinghe and Waters (1983) who investigate the location of solid waste transfer stations in Calgary, Alberta. They first determine the sizes of the collection areas, then position them in the plane, and finally determine the centroids of the collection areas as transfer points. Love et al. (1985) employ a variant of the ℓ_p distance function in order to locate a trucking terminal.

Since most location problems are strategic problems as discussed above, it is no surprise that many applications have more than one objective function. Huxley (1982) locates a church camp in Spain, and his objective is a linear combination of the proximity to a large population, and the average distance between the camp and potential visitors. Köksalan et al. (1995) locate p breweries in Turkey, minimizing the distribution and the inventory cost. Sankaran and Srinivasa Raghavan (1997) locate bottling plants for liquefied petroleum in India. Their mixed-integer linear programming problem minimizes the construction and the operating costs of the plants. Psaraftis et al. (1986) locate equipment for the removal of oil spills, while minimizing the facility, transport, and cleanup costs in addition to the expected costs of the damage done by the spill.

A natural combination of objectives includes components of location and transportation issues in location-routing problems. Nambiar et al. (1981, 1989) describe a problem that deals with the location of collection stations for natural rubber in Malaysia. Their objective includes locating the plants, allocating collection stations to factories, and routing. The problem is similar to that of Jacobsen and Madsen (1980) and their problem of locating transfer points in a distribution system for newspapers in Western Denmark. The authors solve their problem in three stages: they first determine the number of local transfer points, then optimize the tours from the printers to the transfer point, and then they finally optimize the tour from the transfer stations to the retailers. Kirca and Erkip (1988) locate transfer stations for solid waste management, minimizing collection and transportation costs.

Kimes and Fitzsimmons (1990) set up a rule that allows the decision makers of a medium-sized hotel chain in the United States to determine whether or not a given location is likely going to be profitable. The main tool in the study is regression analysis. Schniederjans *et al.* (1982) apply goal programming to determine the locations of trucking terminals in the United States. An interesting feature of this analysis is that it does not just include the usual "hard" data such as costs, but also proxy expressions for the satisfaction of potential customers and drivers. Hopmans (1986) tackles the difficult problem of locating bank branches in the Netherlands by incorporating elements of spatial interaction models, including attraction functions. Barda *et al.* (1990) use tools from decision analysis (*ELECTRE*) to locate thermal power plants in Algeria. For a discussion of tools from decision analysis, readers are referred to part I of this volume.

In addition to the physical location problems described above, there are also nonstandard problems, which are characterized by the fact that the customers and facilities are not physical entities that are actually transported from one place to another.

One of these problems is the *brand positioning problem*, a model that is well known in marketing. The basic setting includes a *d*-dimensional space in which each dimension represents an important feature of the product type under consideration. This gives the *feature space* its name. In order to even consider this setting for a given problem, it is required that all product features be measured on some quantitative scale, and that the level that is considered best by the customers be finite. Given that these conditions are satisfied, we can then map each existing brand of a product in the feature space. Similarly, each customer can also be represented in this space by his "most preferred point." It is apparent that while the mapping of customers in the feature space may be theoretically possible, the huge number of potential customers renders such an approach infeasible. Instead, the modeler will typically consider groups of customers, which are then represented by an average most preferred point. Clearly, aggregation errors will be present in this case.

The next issue concerns the behavior of consumers. Given a customer's most preferred point and the points of all brands, which brand will the customer purchase? The idea here is that customers will generally buy the brand that is closest to their own most preferred point. This requires an appropriate distance function. As Eiselt (1989) pointed out, ℓ_1 distances use the underlying philosophy that "differences build up," in that they measure the distance between two points as the sum of distances of the individual dimensions. On the other hand, the ℓ_∞

distance uses the philosophy that the difference between two points is just as large as the largest distance between any of its individual features. Finally, the Euclidean distance is a combination of these. Which distance function is appropriate depends, of course, on the specific case. Once an appropriate metric is defined, we can determine a tessellation of space, such that a *Voronoi set* V_i is associated with each brand P_i, such that the distance between any point $x \in V_i$ and P_i is no larger than the distance between x and any point P_k with $k \neq i$. The collection of all sets V_i is then called a *Voronoi diagram*. Shamos and Hoey (1975) were the first to describe an efficient $O(n \log n)$ algorithm for the construction of a Voronoi diagram in two dimensions. Their paper also uses Voronoi diagrams for a number of related problems in computational geometry. Given the Voronoi sets for all given brands, it is then possible to determine the sales of each of the brands.

The resulting optimization problem is then as follows. Each of the decision makers who has jurisdiction over a brand can then optimize the location of his own brand, i.e., relocate it in the feature space, so as to maximize the customers that are "captured" by the brand. (The concept of *market capture* was first introduced by ReVelle, 1986). This will have to be done in a competitive context, as the managers of the other brands will also redesign their products for the largest market appeal. Okabe and Suzuki (1987) have considered problems with up to 256 brands, for which they sequentially optimized the locations of each of the brands, one brand at a time.

A similar problem occurs in political science. Rather than products, we have political candidates, and the consumers are now potential voters. The feature space becomes an *issue space*, in which all-important political issues are included. Again, a voter will presumably vote for the candidate who best represents his own views, in the sense that the distance between candidate and voter is as small as possible. The candidate (who supposedly has no opinion of his own or chooses to ignore it) will then relocate, i.e., represent a view that allows him to capture as many voters as possible. The quantification of the issues under consideration is even trickier in this application than in the brand positioning problem. However, while the latter must include prices, the political model does not. There are some excellent pertinent contributions in the volume edited by Niemi and Weisberg (1976).

CHAPTER 2 LOCATION MODELS ON NETWORKS

In this chapter we consider location problems that are defined on a network $G = (N, A)$ with the set of nodes N and the set of directed or undirected arcs A. Typically, but not necessarily, customers are located at the nodes of the network and the demand of all customers at a node n_i is denoted by its "weight" w_i. This demand and node may represent aggregated data of a smaller region or town. Similarly, the distance, unit cost, time or other disutility associated with a trip from node n_i to a node n_j along arc a_{ij} is c_{ij}. The basic problem is to locate some number of facilities (their number can be specified exogenously or determined endogenously) in order to satisfy a given objective.

As an example, consider the network in Figure II.1 with 12 nodes and 17 arcs, where the single-digit number next to an arc represents the distance of the arc, and the double-digit numbers near the nodes indicate their respective weights. Many models assume that transportation movements take place on the shortest connection between nodes, thus necessitating the application of shortest path algorithms; see, e.g., Eiselt and Sandblom (2000) or Ahuja et al. (1993). In general, we will denote by $\mathbf{D} = (d_{ij})$ the matrix of shortest path lengths.

It is easy to envisage that any network location problem in which facilities can locate at a discrete set of locations only, can be formulated by defining binary variables y_j for a set of potential facility locations, such that $y_j = 1$, if a facility is located at the j-th potential site and 0 otherwise. This reduces location problems on networks to zero-one integer programming problems; see, e.g., Eiselt and Sandblom (2000). Similarly, it can be shown that some well-known zero-one programming problems appear as subproblems in a variety of location models. This latter occurrence places these models in the **NP**-hard category, thus requiring either specialized and highly efficient exact solution techniques or good heuristics. In order to avoid describing similar solution methods for a variety of problems, we avoid such repetition by discussing the different models first and describing their solution techniques towards the end of this chapter.

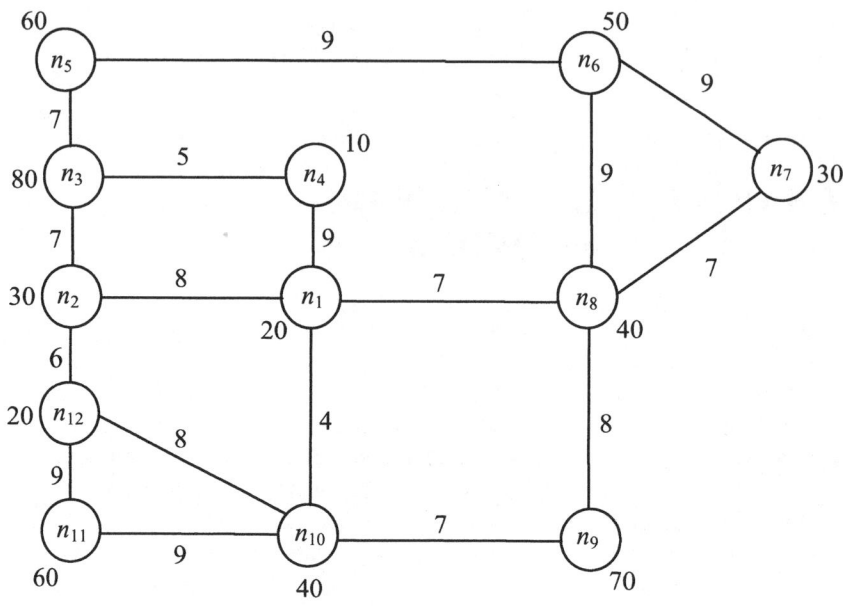

Figure II.1

2.1 Covering Models

The idea behind covering models is to locate facilities that provide some service required by customers. If customers are positioned within a certain predefined critical distance D from any of the facilities, then they are considered served or "covered." Two objectives for the location of facilities are to either cover all customers in the network with the smallest number of facilities or, alternatively, to cover as many customers as possible with a given number of facilities. Typical examples of applications of covering models are found when emergency facilities are to be located. For example, the City of Los Angeles operates 104 fire stations and over 420 types of equipment, such as engines, hook and ladder trucks, staff cars, ambulances, and hazardous materials vehicles, at an annual cost of over $250 million. Here, the objective is to maximize the protection. However, measuring protection is very difficult. In order to find an expression for "protection," we would need to know the value of responding to a fire from different distances or times. For example, what is the difference in protection by increasing the response distance to a neighborhood by a mile? We can safely assume that an increase in the time to respond to an area for fire protection will mean that the fire has a greater chance to spread and it may reduce the chance to save property or a life if one is in jeopardy. Ideally, such information is needed to determine the difference in the effectiveness of one fire service configuration vis-à-vis another. Unfortunately, measurements of the value of protection are nearly impossible to make. Instead, standards of service have been suggested by the Insurance Services Office that, when met, virtually guarantee an adequate level of protection or, at least, the lowest premiums for fire protection. ReVelle *et al.* (1977) report standards for the purpose of rating or scoring the protection capability of cities. Factors other than facility-customer distances that are also included in their rating system include the training and number of personnel, the type of fire prevention program, the availability of water and location of hydrants, to name a few. Such standards are, of course, surrogate measures of performance since direct value measurements cannot be made. For fire services the most commonly used surrogate measure of performance is whether service can be provided to areas within a prespecified distance or time. For example, if fire services are located within 1.5 miles of a neighborhood, then it can be assumed that a timely response to an emergency can be made or, in terms of covering models, that the facility "covers" a demand point. Although we may not know the exact value of this level of service in terms of the potential number of saved lives and property damage prevented, we do know that providing this level of service will make it possible to save lives and protect property.

2.1.1 The Location Set Covering Problem

One of the first models that was developed to site emergency service facilities that incorporates a maximum service distance standard is the *Location Set Covering Problem* (*LSCP*) introduced by Toregas *et al.* (1971). The objective of the *LSCP* is to locate the smallest number of facilities such that each demand node is covered by one or more facilities. If we assume that the cost to purchase land and build a facility is roughly the same for all nodes (or is small in comparison to maintaining each of the needed fire crews), then the objective of using the least number of facilities is equivalent to minimizing the cost of providing service to all demand.

Throughout this chapter, the set I, $|I| = m$, will denote the set of given customer locations, and the set J, $|J| = n$ represents the set of potential facility locations. Furthermore, we assume that $I, J \subseteq N$. If this condition were not initially satisfied, we could create a node where customers are located and/or a potential facility location exists), and $I \cup J \subseteq N$. Define then d_{ij} as the distance or time between the demand area represented by node n_i and the facility site at node n_j. In addition, let the service standard D denote the maximal service distance or time as specified by the decision maker, and define $N_i = \{j \in J: d_{ij} \leq D\}$ as the set of potential facility locations that are able to cover a customer at node n_i. We can then define binary variables y_j which assume a value of 1, if a facility is located at node n_j, and 0 otherwise. We are then able to formulate the *LSCP* as the following integer programming problem.

P_{LCSP}: Min $z = \sum_{j \in J} y_j$

s.t. $\sum_{j \in N_i} y_j \geq 1 \quad \forall i \in I$

$y_j = 0 \vee 1 \quad \forall j \in J,$

or in vector notation we can describe the above model as

Min $z = \mathbf{ey}$, s.t. $\mathbf{Ay} \geq \mathbf{e}, \mathbf{y} \in \{0, 1\}^n$

with appropriately defined "coverage matrix" $\mathbf{A}$.

In the above formulation, the objective function denotes the number of facilities that are located, and the constraints ensure that for each demand point, there exists at least one facility to cover it. The formulation contains exactly one constraint for each demand point and one decision variable for each site. All coefficients are zero or one, including the right hand side, making the formulation a special case of the classical set covering problem in integer programming. Because there sometimes exist many alternate optima to this model, extended models have been developed to help identify specific differences between different but optimal *LSCP* solutions; see, e.g., Daskin and Stern (1981) or Hogan and ReVelle (1986).

It is easy to demonstrate that complete enumeration, while possibly appealing because of its conceptual simplicity, is not a practical solution method. If no single facility site can cover all nodes, then we would need to enumerate all possible two-site patterns to see if any two-site patterns could cover all nodes. If no two-site patterns cover all nodes, then the search would be extended to three-site patterns and continue in this manner until a solution was found that covers all nodes. For a small problem with 100 nodes, we may have to enumerate up to $\binom{100}{1} + \binom{100}{2} + \binom{100}{3} + \binom{100}{4} + \ldots + \binom{100}{100}$ solutions. Since even $\binom{100}{10} = (1.73)10^{13}$, complete enumeration is not feasible for any but the smallest problem. As a result, specialized optimization models must be applied. It is possible to solve an *LSCP* problem in a straightforward manner using an integer-linear programming package. However, it is almost always advisable to employ a special reduction algorithm that often reduces or even eliminates the need to use integer programming. This reduction algorithm is efficient, easily programmed, and has been applied in a number of cities to evaluate the location of fire services; see, e.g., ReVelle *et al.* (1977).

The principles of the reduction technique are easily described. Consider a binary coverage matrix $\mathbf{A} = (a_{ij})$, where $a_{ij} = 1$ indicates that a customer at node n_i can be covered by a facility at node n_j, and $a_{ij} = 0$ specifies that a customer at a node n_i is

Chapter 2: Location Models on Networks

not within covering distance of a potential facility at node n_j. The technique is based on three principal observations:

- *Dominated Row.* A row $\mathbf{a}_{i\bullet}$ dominates a row $\mathbf{a}_{k\bullet}$ ($\mathbf{a}_{i\bullet} \succ \mathbf{a}_{k\bullet}$) if $a_{ij} \leq a_{kj} \ \forall \ j$. A dominated row can be deleted.
- *Dominated Column.* A column $\mathbf{a}_{\bullet j}$ dominates a column $\mathbf{a}_{\bullet \ell}$ ($\mathbf{a}_{\bullet j} \succ \mathbf{a}_{\bullet \ell}$), if $a_{ij} \geq a_{i\ell} \ \forall \ i$. A dominated column can be deleted.
- *Essential Column.* A column $\mathbf{a}_{\bullet \ell}$ is essential, if there exists a row $\mathbf{a}_{i\bullet}$, such that $\mathbf{a}_{i\bullet} = \mathbf{e}_\ell$. A facility must then be located at the node n_ℓ.

The rationale behind the three rules is as follows. If row i dominates row k, then a customer located at n_i can be covered by some, but possibly not all, of those facilities covering customers at n_k. This means that customers at n_i are more difficult to cover than those at n_k. In other words, if a facility can cover the customers at the node n_i, then the same facility will also cover all customers at the node n_k; hence, row $\mathbf{a}_{k\bullet}$ can be deleted. On the other hand, if column j dominates column ℓ, then a facility located at node n_j can cover the same customers as a facility at n_ℓ could, plus possibly some additional customers. Consequently, column $\mathbf{a}_{\bullet \ell}$ can be deleted. Finally, if a row, say $\mathbf{a}_{i\bullet}$, is a unit vector with the "1" in the ℓ-th position, then, in order to cover the customer at n_i at all, a facility must be located at n_ℓ. Doing so leads to the elimination of column $\mathbf{a}_{\bullet \ell}$ as well as all rows $\mathbf{a}_{k\bullet}$ for which $a_{k\ell} = 1$.

Utilizing these three principles, Toregas and ReVelle (1973) developed a special reduction algorithm for the *Location Set Covering Problem.* Note the strong similarity of this procedure and that of iterated dominance discussed in Chapter 3 of Part I. Toregas and ReVelle's algorithm can be described as follows.

LSCP Reduction Algorithm

Step 1: Eliminate all dominated rows.
Step 2: Eliminate all dominated columns.
Step 3: Identify all essential columns, locate facilities at them, and remove all rows covered by essential columns.
Step 4: Repeat the above procedure until either no matrix is left or no removals are made in one complete cycle of steps.

If the reduction algorithm terminates with no matrix remaining, an optimal solution has been identified, the problem is solved and there is no need for a branch and bound method. If the reduction algorithm terminates with a matrix remaining, this remaining matrix is called cyclic and an integer programming method will have to be applied to it. Because the reduction technique can significantly reduce the size of the problem needing to be solved by integer programming, it is almost always used as a first step in solving large location set covering problems.

Example: Consider the network in Figure II.1 and ignore the weights at the nodes. Given a covering distance of $D = 12$, the covering matrix is

	n_1	n_2	n_3	n_4	n_5	n_6	n_7	n_8	n_9	n_{10}	n_{11}	n_{12}
n_1	1	1	0	1	0	0	0	1	1	1	0	1
n_2	1	1	1	1	0	0	0	0	0	1	0	1
n_3	0	1	1	1	1	0	0	0	0	0	0	0
n_4	1	1	1	1	1	0	0	0	0	0	0	0
n_5	0	0	1	1	1	1	0	0	0	0	0	0
n_6	0	0	0	0	1	1	1	1	0	0	0	0
n_7	0	0	0	0	0	1	1	1	0	0	0	0
n_8	1	0	0	0	0	1	1	1	1	1	0	0
n_9	1	0	0	0	0	0	0	1	1	1	0	0
n_{10}	1	1	0	0	0	0	0	1	1	1	1	1
n_{11}	0	0	0	0	0	0	0	0	0	1	1	1
n_{12}	1	1	0	0	0	0	0	0	0	1	1	1

It is apparent that $a_{9\bullet} \succ a_{1\bullet}$, $a_{3\bullet} \succ a_{4\bullet}$, $a_{7\bullet} \succ a_{6\bullet}$, $a_{7\bullet} \succ a_{8\bullet}$, $a_{9\bullet} \succ a_{8\bullet}$, $a_{11\bullet} \succ a_{10\bullet}$, $a_{12\bullet} \succ a_{10\bullet}$, and $a_{11\bullet} \succ a_{12\bullet}$. Consequently, we can delete rows 1, 4, 6, 8, 10 and 12. The reduced matrix is then

	n_1	n_2	n_3	n_4	n_5	n_6	n_7	n_8	n_9	n_{10}	n_{11}	n_{12}
n_2	1	1	1	1	0	0	0	0	0	1	0	1
n_3	0	1	1	1	1	0	0	0	0	0	0	0
n_5	0	0	1	1	1	1	0	0	0	0	0	0
n_7	0	0	0	0	0	1	1	1	0	0	0	0
n_9	1	0	0	0	0	0	0	1	1	1	0	0
n_{11}	0	0	0	0	0	0	0	0	0	1	1	1

The following relations between columns in the reduced matrix are now observed: $a_{\bullet 1} \succ a_{\bullet 9}$; $a_{\bullet 10} \succ a_{\bullet 1}$; $a_{\bullet 3} \succ a_{\bullet 2}$; $a_{\bullet 4} \succ a_{\bullet 2}$; $a_{\bullet 3} = a_{\bullet 4}$; $a_{\bullet 3}, a_{\bullet 4} \succ a_{\bullet 5}$; $a_{\bullet 6} \succ a_{\bullet 7}$; $a_{\bullet 8} \succ a_{\bullet 7}$; $a_{\bullet 8} \succ a_{\bullet 9}$; $a_{\bullet 10} \succ a_{\bullet 9}$; $a_{\bullet 10} \succ a_{\bullet 11}$; $a_{\bullet 10} \succ a_{\bullet 12}$; and $a_{\bullet 12} \succ a_{\bullet 11}$, so that the columns 1, 2, 5, 7, 9, 11, and 12 can be deleted. The reduced matrix is then

	n_3	n_4	n_6	n_8	n_{10}
n_2	1	1	0	0	1
n_3	1	1	0	0	0
n_5	1	1	1	0	0
n_7	0	0	1	1	0
n_9	0	0	0	1	1
n_{11}	0	0	0	0	1

Chapter 2: Location Models on Networks

Here, row 11 is a unit vector with $\mathbf{a}_{11\bullet} = \mathbf{e}_{10}$, so that column 10 is an essential column and a facility must be located at n_{10}. This facility covers customers at the nodes n_2, n_9, and n_{11}, so that column 10 and rows 2, 9, and 11 can be deleted. This results in the following reduced matrix:

	n_3	n_4	n_6	n_8
n_3	1	1	0	0
n_5	1	1	1	0
n_7	0	0	1	1

In this reduced matrix, $\mathbf{a}_{3\bullet} \succ \mathbf{a}_{5\bullet}$ and hence row 5 is dominated and can be deleted. This leaves the reduced matrix

	n_3	n_4	n_6	n_8
n_3	1	1	0	0
n_7	0	0	1	1

At this point, it is apparent that one facility must be located at either n_3 or n_4, and another must be located at n_6 or n_8. Together with the previously located facility at n_{10}, we obtain four alternative optimal solutions $\{n_3, n_6, n_{10}\}$, $\{n_3, n_8, n_{10}\}$, $\{n_4, n_6, n_{10}\}$, and $\{n_4, n_8, n_{10}\}$. In all cases, three facilities are sufficient to cover all customers, given the covering distance $D = 12$.

Notice that the reduction technique was sufficient to solve the problem and no formal integer programming procedure was required. Given the existence of alternative optimal solutions, secondary criteria could be used. For instance, the last two solutions cover a customer at least barely, i.e., they require the full covering distance of $D = 12$. This may cause a decision maker to favor one of the first two solutions.

Other solution approaches include the use of techniques that can solve general set covering problems and specialized solution procedures that have been developed for related location models; see Nemhauser and Wolsey (1988).

2.1.2 The Maximal Covering Location Problem

Suppose now that the location set covering problem resulted in a solution that requires a number of facilities that cannot be afforded by the decision maker, such as a municipality or city. Clearly some compromise must be made. One such compromise would be to relax the distance standard by increasing the maximal service distance D, thus decreasing the number of facilities needed for complete coverage. Actually, if D is large enough, only one facility will be needed for complete coverage. By varying the service distance and solving a series of *LSCP* problems, one can generate a tradeoff curve that is negatively sloped in a diagram whose axes represent the covering distance D and the number of facilities required to cover all customers. Another possible compromise is to relax the coverage standard and allow less than complete coverage to occur. In this case we may focus on how much coverage we can provide for a given level of resources. This is the basis of the Maximal Covering Location Problem (*MCLP*). Simply stated, the *MCLP* seeks the maximum amount of coverage (in terms of population, property values, or similar parameters), given a covering distance and a specific number of facilities that can be used. To formulate this model we need some additional variables and parameters.

For that purpose, let p denote the number of facilities that are to be located. Moreover, define additional zero-one variables x_i that assume a value of one, if a customer at node n_i is covered by at least one facility, and zero otherwise. With demands (or populations, or total property values) of w_i at node n_i, Church and ReVelle (1974) have formulated what is now known as the *Maximal Covering Location Problem* (*MCLP*).

$$P_{MCLP}: \text{Max } z = \sum_{i \in I} w_i x_i$$

$$\text{s.t.} \quad x_i \leq \sum_{j \in N_i} y_j \ \forall \ i \in I$$

$$\sum_{j \in J} y_j = p$$

$$y_j = 0 \vee 1 \ \forall j \in J$$

$$x_i = 0 \vee 1 \ \forall i \in I$$

or, in vector notation,

$$\text{Max } z = \mathbf{wx}, \text{ s.t. } \mathbf{Ay} - \mathbf{ex} \geq \mathbf{0}, \mathbf{y} \in \{0, 1\}^n, \mathbf{x} \in \{0, 1\}^m.$$

This formulation maximizes the demand covered by the location of p facilities. The first set of constraints allows the decision variable x_i to be equal to one only if at least one facility has been located among the set of sites that can provide coverage to node n_i. If no sites are used among the set N_i, then x_i is forced to zero, indicating that node n_i is not covered. The remaining structural constraint establishes that exactly p facilities must be located.

An equivalent formulation is obtained by replacing the variables x_i by their complements $\bar{x}_i = 1 - x_i$, so that $\bar{x}_i = 1$ indicates that the customer at node n_i is not covered. This results in the alternative formulation

$$P'_{MCLP}: \text{Min } \bar{z} = \sum_{i \in I} w_i \bar{x}_i$$

$$\text{s.t.} \sum_{j \in N_i} y_j + \bar{x}_i \geq 1 \ \forall \ i \in I$$

$$\sum_{j \in J} y_j = p$$

$$y_j = 0 \vee 1 \ \forall j \in J$$

$$\bar{x}_i = 0 \vee 1 \ \forall i \in I,$$

or, in vector notation,

$$\text{Min } \bar{z} = \mathbf{w}\,\bar{\mathbf{x}}, \text{ s.t. } \mathbf{Ay} + \mathbf{e}\bar{\mathbf{x}} \geq \mathbf{1}, \mathbf{y} \in \{0; 1\}^n, \bar{\mathbf{x}} \in \{0; 1\}^m.$$

This alternate formulation involves minimizing the amount of population not covered, given the placement of p facilities. Both formulations include $2n$ variables and $(n + 1)$ constraints and are thus fairly compact. Again, dominated columns can be eliminated as a first step in solving this problem. This alternative formulation has proved itself superior in computational experiments. Even though general integer linear programming techniques are most commonly used to solve an *MCLP*, other solution procedures, both heuristic and optimal procedures, have been developed by authors such as Weaver and Church (1983).

This model has been applied in a number of different settings. Eaton *et al.* (1981) and Eaton *et al.* (1985) report an application of *MCLP* regarding the location of ambulances in Austin, Texas, and the location of health centers in Colombia.

We will return to covering problems in Section 2.3.1 of this part, where we demonstrate that the maximal covering problem is a special case of another important class of location models, called *p*-median problems.

We would like to conclude this section with a number of extensions of the basic covering models discussed above. One possibility is to attempt to cover at least a certain proportion of the population within a prespecified distance or time. A typical pertinent standard requires that 90% of the population in an area be accessible within 8 minutes of the facility. An important issue is potential congestion within the service, i.e., what happens if a service call arrives while the unit is presently busy? A possible way to deal with such situations is the introduction of backup coverage. The obvious question then concerns the tradeoff between primary coverage for some people versus backup coverage for others. Other extensions are the probabilistic coverage models which include maximum expected coverage as studied by Daskin (1983), so-called *alpha-reliability coverage models* of ReVelle and Hogan (1989) and ReVelle and Marianov (1991), and the reliability coverage model of Ball and Lin (1993). Such models are designed to address the chances that a service facility may be busy and not available. In the above example of a station with an ambulance, the computation of coverage requires the estimation of the availability of each unit and the added level of coverage provided by an additional unit located at any specific location. Such location models rely on an endogenous estimation approach so that a model can optimally locate a set of facilities which can maximize the estimated level of coverage.

The second type of extended coverage model is the *weighted benefit maximal coverage model* of Church and Roberts (1983). This model involves discretizing the distance from a facility into a set of service ranges and assigning a relative benefit of providing service within that distance range. This makes it possible to account for the service response provided to those that are beyond the desired maximal service distance, as well as account for the differences that might be measured for having desirable service within 5 minutes instead of 8 minutes. Most coverage models utilize only one implied weighted benefit and one coverage distance. The weighted benefit maximal coverage model uses a graduated set of service standards and benefits.

The third type of extended coverage model that is important to mention is the class of *capacitated coverage models*. Such models were originally defined as a family of problems by Current and Storbeck (1988). Many of the models in this class can be shown to be special cases of capacitated warehouse location problems. Such problems have presented a challenge to those interested in solving location problems optimally.

2.2 Center Problems

Another classical problem type comprises the so-called *center problems*. The basic idea of all center problems is that they locate facilities so as to minimize the longest distance between a customer and his closest facility. The underlying logic of center problems is based on Rawls's "Theory of Justice" (1971), according to which the quality of a solution depends on the least well-served entity. One of the problems associated with the concept of centers is their exclusive focus on the customer with the longest facility – customer distance. As in the Wald criterion in games against nature, this can lead to highly undesirable situations. Harsanyi (1975) provides a thoughtful response to Rawls's theory, in favor of minisum problems.

The 1-*node center* of a graph is defined as a node that minimizes the longest distance between itself and any other node. A 1-*general center* is any node whose furthest point on the graph (a node or a point on an edge) is as close as possible. Whereas the *node center* is associated with keeping the furthest node-to-node distance as small as possible, the *general center* is associated with minimizing the distance between a node and the point furthest from it, which is likely to be on an arc. The *1-absolute center* is a point on a network (a node or a point on an edge) whose furthest node is as close as possible and, finally, the *1-general absolute center* is defined as a point on the graph whose most distant point is the closest possible.

The concept of centers applies to multi-facility frameworks as well. For example, the *p*-absolute center problem involves the location of *p* facilities on a graph or network so as to minimize the furthest distance that separates any node and its nearest facility. While all center problems come in weighted and unweighted versions (where in the former a weighted maximal distance is to be minimized), we will consider only unweighted problems in this book. Here, the relation between covering models and center problems becomes apparent: whereas the location set covering problem attempts to locate the least number of facilities that can satisfy coverage within a given distance standard, center problems minimize the distance standard by determining the placement of a fixed number of facilities. Special algorithms have been developed for center problems; see, for example, Handler (1974), Minieka (1977 and 1981), and Daskin (1995). For simplicity we will assume throughout this section that the underlying graph is undirected.

2.2.1 1-Center Problems

The simplest center problem is the 1-node center problem. Letting $I = J$, delineating d_{ij} again as the shortest distance between nodes n_i and n_j, and defining variables $y_j = 1$, if the single facility is to be located at the node n_j, and 0 otherwise, the problem can be formulated as

$$P: \text{Min } z = \max_{i \in I} \{d_{ij} y_j\}$$

$$\text{s.t.} \sum_{j \in J} y_j = 1$$

$$y_j = 0 \vee 1 \; \forall \, j \in J.$$

What makes this problem so simple is the restriction of potential facility location to nodes, thus making complete enumeration possible. In particular, if $\mathbf{D} = (d_{ij})$ is the matrix of shortest distances, we can compute $d_i = \max_j \{d_{ij}\} \, \forall \, i$, and then the node n_k with $k = \arg\min_i \{d_i\}$ is the 1-node center of the given graph and d_k is the longest distance from n_k to any other node in the network. As an example, consider the network in Figure II.2 where the numbers next to the edges indicate the distances between adjacent nodes along the undirected arcs.

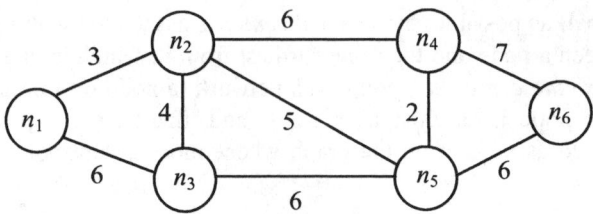

Figure II.2

The matrix of shortest paths is

$$D = \begin{bmatrix} 0 & 3 & 6 & 9 & 8 & 14 \\ 3 & 0 & 4 & 6 & 5 & 11 \\ 6 & 4 & 0 & 8 & 6 & 12 \\ 9 & 6 & 8 & 0 & 2 & 7 \\ 8 & 5 & 6 & 2 & 0 & 6 \\ 14 & 11 & 12 & 7 & 6 & 0 \end{bmatrix}.$$

The column maxima are then $(d_i) = [14, 11, 12, 9, 8, 14]^T$ with the unique minimum of 8 at node n_5. Hence, n_5 is the 1-node center and the longest distance between n_5 and any other node (here n_1) is 8. The complexity of this brute force search method is $O(n^2)$, given that the matrix of shortest path distances is known; otherwise, it increases to $O(n^3)$.

Relax now the restriction that the facility must be located at a node of the network. For the resulting 1-absolute center problem, it is readily apparent that an optimal location is not necessarily located at a node. As a primitive example, consider a graph that consists of two nodes connected by a single edge, where the 1-absolute center is located at the center of the edge halfway between the two nodes.

A simple extension of the result on a single edge is applied to the problem of finding 1-absolute centers in tree networks. Such problems can easily be solved by finding the longest path in the tree; the 1-absolute center is then located at the center of that path. This can be accomplished by the following simple algorithm.

1-Absolute Center on Tree Networks

Step 1: Starting with any node n_i, determine the node farthest from it. Denote this node by n_1.

Step 2: Starting with n_1, determine the node farthest from it. Denote this node by n_2. The 1-absolute center is then located halfway between n_1 and n_2 on the unique path connecting these two nodes.

The above algorithm can be implemented very efficiently by using a labeling technique similar to that in Dijkstra's method for the determination of shortest paths (see, e.g., Eiselt and Sandblom (2000)). Given appropriate data structures, this task can be accomplished in $O(n)$ time.

Example: Consider the network shown in Figure II.3 where the numbers next to the edges denote their lengths.

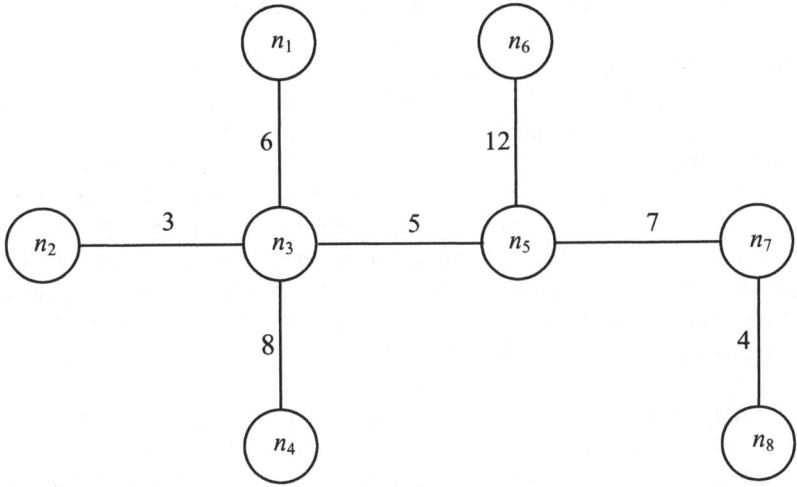

Figure II.3

Starting arbitrarily at the node n_5 with the label "0", we can then label n_3 with 5, n_7 with 7, and n_6 with 12. Continued labeling from n_7 results in a label of 11 at n_8. From n_3, we then label n_1 with 11, n_2 with 8, and n_4 with 13. Now, all nodes have received a label, and the node with the highest label, here n_4, is the node most distant from n_5. Now all labels are erased and the process starts anew with n_4. At its end, we find that the node n_6 has the highest label, 25. As a result, the 1-absolute center is located at a distance of 12½ from n_4 towards n_6, i.e., on the edge (n_3, n_5) at a distance of 4½ from n_3.

An extension of the above procedure was proved to yield 2-absolute centers on tree networks. For that purpose, the following simple extension of the above method can be employed.

2-Absolute Centers on Tree Networks

Step 1: Determine the 1-absolute center on the tree network.

Step 2: Is the 1-absolute center located at a node?
 If yes: Delete either of the two edges on the longest path that are incident to the node at which the 1-absolute center is located, and go to Step 3.
 If no: Delete the edge that includes the 1-absolute center and go to Step 3.

Step 3: Determine 1-absolute centers in the resulting two tree networks. They are the 2-absolute centers of the original network.

Example: Consider again the graph in Figure II.3. As the 1-absolute center is located on the edge (n_3, n_5), this edge is deleted. The resulting two trees are $T_1 = (N_1, E_1)$ and $T_2 = (N_2, E_2)$ with $N_1 = \{n_1, n_2, n_3, n_4\}$, and $N_2 = \{n_5, n_6, n_7, n_8\}$. The 1-absolute center in T_1 is the point on the edge (n_3, n_4) one unit away from n_3, while its counterpart in T_2 is the point on (n_5, n_6) that is ½ unit away from n_5. These two points are the 2-absolute centers on the original tree. The objective value is max $\{7; 11½\} = 11½$, revealing that the establishment of a second center has resulted in a mere reduction of the objective value from 12½ to 11½, important information to the decision maker who will have to balance this improvement with the costs of the second facility.

Unfortunately, this procedure neither generalizes to p-absolute centers on trees with $p > 2$, nor to 1-absolute centers on general graphs. As far as graphs are concerned, consider some node n_k as well as an (undirected) edge that connects two nodes n_i and n_j with $k \neq i, j$. We will first determine the distance between n_k and all points on the edge (n_i, n_j), and then repeat the process for all nodes n_k, $k \neq i, j$. The maximum of all functions, the "upper envelope," gives the longest distances between all points on the edge (n_i, n_j), and its lowest point will be the best location of a facility on this edge. The location of this point and its objective value are stored, and the procedure is repeated for all edges in the network. The point with the lowest objective value will then be the 1-absolute center of the network. To illustrate the basic idea, consider the edge (n_i, n_j) and some node n_k as shown in Figure II.4.

Without loss of generality, assume that $d_{ki} \leq d_{kj}$. We can now determine the distances between the node n_k and all points on the edge (n_i, n_j)

- via n_i, and
- via n_j.

Chapter 2: Location Models on Networks

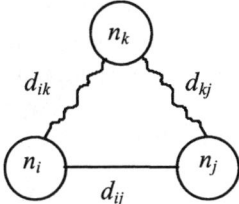

Figure II.4

The distance functions at some point x on the edge (n_i, n_j) are then

- $d_{ki} + x$ via n_i, and
- $d_{kj} + (d_{ij} - x)$ via n_j.

Note that the slopes of the two functions are +1 and –1, respectively. We can now distinguish between two cases.

Case 1: $d_{kj} > d_{ki} + d_{ij}$. The two distance functions in this case are shown in Figure II.5a.

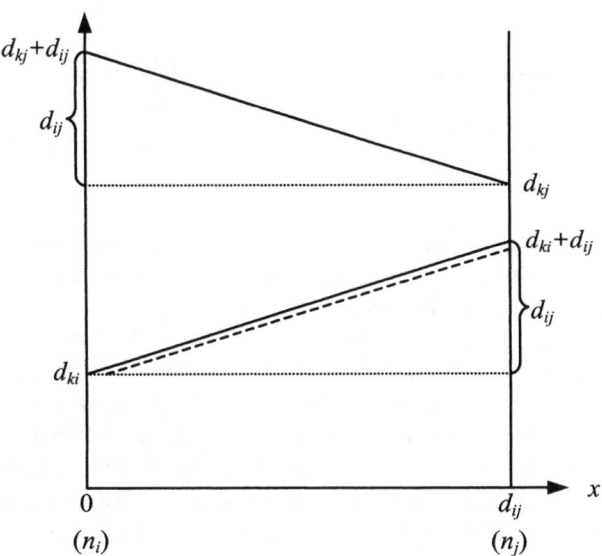

Figure II.5 a

The broken line is the lower envelope of the two functions. This linear function marks the shortest distance between the node n_k and all points on the edge (n_i, n_j). Here, the shortest paths from n_k to all points on that edge all pass by n_i.

Case 2: $d_{kj} \leq d_{ki} + d_{ij}$. The two distance functions are shown in Figure II.5b.

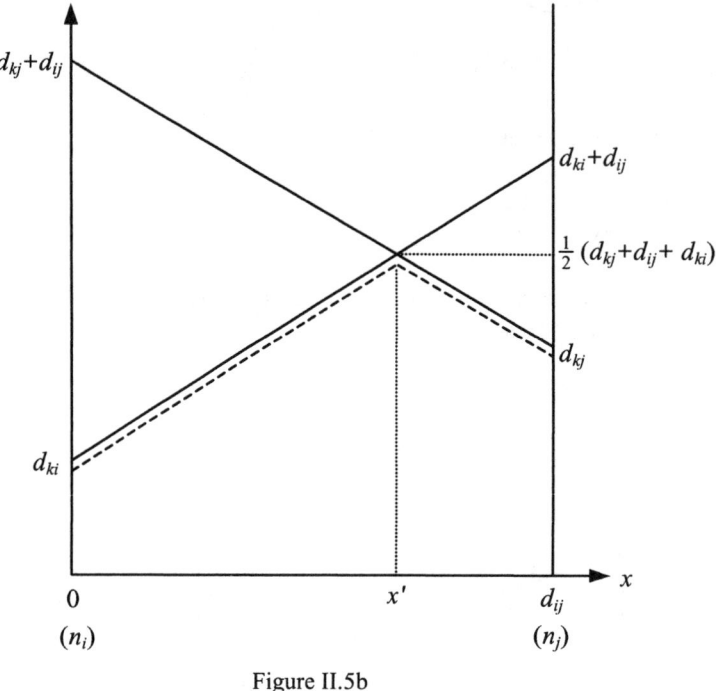

Figure II.5b

Again, the broken line shows the lower envelope of these two distance functions. The function is piecewise linear and concave with a maximum at $x' = \frac{1}{2}(d_{kj} + d_{ij} - d_{ki})$ at which point the distance from n_k is $\frac{1}{2}(d_{kj} + d_{ij} + d_{ki})$.

Lower envelopes, such as the ones shown in Figures II.5a and II.5b, can now be constructed for the edge (n_i, n_j) and all nodes $n_k \in N$. As the center objective is concerned with the most distant customer, the relevant function is then the maximum of all individual distances, i.e., the upper envelope. Unfortunately, the resulting function is typically neither convex nor concave. The objective is now to find the minimum, i.e., the lowest point on this upper envelope. Once this is accomplished, the process is repeated for all edges and, by comparison, the solution with the smallest objective value is chosen as optimum. In summary, the procedure determines

$$\underset{(n_i,n_j)\in E}{\text{Min}} \; \underset{x\in(n_i,n_j)}{\min} \; \underset{n_k}{\max} \; \min\{d_{ki}+x); (d_{kj}+d_{ij}-x)\},$$

or, in other words, the smallest objective value of the lowest points on any upper envelope of lower envelopes.

Example: Consider the network in Figure II.6.

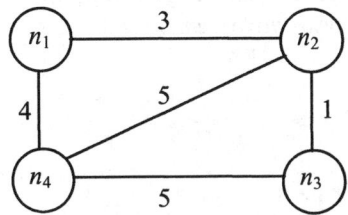

Figure II.6

We will illustrate the process by considering just a single edge. The solid lines in Figure II.7 display the distances between all four individual nodes to points on the edge (n_2, n_4), and the broken line shows the upper envelope.

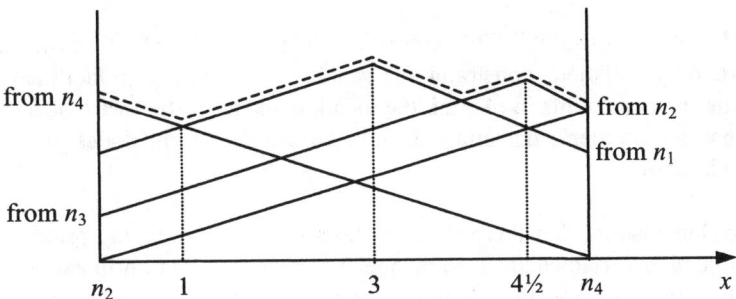

Figure II.7

The upper envelope is a piecewise linear function connecting the points (0, 5), (1, 4), (3, 6), (4, 5), (4½, 5½), and (5, 5). The lowest point on the upper envelope is reached at $x = 1$ distance units from n_2, from where the distance to the farthest two nodes n_1 and n_4 is 4. This point is the optimal location on the edge (n_2, n_4). Repeating this process for all other edges, we obtain objective values of 4 on (n_1, n_2) at n_1, 4 on (n_1, n_4) at n_1, 5½ at $x = ½$ on (n_2, n_3), and 4½ at $x = 4½$ on (n_4, n_3). This leaves n_1 and $x = 1$ on (n_2, n_4) as 1-absolute centers of the network; in both cases the farthest node is 4 distance units away.

2.2.2 *p*-Center Problems

This subsection considers *p*-center problems. For reasons of space, we will restrict ourselves to node *p*-centers. In order to formulate the problem, we need two types of variables: the usual *locational variables* y_j which assume a value of 1, if a facility is located at the node n_i, and 0 otherwise, and *allocation variables* x_{ij} which denote the proportion of the total demand of the customer at node n_i that is satisfied by a facility at node n_j. Let $J \subseteq N$ denote the set of nodes at which a facility could be located. The *p-node center* problem can then be formulated as follows.

$$P: \text{Min } z = \max_i \sum_{j \in J} d_{ij} x_{ij} \tag{1}$$

$$\text{s.t. } \sum_{j \in J} x_{ij} = 1 \quad \forall \, i \in I \tag{2}$$

$$\sum_{j \in J} y_j = p \tag{3}$$

$$x_{ij} \leq y_j \quad \forall \, i \in I; j \in J \tag{4}$$

$$x_{ij} \geq 0 \quad \forall \, i \in I; j \in J \tag{5}$$

$$y_j = 0 \vee 1 \quad \forall \, j \in J. \tag{6}$$

First consider the constraints. The constraints (2) ensure that all customers' demands are fully satisfied, constraint (3) requires that exactly *p* facilities are located, while the constraints in (4) link the location and allocation variables so as to ensure that a customer's demand can only be satisfied from nodes at which facilities are located.

It is easy to demonstrate that there exists at least one solution to the problem in which all allocation variables are zero or one. The reason is that no linear convex combination of the distances to a customer—the sum in (1)—can be smaller than the smallest facility-customer distance. Consequently, the sum in (1) will be the distance between a customer and its closest facility; the longest of all these distances is used as a criterion to be minimized.

The *p*-node center problem was proved to be **NP**-hard by Kariv and Hakimi (1979a). However, there is a reasonably efficient solution technique that is based on binary search. The basic idea is to solve the problem by a sequence of covering problems. Starting with a covering distance *D*—an initial guess—the number of facilities required to cover all customers is determined as *p(D)* by any of the pertinent methods. If $p(D) > p$, then *D* has to be revised upward; if $p(D) \leq p$, then *D* can be decreased. This process continues until a covering distance is found that cannot be reduced any further without requiring additional facilities.

Chapter 2: Location Models on Networks

The process commences with an interval $[\underline{D}; \overline{D}]$ that includes the optimal solution. $\underline{D} = 0$ is an obvious choice, while $\overline{D} = (n-1)\max_{e_{ij} \in E}\{d_{ij}\}$ is an upper bound for the length of the longest path in the network. (By definition, no edge can be longer than the largest length of any edge in the network, and the longest path cannot include more than $(n-1)$ edges). Given a covering distance $\overline{D}$, one facility is sufficient to cover all customers. The algorithm can then be described as follows.

Bisection Search for Node p–Center Problems

Step 1: Determine $D := \lfloor 0.5(\overline{D} + \underline{D})\rfloor$.

Step 2: Solve a set covering problem with the covering distance D. Let $p(D)$ denote the number of facilities required to cover all nodes.

Step 3: Is $p(D) \leq p$?
 If yes: Set $\overline{D} := D$, and go to Step 4.
 If no: Set $\underline{D} := D + 1$, and go to Step 4.

Step 4: Is $\overline{D} = \underline{D}$?
 If yes: Stop, $D = \overline{D} = \underline{D}$ is the optimal objective value of the p-node center problem; the last covering problem provides the locational pattern of the facilities.
 If no: Go to Step 1.

Example: Consider the graph in Figure II.8 and assume that $p = 3$ facilities are to be located.

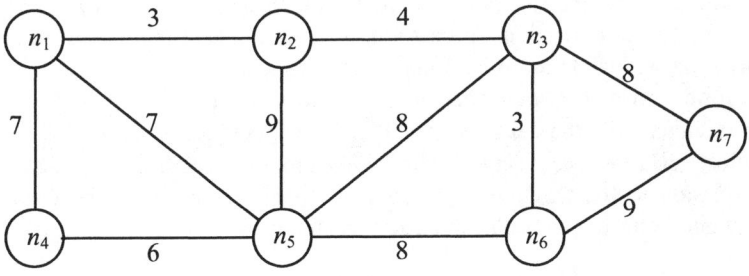

Figure II.8

The algorithm commences with $\underline{D} = 0$ and $\overline{D} = 9(6) = 54$. The first guess is $D = 27$, for which the set covering problem has a solution of $p(D) = p(27) = 1$, i.e., a single facility, located anywhere, can cover all nodes. As $p(D) = 1 < 3 = p$, we set $\overline{D} := 27$.

The results of the subsequent iterations are summarized in Table II.1.

Table II.1

Iteration #	$\underline{D}$	$\overline{D}$	D	$p(D)$	Location
1	0	54	27	1	anywhere
2	0	27	13	1	n_2
3	0	13	6	4	n_7, one at n_4 or n_5
4	7	13	10	2	n_1 and n_3, or n_4 and n_6
5	7	10	8	2	n_1 and n_3
6	7	8	7	3	n_1 and n_7, and one of n_3 and n_6
7	7	7			Stop, optimal.

2.3 Median Problems

As opposed to the minimax problems in the previous section, the common characteristic of all median problems is their minisum objective. As already elaborated upon in Chapter 1, most applications of median problems are found in the private sector. Typically, we can think of the facilities as warehouses from which deliveries are to be made to customers, in which case the warehouse planner's objective is to minimize the total cost of the deliveries. Provided that each delivery requires an individual trip, this cost is the sum of individual delivery costs which, in turn, are expressed as the delivery quantity multiplied by the customer-facility distance or cost. If deliveries are made to more than one customer on a tour, the problem becomes much more difficult, as it now incorporates a location problem and a routing (traveling salesman-style) problem. But even if individual trips to customers are realistic in a given scenario, the assumption of a linear transport cost function implies that each unit of demand is shipped individually. If the delivery vehicle has a capacity of, say κ units, then the cost function will resemble a step function with jumps at 0, κ, 2κ, ... If κ is quite small in relation to the total demand, then a linear function could be used as a reasonably good approximation of the true cost function.

However, the use of minisum objectives (and median problems with them) is not confined to the private sector. Such problems may equally well be applied to the location of public facilities, if it is agreed that aggregated distances are a reasonable measure of the service level. In other words, if a facility is 3 miles away from an area with 100 people, and 5 miles away from an area with 200 users, then the aggregated distance is 3(100) + 5(200) = 1300 miles and it is this distance that provides the yardstick for the evaluation of the solution.

Notice the difference between the minisum and the minimax objectives: on a single edge with one node representing a large agglomeration of demand and the other representing a very small demand (e.g., a town and a "lone guy with his dog in the woods"), the center objective would pull the facility to a location halfway between the two, giving a strong consideration for the single individual. The minisum objective, on the other hand, will locate in town, maximizing the benefit of the majority and leaving the single individual with very poor service. Hybrid models of center and median objectives have been called *centdian* problems (*cent*er + me*dian*). They were first analyzed by Halpern (1976). It is also worth pointing out that as long as the total demand is constant, the minimization of the sum of customer-facility distances is equivalent to minimizing the average customer-facility distance; see, e.g., ReVelle and Swain (1970).

2.3.1 Basic Results and Formulation of the Problem

To stimulate the discussion, consider first the 1-median problem on a general graph where the facility is permitted to locate at any point on the graph. Suppose that the facility is located at some point x on an edge (n_i, n_j), and consider the shortest distance between the facility at x and a customer at some node n_k. The distance function for all points x on (n_i, n_j) is then identical to that in Figures II.5a and II.5b, i.e., the function is either linear with a slope of +1 or −1, or piecewise linear with a single break point, to the left of which the slope is +1, and to the right of which the slope is −1. This property holds for the distance functions to all customers n_k, $k = 1, ..., n$. Regardless of their individual shape, all functions are concave. The minisum objective then adds these functions and, based on the well-known result that the sum of concave (convex) functions is again a concave (convex) function, the weighted or unweighted sum of shortest distances to any point x on the edge (n_i, n_j) is concave. The fact that the minimum of a concave function $f(x)$ on the domain $x \in X$ (here, the domain is the edge (n_i, n_j)) is attained on the boundary of the domain, immediately leads to

Theorem II.1 (Hakimi, 1964): At least one optimal solution of a p-median problem is located at a node.

This theorem implies that, as opposed to center problems where it was necessary to distinguish between node centers and absolute centers, such distinction is not required for median problems. If the decision maker's goal is to find any (of the) optional solution(s), then it is sufficient to consider only node solutions. The consequences of this "node property" or "Hakimi property" are tremendous: rather than having to search an infinite set of points on a graph G, it is sufficient to examine the finite set of node locations. In that sense, Hakimi has done for median problems what Dantzig did for linear programming with his "corner point theorem," that is, limit the search to a finite, albeit astronomically large, set. Hakimi's theorem does not, of course, provide a simple solution. Complete enumeration for a problem with p facilities on a graph with n nodes would require the examination of $\binom{n}{p}$ solutions. For example, locating 5 facilities on a graph with 1,000 nodes has no less than $(8.25)10^{12}$ node solutions.

Using Hakimi's result, we can now follow ReVelle and Swain (1970) and formulate the *p-median problem*. In addition to the usual node weights (demands) w_i of a customer at a node n_i and the shortest distances d_{ij}, we need to define locational zero-one variables y_j that equal one, if a facility is located at node n_j, and zero otherwise. In addition, we define variables x_{ij} as the proportion of customer i's demand that is satisfied by a facility at node n_j. As usual, let I denote the set of nodes with positive customer demand, and let J denote the set of nodes that are potential facility locations. The *p*-median problem can then be formulated as follows.

$$P: \text{Min } z = \sum_{i \in I} \sum_{j \in J} w_i d_{ij} x_{ij}$$

$$\text{s.t. } \sum_{j \in J} x_{ij} = 1 \; \forall \; i \in I \tag{7}$$

$$x_{ij} \leq y_j \; \forall \; i \in I, j \in J \tag{8}$$

$$\sum_{j \in J} y_j = p \tag{9}$$

$$x_{ij} \geq 0 \; \forall \; i \in I, j \in J; \; y_i = 0 \vee 1 \; \forall \; j \in J.$$

The constraints (7) ensure that each customer is allocated to exactly one facility, the constraints (8) guarantee that service from a facility at a node n_j is provided only if there exists a facility at that node, and constraint (9) requires that exactly p facilities are located.

It is easy to show that in this formulation, all variables x_{ij} will be zero or one, even if we do not require them to be. The underlying reason for this behavior is that the minimization objective function will assign each customer to his nearest facility. (In case more than one facility is closest to a customer, alternative optimal solutions exist, some of which are non zero-one). Note that this argument no longer holds once the facilities become capacitated.

The above formulation contains n^2 variables and n^2+1 constraints. Consequently, even a modestly sized network of 100 nodes would have 10,000 variables and

Chapter 2: Location Models on Networks

10,001 constraints. While the number of facilities to be located does not have an influence on the number of variables and constraints of the problem, it greatly influences the difficulty of the problem.

It is, however, possible to write the constraints (8) in a more compact form as

$$\sum_i x_{ij} \leq n y_j \quad \forall \ j \tag{8a}$$

This formulation requires the variable y_j to assume a value of one, if a shipment of any size is to be directed to facility j.

While Kariv and Hakimi (1979b) proved that the general p-median problem on networks is **NP**-hard, they also described some interesting special cases for which polynomial algorithms exist. In particular, they demonstrated that the p-median problem on tree networks can be solved in $O(n^2 p^2)$ time, whereas the 1-median problem on trees is solvable in linear time.

Finally, we wish to point out a relation between the seemingly very different covering problems and p-medians. Church and ReVelle (1976) and Hillsman (1984) have derived these relationships by demonstrating that both the location set covering problem and maximal covering location problem of Sections 2.1.1 and 2.1.2 can be cast into the mold of the p-median model. For the maximal covering location problem, define modified distances d'_{ij}, which are defined as $d'_{ij} = 1$ if $d_{ij} > D$ and 0 otherwise. Using d'_{ij} instead of d_{ij} in the p-median model changes the cost or distance of assignment to 0 if the closest facility is within covering distance and w_i if the closest facility is further than D distance units from n_i. This transforms the objective to a function that minimizes the demand or amount of population that is not assigned to a facility within D distance by the location of the p facilities. This is an equivalent statement of the *MCLP*. Thus, with this simple modification, the *MCLP* is a special case of the p-median model.

For the location set covering problem, we define distances d'_{ij} that are $M >> 0$, if $d_{ij} > D$, zero, if $d_{ij} \leq D$, and one, if $i = j$. Furthermore, we choose a sufficiently large value of p, so that a sufficient number of facilities are available to cover all demand. In addition to the existing potential facility locations, we introduce p pseudo sites to the problem where the cost of self assignment for the pseudo sites is zero, i.e., $d'_{jj} = 0$ for $j = n+1, \ldots, n+p$, and where the pseudo sites cannot serve any demand; i.e., $d'_{ij} = M \ \forall \ i = 1, \ldots, n; j = n+1, \ldots, n+p$. All potential facility locations, pseudo and non-pseudo sites, have weights $w_i = 1 \ \forall \ i \in I$.

Adding constraints $x_{jj} = y_j$ to the standard p-median formulation does not change the solution, as the requirement of a site to serve its own customer is costless due to the zero distance between a facility and the customer at the same site. With these new constraints, the pseudo sites, and the redefined distances, the optimizer will now act as follows. First, the minimization objective in the *p*-median model will ensure that only assignments with $d_{ij} \leq D$ are made, as all other assignments are extremely costly. On the other hand, each facility will carry costs of 1, if it is at a regular node, and zero if it is at a pseudo node. The reason for this is that self-assignments are enforced by the added constraints. As facilities located at pseudo nodes carry zero costs, the objective will minimize the number of facilities to be located, which is the objective of the location set covering model.

In the remainder of this section, we first discuss the easy special case in which $p = 1$ in Section 2.3.2, followed by exact and heuristic solution techniques for general *p*-median problems as described in Section 2.3.3.

2.3.2 1-Median Problems

This section discusses the special case in which only a single facility must be located. Due to Hakimi's theorem we know that at least one optimal solution can be found at a node, so that complete enumeration of the objective values at all *n* nodes is indeed feasible. An illustration of the procedure is provided in the following

Example: Consider the graph in Figure II.9, where the single-digit numbers next to the edges denote the direct distances between adjacent nodes, and the double-digit numbers next to the nodes are the demands at the nodes.

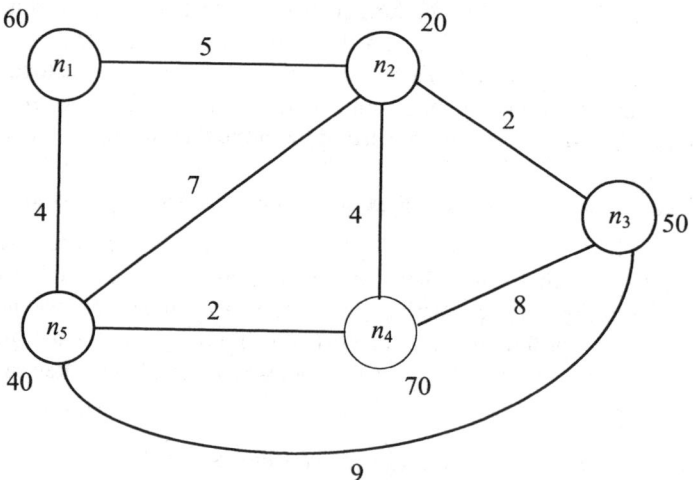

Figure II.9

Chapter 2: Location Models on Networks

The matrix of shortest path distances is then

D:

	n_1	n_2	n_3	n_4	n_5
n_1	0	5	7	6	4
n_2	5	0	2	4	6
n_3	7	2	0	6	8
n_4	6	4	6	0	2
n_5	4	6	8	2	0

Given customer demands of $\mathbf{w} = [60, 20, 50, 70, 40]$, the vector $\mathbf{wD} = [1,030, 920, 1,200, 820, 900]$. In its j-th component, the value of the vector $\mathbf{wD}$ is $\sum_i w_i d_{ij}$, which is nothing but the total transportation cost, given that a facility is located at n_j. Elementary comparison and choice of the location with the lowest cost will then result in an optimal location. Here, the optimum is located at n_4 with an objective value of $z = 820$. As far as computational complexity goes, the determination of the matrix of shortest distances is $O(n^3)$ (e.g., by the Floyd-Warshall algorithm, see, e.g., the Procedure c6 in the Introduction or Eiselt and Sandblom, 2000) and the matrix multiplication is accomplished in $O(n^2)$ time, whereas the determination of the minimum can be done in linear time; hence the complexity is $O(n^3)$ if $\mathbf{D}$ is not known, and $O(n^2)$ if it is.

The problem is even simpler if we were to locate a single facility on a tree network T. This problem was independently solved by Hua Lo-Keng and others (1962) and Goldman (1971). The main difference between general graphs and tree networks is that in the latter, there exists exactly one path between any two points on the tree. This is the major simplifying feature inherent in trees. It is also the reason why in tree networks, the distances between nodes are irrelevant for the determination of the optimal facility location. In particular, Zelinka (1968) already proved that the 1-median on a tree is a node whose removal subdivides the tree into subtrees such that the weight of the largest subtree does not exceed half the weight of the entire tree. Here, the weight of a (sub-) tree is defined as the sum of the (sub-) tree's nodes' weights. Notice the similarity of this result and that for the 1-median on a line.

Goldman's algorithm is implemented in the form of a *folding algorithm*. This means that initially, a pendent node, i.e., a node with degree "1," is examined as to whether or not it satisfies the condition for a median. If so, a solution is found; otherwise, the node is folded onto its unique neighbor and its weight is added to that of the neighbor. Folding a pendent node onto its neighbor is accomplished by adding the weight of the pendent node to its neighbor, and deleting the pendent node and the edge incident to it. This process is repeated at most n times until a solution is found. Given appropriate data structures, Goldman's algorithm can be implemented in linear time.

Example: As an illustration of the optimality condition discussed above, consider the tree network in Figure II.10 with the weights (demands) being indicated next to the nodes.

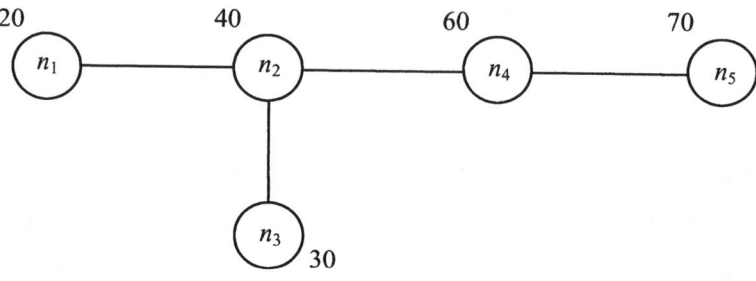

Figure II.10

The total weight of the tree equals $20 + 40 + 30 + 60 + 70 = 220$. Consider n_2 as a potential median. Its removal subdivides the tree network into subtrees with sets of nodes $\{n_1\}$, $\{n_3\}$, and $\{n_4, n_5\}$ whose weights are 20, 30, and 130, respectively. As the weight of the last subtree exceeds $220/2 = 110$, the node n_2 is not a 1-median. Considering n_4 instead, the subtrees are $\{n_1, n_2, n_3\}$ and $\{n_5\}$, whose weights are 90 and 70, respectively, thus satisfying the condition for a median.

2.3.3 *p*-Median Problems

The *p*-median problems investigated in this section are probably among the most researched problems in all of location theory. We will first describe an exact method that uses Lagrangean relaxation. Clearly, as the problem is **NP**-hard, there is a limit on the size of problem that can be solved to optimality. This is followed by some of the best-known heuristic algorithms that can either be used by themselves or as subalgorithms to establish lower bounds.

There has been considerable interest in the development of procedures to solve specific network location models. The techniques that are commonly used include linear programming with branch and bound, special relaxation techniques based on the application of Lagrange multipliers and heuristics based on substitution/movement strategies, meta-heuristics like tabu search, genetic algorithms, simulated annealing, and heuristic concentration. Several of the network location-allocation models have been formulated within the context of integer-linear programming problems. This means that one possible solution approach is the application of general purpose linear programming packages which feature a branch and bound algorithm. After a relaxed linear program is solved, the branch and bound routines can be used to resolve any fractional variables. In fact, all of the models introduced in this chapter have been solved in this way. Unfortunately, there are practical limits as to the size of the problem that can be approached with linear programming. As an example, the p-median problem requires n^2 variables and $O(n)$ constraints where n is the number of nodes on the network. Even though special techniques have been developed to generate a frugal number of constraints for a p-median problem implementation (see Rosing et al. (1979)), still only modestly sized p-median problems with less than 150 nodes have been solved with this straightforward approach.

On the other hand, linear programming with branch and bound is a very competitive approach to solving a number of the covering problems that have been developed, since the formulations are typically much smaller for networks of similar sizes. There are some types of covering problems, however, that cannot be easily solved in this fashion. Suffice it to say that integer linear programming has remained a standard technique to solve a number of network location problems as long as the resulting formulation is moderate in size.

For larger problems, specialized Lagrangean procedures have proven to be particularly robust; see Erlenkotter (1978), Christofides and Beasley (1982), Hanjoul and Peeters (1985), and Weaver and Church (1985). A concise description of Lagrangean relaxation for general integer programming problems can be found in Eiselt and Sandblom (2000). The most popular procedure employs a specially formulated Lagrangean dual that provides a valid bound on the original primal problem. The basic idea is to use the bound along with information on the violated constraints to generate improved bounds. If one can at the same time generate good feasible primal solutions, then it may be possible to converge to a tight lower bound on a feasible solution, and thereby confirm that a solution is appropriately close to optimality. This procedure will be discussed in greater detail in the next section. Such a process has been successfully applied to capacitated location models, p-median location models, simple plant location models and the maximal covering location problem.

One of the most effective procedures developed to date is based on the use of Lagrangean relaxation as demonstrated by Weaver and Church (1985). The basic premise is that one or more constraints are relaxed using a classical application of Lagrange multipliers. Due to the relaxation of one or more constraints, the new model has a larger solution space which includes that of the original problem. Consequently, for a minimization problem, an optimal solution to a given Lagrangean relaxation will be a lower bound for the original primal problem. If such a valid lower bound is close in value to a known feasible primal solution, then we can be certain that the best known primal solution is either optimal or very close to optimal. There are three major questions that need to be addressed in designing a Lagrangean-based solution approach (for further details see Fisher (1981) or Eiselt et al. (1987)):

- Which constraints should be relaxed? The relaxed problem should be significantly easier to solve than the original problem, but retain enough of the structure to yield "tight" bounds on the primal problem.
- What multipliers generate good bounds on the primal problem?
- Given a solution to the relaxed problem, how can we deduce a good feasible solution to the original problem?

In order to address the above questions and demonstrate the specifics associated with this process, consider the *p*-median location problem. We relax the assignment constraints $\sum_{j \in J} x_{ij} = 1 \; \forall \; i \in I$ in the formulation presented above. Alternatively, other constraints may be relaxed instead. For a pertinent discussion of these issues, readers are referred to Daskin (1995). By defining one Lagrange multiplier, λ_i, for each of the assignment constraints, we can establish the following Lagrangean relaxation (or dual problem):

$$P: \text{Min } z_D = \sum_{i \in I} \sum_{j \in J} w_i d_{ij} x_{ij} + \sum_{i \in I} \lambda_i \left(1 - \sum_{j \in J} x_{ij} \right)$$

$$\text{s.t. } x_{ij} \leq y_j \; \forall \; i \in I, j \in J$$

$$\sum_{j \in J} y_j = p$$

$$x_{ij} = 0 \vee 1, \; \forall \; i \in I, \; j \in J$$

$$y_j = 0 \vee 1 \; \forall j \in J$$

Given a set of multipliers $\lambda_i \geq 0$, the above relaxed problem will yield an objective function value that is a lower bound on any feasible solution to the original *p*-median formulation. We can rearrange the terms of the objective function, so that the problem reads

$$\text{Min } z_D = \sum_{i \in I} \sum_{j \in J} (w_i d_{ij} - \lambda_i) x_{ij} + \sum_{i} \lambda_i$$

subject to all constraints of the relaxation.

We are now able to compute the contribution that results from locating a site at node n_j. Locating at n_j implies that $y_j = 1$. By virtue of the constraints $x_{ij} \leq y_j$, we obtain $x_{ij} \leq 1 \; \forall \; i$. In order to minimize the objective in the relaxed problem, we will set $x_{ij}^* = 1 \; \forall \; i: w_i d_{ij} < \lambda_i$ and $x_{ij}^* = 0$ otherwise. Hence, the contribution of locating a facility at n_j is $\Delta_j = \sum_{i \in I} \min \{0; w_i d_{ij} - \lambda_i\}$ and the objective of the relaxed problem can be written as

Chapter 2: Location Models on Networks

$$\text{Min } z_D = \sum_{j \in J} \Delta_j y_j + \sum_{i \in I} \lambda_i$$

subject to $\sum_{j \in J} y_j = p$ and the zero-one constraints. For any given set of multipliers λ_i, $i \in I$, the second sum in the objective is a constant and the remaining problem is trivial to solve: all that needs to be done is to choose the p smallest Δ_j values; their corresponding variables y_j are set equal to one. Denote the solution by $\bar{y}_j \; \forall j \in J$, then the value of the dual objective function is

$$z_D^* = \sum_{j \in J} \Delta_j \bar{y}_j - \sum_{i \in I} \lambda_i .$$

Given $\bar{y}$, the primal objective value $\bar{z}$ of the current solution is determined by setting $\bar{x}_{ij} = 1$, if $\bar{y}_j = 1$ and $w_i d_{ij} = \min_{\ell:\bar{y}_\ell=1}\{w_i d_{i\ell}\}$, so that $\bar{z} = \sum_{i \in I}\sum_{j \in J} w_i d_{ij} \bar{x}_{ij}$.

Clearly, as $\bar{x}_{ij}$ determines a feasible solution, $\bar{z}$ constitutes an upper bound UB on the optimal value of the objective function. Similarly, the dual solution is established by $x_{ij}^* = 1$, if $\bar{y}_j = 1$ and $w_i d_{ij} < \lambda_i$ and, as the Lagrangean Relaxation relaxes some constraints, its objective value z_D^* constitutes a lower bound LB for the optimal solution. In case $\bar{z} = z_D^*$, and optimal solution is found. Otherwise, we need to determine an improved solution by changing the values of the multipliers λ_i, $i \in I$. This is accomplished by first determining the step length

$$t = \frac{\alpha(UB - LB)}{\sum_i \left(\sum_j x_{ij}^* - 1\right)^2}$$

with a parameter $\alpha \in [0; 2]$. The modified values of the multipliers are then determined by setting $\lambda_i := \max\left\{0; \lambda_i - t\left(\sum_j x_{ij}^* - 1\right)\right\}$. The process is then repeated with the new set of multipliers. This process is summarized in the following algorithm:

Lagrangean Relaxation for the p-Median Problem

Step 1: Determine the matrix $\mathbf{G} = (g_{ij}) = (w_i d_{ij})$ and initialize the process with some set of Lagrangean multipliers λ_i, $i \in I$, and $UB = \infty$ and $LB = 0$.

Step 2: Determine the matrix $\Delta = (\Delta_{ij})$, so that $\Delta_{ij} = \min \{0; g_{ij} - \lambda_i\}$ and $\Delta_j = \sum_{i \in I} \Delta_{ij} \ \forall \ j$. Define a set $P = \{j\}$, such that $|P| = p$ and $\Delta_j \geq \Delta_k \ \forall \ j \in P, \ k \notin P$.

Step 3: The current solution is $\bar{y}_j = 1 \ \forall \ j \in P$ and 0 otherwise.

Set $x^*_{ij} = 1 \ \forall \quad \bar{y}_j = 1$ and $w_i d_{ij} < \lambda_i$.

Compute: $z^*_D = \sum_{j \in J} \Delta_j \bar{y}_j - \sum_{i \in I} \lambda_i$ and $\bar{z} = \sum_{i \in I} \sum_{j \in J} w_i d_{ij} \bar{x}_{ij}$, where

$\bar{x}_{ij} = 1$, if $\bar{y}_j = 1$ and $w_i d_{ij} = \min_{\ell : \bar{y}_\ell = 1} \{w_i d_{i\ell}\}$.

Set $UB := \min \{UB; \bar{z}\}$ and $LB := \max \{LB; z^*_D\}$.

Step 4: Is $UB = LB$?

If yes: Stop, the current solution $(\bar{y}, \bar{x})$ with objective value $\bar{z}$ is optimal.

If no: Go to Step 5.

Step 5: Determine $t = \dfrac{\alpha(UB - LB)}{\left(\sum_i \left(\sum_j x^*_{ij} - 1\right)\right)^2}$, and the new multipliers

$\lambda_i := \max \left\{0; \lambda_i - t\left(\sum_j x^*_{ij} - 1\right)\right\}$. Go to Step 2.

Example: Consider the network in Figure IV.11, where the single-digit numbers indicate distances, and the double-digit numbers weights.

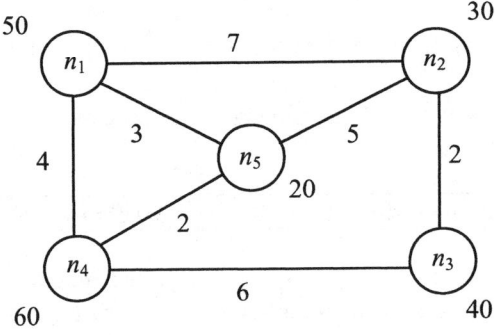

Figure II.11

Chapter 2: Location Models on Networks

The objective is to optimally locate $p = 2$ facilities. The matrix of shortest distances is

$$\mathbf{D} = \begin{bmatrix} 0 & 7 & 9 & 4 & 3 \\ 7 & 0 & 2 & 7 & 5 \\ 9 & 2 & 0 & 6 & 7 \\ 4 & 7 & 6 & 0 & 2 \\ 3 & 5 & 7 & 2 & 0 \end{bmatrix}$$

and the matrix

$$\mathbf{G} = \begin{bmatrix} 0 & 350 & 450 & 200 & 150 \\ 210 & 0 & 60 & 210 & 150 \\ 360 & 80 & 0 & 240 & 280 \\ 240 & 420 & 360 & 0 & 120 \\ 60 & 100 & 140 & 40 & 0 \end{bmatrix},$$

where $g_{ij} = w_i d_{ij} \ \forall \ i, j$. Arbitrarily set $\lambda_1 = 200$, $\lambda_2 = \lambda_3 = \lambda_4 = \lambda_5 = 100$, and let $\alpha = 1$. Now the matrix

$$\Delta = \begin{bmatrix} -200 & 0 & 0 & 0 & -50 \\ 0 & -100 & -40 & 0 & 0 \\ 0 & -20 & -100 & 0 & 0 \\ 0 & 0 & 0 & -100 & 0 \\ -40 & 0 & 0 & -60 & -100 \end{bmatrix}$$

with $\Delta_{ij} = \min\{0; g_{ij} - \lambda_i\}$, and the column sums are

$(\Delta_j) = [-240 \ -120 \ -140 \ -160 \ -150]$.

Now $P = \{1, 4\}$, so that the current solution is $\bar{y}_1 = \bar{y}_4 = 1$ and $\bar{y}_2 = \bar{y}_3 = \bar{y}_5 = 0$. This implies the primal solutions $\bar{x}_{ij}$ and dual solutions $\bar{x}_{ij}^*$ as shown below. Again, $\bar{x}_{ij} = 1$, if $y_j = 1$ and $y_{i\ell}$ is the row minimum, and $x_{ij}^* = 1$, if $y_j = 1$ and $\Delta_{ij} < 0$.

$$\bar{x}_{ij} = \begin{bmatrix} 1 & 0 & 0 & 0 & 0 \\ 1 & 0 & 0 & 0 & 0 \\ 0 & 0 & 0 & 1 & 0 \\ 0 & 0 & 0 & 1 & 0 \\ 0 & 0 & 0 & 1 & 0 \end{bmatrix} \text{ and } x_{ij}^* = \begin{bmatrix} 1 & 0 & 0 & 0 & 0 \\ 0 & 0 & 0 & 0 & 0 \\ 0 & 0 & 0 & 0 & 0 \\ 0 & 0 & 0 & 1 & 0 \\ 1 & 0 & 0 & 1 & 0 \end{bmatrix}.$$

Then $\bar{z} = \mathbf{G}\bar{x} = 0 + 210 + 240 + 0 + 40 = 490$, and $z_D^* = \sum_j \Delta_j y_j + \sum_i \lambda_i = -240 - 160 + 600 = 200$, so that $UB = 490$ and $LB = 200$. The current solution is not optimal and the multipliers need to be adjusted. Here, $t = \dfrac{1\,(490-200)}{0+1+1+0+1} \cong 97$. The numbers in the denominator are based on the dual solution x_{ij}^* that, at present, provides single assignments in rows 1 and 4, no assignments in rows 2 and 3, and a double assignment in row 5. This solution deviates from the required single assignments by 0, 1, 1, 0, and 1 elements. The new multipliers are then calculated as $\lambda_1 = 200$, $\lambda_2 = \lambda_3 = 100 - 97(-1) = 197$, $\lambda_4 = 100$, and $\lambda_5 = 100 - 97(1) = 3$. These multipliers allow us to compute the new matrix $\mathbf{\Delta}$, which is shown below.

$$\Delta = \begin{bmatrix} -200 & 0 & 0 & 0 & -50 \\ 0 & -197 & -137 & 0 & -47 \\ 0 & -117 & -197 & 0 & 0 \\ 0 & 0 & 0 & -100 & 0 \\ 0 & 0 & 0 & 0 & -3 \end{bmatrix},$$

whose column sums are

$(\Delta_j) = [-200 \quad -314 \quad -334 \quad -100 \quad -100]$.

Now $P = \{2, 3\}$, so that the new solution is $\bar{y}_2 = \bar{y}_3 = 1$ and $\bar{y}_1 = \bar{y}_4 = \bar{y}_5 = 0$. The process continues from here for a number of iterations. The main results are summarized in the Table II.2, where # is the iteration counter, and $\bar{y}_j = 1$ indicates where the facilities are located.

Lagrangean relaxation with subgradient optimization has proved to be capable of solving instances of the p-median of up to 1,000 nodes. Furthermore, other similarly structured location problems such as the simple plant location problem have been solved with Lagrangean relaxation as well. However, at times, it is required to use a branch and bound subalgorithm in case a problem is to be solved to optimality.

Table II.2

#	Multipliers λ_i	$\bar{y}_j = 1$	$\bar{z}$	z_D^*	UB	LB
1	200, 100, 100, 100, 100	(1, 4)	490	200	490	200
2	200, 197, 197, 100, 3	(2, 3)	810	49	490	200
3	258, 139, 139, 158, 61	(1, 4)	490	259	490	259
4	200, 197, 197, 158, 3	(2, 3)	810	107	490	259
5	246, 151, 151, 204, 49	(1, 4)	490	296	490	296
6	181, 216, 216, 204, 49	(2, 3)	810	152	490	296
7	220, 177, 177, 243, 88	(4, 5)	540	286	490	296
8	177, 177, 226, 194, 39	(2, 3)	810	141	490	296
9	210, 138, 187, 233, 78	(3, 4)	300	300	300	300

Due to the computational complexity of the p-median problems, it is no surprise that a variety of heuristic methods have been developed for the solution. They range from simple greedy algorithms to genetic algorithms, systematic search and improvement processes, semi-greedy methods, greedy randomized adaptive search processes, tabu search, simulated annealing, random sampling, and heuristic concentration. For an introduction to heuristic methods in general, readers are referred to Eiselt and Sandblom (2000).

As usual, any heuristic method falls into one of two classes: construction and improvement heuristics. Simply speaking, construction methods establish a solution, while improvement methods start with a given solution, and attempt to improve it as much as possible. In this section, we will describe one construction and two improvement heuristics.

The simplest construction heuristic is of the greedy type. The basic idea is locate facilities sequentially, so that in each step, a facility is located so as to minimize the total transportation cost, assuming that the present facility is the last one to be located. The latter assumption is typically for all greedy algorithms: there is no foresight beyond the present step. In order to formally describe the method, assume that $J \subset N$ is the set of nodes that has already been located. The method can then be described as follows.

The *Greedy* Algorithm for *p*-Median Problems

Step 1: Is $|J| = p$?
  If yes: Stop. The current location of the p facilities represents the solution.
  If no: Set $\ell := 1$ and go to Step 2.

Step 2: Is $\ell = n+1$?

If yes: Determine $k = \arg\min_\ell \{\Delta_\ell\}$ and locate a facility at the node n_k.

Go to Step 1.

If no: Go to Step 3.

Step 3: Determine $\Delta_\ell = \sum_{i=1}^{n} w_i \min\{\min_{j\in J}\{d_{ij}\}; d_{i\ell}\} \ \forall \ \ell \notin J$,

(if $\ell \in J$, define $\Delta_\ell \to \infty$), set $\ell := \ell + 1$, and go to Step 2.

In order to illustrate the above *Greedy Algorithm*, consider the following

Example: Consider again the graph in Figure II.9 in Section 2.3.2, and suppose now that $p = 3$ facilities are to be located. For convenience, we restate the matrix of shortest distances and vector of weights and $\mathbf{w} = [60, 20, 50, 70, 40]$.

D:

	n_1	n_2	n_3	n_4	n_5
n_1	0	5	7	6	4
n_2	5	0	2	4	6
n_3	7	2	0	6	8
n_4	6	4	6	0	2
n_5	4	6	8	2	0

The first iteration of the *Greedy Algorithm* is identical to the exact (complete enumeration) method of Section 2.3.2 to locate 1-medians. We obtain $\Delta = (\Delta_\ell) = \mathbf{wD} = [1{,}030, 920, 1{,}200, 820, 900]$, so that the first facility is located at the node n_4.

As $|J| = 1 < 3 = p$, the process continues. Now, $\Delta = (\Delta_\ell) = [460, 480, 480, \infty, 620]$, so that the second facility is located at n_1. In the next iteration, $\Delta = (\Delta_\ell) = [\infty, 180, 120, \infty, 380]$, so that the third and last facility is located at n_3. Now the process terminates, having located facilities at n_1, n_3, and n_4, so that the sum of transportation costs are 120, a value equal to the smallest Δ_ℓ value in the last iteration.

Any construction heuristic is typically followed by the application of an improvement heuristic. In the case of *p*-median problems, among the most famous such techniques is the location-allocation heuristic. It was proposed by Maranzana (1964) for network location and independently by Cooper (1964) for the location of facilities in continuous space. Its principle is simple: it alternates between the location of the facilities and allocation of the demand to them. This method has become the basis for many heuristics in location science. It is initialized with any arbitrary location of the *p* facilities, i.e., an "educated guess". It then proceeds as follows.

Chapter 2: Location Models on Networks

Location-Allocation Heuristic

Step 1: (*Allocation*) Assign each customer to its closest facility. This results in sets of customers $S_1, S_2,..., S_p$, so that all customers in S_k are served by the k-th facility.

Step 2: (*Location*) For each $k = 1,..., p$, optimize the location of the k-th facility among all customers in S_k.

Step 3: Go to Step 1 and repeat the process until location and allocation procedures have converged.

Example: Consider again the graph in Figure II.9, assuming now that $p = 2$ facilities are to be located. Suppose that the facilities are initially located at n_1 and n_2. Given these locations, $S_1 = \{n_1, n_5\}$ and $S_2 = \{n_2, n_3\ n_4\}$, i.e., the facility at n_1 serves customers at the nodes n_1 and n_5, while the facility at the node n_2 serves the customers at n_2, n_3, and n_4. This configuration has an objective function value of $z = 540$. The shortest distance matrix and weight vector for the facility at n_1 are

$$\mathbf{D}_1 = \begin{bmatrix} 0 & 4 \\ 4 & 0 \end{bmatrix}$$

and $\mathbf{w}_1 = [60, 40]$, so that $\mathbf{w}_1\mathbf{D}_1 = [160, 240]$ with the median at n_1. For the second facility we obtain

$$\mathbf{D}_2 = \begin{bmatrix} 0 & 2 & 4 \\ 2 & 0 & 6 \\ 4 & 6 & 0 \end{bmatrix},$$

$\mathbf{w}_2 = [20, 50, 70]$, and $\mathbf{w}_2\mathbf{D}_2 = [380, 460, 380]$, so that there are two medians, at n_2 and n_4. If we were to arbitrarily choose n_2, then the new facility locations would be n_1 and n_2. Since this solution equals the locations in the previous step, the method has converged to a solution with a value of the objective function of $160 + 380 = 540$.

If we were to choose n_4 instead of n_2, then the new solution locates the facilities at n_1 and n_4. Allocating customers in the next iteration yields $S_1 = \{n_1\}$ and $S_2 = \{n_2, n_3, n_4, n_5\}$. Locating the facilities optimally within the respective sets results in locations at n_1 and n_4, so that the method has again converged. The objective value here equals 460, a better value than if we had broken the tie in favor of n_1. This suggests the use of a "multi-start method," in which a variety of initial solutions are improved by a heuristic such as the location-allocation technique, and the best solution among those obtained is then chosen. However, it must be noted that the method is still a heuristic which may not find an optimal solution. For instance, in this example the pair of locations at n_1 and n_4 with $z = 460$ represents a good solution, but it is not optimal. The optimal solution locates facilities at n_3 and n_5 with $z = 420$.

The third heuristic to be discussed in this section was originally described by Teitz and Bart (1968) for solving the p-median model, but its use nowadays is far beyond this particular model. Their *vertex substitution process*, as applied to the p-median problem, is based on the following ideas. Again, it commences with some initial locational pattern of the p facilities, and then it systematically considers replacements for one of the current (tentative) facility locations. If a replacement produces an improvement in the objective function, the replacement is made and the search continues with the new pattern; otherwise other potential replacements are examined. This process terminates when no replacement or substitution in the pattern exists that produces an improvement.

The algorithm is initialized with a tentative set of facility locations at vertices in the set J, whose objective value is z_J. Initially, the set $H \subseteq J$ of nodes that have already been considered for replacement is set $H := \emptyset$. The algorithm can then be described as follows.

Vertex Substitution Method

Step 1: Is $H = J$?
If yes: Stop. The current facility locations represent the solution to the problem.
If no: Go to Step 2.

Step 2: Choose some facility $n_\ell \in J \setminus H$. Calculate
$$z_{\mu\ell} = \sum_i \min_{n_j \in J \setminus \{n_\ell\} \cup \{n_\mu\}} \{d_{ij}\} \ \forall \ n_\mu \in N \setminus J, \text{ and let}$$
$$k = \arg\min_\mu \{z_{\mu\ell}\}.$$

Step 3: Is $z_{k\ell} < z_\mu$?
If yes: Set $J := J \setminus \{n_\ell\} \cup \{n_k\}$, $z_J := z_J + z_{k\ell}$, and go to Step 1.
If no: Set $H := H \cup \{n_\ell\}$, and go to Step 2.

The vertex substitution method can be illustrated by the following

Example: Consider again the graph in Figure II.9 and assume that $p = 2$ facilities are to be located. Suppose that the initial solution includes $J = \{n_1, n_2\}$, so that $z_J = 540$. Since none of the nodes has been considered for replacement so far, $H = \emptyset$.

In the first iteration, we choose $\ell = 1$, i.e., we consider relocating the facility that is currently located at the node n_1. We can then calculate $z_{31} = 820$, $z_{41} = 480$, and $z_{51} = 480$. Arbitrarily breaking the tie between z_{41} and z_{51}, we choose $k = 1$. Since $z_{41} = 480 < 540 = z_J$, we replace n_1 by n_4, so that $J = \{n_2, n_4\}$ with $z_J = 480$. The second iteration commences again with $H = \emptyset$. Choosing $\ell = 2$, we can then calculate $z_{12} = 460$, $z_{32} = 480$, and $z_{52} = 620$, so that $k = 1$. Since $z_{12} = 460 < 480 = z_J$, we replace n_2 by n_1 and set $J := \{n_1, n_4\}$ with $z_J = 460$. The third iteration again starts with $H = \emptyset$, and we choose $\ell = 1$. This leads to $z_{21} = 480$, $z_{31} = 480$, and $z_{51} = 620$, so that $k = 1$. However, $z_{21} = 480 > 460 = z_J$, so that $H = \{n_1\}$. With the set $J := \{n_1, n_4\}$ unchanged, we choose $\ell = 4$, leading to $z_{24} = 540$, $z_{34} = 620$, and $z_{54} = 590$, so that $k = 2$. Again, $z_{24} = 540 > 460 = z_J$, so that we redefine $H := \{n_1, n_4\}$. Now $H = J$, and the algorithm terminates with the two facilities being located at n_1 and n_4, with an objective value of $z = 460$. Notice that again a reasonably good, but not an optimal, solution has been found.

Any improvement heuristic may get trapped at a local optimum, and the vertex substitution method is no exception. One recent attempt to break out of such a local minimum was proposed by Rosing and ReVelle (1997) in what they call a *heuristic concentration* method. The technique first employs a heuristic such as vertex substitution to generate a number of solutions from which a site list is developed. This site list is likely to be a small subset of the original set of sites. Sites in such a list can then easily be incorporated into an integer programming formulation to pick the best pattern from among the smaller list of sites. Its solution will be no worse than the best of the solutions found when generating the list. This two step process was shown to be quite robust in identifying optimal median solutions.

2.4 Simple and Capacitated Plant Location Problems

In the models discussed in the previous sections, the number of facilities was either prespecified (as in the maximal covering location problem, p-center, and p-median problems) or was to be minimized (as in the location set covering problem). All those models have in common that they consider all potential sites equally costly. This assumption is dropped in the simple plant location problem (*SPLP*) and its variant, the capacitated plant location problem (*CPLP*). Its objective is to locate an unspecified number of facilities (often referred to as "plants" or "factories"), so as to meet all demand while minimizing the sum of site-related and transportation costs. The model makes a number of assumptions, the most important of which are as follows:

- The facility can be supplied with unlimited resources whose prices do not vary by source.
- The transportation costs from factories to markets are linear, i.e., there are no economies of scale.
- The production costs at a facility are linear in the quantity produced once an initial fixed cost has been incurred.
- Demand is known and does not vary with changes in the delivered price.
- There is no capacity limitation on the quantity produced at a factory (for the *SPLP*).
- There is a fixed cost for purchasing and developing a site if selected for a facility.

Given the above assumptions we can define the following parameters. Denote by d_{ij} the unit transportation cost from a facility at node n_i to a customer at node n_j, a customer at node n_j demands w_j units of the product, the fixed cost for opening a facility at node n_i is f_i, and the variable unit production costs are p_i. We then need to define two sets of binary variables: $y_i = 1$, if a facility is located at the node n_i, and 0 otherwise, and $x_{ij} = 1$, if a customer at node n_j is served by a facility located at node n_i, and 0 otherwise. We can then formulate the *simple plant location problem* (*SPLP*) as follows.

$$P_{SPLP}: \text{Min } z = \sum_{i \in I}\sum_{j \in J} w_j d_{ij} x_{ij} + \sum_{i \in I} p_i \sum_{j \in J} w_j x_{ij} + \sum_{i \in I} f_i y_i$$

$$\text{s.t. } \sum_{i \in I} x_{ij} = 1 \ \forall j \in J$$

$$x_{ij} \leq y_i \ \forall \ i \in I, j \in J$$

$$y_i = 0 \vee 1 \ \forall \ i \in I$$

The three main components of the above formulation minimize the transportation cost, the production cost, and the fixed cost of siting the facilities. The above formulation was originally proposed but later abandoned by Balinski (1965). It was analyzed more fully and found highly functional by Morris (1978), who examined the relaxed linear programming solution of 600 random test problems and found that 96% of all solutions to relaxed problems did already satisfy integrality for all variables. Perhaps one of the most efficient programming approaches for solving the simple plant location problem is that of Erlenkotter (1978). Erlenkotter's approach was to use Lagrangean relaxation in a dual ascent approach, embedded within a branch and bound algorithm. Computation times for most problems are very short and his code has been used successfully in problems ranging up to 800 nodes.

The most common variant of the *SPLP* is the capacitated plant location problem (*CPLP*). The only difference between *SPLP* and *CPLP* is that the latter includes capacities for the facilities. In particular, define κ_i as the capacity of a facility located at node n_i. The capacity constraints are then formulated as $\sum_{j \in J} w_j x_{ij} \leq \kappa_i y_i \ \forall i \in I$ which replace the constraints $x_{ij} \leq y_i$ in the *SPLP*. Note that with such capacities, facility sites are constrained in ultimate development and may not be able to handle all nearby demand. An open facility that is the least cost source for a demand at node n_j may not be able to serve any of thé demand at that node. Essentially, the capacitated plant location problem involves an endogenous classical transportation problem that is used to assign supply at located sources to demand points. The complete formulation of the *capacitated plant location problem* (*CPLP*) is then as follows.

$$P_{CPLP}: \text{Min } z = \sum_{i \in I} \sum_{j \in J} w_j d_{ij} x_{ij} + \sum_{i \in I} p_i \sum_{j \in J} w_j x_{ij} + \sum_{i \in I} f_i y_i$$

$$\text{s.t. } \sum_{j \in J} x_{ij} = 1 \ \forall i \in I$$

$$\sum_{j \in J} w_j x_{ij} \leq \kappa_i y_i \ \forall i \in I$$

$$y_i = 0 \vee 1 \ \forall i \in I$$

Whereas in the simple plant location problem, customers are assigned to a source based on the lowest total cost of supply and transport, the capacitated plant location problem may assign customers to a facility that is not the lowest cost source. Furthermore, the capacity constraints change the problem structure considerably. For instance, the linear programming relaxation of the *capacitated plant location problem* almost always yields solutions with fractional variables because the variables y_i have effective fractional lower bounds due to the capacity constraints.

These two models have formed the basis upon which many of the warehouse location and factory optimization models have been developed. Although some of the assumptions under which the above models have been formulated are very restrictive, Geoffrion and McBride (1978) report a number of models which relax these restrictions. A number of solution methods have been developed to solve the above problem structures; see, e.g., Fisher (1981) and Efroymson and Ray (1966). Several of the solution methods discussed in Section 2.3 can be modified to deal with either the capacitated or simple plant location models; see Neebe (1978).

2.5 An Application of the Capacitated Facility Location Problem

In this section, we discuss a variant of the capacitated facility location problem. In particular, the example deals with a scenario that involves a hierarchy of facilities that are to be located. The RJ Reynolds-Nabisco company operates a flour mill in Toledo, Ohio which satisfies 85% of the company's need. Additional flour is purchased from various sources whenever required. Such a strategy makes sense, since the company mill can operate at a high level of capacity, and additional flour is purchased only when given demand fluctuations require it. Also, overall cost might be minimized by keeping a company-owned facility operating at full capacity, allowing for some outside purchases, as compared to having to build up a higher capacity of the company's own mill in order to meet all flour needs, which usually leads to significant slack capacity some of the time. The flour is distributed to 27 baking facilities in the United States. Eleven of these facilities are owned by the company and the remaining 16 baking facilities operate under production agreements. Baked goods ranging from Ritz Crackers© to Cheese Nips© are directly delivered to approximately 110,000 separate delivery points utilizing nine major warehouses and 130 shipping branches. This system can be viewed in terms of a hierarchy of levels that starts with the flour mill (and additional flour sources) and terminates at the many groceries. The top level of the hierarchical system is composed of the flour mill and other market sources of flour. The flour is then shipped to baking facilities on the second level. It is also possible to add suppliers of other major ingredients such as sugar to the first level; however, such raw materials can also be included as a per unit cost of production at each baking facility. The third level comprises the warehouses from where the products are shipped to the bottom level which, in turn, represents the retail outlets that are the customer base for the system.

At each level specific location problems can be identified. For example, at the top level, we need to find the optimal number and location of flour mills to support the baking operations with a combined minimal operation and transportation cost, as well as determine if flour mills should be owned and operated, or if flour should be purchased for all baking operations. Further problems involve determining the best warehouse locations in order to minimize warehousing costs, transportation costs from baking facilities to warehouses, and transportation costs to individual retail locations. Good solutions to each of these subproblems could help make the large hierarchical system more efficient. Each individual level of the system is reminiscent of a capacitated or simple plant location problem. For example, consider the problem of optimally locating a set of company-owned baking facilities. Further assume that production at a baking facility is viewed in terms of a homogeneous product which represents a composite of individual baked goods. For that purpose, we use indices $k \in K$ for the sources of flour, $i \in I$ for baking facilities, and $j \in J$ for warehouses. Furthermore, let d_{ki} (d_{ij}) denote the unit transport cost from the k-th source of flour to the i-th baking facility (from the i-th

Chapter 2: Location Models on Networks

baking facility to the j-th warehouse), let w_j denote the quantity of the product required at the j-th warehouse for further distribution, define f_i as the fixed cost to open a baking facility at node n_i and let p_i denote the unit production cost of a facility at node n_i. We can then define binary variables y_i that assume a value of 1, if a baking facility is located at node n_i and 0 otherwise. Furthermore, define variables x_{ki} as the quantity of flour shipped from the k-th source of flour to the i-th baking facility and variables x_{ij} that denote the quantity of baked goods shipped from the i-th bakery to the j-th warehouse. We can then formulate a baking facility location model as follows.

$$P: \text{Min } z = \sum_{k \in K}\sum_{i \in I}(d_{ki} + p_i)x_{ki} + \sum_{i \in I}\sum_{j \in J}d_{ij}x_{ij} + \sum_{i \in I}f_i y_i$$

$$\text{s.t. } \sum_{i \in I} x_{ij} = w_j \; \forall \; j \in J$$

$$\sum_{k \in K} x_{ki} = \sum_{j \in J} x_{ij} \; \forall \; i \in I$$

$$\sum_{j} x_{ij} \leq \kappa_i y_i \; \forall \; i \in I$$

$$y_i = 0 \vee 1 \; \forall \; i \in I$$

$$x_{ij} \geq 0 \; \forall \; i \in I, j \in J$$

$$x_{ki} \geq 0 \; \forall \; k \in K, i \in I$$

The above formulation minimizes the cost of shipments to and production at the bakeries, the transportation costs of baked goods to the distribution warehouses and the site development costs of the baking facilities. It does so by determining the optimal arrangement of bakeries and the optimal pattern of production and transport of the products from the flour mills to the bakeries and the warehouses. The first set of constraints ensures that the demand at the warehouses is met, the second set equates the supply of flour at each bakery with the quantity of baked goods shipped out from each bakery, and the third set of constraints are the capacity constraints for the bakeries. This model expands on the capacitated facility location problem in that it adds throughput constraints for each baking site. This second set of constraints changes the character of the capacitated plant location model considerably. They ensure that the amount of flour (in units of finished baked goods as well as other ingredients) shipped from all sources to an individual bakery site equals the amount of product (in units of finished baked goods) that is made at and shipped from that bakery site.

CHAPTER 3 CONTINUOUS LOCATION MODELS

Models discussed in this chapter fall into various classes, based on the dimensionality of the space under consideration. The simplest such type of problem involves a line, a one-dimensional market, along which retail market outlets are to be located, or, similarly, a train route along which stations are located. More common and realistic are location problems in the two-dimensional Cartesian plane. It is usually assumed that the plane is unbounded and that the location of demand points and facilities can be described by their coordinates. A few studies have dealt with continuous spaces other than the plane; most prominent among them is the surface of a simple sphere. Locating facilities in three or more dimensions is also possible. An application of a location problem in a d-dimensional space is a clustering problem involving biological genotypes; for details see Cooper (1973). An example of this type of problem outside of clustering and pattern recognition is the positioning of satellites. Typical problems involve the number and position of satellites in order to optimize global surveillance or communication. A microscopic application relates to medicine where a radiological device is implanted to kill a tumor. The best position for such a device can be structured as a three-dimensional location problem.

Throughout this chapter, we assume that there are n customers, whose locations are (a_i, b_i), $i = 1, \ldots, n$, and their respective demands or weights are w_i, $i = 1, \ldots, n$. In general, there are p facilities to be located, whose coordinates are (x_j, y_j), $j = 1, \ldots, p$. In the case of a single facility, we simply use (x, y) as coordinates.

Following the same sequence of problem types as we did in our discussion of network location models in the previous chapter, the first section investigates covering models. While, typically, location problems in continuous spaces are represented by nonlinear optimization problems (where the nonlinearity derives from the—often Euclidean—distance function), it is shown how standard covering problems can be reduced so as to allow a discrete search. The second section discusses minimax problems, and Section 3.3 describes the exact and approximate solution of minisum problems with a variety of distance functions.

3.1 Covering Problems

The justification of covering problems in continuous spaces closely resembles its equivalent in the previous chapter on networks. Again, a distance standard D is set by a planning agency. For instance, city planners may wish to locate fire stations so that all neighborhoods within their jurisdiction are within a prespecified distance, typically 1½ miles, from their nearest fire station. Again, the concept of coverage arises in the form of its two versions: the planar *location set covering problem* (*LSCP*) that locates facilities so as to minimize the number of facilities subject to the constraint that all demand points are covered, and the planar *maximal covering location problem* (*MCLP*) whose objective it is to locate a given number of facilities so as to maximize the number of customers (or the demand) covered by the facilities.

Consider first the maximal covering location problem. The task of locating a given number of p facilities in order to maximize coverage within the prespecified maximal covering distance D, can be restated in geometrical terms as follows: locate p circles of radius D, so that as much of the population is covered as possible. Define P_i as the i-th demand point, the parameter w_i as its weight, and D_i as a circle with radius D and center at P_i. Following Church (1984), define the *circle intersect point set*, or *CIPS*, as the set Q of all demand points plus all points in which some pair D_i, D_j ($i \neq j$) of circles touch or intersect. By construction, Q is a finite set with at most ½$n(n-1) + n$ points. Church then proved

Lemma II.2: There exists at least one optimal solution to the planar maximal covering location problem in which all facilities are located in the set Q.

This concept will be explained in the following

Example: Consider five customers who are represented by points P_1, P_2, ..., P_5 as shown in Figure II.12. The weights of the customers are **w** = [30, 40, 40, 20, 30]. The circles with centers at P_i and radius D include all potential facility locations that will cover P_i. The *CIPS* then comprises the five points P_i and the intersection points of the circles, shown as I_1, I_2, ..., I_{11}. We now examine the coverage that is achieved at each of these points. For instance, a facility that locates at I_7 will capture the customers at P_1 and P_5 for a total captured weight of 30 + 30 = 60. At I_9 a facility would capture customers are P_3, P_4, and P_5 for a total demand of 40 + 20 + 30 = 90, etc. Suppose now that $p = 2$ facilities are to be located. Lemma II.2 then prescribes to choose two out of the 17 points I_k and P_i, so as to maximize the coverage. While this is a combinatorial problem with astronomically many solutions, reasonably-sized problems have been solved. In our example, locating one facility anywhere in the set determined by I_1 and I_2, and another in the triangle spanned by I_8, I_9, and I_{10} covers all customers with a total coverage of 160.

Chapter 3: Continuous Location Models

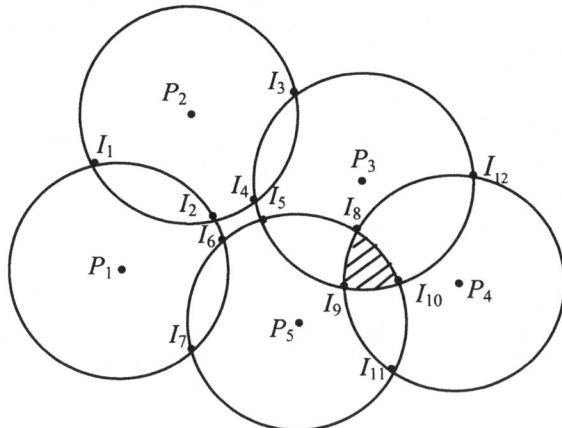

Figure II. 12

It can further be proved that Lemma II.2 also holds for planar location set covering problems. Since a finite set of points exists within which at least one optimal solution exists, then a discrete integer programming solution approach can be used, like that formulated for the network maximum covering location problem in Chapter 2 of this part. Further, by using dominance arguments, the set Q may be reduced to a set Q', called the *reduced circle intersection point set*. In particular, any site in Q which, when compared to another site in Q, covers the same demand, plus perhaps possibly more, is said to dominate the other site. All dominated sites can then be eliminated from Q. The sites in Q' can then be used to solve location set covering and maximum covering location problems. The process of generating the set Q is simple and its computational complexity is $O(n^2)$. The process of performing reductions has a similar computational complexity, and the reduced set Q' often includes no more than $3n$ or even less potential facility locations.

A realistic type of problem that uses the concept of coverage arises in the location of radio transmission towers. The idea is to locate towers of a certain height that can transmit signals to as many customers as possible. If the customers are police officers who patrol a certain area, then the customer locations are not a discrete set, since all points along the roads or streets are to be patrolled. In essence, this type of problem includes two decision variables for each transmission tower: its location and its height. Lee (1991) describes such a model. For simplicity, assume that for the height of the tower, there is only a finite number of discrete choices. In order to formalize the problem, let f_{jk} denote the cost of constructing a tower of height k at point j, and let w_i denote the value of serving customer i. Similarly, define $N_i := \{(j, k)\}$ as the set of sites j at which a tower of height k can communicate with a customer at point i. The binary decision variables are then defined as y_{jk} that equal 1, if a tower of height k is located at a point j, and 0 otherwise. Furthermore, let x_j equal 1, if demand point i is *not* served by any of the towers. Then the bi-objective *communications design problem* can be written as the vector optimization problem

$$P: \text{Min } z_1 = \sum_{i \in I} w_i x_i$$

$$\text{Min } z_2 = \sum_{i \in I} \sum_{(j,k) \in N_i} f_{jk} y_{jk}$$

$$\text{s.t.} \quad \sum_{(j,k) \in N_i} y_{jk} + x_i \geq 1 \; \forall \; i \in I$$

$$y_{jk} = 0 \vee 1 \; \forall \; (j, k) \in N_i, \; \forall \; i \in I$$

$$x_i = 0 \vee 1 \; \forall \; i \in I.$$

The first objective minimizes the value of customers that are not covered (or, in case of equal values of all customers, the number of customers that are not covered by any of the towers). The second objective minimizes the construction costs of the transmission towers. The structural constraints link the two types of variables. If the sum on the left-hand side equals zero, then a customer is not covered and, in order to satisfy the inequality, the variable x_i is forced to one, indicating non-coverage. Similarly, if the sum is at least one, then x_i can assume any nonnegative value. Due to the first objective function, x_i will then be zero, indicating that the customer is indeed covered. Finally, since the sum is integer, the constraints $x_i = 0 \vee 1$ can be relaxed to $x_i \geq 0$; these variables will automatically be integer in any optimal solution to the problem.

While the problem is easily defined, the number of potential sites is typically very large. It would also be beneficial not to discretize the tower heights. This would, however, dramatically complicate the problem.

3.2 Single-Facility Minimax Problems

The problems in this chapter have in common that they locate a single facility given some measure of centrality. Single-and multiple-facility minimax and minisum have one characterization of the optimal solution in common. In order to explain this, define the *convex hull* of a set of given points as the intersection of all sets that include these points. Then there exists at least one optimal solution in which all facilities are located within the convex hull.

Consider now a specific scenario; in particular, the problem of locating an unweighted center in the plane given ℓ_1 distances. It turns out that this model has a closed-form solution. Without loss of generality, assume that the coordinates of all

Chapter 3: Continuous Location Models

customers are nonnegative. Clearly, this implies that the coordinates of the center will also be nonnegative. The single-facility minimax problem can then be written as

P: Min $z = \max_{i \in I} \{|a_i - x| + |b_i - y|\}$, or, equivalently,

P: Min z

s.t. $z \geq |a_i - x| + |b_i - y| \quad \forall\, i \in I$.

Considering possible positive and negative values of the absolute values, this problem can be rewritten as the standard linear programming problem

P: Min z

s.t. $z \geq a_i - x + b_i - y \quad \forall\, i \in I$
$z \geq -a_i + x + b_i - y \quad \forall\, i \in I$
$z \geq a_i - x - b_i + y \quad \forall\, i \in I$
$z \geq -a_i + x - b_i + y \quad \forall\, i \in I$
$x, y, z \geq 0$.

Rearranging the terms leads to

P: Min z
s.t. $x + y + z \geq \max_{i \in I} \{a_i + b_i\}$
$-x + y + z \geq \max_{i \in I} \{-a_i + b_i\}$
$x - y + z \geq \max_{i \in I} \{a_i - b_i\}$
$-x - y + z \geq \max_{i \in I} \{-a_i - b_i\}$
$x, y, z \geq 0$.

Define now $\alpha_1 = \max_{i \in I} \{a_i + b_i\}$, $\alpha_2 = \max_{i \in I} \{-a_i + b_i\}$, $\alpha_3 = \max_{i \in I} \{a_i - b_i\}$, and $\alpha_4 = \max_{i \in I} \{-a_i - b_i\}$. With $\alpha_5 = \max\{(\alpha_1 + \alpha_4), (\alpha_2 + \alpha_3)\}$, we can then state

Lemma II.3: The 1-facility minimax location problem in the plane with ℓ_1 distances has optimal solutions

$\bar{x} = \frac{1}{2}(\alpha_3 - \alpha_4)$, $\bar{y} = \frac{1}{2}(-\alpha_3 - \alpha_4 + \alpha_5)$, and
$\bar{x} = \frac{1}{2}(\alpha_1 - \alpha_2)$, $\bar{y} = \frac{1}{2}(\alpha_1 + \alpha_2 - \alpha_5)$, with $\bar{z} = \frac{1}{2}\alpha_5$ in both cases.

Proof: We first demonstrate that ½ α_5 is a lower bound on the objective function, and we then show that the above solutions are feasible. The first task is easily accomplished. We add the first and the fourth constraint, as well as the second and the third constraints. Dividing each of the resulting inequalities by 2, we obtain $z \geq$ ½ $(\alpha_1 + \alpha_4)$ and $z \geq$ ½ $(\alpha_2 + \alpha_3)$. Since both constraints have to be satisfied, $z \geq$ max $\{½ (\alpha_1 + \alpha_4); ½ (\alpha_2 + \alpha_3)\}$ = ½ α_5 is a lower bound on the value of the objective function z. This completes the first part of the proof.

We now show that $\bar{x}$ = ½ $(\alpha_3 - \alpha_4)$, $\bar{y}$ = ½ $(-\alpha_3 - \alpha_4 + \alpha_5)$, and $\bar{z}$ = ½ α_5 is a feasible solution. There are two cases.

Case 1: Let $\alpha_1 + \alpha_4 \geq \alpha_2 + \alpha_3$, so that $\alpha_5 = \alpha_1 + \alpha_4$.
Then $\bar{x}$ = ½ $(\alpha_3 - \alpha_4)$, $\bar{y}$ = ½ $(\alpha_1 - \alpha_3)$, and $\bar{z}$ = ½ $(\alpha_1 + \alpha_4)$.
The individual constraints are then

$$\bar{x} + \bar{y} + \bar{z} = ½ (\alpha_3 - \alpha_4) + ½ (\alpha_1 - \alpha_3) + ½ (\alpha_1 + \alpha_4) = \alpha_1,$$
$$-\bar{x} + \bar{y} + \bar{z} = -½ (\alpha_3 - \alpha_4) + ½ (\alpha_1 - \alpha_3) + ½ (\alpha_1 + \alpha_4)$$
$$= \alpha_1 - \alpha_3 + \alpha_4 \geq \alpha_2 + \alpha_3 - \alpha_3 = \alpha_2,$$
$$\bar{x} - \bar{y} + \bar{z} = ½ (\alpha_3 - \alpha_4) - ½ (\alpha_1 - \alpha_3) + ½ (\alpha_1 + \alpha_4) = \alpha_3, \text{ and}$$
$$-x - \bar{y} + \bar{z} = -½ (\alpha_3 - \alpha_4) - ½ (\alpha_1 - \alpha_3) + ½(\alpha_1 + \alpha_4) = \alpha_4;$$

thus establishing the feasibility of the proposed solution.

Case 2: Let $\alpha_2 + \alpha_3 \geq \alpha_1 + \alpha_4$, so that $\alpha_5 = \alpha_2 + \alpha_3$.
Then $\bar{x}$ = ½ $(\alpha_3 - \alpha_4)$, $\bar{y}$ = ½ $(\alpha_2 - \alpha_4)$, and $\bar{z}$ = ½ $(\alpha_2 + \alpha_3)$.
The individual constraints are then

$$\bar{x} + \bar{y} + \bar{z} = ½ (\alpha_3 - \alpha_4) + ½ (\alpha_2 - \alpha_4) + ½ (\alpha_2 + \alpha_3)$$
$$= \alpha_2 + \alpha_3 - \alpha_4 \geq \alpha_1 + \alpha_4 - \alpha_4 = \alpha_1,$$
$$-\bar{x} + \bar{y} + \bar{z} = -½ (\alpha_3 - \alpha_4) + ½ (\alpha_2 - \alpha_4) + ½ (\alpha_2 + \alpha_3) = \alpha_2,$$
$$\bar{x} - \bar{y} + \bar{z} = ½ (\alpha_3 - \alpha_4) - ½ (\alpha_2 - \alpha_4) + ½ (\alpha_2 + \alpha_3) = \alpha_3, \text{ and}$$
$$-x - \bar{y} + \bar{z} = -½ (\alpha_3 - \alpha_4) - ½ (\alpha_2 - \alpha_4) + ½(\alpha_2 + \alpha_3) = \alpha_4;$$

and again, feasibility of the solution has been established. In conjunction with the bounding, this implies the optimality of the solution. The optimality of the second solution in the lemma is proved similarly. The optimality of all linear convex combinations of the two optimal points can be proved directly as well, but it follows from standard results in linear programming.

Example: As an illustration, consider six customers located at the points (0, 0), (3, 0), (6, 1), (0, 2), (1, 5), and (4, 6). Here, α_1 = max $\{(0 + 0); (3 + 0); (6 + 1); (0 + 2); (1 + 5); (4 + 6)\}$ = 10 and, similarly, $\alpha_2 = 4$, $\alpha_3 = 5$, $\alpha_4 = 0$, and $\alpha_5 = 10$. This results in centers being located at the two points $(\bar{x}, \bar{y}) = (2½, 2½)$ and (3; 2), as well as at all points on the line segment between them. All points have an objective value of $\bar{z} = 5$. The diamonds in Figure II. 13 with the solid and broken lines are contour lines that include all points that are $\bar{z} = 5$ units away from the points (2½, 2½) and (3, 2), respectively. In both cases, the customers P_1, P_3, P_5, and P_6 are located on the contour line, indicating that they are critical, in the sense that they are farthest away from the facility.

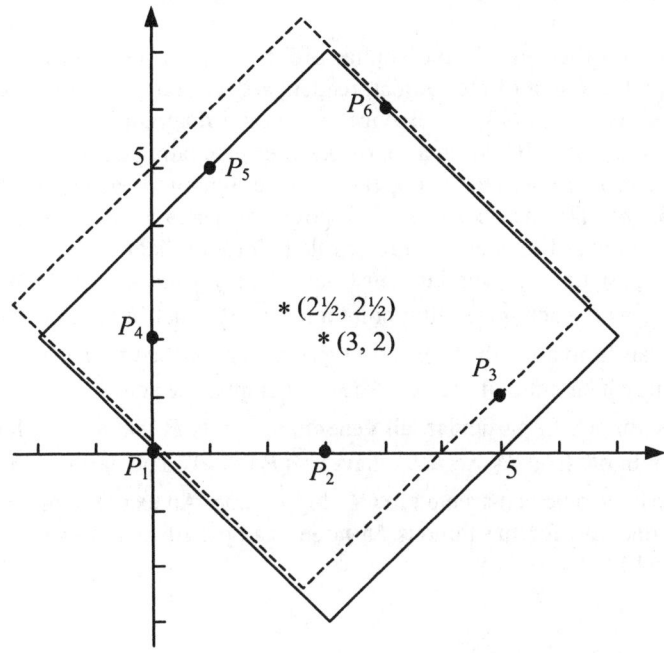

Figure II. 13

Extending this problem to weighted problems is possible, whereas extensions to problems with multiple facilities are considerably more difficult. No solution procedure has been developed that can solve large problems to optimality within a reasonable amount of computational effort. For a discussion, readers are referred to Love *et al.* (1988).

Consider now the same minimax problem in the plane, but with Euclidean distances. Again, we assume that all demand points (a_i, b_i) carry the same (unit) weight, so that the problem is to locate a single facility at (x, y) so as to minimize the maximal distance between the new point and any of the given points. The 1-center problem with Euclidean distances can then be written as

$$\text{P:} \quad \text{Min } z = \max_{i \in I} \{(a_i - x)^2 + (b_i - y)^2\}^{\frac{1}{2}}.$$

Graphically, the objective is to find a point in the plane, so that the radius of a circle, centered at that point and covering all demand points, is minimized. As Shamos and Hoey (1975) pointed out, this is nothing but the well-known "smallest enclosing circle" problem in computational geometry. It turns out that the solution to this problem, unlike the solution to the minimax problem with ℓ_1 distances, is unique. For a proof, see Eiselt *et al.* (1997).

The basic tool used in this and similar problems is the *Voronoi diagram*. For an exhaustive discussion of the subject, readers are referred to Okabe *et al.* (1992) or Okabe and Suzuki (1997). In essence, a Voronoi diagram is a tessellation, i.e., a subdivision of space based on a set of given or generating points P_i, $i \in I$. Given a metric, a Voronoi diagram is constructed by assigning to each generating point P_i a *Voronoi set* $V(P_i)$ that consists of all points closer to P_i than to any other of the generating points. It is well established that Voronoi diagrams can be constructed in $O(n \log n)$ time. A similar construct is the *farthest-point Voronoi diagram* which assigns to each generating point a set $\overline{V}(P_i)$ that includes all points farther from P_i than from any other generating point. As opposed to the standard Voronoi diagram in which none of the sets $V(P_i)$ is empty, the sets $\overline{V}(P_i)$ of some points P_i may be empty. In particular, all generating points P_i that are not located on the convex hull of $\{P_1, P_2, ..., P_n\}$ have $\overline{V}(P_i) = \varnothing$. The farthest-point Voronoi diagram can also be constructed in $O(n \log n)$ time. An example of a farthest-point Voronoi diagram for the famous 55-node example of Swain (1974) is shown in Figure IV.14.

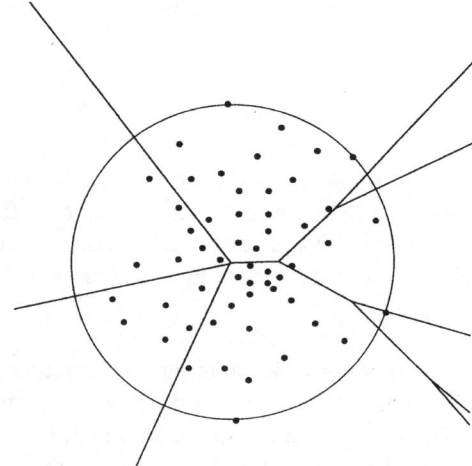

Figure II.14

Once this diagram is known, the center problem can easily be solved. Its solution is based on the fact that the location of the 1-center is either at the center of the line segment connecting the two generating points that are farthest apart, or at any of the vertices of the farthest-point Voronoi diagram.

Another, more traditional, approach to solving the 1-center problem is the algorithm by Elzinga and Hearn (1972). It is initialized with any two generating points A and B, and a tentative center location at X halfway between A and B. The current estimate of the critical distance is $\bar{d} = d_{AX} = d_{BX}$. The algorithm can then be stated as follows.

The Elzinga-Hearn Algorithm for 1-Centers with Euclidean Distances

Step 1: Are all points P_i inside a circle with center at X and radius $\bar{d}$?
If yes: Stop, X is the optimal 1-center.
If no: Choose C as any point P_i outside the circle and go to Step 2.

Step 2: Determine the angles of the triangle Δ_{ABC}. Are any angles of Δ_{ABC} right or obtuse?
If yes: Delete the point at the vertex with the right or obtuse angle,
rename the remaining points A and B, compute X and $\bar{d}$ and go to Step 1.
If no: Go to Step 3.

Step 3: Determine the circumcircle defined by the triangle Δ_{ABC}, denote its center by X and determine the distance $\bar{d} = d_{AX} = d_{BX} = d_{CX}$. Does the circle centered at X with radius $\bar{d}$ cover all points $P_1,..., P_n$?
If yes: Stop, X is the optimal 1-center.
If no: Go to Step 4.

Step 4: Let D be any point P_i outside the circle. Denote by E the point among A, B, and C that is farthest from D. There is a unique line through X and E. Let F be the point among A, B, and C that is in the same half plane as D. Denote by G the point among A, B, and C that is in the other half plane. Redefine E, G, and D as A, B, and C and go to Step 2.

Example: Consider the same example solved above for ℓ_1 distances, in which there are $n = 6$ customers at points $P_1 = (0, 0)$, $P_2 = (3, 0)$, $P_3 = (6, 1)$, $P_4 = (0, 2)$, $P_5 = (1, 5)$, and $P_6 = (4, 6)$. Arbitrarily choose $A = P_1 = (0, 0)$ and $B = P_2 = (3, 0)$, then $X = (1\frac{1}{2}, 0)$ and $\bar{d} = 1\frac{1}{2}$. Clearly, the circle centered at X with diameter $\bar{d}$ does not cover all generating points (in fact it covers only P_1 and P_2). As a point that is not covered we arbitrarily choose P_5. The triangle $\Delta_{ABC} = \Delta_{P_1 P_2 P_5}$ has no right or obtuse angles. Its circumcircle has a center at $X = (1.5, 2.3)$ and radius $\bar{d} = 2.7459$. This circle does not cover P_3. Hence we set $D = P_3$, the farthest point from P_3 is $E = P_5$. The line through X and E has $F = P_2$ on the same side as $D = P_3$, while $G = P_1$ is on the other side of the line. Hence the new triangle consists of the points P_1, P_3, and P_5 which are now defined as A, B, and C, respectively.

The triangle $\Delta_{ABC} = \Delta_{P_1P_3P_5}$ has a circumcircle with center at $X = (2.7415, 2.0517)$ and radius $\bar{d} = 3.4242$. This circle covers all nodes except P_6. Hence we set $D = P_6$, and the farthest point from P_6 is $E = P_1$. The line through X and P_1 has $F = P_5$ on the same side as $D = P_6$, while $G = P_3$ is on the other side. This gives us the new triangle with points P_1, P_3, and P_6 which are now redefined as A, B, and C. The new triangle $\Delta_{ABC} = \Delta_{P_1P_3P_6}$ has a circumcircle with center at $X = (2.65625, 2.5625)$ and radius $\bar{d} = 3.6908$. This circle covers all points, hence X is the optimal location of the 1-center.

3.3 Minisum Problems

The models discussed in this section have in common that they use the average, rather than the largest, distance between a facility and its customers as the criterion for optimization. The basic setting is the same as in previous chapters with n customers, located at the points (a_i, b_i), $i = 1,..., n$ and demanding w_i units each. The task is then to locate p facilities at coordinates (x_j, y_j), $j = 1, ..., p$, so as to minimize the average, or, equivalently, the sum of weighted customer-facility distances. This minisum problem is often referred to as a *Weber problem* or, less frequently, a *Steiner-Weber problem*, in reference to Weber's (1909) seminal work, based on Steiner's (1835) contribution. The following sections investigate this problem for various distance functions.

3.3.1 Single-Facility Problems

This section deals with location problems with minisum objective. As discussed in the chapter on network location problems, most location problems that arise in the private sector of the economy fall into this category. However, as opposed to network problems, here we face the difficulty of choosing the appropriate distance function. The sections below discuss minisum problems with different ℓ_p distances. Formally, define d_i as the distance between the i-th customer and the new facility. Then the objective can be written as

$$\text{Min } z = \sum_{i=1}^{n} w_i d_i \qquad (1)$$

Chapter 3: Continuous Location Models

Consider now the 1-minisum location problems with rectilinear or Manhattan distances. The model goes back to the work by Wersan et al. (1962) who used it in the context of locating a municipal incinerator. The authors formulated the model as a linear programming problem. Possibly the most elegant description and proposed solution process are due to Vergin and Rodgers (1967). With these distances, the general objective in (10) reduces to

$$\text{Min } z = \sum_{i=1}^{n} w_i [|a_i - x| + |b_i - y|] = \sum_{i=1}^{n} w_i |a_i - x| + \sum_{i=1}^{n} w_i |b_i - y|. \quad (2)$$

It is apparent that this objective is separable. This allows us to optimize in one dimension at a time. Without loss of generality, consider the abscissa along which the n customers are lined up. We notice that the objective function (2) is piecewise linear, hence it suffices to consider the break points, which occur only at points where customers are located. We first rearrange the coordinates a_i in nondecreasing order, merging equal values and aggregating the corresponding weights w_i. The objective values at some adjacent points i^* and (i^*+1) are then

$$z_{i^*} = \sum_{i=1}^{i^*-1} w_i d_{ii^*} + \sum_{i=i^*+1}^{n} w_i d_{ii^*} \quad \text{and} \quad (3a)$$

$$z_{i^*+1} = \sum_{i=1}^{i^*} w_i d_{i,i^*+1} + \sum_{i=i^*+2}^{n} w_i d_{i,i^*+1}. \quad (3b)$$

The slope along the segment between i^* and (i^*+1) is then

$$S_{i^*,i^*+1} = d_{i^*,i^*+1}^{-1}(z_{i^*+1} - z_{i^*}) \quad \text{or}$$

$$d_{i^*,i^*+1}^{-1} \left[\sum_{i=1}^{i^*} w_i d_{i,i^*+1} - \sum_{i=1}^{i^*-1} w_i d_{ii^*} + \sum_{i=i^*+2}^{n} w_i d_{i,i^*+1} - \sum_{i=i^*+1}^{n} w_i d_{ii^*} \right] \quad (4)$$

As $d_{i,i^*+1} = d_{ii^*} + d_{i^*,i^*+1}$ for $i = 1,..., i^*$ and $d_{i,i^*+1} = d_{ii^*} + d_{i^*,i^*+1}$ for $i = i^* + 1,..., n$, and $d_{i^*i^*} = 0$, we obtain a slope of

$$S_{i^*,i^*+1} = \sum_{i=1}^{i^*} w_i - \sum_{i=1^*+1}^{n} w_i. \quad (5)$$

Notice in relation (5) that the distances have cancelled out and are immaterial for our optimization. Furthermore, the function is nondecreasing which, coupled with piecewise linearity, implies convexity. This allows us to find an optimal location at i^* by computing slopes that satisfy the conditions

$$S_{i^*-1,i^*} \leq 0 \leq S_{i^*,i^*+1}. \tag{6}$$

This is the point where the objective has reached its minimum and its slope changes sign. Using (5), relation (6) can be rewritten as

$$\sum_{i=1}^{i^*} w_i \geq \tfrac{1}{2} \sum_{i=1}^{n} w_i \tag{7}$$

and

$$\sum_{i=1}^{i^*-1} w_i \leq \tfrac{1}{2} \sum_{i=1}^{n} w_i \tag{8}$$

In other words, adding up weights from either side, the optimal location of the facility is at a point i^*, where the sum of weights reaches or exceeds half the total weight for the first time. This is the source of the term "median," whose meaning is the same as in statistics.

The one-dimensional problem can also be viewed as a problem on graphs, with the customers at the nodes, and the connections between them as undirected edges. The result is a simple path, which, in turn is a special case of a tree, so that the optimality condition of Section 2.3.2 applies. Recall that the theorem states that a node is the median, if the weight of neither of its (here: two) subtrees exceeds half the total weight of the tree, which leads directly to relations (7) and (8).

Example: Consider a 1-minisum problem with rectilinear distances, $n = 11$ customers, and a total demand of $\sum_{i=1}^{11} w_i = 165$.

Table II.3

Customer #	1	2	3	4	5	6	7	8	9	10	11
a_i	10	20	33	25	13	10	15	30	43	30	20
b_i	10	13	15	25	20	27	30	30	23	8	0
w_i	10	10	30	5	20	15	15	15	15	10	20

The distribution of customers in the plane and their associated weights are shown in Table II.3. Ordering the customers with respect to nondecreasing x-coordinates, we obtain the ordered customer set (1, 6, 5, 7, 2, 11, 4, 8, 10, 3, 9). Adding up weights in this order we obtain 10, 25, 45, 60, 70, 90 > 82½ = $\tfrac{1}{2}\sum_{i=1}^{11} w_i$, i.e., at customer 11 we exceed half the total weight on the x-coordinate for the first time. Ordering now customers with respect to their y-coordinates results in the ordered customer set (11, 10, 1, 2, 3, 5, 9, 4, 6, 7, 8). Adding the weights in this order yields 20, 30, 40, 50, 80, 100 at which point half the total weight is again exceeded for the first time. This occurred at customer 5. Hence, the (unique) optimal solution is located at the x-coordinate of customer 11 and the y-coordinate of customer 5, i.e., at $(\bar{x}, \bar{y}) = (20, 20)$. The objective value is then computed by determining the distances between (20, 20) and all customers. Here, $\bar{z} = 10(20) + 10(7) + 30(18) + 5(10) + 20(7) + 15(17) + 15(15) + 15(20) + 15(26) + 10(22) + 20(20) = 2,790$.

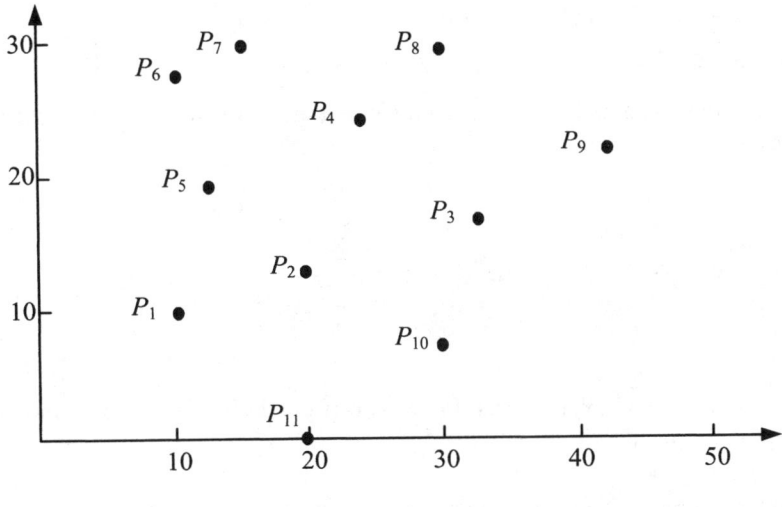

Figure II. 15

Consider now single-facility minisum problems with squared Euclidean distances $d_{ix} = (a_i - x)^2 + (b_i - y)^2$. While such distances are a theoretical construct (they are defined as the square of a one-dimensional measure), they do have a number of desirable properties. One such property is that squaring distances will exaggerate longer distances which, in turn, are distances the minimization objective will naturally try to avoid. Thus, while minimizing the sum of a distance measure, this distance measure attempts to keep the longest distance short. In that sense, it combines the features of both the minisum and the minimax objective. While transportation costs as function of distance can rarely be approximated by squared distances, the main appealing feature of squared Euclidean distances is the fact that—as opposed to the Euclidean distances discussed below—the optimization function is separable. In particular, the objective is

$$\text{Min } z = \sum_{i=1}^{n} w_i \left[(a_i - x)^2 + (b_i - y)^2\right] = \sum_{i=1}^{n} w_i (a_i - x)^2 + \sum_{i=1}^{n} w_i (b_i - y)^2. \qquad (9)$$

It is apparent from relation (8) that the choice of x-coordinate does not influence the choice of y-coordinate. In other words, we can optimize in both directions separately. To do so, we determine the partial derivatives of the objective (9) with respect to x and y, respectively. Setting them equal to zero, we obtain

$$\frac{\partial z}{\partial x} = \sum_{i=1}^{n} -2w_i(a_i - x) = 0, \text{ and}$$

$$\frac{\partial z}{\partial y} = \sum_{i=1}^{n} -2w_i(b_i - y) = 0. \qquad (10)$$

After a few standard algebraic transformations, we obtain the closed-form solutions

$$\bar{x} = \frac{\sum_{i=1}^{n} w_i a_i}{\sum_{i=1}^{n} w_i} \quad \text{and} \quad \bar{y} = \frac{\sum_{i=1}^{n} w_i b_i}{\sum_{i=1}^{n} w_i} \qquad (11)$$

Note that the point $(\bar{x}, \bar{y})$ is also the center-of-gravity of the given points (a_i, b_i) and their weights w_i.

Example: Consider a problem with customers located at (0, 0), (6, 0), and (0, 2), whose demands are 1, 3, and 2, respectively. The optimal location of the 1-minisum problem with squared Euclidean distances is then $\bar{x} = 18/6 = 3$ and $\bar{y} = 4/6 = 0.67$. Notice the similarity between this solution and that of the 1-minisum problem with Euclidean distances.

Another version of the single-facility minisum problem uses Euclidean distances. In such a case $d_{ix} = [(a_i - x)^2 + (b_i - y)^2]^{1/2}$, so that the objective function can be written as

$$\text{Min } z = \sum_{i=1}^{n} w_i \left[(a_i - x)^2 + (b_i - y)^2\right]^{1/2} \qquad (12)$$

Taking partial derivatives with respect to x and y, respectively, results in the first-order optimality conditions

Chapter 3: Continuous Location Models

$$\frac{\partial z}{\partial x} = \sum_{i=1}^{n} \tfrac{1}{2} w_i \, [(a_i - x)^2 + (b_i - y)^2]^{-\frac{1}{2}} \, 2(a_i - x)(-1) = 0 \qquad (13)$$

and

$$\frac{\partial z}{\partial y} = \sum_{i=1}^{n} \tfrac{1}{2} w_i \, [(a_i - x)^2 + (b_i - y)^2]^{-\frac{1}{2}} \, 2(b_i - y)(-1) = 0 \qquad (14)$$

A few routine calculations transform relation (13) to

$$\sum_{i=1}^{n} \frac{w_i a_i}{d_{ix}} = x \sum_{i=1}^{n} \frac{w_i}{d_{ix}} \qquad (15)$$

and similar for relation (14). As the variable x appears in the denominator as part of d_{ix}, it is apparent that relation (15) does not have a closed-form solution. Dividing by the sum on the right-hand side, we obtain

$$x = \frac{\displaystyle\sum_{i=1}^{n} \frac{w_i a_i}{d_{ix}}}{\displaystyle\sum_{i=1}^{n} \frac{w_i}{d_{ix}}} \qquad (16)$$

and similar for y. Weiszfeld (1937) was the first to suggest using relation (16) in the context of an iterative algorithm by starting with an initial guess (x^0, y^0), using these values to compute the right-hand side of relation (16), thus obtaining a new value x^1 (and similar for y^1), and using the new solution (x^1, y^1) as new guesses. The Weiszfeld method is initialized with an initial guess (x^0, y^0), an iteration counter $t = 0$, and some stop criterion. The procedure can then be summarized as follows.

The Weiszfeld Method

Step 1: Is the stop criterion satisfied?
If yes: Stop, the current solution (x^t, y^t) is sufficiently close to optimality.
Use it to compute the objective value $z^t = \sum_{i=1}^{n} w_i \, [(a_i - x^t)^2 + (b_i - y^t)^2]^{\frac{1}{2}}$.

If no: Go to Step 2.

Step 2: Compute

$$x^{t+1} = \frac{\sum_{i=1}^{n} \frac{w_i a_i}{\sqrt{(a_i - x^t)^2 + (b_i - y^t)^2}}}{\sum_{i=1}^{n} \frac{w_i}{\sqrt{(a_i - x^t) + (b_i - y^t)^2}}}, \text{ and} \qquad (17a)$$

$$y^{t+1} = \frac{\sum_{i=1}^{n} \frac{w_i b_i}{\sqrt{(a_i - x^t)^2 + (b_i - y^t)^2}}}{\sum_{i=1}^{n} \frac{w_i}{\sqrt{(a_i - x^t) + (b_i - y^t)^2}}} \qquad (17b)$$

Set $t := t + 1$ and go to Step 1.

A number of different stop criteria can and have been used in this method. Typical examples are an upper bound on the number of iterations, the distance between two successive points (x^{t-1}, y^{t-1}) and (x^t, y^t), and the improvement of the objective function $z^{t-1} - z^t$. Each or any combination may be employed in Weiszfeld's algorithm.

Example: As an illustration of Weiszfeld's method, consider the previous small problem with $n = 3$ customers located at $(0, 0)$, $(6, 0)$, and $(0, 2)$, respectively. The respective customer demands are 1, 3, and 2. To illustrate the generally fast convergence of the Weiszfeld algorithm, we initialize the process with the point $(x^0, y^0) = (10, 10)$. Clearly, this point is outside the convex hull of the demand points and far from any potential optimum. The first iteration computes

$$x^1 = \frac{\frac{1(0)}{\sqrt{(0-10)^2 + (0-10)^2}} + \frac{3(6)}{\sqrt{(6-10)^2 + (0-10)^2}} + \frac{2(0)}{\sqrt{(0-10)^2 + (2-10)^2}}}{\frac{1}{\sqrt{(0-10)^2 + (0-10)^2}} + \frac{3}{\sqrt{(6-10)^2 + (0-10)^2}} + \frac{2}{\sqrt{(0-10)^2 + (2-10)^2}}}$$

$\cong 3.3066$, and

$$y^1 = \frac{\frac{1(0)}{\sqrt{(0-10)^2 + (0-10)^2}} + \frac{3(0)}{\sqrt{(6-10)^2 + (0-10)^2}} + \frac{2(2)}{\sqrt{(0-10)^2 + (2-10)^2}}}{\frac{1}{\sqrt{(0-10)^2 + (0-10)^2}} + \frac{3}{\sqrt{(6-10)^2 + (0-10)^2}} + \frac{2}{\sqrt{(0-10)^2 + (2-10)^2}}}$$

$\cong 0.6180$.

Another iteration results in $(x^2, y^2) = (3.356, 0.575)$, and a third locates the facility at $(x^3, y^3) = (3.410, .562)$. We notice that while the changes in the x- and y-values were very significant in the first step, they were already very small in the second iteration.

It is possible that at some iteration t, the iterate (x^t, y^t) equals one of the given demand points. In such a case, the distance between the customer and the current location of the firm is zero. As this distance appears in the denominator of relations (17a) and (17b), these relations are no longer defined. As a remedy, a perturbation technique is usually employed. It adds a small positive constant ε to the distance, thus avoiding that any denomination equals zero. This technique is sometimes referred to as *hyperboloid approximation*.

One can easily picture the process of the Weiszfeld algorithm using a mechanical analog. Such a mechanical analog was originally proposed by Varignon (hence the name "Varignon apparatus", or "Varignon frame") and rediscovered by Miehle (1958). This analog is depicted in Figure II.16 where a board has holes at the customer locations. Through each such hole a string is fed, and all strings are tied together in a knot. At the other end of the strings weights are attached that are equivalent to the weight or demand of the customer to whom the hole belongs through which the string is fed. Assuming the absence of friction, the knot will be pulled by the weights to a position of equilibrium. This is the solution to the 1-minisum problem discussed in this section.

Although the Weiszfeld algorithm usually takes only a few iterations, Drezner (1992) has shown that there are cases when the process can consume thousands of iterations. As an example, Drezner considers the three points (0, 0), (30, 0), and (0, 40) with weights 5, 3, and 4 respectively. Using a convergence tolerance of .00001, the Weiszfeld algorithm takes 2,230 iterations to converge. A second simple pathological problem consists of the four points (0, 0), (1, 0), (0, 1), (1, 1), and (100, 100) with weights 1, 1, 1, 1, 4 respectively. Here, the algorithm takes over 100,000 iterations to converge, hence the obvious need to accelerate it. Ostresh (1978) has developed an approach that uses a variable step size that can aid in speeding up the Weiszfeld algorithm. In iteration $t + 1$, it computes a solution (x^{t+1}, y^{t+1}) which is subsequently revised to $\tilde{x}^{t+1} = x^t + \lambda(x^{t+1} - x^t)$ and similarly for $\tilde{y}^{t+1}$. The parameter λ is the step length. If $\lambda = 1$, then $\tilde{x}^{t+1} = x^{t+1}$ and there is no change in the process. If $\lambda > 1$, then the move to the next trial solution is extended by the factor λ. Ostresh demonstrated that if $1 < \lambda < 2$, then the new process does converge. Drezner found in his study that the number of iterates tends to decrease as n increases. Typically, the iterations are reduced most for values of λ close to 2. Unfortunately, there are pathological examples that resist even this acceleration device.

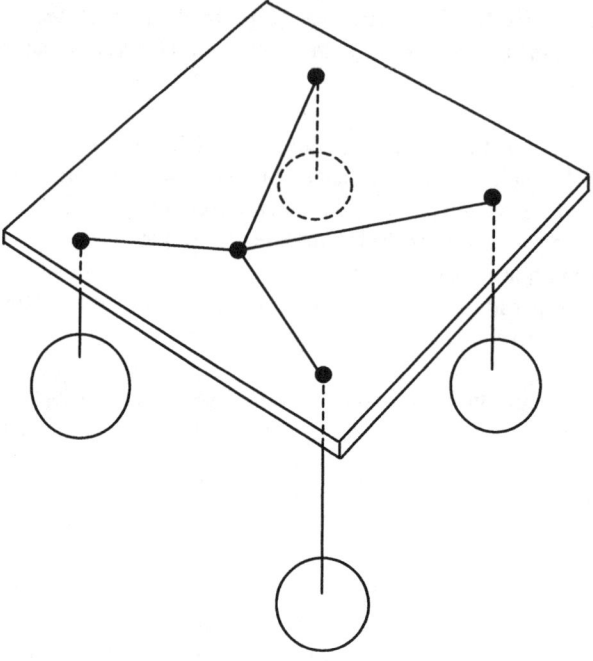

Figure II.16

3.3.2 Multi-Facility Problems

This section extends the discussion of the previous sections to multiple facilities. Here, we have to distinguish between two very different models. First, there are multiple facility problems with fixed interactions. These models assume that the customers are served from a specific firm regardless of its location. However, there may be interactions between the facilities themselves. On the other hand, in *location-allocation* models the location of the facilities is only one problem, the other is to allocate the customers to a facility (typically the one closest to it).

First assume that the interactions (i.e., the magnitude of the shipments) between customers and facilities are fixed. If there is no interaction between the facilities, then the resulting location problem lacks the "allocation" feature of the usual location-allocation problems, and thus decomposes into p single Weber problems, making the problem easy to solve. However, introducing fixed interactions between facilities makes the problem again difficult. Formally, define w_{ij}^{cf} as the interaction between customer i and facility j, and w_{jk}^{ff} as the interaction between facilities j and k. Similarly, define d_{ij}^{cf} as the distance between customer i and facility j, and d_{jk}^{ff} the distance between facilities j and k. The objective of the problem can then be formulated as

$$\text{P: Min } z = \sum_{i \in I} \sum_{j \in J} w_{ij}^{cf} d_{ij}^{cf} + \sum_{j \in J} \sum_{k \in J} w_{jk}^{ff} d_{jk}^{ff}$$

where the variables that express the facility locations (x_j, y_j) are included in the distances d_{ij}^{cf} and d_{jk}^{ff}. Given rectilinear distances, the objective is again decomposable, i.e., the optimal facility locations for the x- and y-coordinates can be optimized separately. For the x-coordinate, define variables $x_1, x_2,..., x_p$ as the x-coordinates for the p facilities, as well as variables α_{jk} and β_{jk} that measure the distances between facilities j and k along the x-axis, and variables γ_{ij} and δ_{ij} that measure the distances between the i-th customer and the j-th facility. Two variables for each distance are required as the coordinate differences may be positive or negative, and we need to convert this to a nonnegative value. The problem (for the x-coordinate only) can then be formulated as

$$\text{P: } \quad \text{Min } z_x = \sum_{j=1}^{p-1} \sum_{k=j+1}^{p} w_{jk}^{ff}(\alpha_{jk} + \beta_{jk}) + \sum_{i=1}^{p} \sum_{j=1}^{n} w_{ij}^{cf}(\gamma_{ij} + \delta_{ij})$$

s.t. $x_j - x_k - \alpha_{jk} + \beta_{jk} = 0 \ \forall \ 1 \leq j < k \leq p$

$x_j - \gamma_{ij} + \delta_{ij} = a_i \ \forall \ i = 1,..., n; j = 1,..., p$

$x_j \in \mathbb{R} \ \forall \ j = 1,..., p$

$\alpha_{jk}, \beta_{jk}, \gamma_{ij}, \delta_{ij} \geq 0 \ \forall \ i, j, k$

As an example, suppose that customers are located at (0, 0), (6, 1), (2, 4), and (1, 2); i.e., $a_i = 0, 6, 2, 1$ for $i = 1,..., 4$ and $b_i = 0, 1, 4,$ and 2 for $i = 1,..., 4$. Two facilities are to be located, so that $w_{i1}^{cf} = [2, 0, 0, 4]$, i.e., the interaction between customer 1 and facility 1 has a value of 2, and the interaction between customer 4 and facility 1 has a value of 4. Similar, the interactions between the customers and the second facility are $w_{i2}^{cf} = [3, 6, 1, 0]$. Finally, the interaction between the facilities is $w_{12}^{ff} = 2$. The problem for the x-coordinate is then

P': Min $z_x = 2(\alpha_{12} + \beta_{12}) + 2(\gamma_{11} + \delta_{11}) + 4(\gamma_{41} + \delta_{41}) + 3(\gamma_{12} + \delta_{12})$
$\qquad + 6(\gamma_{22} + \delta_{22}) + 1(\gamma_{32} + \delta_{32})$

s.t. $\quad x_1 - x_2 - \alpha_{12} + \beta_{12} = 0$
$\quad\quad x_1 \quad\quad - \gamma_{11} + \delta_{11} = 0$
$\quad\quad x_1 \quad\quad - \gamma_{41} + \delta_{41} = 1$
$\quad\quad x_2 - \gamma_{12} + \delta_{12} = 0$
$\quad\quad x_2 - \gamma_{22} + \delta_{22} = 6$
$\quad\quad x_2 - \gamma_{32} + \delta_{32} = 2$
$\quad\quad x_1, \; x_2 \quad\quad\quad \in \mathbb{R}$
$\quad\quad \alpha_{12}, \beta_{12}, \gamma_{ij}, \delta_{ij} \geq 0 \; \forall \; i, j$

The formulation that optimizes the y-coordinate is very similar, except that it replaces x_j by y_j $\forall$ j, and the right-hand side values of the second set of constraints are b_i rather than a_i. The problem has multiple optimal solutions. One of these solutions is $\bar{x}_1 = 1$, $\bar{x}_2 = 2$, $\bar{\beta}_{12} = 1$, $\bar{\gamma}_{11} = 1$, $\bar{\gamma}_{12} = 2$, $\bar{\delta}_{22} = 4$, and $\bar{\gamma}_{32} = 4$ with $\bar{z}_x = 34$. The solution for the y-coordinate is $\bar{y}_1 = 2$, $\bar{y}_2 = 1$, $\bar{\alpha}_{12} = 1$, $\bar{\gamma}_{11} = 2$, $\bar{\gamma}_{12} = 1$, and $\bar{\delta}_{32} = 4$ with $\bar{z}_y = 12$. In other words, one of the optimal solutions is to locate our facility at (1, 2), and the second at (2, 1). The sum of weighted rectilinear distances is then $\bar{z} = \bar{z}_x + \bar{z}_y = 34 + 12 = 46$.

In general, the linear programming problem solved above has $(p + 2n)p$ variables and $½p(p - 1) + pn$ constraints. Some interesting applications of this problem are described by Love and Yerex (1976). While the multi-facility location problem with fixed interactions is, at least in principal, easy to solve as long as rectilinear distances are employed, many of the problem's extensions are difficult.

The best known minisum facility problem in the plane with multiple facilities is the *Multifacility Weber problem*, or the *Multi-Weber problem* for short. For an excellent historical account of the problem, see Wesolowsky (1993). The Multifacility Weber problem allows no interactions between facilities, but locates facilities so as to minimize the total weighted distance of assignment, where each demand point is assigned to its closest facility. The locations are allowed to be anywhere on the continuous surface. Virtually all research on Multi-Weber problems has involved the assumption that the location surface can be represented by a Cartesian plane or a spherical surface.

Define now (x_j, y_j) as the coordinates of the (unknown) j-th facility location, and define additional variables u_{ij} that assume a value of 1, if the i-th customer is served by the j-th facility, and 0 otherwise. With d_{ij} denoting the distance between the i-th customer and the j-th facility, the *Multi-Weber Problem* can be formulated as

Chapter 3: Continuous Location Models   231

$$P: \quad \text{Min } z = \sum_{i \in I} \sum_{j \in J} w_i u_{ij} d_{ij}$$

$$\text{s.t.} \sum_{j \in J} u_{ij} = 1 \ \forall \ i$$

$$u_{ij} = 0 \vee 1 \ \forall \ i \in I, j \in J.$$

In this formulation, the weighted distance between a customer and a facility is included in the objective function, only if $u_{ij} = 1$, i.e., the i-th customer is served by the j-th facility. The structural constraints ensure that each customer is served by exactly one facility.

At first glance, this model appears to be an equivalent of the network p-median problem in continuous space. There are some important differences, though. Firstly, the Multi-Weber problem is a nonlinear programming problem as it involves the product of the variables x_j and u_{ij}, and of y_j and u_{ij}, respectively. Secondly, there is no direct equivalent of Hakimi's theorem that allows searching for optimal locations on a finite, albeit astronomically large, set. However, it is possible to use the result that the optimal solution will consist of allocations of customers that are p subsets of the set of customers I, which are mutually exclusive and collectively exhaustive. This idea is applied in the approach described below.

In general, it can be said that location-allocation models are difficult, and most exact solution methods have been applied to problems with less than 100 demand points.

One exact solution method for Multi-Weber problems was suggested by Rosing (1992) and Rosing and Harris (1990). The basic idea of the approach is as follows. Wherever the facilities will be located at optimum, their locations will result in a partition of the set of customers, such that each customer is assigned to its closest facility, all customers assigned to the same facility define one customer set, and within each such customer set the facility is located so that it minimizes the sum of weighted distances to the customers in the set it is assigned to. Clearly, the latter feature makes each facility locate at the optimal Weber point within its customer set. The idea is now to enumerate all (or, if Rosing's method is to be used as a heuristic, some subset of) the partitions (and their customer sets) possible in the problem. Defining variables v_j which assume a value of one, if the j-th customer set is included in the solution, and zero otherwise. Furthermore, let c_j denote the sum of the weighted distances from all customers in the j-th customer set to the facility serving that set, i.e., $c_j = \sum_i w_i d_{ij}$ with d_{ij} denoting the (Euclidean) distance between the i-th customer and the j-th facility. Clearly, determining c_j requires the solution of a single-facility Weber problem for each customer set under consideration. Define p as the number of facilities to be located and let N_i

comprise all customer sets that include customer i. Then the Multi-Weber problem can be formulated as

$$P: \quad \text{Min } z = \sum_j c_j v_j$$

$$\text{s.t. } \sum_j v_j = p$$

$$\sum_{j \in N_i} v_j \geq 1 \quad \forall \, i = 1,\ldots, n$$

$$v_j = 0 \vee 1 \quad \forall \, j.$$

The objective function adds the weighted customer-facility distances of all partitions included in the solution, the first constraint ensures that exactly p partitions (and facilities to serve them) are included in the solution, and the remaining structural constraints guarantee that each customer is included in at least one partition. This reduces Rosing's formulation to a constrained set covering model. It should be noted that at optimum, each customer will be assigned to exactly one partition. The proof is by contradiction. If this property were violated, then there would exist at least one customer assigned to two or more customer sets. In such a case, one of these customer sets could be reduced in size without leaving a customer without service. For the reduced customer set, a facility location with a smaller c_j value can be computed, thereby creating a better solution. Hence a solution with overlapping customer sets cannot be optimal. Note that this property holds only in the exact case, i.e., if *all* partitions and their customer sets are included in the model.

The major drawback of Rosing's approach is that the number of potential partitions is extremely large. However, some partitions can be discarded from consideration. For example, if S_1, S_2, S_3, and S_4 are arbitrary customer sets, such that $S_1 \cup S_2 = S_3 \cup S_4$, then only one of the two pairs may be included. Thus we may discard from consideration the pair whose c_j value is higher. Rosing (1992) and Rosing and Harris (1990) have shown a number of possible ways to implicitly enumerate all possible partitions and customer sets, represent those that are not eliminated by rejection rules, and then use the above covering model to solve the Multi-Weber problem. Rosing has applied this approach to problems of up to 60-100 nodes. Beyond that size, exact techniques have yet to be developed.

Example : Consider four customers located at (0, 1), (2, 1), (2, 0), and (0, 0); these customers will be numbered 1, 2, 3, and 4. For simplicity, all customers are assumed to have unit weights. The customer sets are then {1}, {2}, {3}, {4}, {1, 4}, {2, 3}, {1, 2}, {3, 4}, {1, 2, 3}, {1, 2, 4}, {1, 3, 4}, and {2, 3, 4}. Note that {1, 3} and {2, 4} are not included as they would result in noncontiguous regions. Numbering the customer sets $v_1, v_2,\ldots, v_{12}$, the sum of customer-facility distances

Chapter 3: Continuous Location Models   233

for these sets are $c_j = 0$ for $j = 1,..., 4$; $c_j = 1$ for $j = 5, 6$; $c_j = 2$ for $j = 7, 8$; and $c_j \cong 2.9$ for $j = 9,..., 12$. Given that $p = 2$ facilities are to be located, the problem can be written as

P: Min $z = \quad v_5 + v_6 + 2v_7 + 2v_8 + 2.9v_9 + 2.9v_{10} + 2.9v_{11} + 2.9v_{12}$

$$\begin{array}{l} \text{s.t.} \quad v_1 + v_2 + v_3 + v_4 + v_5 + v_6 + v_7 + v_8 + v_9 + v_{10} + v_{11} + v_{12} = 2 \\ \quad v_1 \qquad\qquad\quad + v_5 \qquad + v_7 \qquad + v_9 + v_{10} + v_{11} \qquad\quad \geq 1 \\ \qquad v_2 \qquad\qquad\qquad + v_6 + v_7 \qquad + v_9 + v_{10} \qquad\quad + v_{12} \geq 1 \\ \qquad\quad v_3 \qquad\qquad\quad + v_6 \qquad + v_8 + v_9 \qquad\qquad + v_{11} + v_{12} \geq 1 \\ \qquad\qquad v_4 + v_5 \qquad\qquad + v_8 \qquad\quad + v_{10} + v_{11} + v_{12} \geq 1 \\ \quad v_1,\ v_2,\ v_3,\ v_4,\ v_5,\ v_6,\ v_7,\ v_8,\ v_9,\ v_{10},\ v_{11},\ v_{12} = 0 \vee 1. \end{array}$$

The unique optimal solution of this problem is $\overline{v}_5 = \overline{v}_6 = 1$ and $\overline{v}_j = 0$ otherwise, with an objective value $\overline{z} = 2$. In terms of the original problem, the solution indicates that one of the facilities will serve customers 1 and 4, while the other facility will serve customers 2 and 3. The two facilities will be located at $(0, ½)$ and $(2, ½)$, respectively, so that the distances are halfway between each facility-customer pair for a total distance of 2.

The remainder of this section is devoted to heuristic methods. As usual, heuristic solution methods are categorized with respect to their function as construction heuristics and improvement heuristics. In order to apply a Greedy-style construction heuristic, we need a technique that solves a conditional location problem. Such a technique will optimally locate a single facility in the presence of $s \in [1; p[$ already existing facilities. Given such a method, a Greedy method could then locate the first facility with any of the pertinent single-facility methods described in the previous section. It would then be applied repeatedly, locating one additional facility in each step.

Another possible construction heuristic exploits the similarity of location and clustering methods. The idea is to form p clusters that are similar in the sense that each cluster includes points (i.e., customers) that are located close to each other. Customers in the same cluster are connected by one or more edges. A simple heuristic could be as follows.

A Clustering Heuristic

Step 1: Calculate the distances between all pairs of customers and put them in nondecreasing order. Set the number of edges introduced so far to $k := 0$.

Step 2: Are all customers assigned to clusters?
> If yes: Stop. Determine the optimal location within each cluster by one of the pertinent methods.
> If no: Set $k =: k+1$, and go to Step 3.

Step 3: Choose the k-shortest connection and assign the two customers that it connects to the same cluster (a new cluster if both of them were unassigned so far). Go to Step 2.

Example 1: Consider six customers whose locations and weights are shown in Table II.4.

Table II.4

Customer	P_1	P_2	P_3	P_4	P_5	P_6
a_i	0	0	2	6	6	8
b_i	0	5	3	0	3	3
w_i	50	40	30	60	30	50

Assume now that $p = 2$ facilities are to be located, i.e., we need to establish two clusters.

For simplicity, we will use squared Euclidean distances. These distances between the customers are summarized in the direct distance matrix.

$$\begin{bmatrix} 0 & 25 & 13 & 36 & 45 & 73 \\ 25 & 0 & 8 & 61 & 40 & 68 \\ 13 & 8 & 0 & 25 & 16 & 36 \\ 36 & 61 & 25 & 0 & 9 & 13 \\ 45 & 40 & 16 & 9 & 0 & 4 \\ 73 & 68 & 36 & 13 & 4 & 0 \end{bmatrix}$$

The first six pairs of points that are closest to each other are (P_5, P_6), (P_2, P_3), (P_4, P_5), (P_4, P_6), (P_1, P_3), and (P_3, P_5). The first edge connects P_5 and P_6 in one cluster C_1, the second edge path P_2 and P_3 in one cluster C_2, the third edge connects P_4 and P_5, so that cluster C_1 now includes P_4, P_5, and P_6. The fourth edge connects P_4 and P_6 which has no effect as P_4 and P_6 are already in one cluster, and the fifth edge adds P_1 to C_2. At this point there are $p = 2$ connected components, and the procedure terminates with $C_1 = \{P_4, P_5, P_6\}$, and $C_2 = \{P_1, P_2, P_3\}$. It is only at this point that the locations are determined. The optimal locations within the two clusters are then $(940/140, 240/140) = (6.7143, 1.7143)$, and $(60/120, 290/120) = (0.5, 2.4167)$. At this point, an improvement heuristic may be used.

Chapter 3: Continuous Location Models

As in the case of network locations, many improvement heuristics can be devised. One such method dates back to the 1960s, when Cooper (1963) proposed several solution procedures for planar location models. One of those methods is known as Cooper's *Alternate Method*, developed to solve Multi-Weber models. This heuristic algorithm alternates between location and allocation phases, and it is said to converge if two successive locations or allocations are identical. In the location phase, a problem is solved that locates a single facility in each known cluster of customers. The allocation phase takes a given set of facility locations and assigns each customer to his closest facility. Starting with an arbitrary clustering of customers or set of facility locations, these two phases alternate. Similar heuristics were devised by Schultz in the sanitary engineering literature; see Helms and Clark (1971), and Maranzana (1964) in the context of warehouse location.

It is easy to prove that the approach will monotonically converge. Unfortunately, it most often stops at a local, but not a global, optimum. The application of such a technique should always involve the restarting of the heuristic a number of times with different initial solutions to ensure that a near-optimal solution is found.

Example 2: Consider again the scenario in Example 1, and suppose that the clusters with $C_1 = \{P_1, P_4\}$, and $C_2 = \{P_2, P_3, P_5, P_6\}$ have been generated. Given this allocation of P_1 and P_4 to facility 1, and P_2, P_3, P_5, and P_6 to facility 2, we now apply the location phase by optimally locating a single facility in each cluster. Given the squared Euclidean metric, the optimal locations are determined to be at $(360/110, 0) = (3.2727, 0)$ for C_1, and $(640/150, 530/150) = (4.2667, 3.5333)$ for C_2. At this point we enter the allocation phase, which assigns a customer to his closest facility. Here, this allocation step again assigns the customers P_1 and P_4 to facility 1 and the customers P_2, P_3, P_5, and P_6 to facility 2, so that the method has converged. The value of the objective function is then $\bar{z} = 61.4376$.

Another well-known improvement heuristic is the *vertex substitution method*. Suitably adapted, it can also be used for the Multi-Weber problem. Such an adaptation could proceed as follows. Given clusters, remove one customer from a cluster and tentatively assign him to the other clusters, one at a time. For each such assignment, the new objective value is calculated. This process is then repeated for all customers. A customer is then reallocated if his reallocation results in the largest savings. If no savings are possible, the procedure terminates. It should be pointed out that the process can be modified, so that any reassignment that results in cost savings is immediately made, rather than calculating the savings for all possible reassignments of a single customer, and then making the change with the lowest value of the objective function.

Example 3: Consider again the scenario described in Example 1, and suppose again that the clusters $C_1 = \{P_1, P_4\}$, and $C_2 = \{P_2, P_3, P_5, P_6\}$ have been determined "somehow." For these clusters, the optimal locations are again

(3.2727, 0) for C_1, and (4.2667, 3.5333) for C_2, and the objective value is $\bar{z} = 61.4376$.

Tentatively remove now P_1 from cluster C_1 and allocate it to C_2. The optimal locations (given again the squared Euclidean metric) are then (6, 0) for $C_1 = \{P_4\}$, and (640/200, 530/200) = (3.2, 2.65) for $C_2 = \{P_1, P_2, P_3, P_5, P_6\}$ for an objective value of $\bar{z} = 0 + 65.7125 = 65.7125$. There is no improvement here, so the method continues.

In the next step, we remove customer P_2 from his cluster, and tentatively assign him to cluster C_1. This results in clusters $C_1 = \{P_1, P_2, P_4\}$, and $C_2 = \{P_3, P_5, P_6\}$. The optimal locations within these clusters are (360/150, 200/150) = (2.4, 1.3333), and (640/110, 330/110) = (5.8182, 3), with a value of the objective function of $\bar{z} = 41.48 + 19.3720 = 60.8520$. As this exchange results in a reduction of the value of the objective function, the move is made, and the process continues.

CHAPTER 4   OTHER LOCATION MODELS

In this chapter we consider a variety of location problems that do not fit into the categories of covering, center, and median problems. One strand of research deals with undesirable facilities. As individuals are less likely to tolerate new facilities constructed nearby them (which has an interesting correlation with the increase of compensation settlements in and out of court), locations of undesirable facilities have become increasingly scrutinized and analyzed. Along similar lines, objectives that include some component purportedly designed to measure "fairness" have also been introduced. Other location problems that are generating interest include the location of hubs, competitive location problems, and the location of facilities that are too large to be represented as points in some given space.

4.1   The Location of Undesirable Facilities

In the discussion in the previous chapters we have assumed that proximity of a facility to customers is a virtue. Put differently, one of our assumptions was that the spatial separation between facilities and customers represented a disutility, such as a cost or distance. Clearly, not all facilities fall into this category. Many facilities are noxious or obnoxious in the sense that they either emit poisonous substances or are simply a nuisance due to their noise, air, or water pollution. While there is a significant difference between noxious (i.e., toxic) facilities and those that are simply obnoxious (i.e., annoying), there is no difference in the models that are used to locate them. In order to unify the terminology, we will simply call them *undesirable facilities*. The first to discuss undesirable facilities were Goldman and Dearing (1975). Until then, all facilities that were located had been assumed desirable in the sense that the closer they were to the customer, the better the solution would be. This desirability was now put into question. Clearly, power plants, landfills, and ammunitions dumps could hardly be called desirable. However, if the decision maker wants to minimize the cost of delivering a service from the facility to its customers, then proximity is desirable, and the usual minisum objectives apply. Most often it is not the planner who finds the facilities objectionable, but the general public who has to live with them. Their objective is then to put as much distance between themselves and the facilities to be located. This discussion demonstrates that it is not so much the fact that the facility is desirable or undesirable (even though that is what we will call it) which determines the objective of the model, but rather the decision maker's utility that will indicate whether proximity of a facility to the customers is desirable or undesirable. It also demonstrates that in the location of undesirable facilities, a main feature is that there are multiple "players" or stakeholders involved, each with his own objective. For instance, the planner of a power plant, an arguably undesirable facility, would most likely want to locate the facility as close to the population concentration as possible in order to avoid an extensive grid of transmission lines (a typical minisum cost objective), while another player, *viz.*, the public who lives in the vicinity of the plant, would wish it to be as far away as possible from them (a maxisum or maximin objective). It is apparent that location problems with multiple objectives are much more difficult to solve than single-objective problems. For a discussion of multi-objective problems, readers are referred to part I of this volume as well as Eiselt *et al.* (1987). For a classification of location models involving undesirable facilities and pertinent references until that time, see Erkut and Neuman (1989).

To add to the difficulty, desirable and undesirable facilities are but extremes of the continuum. A typical facility has both desirable and undesirable features. Take, for instance, a supermarket. On the one hand, customers would find proximity desirable as it will be convenient for them to be able to purchase groceries nearby. On the other hand, customers will soon realize the undesirable side of the business: deliveries are typically made around 5 or 6 a.m., which is a noisy procedure. As a result, many models are formulated with multiple objectives, thus acknowledging positive and negative aspects of the facility. The communication design problem in Section 3.1 of this part is one example of such a formulation. Deeney (1999) presents a novel approach to the problem.

It is also worth pointing out that there are relatively few applications for the location of undesirable facilities on networks, as network distances are rarely appropriate to measure utilities of spatial separation. In other words, odors, radiation, noise, and other kinds of pollution most frequently travel along straight lines, making continuous spaces more amenable to the location of undesirable facilities. Furthermore, the "Not In My Back Yard" (NIMBY) or, stronger still, the "Build Absolutely Nothing Anywhere Near Anything" (BANANA), syndrome is an important issue in many location problems. One model that addresses these concerns on networks is the *p-maxian* or *p-anti-median* model. Its structure (but not its formulation) is identical to that of the *p*-median, except that the objective involves the maximization, rather than minimization, of the sum of weighted distances. Without any further restrictions, the p-maxian model would locate all p facilities at the same point. (Actually, many regional planners use this concept by defining use zones for polluting facilities, thus having them aggregated, rather than dispersed throughout the region).

Assume now that each node in the network can be home to only one facility. As usual, let I denote the set of nodes with positive customer demand, and J the set of nodes that are potential facility locations. We can then define zero-one variables y_j, so that y_j equals 1, if a facility is located at the node n_j, and zero otherwise. In addition, we define zero-one variables x_{ij}, so that $x_{ij} = 1$, if a customer at node n_i is served by a facility at n_j, and zero otherwise. As usual, each customer has a demand or weight of w_i, d_{ij} denotes the shortest distance between node n_i and node n_j, and p is the number of facilities that are to be located. Finally, the set Q_{ij} is defined as $Q_{ij} = \{k: d_{ik} \leq d_{ij}\} \; \forall \; i \in I, j \in J$, i.e., the set of facilities that are no further from customer i than facility j. We can then formulate the p-maxian problem as follows.

$$P: \text{Max } z = \sum_{i \in I} \sum_{j \in J} w_i d_{ij} x_{ij}$$

$$\text{s.t.} \sum_{j \in J} x_{ij} = 1 \; \forall \; i \in I$$

$$x_{ij} \leq y_j \; \forall \; i \in I, j \in J$$

$$\sum_{j \in J} y_j = p$$

$$\sum_{k \in Q_{ij}} x_{ik} \geq y_j \; \forall \; i \in I, j \in J \qquad (1)$$

$$x_{ij} = 0 \vee 1 \; \forall \; i \in I, j \in J$$

$$y_i = 0 \vee 1 \; \forall \; j \in J.$$

Apart from the maximization objective and the constraints (1), the formulation is no different from the standard p-median problem discussed in Chapter 2. However, without these "closest facility constraints," the problem would allocate a customer not to his closest facility as required (as the minimization objective in the standard p-median problem does automatically), but the maximization objective in this problem would allocate a customer to the facility farthest from him. In order to avoid this, additional constraints are required. In particular, the constraints (1) are redundant, whenever a facility does not exist at node n_j, since y_j is then zero. Consider now the case in which facilities exist at, say, nodes n_ℓ and n_j. Without loss of generality assume that $d_{i\ell} < d_{ij}$. Then the left-hand side of the constraint (1) for i and ℓ is $\sum_{k \in Q_{ij}} x_{ik} = x_{i\ell} \geq y_\ell = 1$, thus implying that $x_{i\ell} = 1$. This is the desired result, as the facility at n_ℓ is closer to the customer at n_i than any other facility. A similar formulation is used by Daskin (1995). One of the drawbacks of this formulation is its large size. The problem formulation includes no less than $(n^2 + n)$ variables, and $(2n^2 + n + 1)$ constraints.

This formulation will be illustrated in the following

Example: Consider the graph shown in Figure II.17.

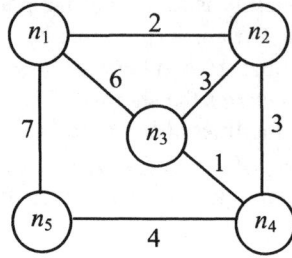

Figure II.17

The sets Q_{ij} can then be shown in the following matrix:

$$(Q_{ij}) = \begin{bmatrix} \{1\} & \{1,2\} & \{1,2,3,4\} & \{1,2,3,4\} & \{1,2,3,4,5\} \\ \{1,2\} & \{2\} & \{1,2,3,4\} & \{1,2,3,4\} & \{1,2,3,4,5\} \\ \{1,2,3,4,5\} & \{2,3,4\} & \{3\} & \{3,4\} & \{1,2,3,4,5\} \\ \{1,2,3,4,5\} & \{2,3,4\} & \{3,4\} & \{4\} & \{2,3,4,5\} \\ \{1,2,3,4,5\} & \{1,2,3,4,5\} & \{3,4,5\} & \{4,5\} & \{5\} \end{bmatrix}.$$

For simplicity, we will only write the constraints that relate to the customer at the node n_3.

P: Max $z = \ldots + w_3 (d_{31}x_{31} + d_{32}x_{32} + d_{33}x_{33} + d_{34}x_{34} + d_{35}x_{35}) + \ldots$

s.t.
$$\begin{aligned}
x_{31} + x_{32} + x_{33} + x_{34} + x_{35} &= 1 \\
x_{31} &\leq y_1 \\
x_{32} &\leq y_2 \\
x_{33} &\leq y_3 \\
x_{34} &\leq y_4 \\
x_{35} &\leq y_5 \\
x_{31} + x_{32} + x_{33} + x_{34} + x_{35} &\geq y_1 \\
x_{32} + x_{33} + x_{34} &\geq y_2 \\
x_{33} &\geq y_3 \\
x_{33} + x_{34} &\geq y_4 \\
x_{31} + x_{32} + x_{33} + x_{34} + x_{35} &\geq y_5
\end{aligned}$$

Chapter 4: Other Location Models

For locations of facilities at, say, nodes n_2 and n_4, we obtain

P: Max $z = \ldots + w_3 (5x_{31} + 3x_{32} + 0x_{33} + 1x_{34} + 5x_{35}) + \ldots$
s.t.
$$x_{31} + x_{32} + x_{33} + x_{34} + x_{35} = 1$$
$$x_{31} \leq 0 \text{ (forcing } x_{31} = 0)$$
$$x_{32} \leq 1 \text{ (redundant)}$$
$$x_{33} \leq 0 \text{ (forcing } x_{33} = 0)$$
$$x_{34} \leq 1 \text{ (redundant)}$$
$$x_{35} \leq 0 \text{ (forcing } x_{35} = 0)$$
$$x_{31} + x_{32} + x_{33} + x_{34} + x_{35} \geq 0 \text{ (redundant)}$$
$$x_{32} + x_{33} + x_{34} \geq 1$$
$$x_{33} \geq 0 \text{ (redundant)}$$
$$x_{33} + x_{34} \geq 1 \ (x_{34} = 1 \text{ as } x_{33} = 0)$$
$$x_{31} + x_{32} + x_{33} + x_{34} + x_{35} \geq 0 \text{ (redundant).}$$

With $x_{34} = 1$, all other variables $x_{31} = x_{32} = x_{33} = x_{35} = 0$, and the distribution variables indicate that the customer at node n_3 will be served by his closest facility at node n_4.

Already in the late 1970s, Church and Garfinkel (1978) demonstrated that the Hakimi property (Theorem II.1 in Chapter 2) does not hold even for 1-maxians, but that there is a finite set of points among which an optimal solution can be found. More specifically, they prove that there exists at least one optimal location on a cycle in the network or at a leaf node (implying that in tree networks at least one optimal location is at a leaf node).

From a technical point of view, the idea is to solve a maxsummin (max-sum-min) problem which maximizes the (weighted) sum of minimal customer-facility distances. An important underlying assumption is that a customer is only and exclusively affected by the facility closest to it. Typically, this is not true if we deal with pollution, in which case a customer is affected by *all* facilities. The degree of pollution is then a function of the distance (many types of pollution decrease with the square of the distance).

In order to demonstrate that the Hakimi property does not hold in general in maxian (i.e., maxsummin) problems, consider the network in Figure II.18 with the single-digit numbers next to the edges indicating the distances, and the double-digit numbers next to the nodes denoting the demands or weights of the customers.

At the nodes n_1, n_2, and n_3, the sum of weighted distances is 240, 340, and 170, respectively, making n_2 the best choice among the nodes. Consider now the edge (n_2, n_3) and denote by x any point on that edge at a distance of x from n_2. The shortest distance between node n_1 and any point x on (n_2, n_3) is then $(6 + x)$ for $x \leq$ ½ (reached via n_2) and $(7 - x)$ for $x \geq$ ½ (via n_3). In other words, the point $x =$ ½ is a break point on the edge (n_2, n_3), and the sum of weighted distances is $z = 345$. It can be shown in general that the sum of weighted distances is a piecewise linear,

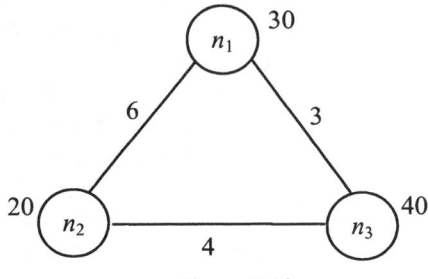

Figure II.18

concave function along each edge, so that, in order to find the maximum of that function, it is sufficient to compute the functional values at the break points on all edges. In this simple example, each edge has but a single break point and their respective values are $z = 415$ for $x = 3$½ distance units from n_1 on (n_1, n_2), $z = 345$ at $x =$ ½ distance unit from n_2 on (n_2, n_3), and $z = 245$ at $x =$ ½ distance unit from n_1 on (n_1, n_3). These three break points together with the three nodes constitute the set of possible optima. Enumeration reveals that the objective value is maximal with $z = 415$ at the break point on (n_1, n_2) which is the unique optimum. In general, the p-maxian problem on networks is **NP**-hard; see, e.g., Erkut *et al.* (1990).

While networks are the space of choice when it comes to transportation, they are not that useful to measure the spread of pollution and other undesirable effects. Modeling problems with undesirable facilities in the plane is often a better way to model the undesirable effects of a facility. Suppose that a facility is to be located in the presence of n customers. It is then apparent that regardless of the metric applied the maxisum objective will push the facility towards infinity. This implies that—as opposed to the usual minisum objectives—we need a bounded set on which to locate the facility.

For the case of maximizing the shortest distance in the presence of ℓ_2 distances, Shamos and Hoey (1975) have shown the equivalence of the maximin location problem and that of finding the center of the *largest empty circle*; i.e., the circle with the largest diameter whose center is located in the feasible set and whose disk it encloses does not include any customer points. They also show that in a Voronoi diagram, the optimal maximin location is either at a Voronoi point or at the intersection of a Voronoi edge and the convex hull of the feasible set of locations.

4.2 p-Dispersion Problems

A somewhat different location problem is a maxisum problem known as the *p-dispersion* problem. Its objective is to locate *p* facilities on a network so that the minimum separation distance between any two pairs of open facilities is maximized. Notice that customers play no role in this model. This problem has been suggested for the case where facilities pose a threat to each other, such as ammunition dumps, but it is also applicable to (slightly) more benign scenarios such as the location of franchises. One of the objectives of a franchiser will be to locate the facilities, so that none of the franchises cannibalizes too much demand from another franchise. In other words, the objective is to spread out the facilities to the greatest extent possible so as to avoid interference between franchises, thus allowing each franchise to prosper.

Defining zero-one variables y_ℓ which assume a value of one, if a facility is located at node ℓ, and zero otherwise, the *p*-dispersion location problem can be written as

P: Max $z = \min_{\substack{i:y_i=1 \\ j:y_j=1}} \{d_{ij}\}$

s.t. $\sum_{\ell \in J} y_\ell = p$

$y_\ell = 0 \vee 1 \ \forall \ \ell \in J.$

The problem can be rewritten as a zero-one knapsack problem with a minimax objective, or, alternatively, as a regular zero – one programming problem as

P: Max z

s.t. $z \leq d_{ij} + M(2 - y_i - y_j) \ \forall \ i, j \in J$

$\sum_{j \in J} y_j = p$

$y_j = 0 \vee 1 \ \forall \ j \in J.$

Consider the first set of constraints. For those pairs of (i, j) that have facilities located at both of them (i.e., $y_i = y_j = 1$), the right-hand side value is d_{ij}, whereas for all other (i, j) pairs, the right-hand side is a function of M and thus redundant. Consequently, the value of z on the left-hand side is bounded only by those pairs of nodes that both have facilities located at them. This will restrict the value of z to the smallest distance between any pair of nodes with facilities located at them. Maximizing z will then find the largest possible such value, which is the objective shown above. A similar formulation of the *p*-dispersion problem was provided by Kuby (1987).

4.3 Location Models with "Equity" Objectives

Problems with "equity" objectives have been discussed for a long time in the social sciences. An important strand of research in this context deals with income inequalities; see, e.g., Dalton (1920). Problems with objectives that attempt to distribute disutilities as evenly as possible among the elements of a given set were referred to as "balancing problems" by Eiselt and Laporte (1995). Their basis, the concept of "equity," is defined in the American Heritage Dictionary as "Something that is just, impartial, and fair." As fairness is as fuzzy a concept as is equity, most authors employ the simplistic concept of equality instead. Such a concept may, for instance, be justified when public decision makers locate facilities that should be "equally" accessible by all (potential) users.

One of the main questions is how (in-) equality of access is to be measured. Marsh and Schilling (1991, 1994) and Eiselt and Laporte (1995) provide a large number of possible measures. Here, we will illustrate some of the more important measures. To formalize, consider the problem of locating a single facility, and define again d_i as the distance between the i-th customer and the new facility, $i = 1, ..., m$. Furthermore, let $\bar{d}$ denote the average distance between the customers and the new facility. We can then devise the following measures of inequality that may be minimized:

(1) the *center* $z_c = \max_i \{d_i\}$,

(2) the *range* $z_r = \max_i \{d_i\} - \min_i \{d_i\}$,

(3) the *total absolute deviation* $z_t = \sum_i |d_i - \bar{d}|$,

(4) the *total variation* $z_v = \sum_i (d_i - \bar{d})^2$, and

(5) the *Gini index* $z_G = \dfrac{\sum_i \sum_\ell |d_i - d_\ell|}{2m^2 \bar{d}}$.

The concept of a center was already discussed at length in the previous chapters of this part. The range is a similar measure that looks at the two extremes of the customer-facility distances, rather than just the longest as the center does. The problems with the range are that, similar to the center objective, it examines only the extremes, and also, in trying to minimize the width of the interval, attempts not only to find locations in which the longest distances are as short as possible, but it will also make the short (and very advantageous) distances longer. The mean absolute deviation and the variance (here expressed in absolute terms rather than divided by the constant m) both incorporate not only the extremes of the distribution, but all of its elements.

The Gini index was developed on the basis of the Lorenz curve, which is a common feature in the study of income distributions. The idea is as follows. On the abscissa, we plot the proportion of the total population, while the ordinate expresses the proportion of the total income. To any point, say x, on the abscissa, the value of y that corresponds to it is the proportion of the total income that the bottom x percent of the population has. If, for example, sixty percent of the population were to be in command of thirty percent of the income, then the point (0.6, 0.3) would be a point on the Lorenz curve. If, in the extreme, all incomes were equal, then the Lorenz curve coincides with the diagonal in the unit square from (0, 0) to (1, 1). On the other hand, if the richest ε percent ($\varepsilon \to 0$) of the population were to have 100 percent of the income, then the Lorenz curve would consist of two linear pieces, one from (0, 0) to (1, 0), and the other from (1, 0) to (1, 1). In other words, the degree to which the Lorenz curve equals or deviates from the diagonal is an expression for the equality of incomes. The Gini index is then calculated as the area between the Lorenz curve and the diagonal, in relation to the entire area below the diagonal (which equals ½). Replacing the proportion of the total population on the abscissa by the proportion of customers, and replacing the proportion of the total income by the proportion of the total customer-facility distances, we can then plot the Lorenz curve with points (x, y) as the proportion of customers x whose customer-facility distances are y percent of the sum of customer-facility distances.

In order to explain the concepts above, consider the following

Example: Three customers are located at the points (0, 0), (6, 0), and (0, 2). Suppose that two possible facility locations have been proposed, (1, 1) and (3, 0). These locations are to be evaluated with respect to the criteria introduced above. The ℓ_1 distances between (1, 1) and the three customers are 2, 6, and 2, respectively, with an average distance of 3.3333. Similarly, the ℓ_1 distances between the potential facility location at (3, 0) and the customers are 3, 3, and 5, respectively, with an average distance of 3.6667. The balancing measures for the two potential facility locations are summarized in Table II.5.

Table II.5

Measure	Facility at (1, 1)	Facility at (3, 0)
z_c	6	5
z_r	4	2
z_t	5.3333	2.6667
z_v	10.6667	2.6667
z_G	4/15 = .2667	4/33 = .1212

It is apparent that while locating a facility at (1, 0) is more efficient (as the average distance to a customer is shorter than from the alternative location at (3, 0)), all balancing criteria prefer locating the facility at (3, 0).

Constructing the Gini index, we first construct the Lorenz curve, whose axes are "% of observations", and "% of total distance". For a facility location at (1, 1), the sum of distances is 10. The first customer, who accounts for 1/3 of all customers, is 2 units away from the facility, which is 2/10 of the total, hence the first point on the Lorenz curve (other than (0, 0)) is (1/3, 2/10). The second-closest customer is also 2 units away from the facility, so that 2/3 of the (closest) customers account for 4/10 of the total distance, giving the point (2/3, 4/10) on the Lorenz curve. Finally, considering all customers, 100% of all customers account for 100% of the total distance, so that the point (1, 1) results. We now have a Lorenz curve that consists of only two linear pieces, one from (0, 0) to (2/3, 2/5), and a second from (2/3, 2/5) to (1, 1). The triangle above these two linear pieces but below the line that connects (0, 0) and (1, 1) has an area of 2/15. Since the total area of the triangle under the diagonal is ½, the triangle with area 2/15 therefore makes up 4/15 of the triangle under the diagonal; hence $z_G = 4/15$.

When choosing a balancing measure, decision makers want to know whether or not their chosen measure satisfies some reasonable properties. One of these axioms is the so-called *principle of transfers* that was first mentioned by Pigou (1912) and Dalton (1920) for income distributions. The principle is said to be satisfied if the transfer from a "rich" to a "poor" individual decreases inequality. In the context of location, it requires that moving closer to a distant facility (and away from a closer facility) will decrease inequality. Another requirement is scale invariance. It simply states that the unit of measurement should have no bearing on the measure. A number of measures in the context of location were set up by Eiselt and Laporte (1998).

Finally, we would like to provide an example that balancing objectives should not be used by themselves without an efficiency objective to accompany them. Suppose that three customers are located at, say, (0, 0), (10, 0), and (5, –0.1). With any objective that tries to make all customer-facility distances equal, the optimal location will be at the center of the unique circle that has all three customers located on its circumference. This center is located at (5, 124.95), so that the facility is located 125.05 distance units away from each of the customers, thus achieving perfect equality. However, if the facility were to move straight downwards, it would get closer to each of the three customers, so that all of them will win—but the solution would no longer be perfectly equal. Hence, a balancing objective without an efficiency component will tend to produce nonsensical solutions.

4.4 Hub Location Problems

One of the success stories of the century is the development of the overnight package express business. The flow of packages between any pair of cities, say, Reno, Nevada, and Charleston, South Carolina, is usually quite small. However, the total potential traffic to/from Reno or Charleston from/to other cities may be sizeable. The design of the Federal Express system started with the location of one hub which could take advantage of this property. Essentially all of the packages leaving Reno are put on a plane heading for Memphis which was chosen as the *hub*. The same applies to all other cities. The routes from various cities to Memphis form a series of *spokes*, sometimes made by several connecting flights. In Memphis, all of the airplanes are unloaded and the packages are sorted for their destinations. Then the packages are reloaded on the planes returning out along their spoke-like routes, dropping off packages. In this way a package heading from Reno to Charleston takes a route through Memphis. The same was true for packages shipped from Los Angeles to neighboring San Diego. In the beginning, Federal Express had only one hub and virtually every package went via Memphis. Now there exist many smaller hubs as well as the principal hub in Memphis as the system has expanded to accommodate growth in package volume. Hubs in general reduce network construction costs, centralize sorting and freight handling, and take advantage of the economies of scale associated with the consolidation of flows as O'Kelly (1986) points out. The hub system was also embraced by the major airlines for passenger traffic. The typical route requires passengers to first take a flight on a spoke to a hub, followed by a second flight to another hub on a main link, and finally a third flight on another spoke to the final destination. For example, Northwest Airlines' hubs are Detroit, Minneapolis/St. Paul, and Memphis. A traveler located in New York destined for Seattle could then be routed on a spoke connection to Detroit, followed by flights to Minneapolis/St. Paul, and on to Seattle. Clearly, if origin and destination are connected to the same hub such as Atlanta and Dallas are in the case of Northwest Airlines (both are directly connected to Memphis and Minneapolis/St. Paul), only two flights to and from the chosen hub are required.

The formulation uses the idea to locate several hubs, so that each hub will be connected directly with all other hubs. Traffic from node n_i to node n_j is to be routed through two hubs. Given this "hub paradigm," define the parameter $d_{\nu\mu}$ as the travel cost from node n_ν to node n_μ, f_{ij} as the flow from origin n_i to destination n_j, p as the number of hubs to be opened, and $\alpha \in [0; 1]$ as the interhub discount factor that specifies the proportion of cost charged for interhub traffic as opposed to transportation between hub and nonhub. For example, a discount factor of $\alpha = 0.8$ indicates that the traffic between hubs is 20% less costly per unit than between a hub and a source or a destination. We can then define variables $x_{ijk\ell}$ as the fraction of flow from origin $i \in I$ to destination $j \in J$ that is routed via a hub $k \in K$

and a second hub $\ell \in L$ and binary variables y_k that equal 1, if node n_k is chosen as a hub, and 0 otherwise. These definitions allows us to formulate the *p-Hub Network Design Problem* as

$$P: \text{Min } z = \sum_{i,j,k,\ell \in S} f_{ij} [d_{ik} + \alpha d_{k\ell} + d_{\ell j}] x_{ijk\ell}$$

$$\text{s.t.} \sum_{k,\ell \in S_{ij}} x_{ijk\ell} = 1 \; \forall \; j > i$$

$$\sum_{\ell \in S_{ij}} x_{ijk\ell} - y_k \leq 0 \; \forall \; j > 1, k \in K$$

$$\sum_{k \in S_{ij}} x_{ijk\ell} - y_\ell \leq 0 \; \forall \; j > 1, \ell \in L$$

$$\sum_k y_k = p$$

$$y_k = 0 \lor 1 \; \forall \; k \in K.$$

This formulation locates hubs and determines the optimal routing between each pair of nodes. It is assumed here that the interflow matrix between all nodes is symmetric. Given the large number of variables and constraints, this problem is relatively difficult to solve. Extensions to the model include restrictions to single node to hub connections, capacitated hubs, and single and multi hub routing. A number of variants are discussed in Campbell (1996), and Campbell *et al.* (2002).

4.5 Competitive Location Problems

The first competitive location model is due to the economist Hotelling (1929). The space Hotelling locates his duopolists in is a linear market, i.e., a line segment. Typical examples of linear markets include the location of bus stops along a route, the location of service plazas along a toll road, the location of supply points along a construction project, and the location or selection of product niches along a continuum associated with one variable. For example, the "cola wars" between Pepsi and Coke have often been represented along a continuous line representing sweetness. Hotelling's basic model of competition incorporates two competing planners who attempt to locate a single facility each on a linear market of length L. The two facilities are usually called A and B. Demand for a homogeneous good offered by A and B is assumed to be constant and uniformly distributed along the market. Note that such a demand structure is only applicable in cases in which an essential good is concerned; otherwise customer demand would typically be decreasing with increasing facility-customer distance. Hotelling used the example of "main street" to justify these assumptions; subsequent authors such as Alonso

(1964) have coined the phrase of the "ice cream vendor on the beach", a context that has not exactly prompted potential users to flock to employ the model. Each of the two firms has two decision variables: its location and the price it charges. Denote by x_A and x_B the locations of the two facilities and let p_A and p_B be the prices they charge. It is further assumed that both facilities employ mill pricing (i.e., customers pay for the price of the product at the firm and pay an additional transport cost based on the distance between them and the facility they patronize) and customers buy from the firm with the lowest full price (= mill price + transport cost). While the latter assumption is frequently justified as "customer rationality", things are more complicated than that. Such customer behavior requires complete information, rational behavior, *and* the fact that costs are a customer's only criterion to determine patronage. Transport costs are assumed to be linear in the quantities, a troublesome assumption that ignores the possibility of economies of scale and multi-purpose shopping. This scenario is shown in Figure II.19.

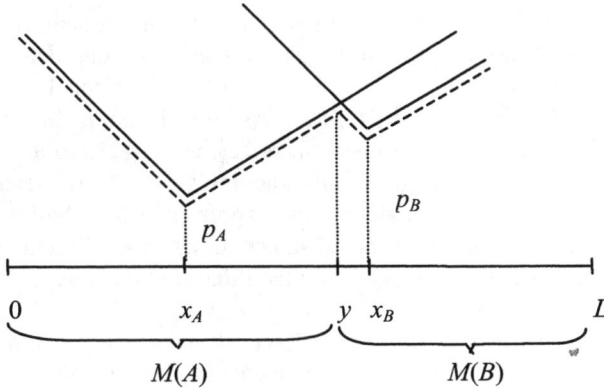

Figure II.19

The market in Figure II.19 is subdivided into three disjoint parts: the area to the left of firm A, called *A's hinterland*, the area to the right of firm B, called *B's hinterland*, and the area between firms A and B, called the *competitive region*. The prices p_A and p_B are charged at the respective facilities and the unit transport cost determine the slope of the V-shaped function above x_A and x_B, respectively. At each point of the market, a customer could purchase the desired goods, but the assumption of lowest full price results in customers paying according to the lower envelope of the two V-shaped full price functions, shown here as the broken line. The point y separates the two market areas $M(A)$ and $M(B)$: to its left, customers patronize firm A and to its right, they satisfy their demand from firm B.

The above discussion enables us to analyze the behavior of the two competitors. First, consider pure location competition. Assume that firm B has temporarily fixed its location and firm A is able to respond to it. For the time being, let $p_A = p_B$. As long as there is a competitive region of positive length (i.e., both firms do not agglomerate), firm A will gain additional market share if it moves towards its competitor. The force that pushes firms together is sometimes referred to as "market share force" as firms gain market share by moving towards each other. There are two main reasons for the existence of this force: one is that a facility does not lose demand in its hinterland as it moves towards its competitor (which is no longer true if the demand is sensitive to the distance between customer and firm as is the case when nonessential goods are considered) and the other is that a hinterland exists (which is the case in simple spaces such as linear markets and tree networks, but not d-dimensional space with $d \geq 2$ or general networks). If the mill prices of the two firms are still fixed, but no longer equal, we encounter a major discontinuity. At first, each facility gains market shares as it approaches its competitor. However, at some point the V-shaped full price function of the lower cost firm (here A) equals that of the more expensive firm (here B) on the right, and A's full price is lower everywhere else. An arbitrarily small distance closer than that, and firm A has a lower full price anywhere, meaning that it has undercut its competitor and now *captures* the entire market. (For details on market capture, readers are referred to ReVelle (1986)). As far as location choice goes, this means that in the case of fixed, but unequal, prices, as a stable solution cannot exist as the firm with the lower price will attempt to locate sufficiently close to its competitor, so that it can undercut and capture it, while the high-price firm will attempt to keep its competitor at a distance. In the case of fixed, and equal, prices, a stable solution can only exist if both facilities agglomerate.

Consider now price competition. For fixed locations, each firm has an incentive to lower its mill price, cut out its opponent and capture the entire market. The incentive to do so is greatest if the two firms are located close to each other. This undercutting does, of course, lower the facilities' profits, so that there is a "pricing force" that makes firms avoid highly competitive locations close to each other and thus pulls firms apart. This is the second major force that governs the process.

There are now two major types of analyses that can be performed. The first is preferred by economists, and it deals with the stability of the competitive situation. The concept applied in these circumstances is the *Nash equilibrium*. Loosely speaking, a Nash equilibrium is a situation out of which none of the players wants to break out unilaterally, as he cannot win by doing so. Most such analyses only analyze whether or not a Nash equilibrium exists in a given situation, thus concluding that the situation is stable in case it does, and that it is not, if no such equilibrium exists. Naturally, matters are more complicated than that. Eiselt and Bhadury (1995) have proposed a measure on a continuum between the extremes "stable" and "unstable." Hotelling's original model was long thought to have such an equilibrium with firms agglomerating at the center (a concept usually referred to as *central agglomeration*, or, in the context of product design, as the *principle of minimum differentiation*), until d'Aspremont *et al.* (1979) demonstrated that no such equilibrium exists. The nonexistence of an equilibrium can be envisaged by considering the forces at work in the model. On the one hand, there is the force of the market share that pulls facilities together, as they gain additional market share in the competitive region (i.e., between the facilities) without losing customers in their hinterlands. On the other hand, price competition pushes facilities apart, as competitive pressures increase the closer the facilities are to each other. Note, however, that the nonexistence result is rather fragile as simple changes to the transport cost function result in vastly different results. For instance, d'Aspremont *et al* in their aforementioned reference demonstrate that in case of quadratic rather than linear transport costs, a Nash equilibrium does exist with both facilities locating at the ends at the market, i.e., $x_A = 0$ and $x_B = L$.

A somewhat related issue is whether or not a pair of locations, be it stable or not, serves the customers well. (Remember that up to now, all planning has been made with the benefit of the firms in mind!) One possible measure would be the average travel distance of a customer which is easily computed to be $¼L$. For the sake of argument, suppose that the two firms were to locate at $x_4 = ¼L$ and $x_3 = ¾L$ instead. The point y that separates the two market areas is at $y = ½L$, and the average distance a customer must travel is now $⅛ L$, i.e., one half of that of central agglomeration. Actually, the solution $x_A = ¼L$ and $x_B = ¾L$ minimizes the average travel distance for customers and is frequently referred to as *social optimum*, as opposed to the *competitive optimum* with $x_A = x_B = ½L$.

The second solution concept is known as a *Stackelberg game*, based on the work of the German economist von Stackelberg (1943). The basic idea is that one firm, say A, acts as leader while the other, say B, is the follower. The leader will then locate, knowing that the follower will subsequently locate. The follower will then take the leader's location as given and optimize the location of its own firm. This planning with foresight was first formally employed by Prescott and Visscher (1977). Hakimi (1983) referred to the leader's location as *centroid*, while the follower's location was called *medianoid*. More specifically, he defined $(r|X_p)$ medianoids as the location of r new facilities that are located by a Stackelberg follower in the presence of p of the leader's firms located at a set of location X_p, so as to maximize the demand captured by its facilities. This type of objective was dubbed maximum capture, or "MAXCAP" by ReVelle (1986). Similarly, the Stackelberg leader faces the task of finding an $(r|p)$ centroid, i.e., optimally (again with respect to market capture) locating p facilities, considering the fact that the follower will locate r facilities later. While the follower's problem is a conditional location problem (i.e., locate r facilities given that p facilities are already located)

Stackelberg solutions of location problems was provided by Eiselt and Laporte (1996), and a framework for competitive location models and an extensive list of more than 100 contributions is found in Eiselt et al. (1993).

In general, medianoid problems, while **NP**-hard (Hakimi, 1983, 1990) can be solved, e.g., by heuristics such as tabu search; see Benati and Laporte (1994). Centroid problems, on the other hand, are considerably harder to solve in practice; in many contexts, nothing is known at all about how to determine centroids. One of the few papers that describe heuristic methods for the determination of centroids is found in Bhadury et al. (2003).

4.6 Locating Extensive Facilities and Routing in Irregular Spaces

The location problems addressed in this section have in common that either the space, in which facilities are located and units are shipped, is irregular, or the facilities are.

First consider the case of "extensive facilities." These are facilities that cannot be represented as a point in some space. This type of location problem is the link between typical location and layout problems. One of the more popular location problems in this category is the location of a line through an area, for which only the beginning and end points are given that the line connects. A typical example is a new road, transmission line, or pipeline that is to be constructed, such that it connects two towns, or places which are already connected with the remainder of the network. As the costs of establishing the connection are usually significant, the objective is typically to minimize the costs.

This "corridor location problem" has been addressed by a number of authors; see, e.g., Huber and Church (1985). Most researchers employ a digitization approach, which can be described as follows. First, the relevant area is subdivided into small cells. Within each cell, the feasibility of establishing the road is determined. Then, the costs of constructing a connection between all adjacent pairs of cells are determined by calculating the costs of connecting their respective centroids. This procedure results in a network, in which the shortest path can be determined between the two prespecified endpoints.

Many approaches have been devised, and most of these are based on the work of McHarg (1969). He developed an approach that can most easily be envisaged by using overhead transparencies. Suppose that a pile of transparencies have been prepared that all outline the same region. Each of these transparencies is designed to show the region with respect to different aspects. For example, the first transparency could show the property ownership, where each cell in the region is colored according to the costs that will be incurred if a right-of-way is established in the cell. Another transparency may show differences in topography, where more difficult areas are shaded in a deeper color. Putting all these transparencies on top of each other and projecting the result on the wall, the combined picture will show some lighter areas, in which routing will be easier, and darker areas, in which routing of the road or transmission line will be difficult and/or expensive. The very same approach is nowadays used in geographical information systems (GIS). At the time of this writing, GIS systems have become very popular with managers in business and government, particularly regional planners. Tremendous advances in the software that is available can now keep track of all sorts of spatial data. In fact, some GIS software now comes with embedded location model algorithms. Church and Sorensen (1994) have identified a number of problems that must be addressed when linking location models with geographical information systems.

One of the problems associated with the use of geographical information systems is the detail that is needed in a particular application. Clearly, if the decision maker wants to locate a retail facility, he will need to know where the potential customers are located—but not where the living room and the bedroom in their houses are located. Hence, if the spatial data are available at such detail, aggregation will be required. While aggregation of spatial data was first mentioned by Goodchild (1979), it still remains an important needed research area; see the work by Hillsman and Rhoda (1978), Current and Schilling (1987, 1990), and Murray and Gottsegen (1997). Another problem with spatial data is the magnitude of data. Relevant problems today are much beyond the capabilities of computers only a few years ago. However, software and hardware has caught up and can nowadays handle the massive amounts of data required by many applications. Another weak point are algorithms: while a number of exact and heuristic methods were known to perform quite well on small to medium-sized problems, some methods are not performing well on large problems. For example, Church and Sorensen (1994) have demonstrated for a real planning problem of modest size, that in order to have a 95% probability of identifying an optimal solution, a commonly employed location heuristic needed to be restarted nearly 800 times. For one 150 node data set representing London, Ontario, Canada, the Teitz and Bart heuristic, described in Chapter 2 of this part, had to be restarted 2,515 times in order to identify the optimal solution 10 times.

Finally, another type of location problem occurs when the facility may well be represented by a point, but the location is not possible everywhere in the plane, and neither is the transportation. If it is not possible to locate anywhere in the plane, we face a location problem with *forbidden areas*. On the other hand, if the transport of goods must detour around obstacles, we talk about a location problem with *barriers*. Barriers are typically rivers that can only be crossed on *gates* such as bridges, military zones that have to be detoured, or wildlife refuges, recreation areas, or national parks that do not allow driving through. The presence of objects such as barriers, gates, and forbidden regions helps add a degree of reality when some geographical features play a major role in the determination of the routes used between the points. For work on location models with forbidden areas and barriers, readers are referred to Klamroth (2002).

CHAPTER 5 LAYOUT MODELS

Typically, *facility* or *plant layout problems* are described in terms of locating items (e.g., machines or workstations or departments) in certain slots, spaces or positions in a plant (or office, school, etc.). As such, layout problems appear very similar to standard location models. There are, however, a number of significant differences between them. First, there are no "customers" in layout problems. Instead, we have "items" that are to be put in available "slots." As an example, stores will be put into (empty) buildings, office machines will be assigned to empty rooms, or pallets with goods are to be positioned in a warehouse.

Secondly, without loss of generality, we assume that the number of items to be located equals the number of available slots. If this assumption is not satisfied *per se*, the missing items or slots can be replaced by dummies (resulting, of course, in empty slots or unassigned items, respectively).

Thirdly, consider the assumption of homogeneity of facilities and products. This assumption is convenient, as it allows flows of materials to be directed to any facility, regardless of their origin. In layout problems this assumption is dropped and we allow heterogeneous facilities and products instead. For instance, the slots can be differently shaped and sized rooms or areas in a room, while the items to be positioned in these slots can be tables, counters, display cabinets, and others, as in the case of a coffee shop. In addition to the heterogeneity of items and slots, there is a flow of people, goods, or other entities between each item and each slot. This interaction will determine in which slots the items should be positioned, i.e., their layout.

To evaluate a given or contemplated facility layout, one may use the total cost of materials handling, including transportation, as a measure. A number of other aspects could also be considered, for instance space utilization, time required for production, convenience for employees, safety and/or aesthetical considerations, etc. Consequently, the facility layout problem is a multicriteria decision problem, difficult to cast in a neat mathematical form.

In the first section of this chapter we will discuss the various aspects of facility layout planning, and in Section 5.2 we simplify the layout problem and provide several formulations for the resulting quadratic assignment problem or *QAP* for short. Section 5.3 is devoted to some special cases of the quadratic assignment problem, and a number of applications are described in Section 5.4. Finally, solution methods, exact as well as heuristic, are discussed in Section 5.5.

For a full treatment of the material covered in this chapter, the reader is referred to Francis *et al.* (1992) as well as Cela (1998) and Rendl (2002).

5.1 Facility Layout Planning

The systematic study of the best layout of a facility is an endeavor of classical industrial engineering. Originally designed as a plant layout problem, the issue was to arrange the components of a factory so as to achieve maximal production efficiency, measured in some quantifiable way. We will use the term "facility" rather than "plant" since the layout problem applies to many other situations than those of a factory, including airports, banks, hospitals, offices, schools, and shopping centers. Applications involving smaller physical dimensions include the layout of information on a desktop computer screen or the visual display for an airplane pilot, and a miniature application is that of placing components on a small circuit board or a microprocessor chip.

As an efficiency measure or objective function in layout problems one might wish to

- minimize the transportation cost of materials,
- minimize the total production cost,
- minimize the total production time,
- maximize the layout flexibility,
- maximize space utilization,
- maximize employee safety and convenience, or
- minimize new investment, subject to some constraints.

The constraints of the problem include limitations of size of the slots, restrictions on the location of items (e.g., dishwashers would not be appreciated in or near the seating area in a restaurant), and the prohibition of the movement of flow in certain areas (e.g., when transporting hazardous materials). In order to produce a facility layout that satisfies given requirements, a number of procedures have been proposed, many of them variations of the original *systematic layout planning* (*SLP*) approach due to Muther (1961). This procedure uses as input the required flow pattern of materials and given locational relationships to develop a *relationship diagram* between the constituent components of the facility.

Considering the required and available space, *a space relationship diagram* can be constructed. Given this diagram, layout alternatives are developed, which are finally evaluated for ultimate selection. In the usual *SLP* terminology, the concept of item is often referred to as activity or operation. For clarity of exposition, the discussion below will use the term *item* or *machine* for the object to be located in the facility, regardless of its actual physical counterpart; similarly, the term *slot* will be used to signify the location or area where the item is to be located.

Movements of people and goods between individual items such as machines or other facilities are, of course, not necessarily following a straight line. People in restaurants will have to walk around tables, employees in an office have to respect existing furniture and doors to go to the photocopier, and fork lifts drive in rectilinear movements around pallets in warehouses. For an in-depth discussion of patterns of the physical flow of materials in a facility, readers are referred to Francis *et al.* (1992).

The flow between items (machines) in a facility will also depend on the type of layout adapted. Following Francis *et al.* (1992), we can identify four basic layout categories:

- *Fixed* (or *static*) *product layout* refers to situations in which the processes are brought to the product. This layout is used when the product is too large and cumbersome to move around; for example when constructing a ship, a large aircraft or a building. Workstations and production centers are then arranged suitably around the product. Typically, the movement of personnel and equipment is relatively more intense than with the other layout types.
- *Production-line layout* is the opposite of the previous layout. Typically, many, but relatively small, products are moved from station to station in a given processing sequence. This layout is used in assembly lines when the flow of materials is large, and personnel and equipment are more or less stationary.
- The *group* (or *cellular*) *layout* is a modified form of the production-line layout. Here, the volumes of individual products are not enough to justify production-line layouts, so that groups or families of products are formed in such a way that a production-line layout makes sense for the family rather than the individual product. The corresponding groups of processes are then called *cells*.
- In *process layout*, all machines involved in a particular process are grouped together. This layout is quite flexible and appropriate when a large range of low-volume products is to be turned out as for example in a job shop. It is also used when there are frequent changes in the volume and/or mix of products. Finally, it is a sensible layout when product or group layouts are unsuitable.

Each of the above four layout types have advantages and drawbacks; for a detailed comparative analysis the reader is referred to the book by Francis *et al.* (1992).

Another systematic layout planning concept is that of the item (machine) *relationship chart* (or *REL chart*), which graphically represents a *closeness rating* of pairs of items. Rather than specifying the actual flow between items, they are usually grouped together in classes as follows. Class A: Absolutely necessary; the two items must be located closely together. Next in closeness ratings follows E: especially important; I: important, O: ordinary importance; U: unimportant; and X: undesirable, the items should preferably be apart. Clearly, the extreme ratings A and U are considered to be the most important to observe.

Example: A small engineering consulting firm operates with two partners P_1 and P_2, an assistant *ASST*, a computer, copying and storage room *CCS*, a reception area *RECEP* and a meeting room *MEET*. The corresponding relationship chart is shown in Tables II6 and II.7 in two different versions. Since the relationships are bi-directional, the matrix format in Table II.6 can be simplified to the format shown in Table II.7, commonly used in the systematic layout planning context.

Table II.6

	P_1	P_2	ASST	CCS	RECEP	MEET
P_1	–	A	E	I	O	U
P_2	A	–	E	I	O	U
ASST	E	E	–	I	E	O
CCS	I	I	I	–	X	O
RECEP	O	O	E	X	–	U
MEET	U	U	O	O	U	–

Table II.7

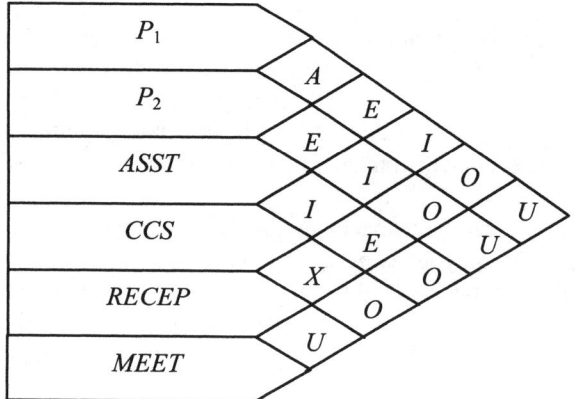

Based on the relationship chart, a *relationship diagram* is developed with the items (machines) as nodes and the relationships as arcs. The relationship chart in this example can be transformed into the relationship diagram displayed in Figure II.20, where "A," "E," "I," and "O" relations are shown by solid lines of decreasing thickness, "U" relations are not indicated, while "X" relations are shown by a broken line.

If the flows between work stations, the number of daily trips required between them, or similar measures were known, we could use only one type of edges but assign appropriate values to them.

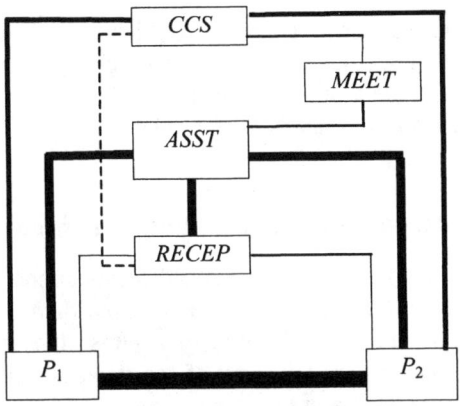

Figure II.20

For an analysis of the flow between the items (machines) in the facility, a number of charts and diagrams depicting the nature, sequencing and timing, as well as type and volume of flow, are used. We will here restrict ourselves to *from-to charts* indicating flow volumes between the items (machines). For planning a layout, one must also have access to the data representing the space requirements of all the items to be located. A more sophisticated version of the relationship diagram is the *space relationship diagram*. It looks similar to relationship diagrams, but here, each item is shown with its particular shape and size that corresponds to its actual appearance. Since different shapes could be contemplated for the various items, alternative space relationships might be constructed. Finally, a *block plan* is designed, which is a tentative layout of the facility, attempting to take all the previous information into consideration.

Example: Consider again the above example of the consulting firm. For accessibility a corridor would be needed, suitably running in the length axis of the facility, separating P_1, P_2 and *CSS* from *RECEP*, *ASST* and *MEET* as indicated in

Figure II.21. The shape of the block plan might also have to be modified to fit the available facility, which, in our example, is the shape of the office floor.

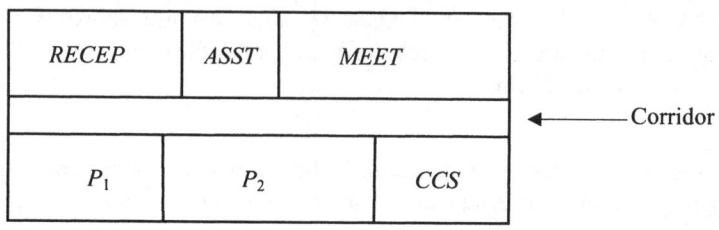

Figure II.21

Once a block plan has been approved (which typically happens after some modifications), a detailed facility design will then be carried out, taking into account the flows in the from-to table. For a detailed treatment, the reader is referred to Wild (1979) and Francis *et al.* (1992).

5.2 Formulations of the Basic Layout Problem

This section describes different mathematical formulations of the layout problem in its simplest form. In this version, we assume that there are n items (machines) T_i, $i = 1, ..., n$ that are to be located in n slots (locations) S_j, $j = 1, ..., n$. Furthermore, each slot S_j can take any of the items T_i and every item must be assigned to exactly one slot, and vice versa, in a one-to-one assignment. There are given flows f_{ik} of a certain commodity that must be sent from item T_i to item T_k, $i, k = 1, ..., n$. The distance (or cost of one unit of flow) between slots S_j and S_ℓ is denoted by $d_{j\ell}$, $j, \ell = 1, ..., n$. (Distances are usually measured between centroids of slots or obvious points of entry, such as doors). Therefore, if item T_i is located in slot S_j and item T_k in slot S_ℓ, the cost of the required flow from T_i at S_j to T_k at S_ℓ would be $d_{j\ell}f_{ik}$. Given the meaning of the parameters $d_{j\ell}$ and f_{ik}, the assumption that they both are nonnegative appears not overly restrictive. The objective of the problem is then to find an assignment of items to slots such that the total cost of all required flows is minimized.

To formulate this problem, define zero-one variables $x_{ij} = 1$, if item T_i is assigned to slot S_j, and zero otherwise; $i, j = 1, ..., n$. Defining $f_{ii} = d_{jj} = 0$ $\forall i, j$, the total cost will then be $z = \sum_i \sum_j \sum_k \sum_\ell d_{j\ell} f_{ik} x_{ij} x_{k\ell}$. Since each item T_i must be assigned to exactly one slot S_j, we must have $\sum_{j=1}^{n} x_{ij} = 1$, $i = 1, ..., n$; and since each slot S_j must be assigned to exactly one machine T_i, we must also have

$\sum_{i=1}^{n} x_{ij} = 1, j = 1, ..., n$. This leads to a nonlinear zero-one integer programming problem, known as the *quadratic assignment problem*

$$QAP: \text{Min } z = \sum_i \sum_j \sum_k \sum_\ell d_{j\ell} f_{ik} x_{ij} x_{k\ell} \quad (1)$$

$$\text{s.t.} \sum_j x_{ij} = 1 \; \forall \; i = 1, ..., n \quad (2)$$

$$\sum_i x_{ij} = 1, \; \forall \; j = 1, ..., n \quad (3)$$

$$x_{ij} = 0 \vee 1 \; \forall \; i, j \quad (4)$$

The quadratic assignment problem *QAP* was first described by Koopmans and Beckmann (1957) and derives its name from the fact that the objective function (1) is quadratic in the variables x_{ij} whereas the constraints (2), (3), (4) are those of the classical linear assignment problem.

The quadratic assignment problem may also include flow-independent annualized costs c_{ij} for assigning item T_i to slot S_j. We would therefore have three $[n \times n]$ − dimensional coefficient matrices: $\mathbf{D} = (d_{j\ell})$ for the distances, $\mathbf{F} = (f_{ik})$ for the flows and $\mathbf{C} = (c_{ij})$ for the assignment costs. The objective function would then be modified to be $z = \sum_i \sum_j \sum_k \sum_\ell d_{j\ell} f_{ik} x_{ij} x_{k\ell} + \sum_i \sum_j c_{ij} x_{ij}$, i.e., having a quadratic and a linear part. In case there are no flows or flow costs between the items, e.g., if $f_{ik} = 0 \; \forall \; i, k$, the quadratic part vanishes and we are left with only the linear part; this special case is nothing but the classical (linear) assignment problem.

In the *balanced* case, the number of items and slots are equal. Quadratic assignment problems are said to be *unbalanced*, if the number of items T_i does not equal the number of slots S_j. As in the linear assignment problem, each unbalanced problem can easily be transformed into a balanced instance by the use of dummies. In particular, if there are more items T_i than slots S_j, we introduce a sufficient number of additional *dummy slots* S_j with distances $d_{j\ell} = 0$ to all existing slots. An item T_i assigned to a dummy slot S_j in the modified problem indicates that in the original problem, this item is not assigned at all. If there are fewer items T_i than slots S_j, we introduce a sufficient number of additional *dummy items* T_i with flows $f_{ik} = 0$ to all existing items. If a dummy item T_i gets assigned to a slot S_j in the modified problem, this slot is vacant in the original problem with no item assigned to it.

There have been many attempts to reformulate the quadratic assignment problem as an integer linear programming problem. Below we will show two of them. One attempt was made by Lawler (1963), who introduces n^4 new binary variables by setting $y_{ijk\ell} = x_{ij} x_{k\ell}$ $\forall\, i, j, k, \ell$. Due to the assignment constraints (2) and (3) it then follows that $\sum_i \sum_j x_{ij} = n$ and $\sum_k \sum_\ell x_{k\ell} = n$, so that $\sum_i \sum_j \sum_k \sum_\ell y_{ijk\ell} = n^2$.

Consider now the constraints $y_{ijk\ell} \leq \frac{1}{2}(x_{ij} + x_{k\ell})$ $\forall\, i, j, k, \ell$. As the variables x_{ij} are binary, the right-hand side of this relation is always less than one, except when x_{ij} and $x_{k\ell}$ both equal one. This implies that the variables $y_{ijk\ell}$ are forced to zero for all cases, except when x_{ij} and $x_{k\ell}$ both equal one, in which case $y_{ijk\ell}$ could be zero or one. As the objective function minimizes an increasing function of $y_{ijk\ell}$, it forces all variables $y_{ijk\ell}$ to zero. However, the constraint $\sum_i \sum_j \sum_k \sum_\ell y_{ijk\ell} = n^2$ requires that $y_{ijk\ell}$ equals one, whenever x_{ij} and $x_{k\ell}$ both equal one. This leads to the following integer linear programming problem:

$$\text{QAPL: Min } z = \sum_i \sum_j \sum_k \sum_\ell d_{j\ell} f_{ik} y_{ijk\ell} \tag{1'}$$

$$\text{s. t. } \sum_j x_{ij} = 1 \quad \forall\, i = 1, \ldots, n \tag{2}$$

$$\sum_i x_{ij} = 1 \quad \forall\, j = 1, \ldots, n \tag{3}$$

$$\sum_i \sum_j \sum_k \sum_\ell y_{ijk\ell} = n^2 \tag{5}$$

$$x_{ij} + x_{k\ell} - 2 y_{ijk\ell} \geq 0 \quad \forall\, i, j, k, \ell = 1, \ldots, n \tag{6}$$

$$y_{ijk\ell} = 0 \vee 1 \quad \forall\, i, j, k, \ell = 1, \ldots, n \tag{7}$$

$$x_{ij} = 0 \vee 1 \quad \forall\, i, j = 1, \ldots, n. \tag{4}$$

This formulation uses $(n^4 + n^2)$ binary variables and $(n^4 + 2n + 1)$ structural constraints.

A modification of the above formulation is due to Christofides *et al.* (1980) who found that it is possible to omit constraint (5) by replacing constraint (6) with

$$y_{ijk\ell} \geq x_{ij} + x_{k\ell} - 1 \quad \forall\, i, j, k, \ell = 1, \ldots, n \tag{6'}$$

Since we minimize and each term of the objective function is nonnegative, the variables $y_{ijk\ell}$ must be as small as possible in any optimal solution. Therefore (6') implies that $y_{ijk\ell} = 1$ if and only if $x_{ij} = 1$ and $x_{k\ell} = 1$ so that (5) follows. Note also

Chapter 5: Layout Models

that the integrality condition (7) on $y_{ijk\ell}$ then becomes redundant and could be replaced by $y_{ijk\ell} \geq 0$.

Another linear formulation is due to Kaufmann and Broeckx (1977). It is more compact than the other two, both with respect to the number of constraints and the number of variables. This formulation uses only n^2 continuous variables, n^2 binary variables and $n^2 + 2n$ constraints and is as follows.

Starting with the quadratic model (1) – (4) above, define $c'_{ij} = \sum_k \sum_\ell d_{j\ell} f_{ik}$ and introduce the variables $w_{ij} = x_{ij} \left(\sum_k \sum_\ell d_{j\ell} f_{ik} x_{k\ell} \right)$. The objective function (1) is then equal to $z = \sum_i \sum_j x_{ij} \left(\sum_k \sum_\ell d_{j\ell} f_{ik} x_{k\ell} \right) = \sum_i \sum_j w_{ij}$. Now consider the problem

$$QAPL' : \text{Min } z' = \sum_i \sum_j w_{ij} \qquad (8)$$

$$\text{s. t. } \sum_j x_{ij} = 1 \quad \forall \, i = 1, ..., n \qquad (2)$$

$$\sum_i x_{ij} = 1 \quad \forall \, j = 1, ..., n \qquad (3)$$

$$c'_{ij} x_{ij} + \sum_k \sum_\ell d_{j\ell} f_{ik} x_{k\ell} - w_{ij} \leq c'_{ij} \quad \forall \, i, j \qquad (9)$$

$$x_{ij} = 0 \vee 1 \quad \forall \, i, j \qquad (4)$$

$$w_{ij} \geq 0 \quad \forall \, i, j.$$

The constraints (9) are valid since they are equivalent to

$\left(\sum_k \sum_\ell d_{j\ell} f_{ik} x_{k\ell} \right) (1 - x_{ij}) \leq \left(\sum_k \sum_\ell d_{j\ell} f_{ik} \right) (1 - x_{ij})$, which are satisfied for all binary x_{ij} and $x_{k\ell}$. This demonstrates the validity of the constraints (9) for any given solution to the quadratic assignment problem.

Conversely, start with the problem $QAPL'$ above. The quadratic formulation (1) – (4) for the quadratic assignment problem is implied if we can show that in $QAPL'$, $w_{ij} = x_{ij} \left(\sum_k \sum_\ell d_{j\ell} f_{ik} x_{k\ell} \right)$ for any optimal solution. This may be derived as follows: suppose that $x_{ij} = 0$. In this case, $w_{ij} = 0$ satisfies constraint (9). Since $w_{ij} =$

0 is feasible, its value is in fact zero in an optimal solution because we minimize the sum of the variables w_{ij}. Now suppose that $x_{ij} = 1$. Constraint (9) implies $w_{ij} \geq \sum_k \sum_\ell d_{j\ell} f_{ik} x_{k\ell}$. We have equality for an optimal solution, again because the sum of the w_{ij} is minimized. Both cases together establish the required equality above for the variables w_{ij}. Consequently, $QAPL'$ is a valid linear formation of the quadratic assignment problem.

Although the sum in (1) contains n^4 terms, only $n^2 - n$ of them are nonzero corresponding to the nonzero entries of the **D** and **F** matrices (recall that they have zero diagonals). On the other hand, the number of feasible solutions to the constraint set (2), (3), (4) is $n!$, so that it is not surprising that the above nonlinear integer programming formulation is extremely difficult to solve computationally as n grows; unfortunately there is no known efficient solution method for this formulation.

For small values of n, the integer linear programming models formulated above may be owed to optimality by exact solution methods, such as branch and bound. These requires the generation of lower bounds for which, typically, continuous relaxation of the problems are employed. This often results in very poor bounds. Computational experience with branch and bound or cutting plane methods is very disappointing. Exact solutions have been found only for problems up to size n=20.

Since the cardinality of the solution set is $n!$, one is led to the observation that every solution can be characterized as a permutation of the set $N = \{1, 2, ..., n\}$. Specifically, let π be a permutation of the set N, i.e., $\pi \in \Pi_n$, where Π_n denotes the set of all permutations of the elements 1, ..., n. In other words, $\pi = [\pi(1), \pi(2), ...\pi(n)]$. By assigning item $T_{\pi(j)}$ to slot S_j $\forall j = 1, ..., n$, we define a particular solution, and each permutation in the set Π_n represents a feasible solution and vice versa. Assigning item $T_{\pi(j)}$ to slot S_j and item $T_{\pi(\ell)}$ to slot S_ℓ, the cost The cost of flow from S_j to S_ℓ is $d_{j\ell} f_{\pi(j),\pi(\ell)}$. The quadratic assignment problem can then be written as

$$QAP': \min_{\pi \in \Pi_n} z(\pi) = \sum_j \sum_\ell d_{j\ell} f_{\pi(j),\pi(\ell)}$$

and the linear part, if applicable, is $\sum_i c_{i,\pi(i)}$.

Example: Consider a quadratic assignment problem with $n = 3$ items and slots, and distance and flow matrices

Chapter 5: Layout Models

$$\mathbf{D} = \begin{bmatrix} 0 & 10 & 12 \\ 15 & 0 & 13 \\ 10 & 11 & 0 \end{bmatrix}, \quad \mathbf{F} = \begin{bmatrix} 0 & 1 & 4 \\ 2 & 0 & 5 \\ 7 & 3 & 0 \end{bmatrix}.$$

Suppose now that some solution assigns item T_1 to slot S_3, item T_2 to slot S_1, and the remaining item T_3 to slot S_2.

For the nonlinear integer programming formulation *QAP*, this assignment corresponds to $x_{21} = x_{32} = x_{13} = 1$ and $x_{ij} = 0$ otherwise. Of the $3^4 = 81$ terms in the objective function, only $3^2 - 3 = 6$ nonvanishing terms corresponding to feasible solutions remain. Specifically, ordering the terms as they appear, reading the F-matrix row-wise, remembering that the diagonal elements are zero, $z = d_{31}f_{12} + d_{32}f_{13} + d_{13}f_{21} + d_{12}f_{23} + d_{23}f_{31} + d_{21}f_{32} = (10)1 + (11)4 + (12)2 + (10)5 + (13)7 + (15)3 = 264$.

Using the permutation formulation *QAP'*, the given solution corresponds to the permutation $\pi = [2, 3, 1]$ and we find $z(\pi) = z(2, 3, 1) = d_{12}f_{23} + d_{13}f_{21} + d_{21}f_{32} + d_{23}f_{31} + d_{31}f_{12} + d_{32}f_{13} = 264$. By exploring all $3! = 6$ feasible solutions to this example, we will find the unique optimal solution $x_{31} = x_{12} = x_{23} = 1$ or $\pi = [3, 1, 2]$ with the objective function value $z = 251$. To put this value in perspective, it is worth mentioning that the *highest* z-value of 271 is obtained for the permutation $\pi = [1, 3, 2]$, indicating a rather narrow spread of the objective values in this particular example.

Using the concept of a permutation matrix, the quadratic assignment problem can also be formulated in an elegant matrix algebraic form, which will be convenient when we discuss solution techniques later on.

Definition II.4: For any permutation $\pi = [\pi(1), \pi(2), ..., \pi(n)] \in \Pi_n$ of the set $N = \{1, 2, ..., n\}$ we define the corresponding $[n \times n]$-dimensional *permutation matrix* $\mathbf{X}_\pi = (x_{ij})$ by $x_{i,\pi(i)} = 1 \; \forall i = 1,..., n$, and $x_{ij} = 0$ otherwise. An element $x_{ij} = 1$ in the permutation matrix indicates the assignment of item T_i to slot S_j.

By definition, each permutation matrix has exactly one unit element in each row and in each column. It is worth noting that there is a one-to-one correspondence between all permutations on one hand and all permutation matrices on the other.

Example: Consider the permutation $\pi = [4, 1, 3, 2]$. Its permutation matrix is

$$\mathbf{X}_\pi = \begin{bmatrix} 0 & 0 & 0 & 1 \\ 1 & 0 & 0 & 0 \\ 0 & 0 & 1 & 0 \\ 0 & 1 & 0 & 0 \end{bmatrix}.$$

Premultiplying any column vector (postmultiplying any row vector) by a permutation matrix is equivalent to rearranging the elements of the column (row) vector.

One can also show that for any permutation matrix $\mathbf{X}_\pi$ and any square matrix $\mathbf{A} = (a_{ij})$ of appropriate dimensions $\mathbf{X}_\pi \mathbf{A} \mathbf{X}_\pi^T = (a_{\pi(i),\pi(j)})$. Recall that the *trace* of a matrix $\mathbf{A}$, $tr(\mathbf{A})$, is defined as $tr(\mathbf{A}) = \sum_i a_{ii}$. One can then easily show that $tr(\mathbf{A}\mathbf{X}_\pi^T) = \sum_j a_{j,\pi(j)}$ and that $tr(\mathbf{X}_\pi \mathbf{A}\mathbf{X}_\pi^T) = \sum_i a_{\pi(i),\pi(i)} = tr(\mathbf{A})$. One can also show that for any square matrix $\mathbf{B}$ of appropriate dimension $tr(\mathbf{A}[\mathbf{X}_\pi \mathbf{B} \mathbf{X}_\pi^T]^T) = \sum_j \sum_\ell a_{j\ell} b_{\pi(j),\pi(\ell)}$. It follows that the quadratic assignment problem in permutation form, *QAP'*, can be written in the algebraic matrix form:

$$QAP'' : \min_{\mathbf{X}_\pi : \pi \in \Pi_n} z(\mathbf{X}_\pi) = tr(\mathbf{D}\mathbf{X}_\pi \mathbf{F}^T \mathbf{X}_\pi^T).$$

If a linear part is present, the objective function

$$z(\pi) = \sum_j \sum_\ell d_{j\ell} f_{\pi(j),\pi(\ell)} + \sum_i c_{i,\pi(i)} \text{ becomes}$$

$$\min z(\mathbf{X}_\pi) = tr([\mathbf{D}\mathbf{X}_\pi \mathbf{F}^T + \mathbf{C}]\mathbf{X}_\pi^T).$$

For further details on the algebraic formulation, see Edwards (1980) and Finke *et al.* (1987).

5.3 Special Cases of the Quadratic Assignment Problem

We will now show how some well known problems can be reformulated as special cases of the quadratic assignment problem.

5.3.1 Triangulation Problems

Consider the following problem. There are n decision alternatives $d_j, j = 1, \ldots, n$ available to the decision makers. The task at hand is to rank these decisions, so as to minimize the disutility of the ranking. For that purpose, define a parameter f_{ik} that denotes the disutility of ranking the i-th decision directly above the j-th decision. These parameters are collected in the matrix **F**. For instance, f_{ik} could symbolize the proportion of stakeholders, who prefer not to rank the i-th product higher than the k-th. In terms of permutations, the task is to rank the alternatives and find the best ordering of the items $\pi(1), \pi(2), \ldots, \pi(n)$. The total disutility of, say, ranking $\pi(1)$ highest is the sum of the disutilities ranking $\pi(1)$ higher than $\pi(2)$, higher than $\pi(3)$, ..., and higher than $\pi(n)$. In terms of the parameters f_{ik}, these costs are $f_{\pi(1),\pi(2)} + f_{\pi(1),\pi(3)} + \ldots + f_{\pi(1),\pi(n)}$. Similarly, the total disutility of having $\pi(i)$ as the i-th element is $f_{\pi(i),\pi(i+1)} + f_{\pi(i),\pi(i+2)} + \ldots + f_{\pi(i),\pi(n)}$.

In order to reformulate the triangulation problem as a quadratic assignment problem, we define the upper triangular matrix of appropriate dimension, viz.,

$$\mathbf{D} = \begin{pmatrix} 0 & 1 & 1 & 1 & & 1 & 1 \\ 0 & 0 & 1 & 1 & \vdots & 1 & 1 \\ 0 & 0 & 0 & 1 & & 1 & 1 \\ \vdots & \vdots & \vdots & \vdots & \ldots & \vdots & \vdots \\ 0 & 0 & 0 & 0 & \vdots & 0 & 1 \\ 0 & 0 & 0 & 0 & & 0 & 0 \end{pmatrix}$$

In order to demonstrate how to write the problem as a triangulation problem, consider the following

Example: A decision maker is faced with $n = 4$ decisions. Suppose it has been decided to rank them in order $[d_2, d_1, d_4, d_3]$. In other words, $\pi(1) = 2, \pi(2) = 1, \pi(3) = 4$, and $\pi(4) = 3$. The total disutility of such a ranking is then

$$\sum_i \sum_{k>i} f_{\pi(i),\pi(k)} = \sum_i \sum_k d_{ik} f_{\pi(i),\pi(k)} = f_{21} + f_{24} + f_{23} + f_{14} + f_{13} + f_{43}.$$

This demonstrates that the best ranking of tasks can be found by solving a quadratic assignment problem with the matrix **D**, making triangulation problems a special case of quadratic assignment problems. Problem instances of triangulation problems of sized up to $n = 70$ can be solved by cutting plane methods.

5.3.2 Traveling Salesman Problems

The traveling salesman problem, defined on a network $G = (N, A, d)$ with node set N, arc set A, and arc values d_{ij}, is the problem of finding a tour along the arcs or edges of the network, so that each node is visited on that tour exactly once, and the total length of the tour, defined as the sum of distances d_{ij}, is minimized. One can think of the problem of finding a tour for a salesman who has to visit all cities in his district, while efficiency will require finding the shortest such tour. A more detailed description can be found in Eiselt and Sandblom (2000). Consider now a permutation $\pi \in \Pi_n$ of n numbers. With each such permutation, we can associate a graph $G = (N, A)$ with node set $N = \{1, 2, ..., n\}$ and arc set $A = \{(i, \pi(i), i \in N\}$. As an illustration, consider the following

Examples: Consider three permutations on four numbers, *viz.*, $\pi = [4, 1, 3, 2]$, $\pi = [3, 4, 1, 2]$, and $\pi = [4, 3, 1, 2]$. The graphs that belong to these permutations are shown in Figures II.22a, b, and c, respectively.

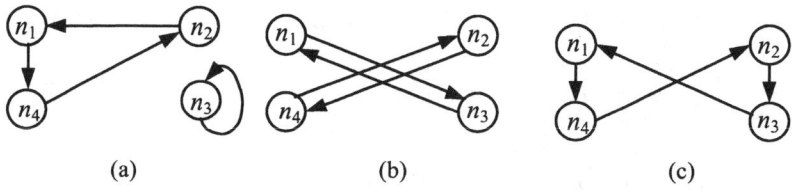

Figure II.22

Whereas the graphs of the first two permutations include two circuits each, the graph in Figure II.22c includes a single circuit, which is a traveling salesman tour. In general, a permutation whose graph contains only one circuit, i.e., a *tour*, is called *cyclic*, and we denote by Π^c the set of all *cyclic permutations*. With this definition, the traveling salesman problem can be formulated as

$$\text{TSP:} \quad \underset{\pi \in \Pi^c}{\text{Min}} \ z = \sum_{i=1}^{n} d_{i, \pi(i)}$$

Consider now a tour, e.g., $(n_1, n_2, ..., n_{n-1}, n_n, n_1)$, which corresponds to the cyclic permutation which has the permutation matrix

Chapter 5: Layout Models   269

$$\mathbf{F} = \begin{bmatrix} 0 & 1 & 0 & 0 & \dots & 0 & 0 \\ 0 & 0 & 1 & 0 & \dots & 0 & 0 \\ 0 & 0 & 0 & 1 & \dots & 0 & 0 \\ . & . & . & . & \dots & . & . \\ . & . & . & . & \dots & . & . \\ 0 & 0 & 0 & 0 & \dots & 0 & 1 \\ 1 & 0 & 0 & 0 & \dots & 0 & 0 \end{bmatrix}$$

The so-called *cycle property* of permutations (see, e.g., Hall, 1959) establishes that for any given cyclic permutation π with permutation matrix $\mathbf{X}_\pi$ and hence any tour, there exists some, not necessarily cyclic, permutation matrix $\mathbf{X}$ such that

$$\mathbf{X}_\pi = \mathbf{X}\mathbf{F}\mathbf{X}^T$$

Therefore,

$$\sum_{i=1}^{n} d_{i,\pi(i)} = tr(d_{i,\pi(j)}) = tr(\mathbf{D}\mathbf{X}_\pi^T) = tr(\mathbf{D}[\mathbf{X}\mathbf{F}\mathbf{X}^T]^T) = tr(\mathbf{D}\mathbf{X}\mathbf{F}^T\mathbf{X}^T)$$

for some permutation matrix $\mathbf{X}$. We can then write the traveling salesman problem as a quadratic assignment problem

$$TSP': \underset{\mathbf{X}\in\Pi_n}{\text{Min}} z = tr(\mathbf{D}\mathbf{X}\mathbf{F}^T\mathbf{X}^T)$$

For a full discussion of traveling salesman problems, see Eiselt and Sandblom (2000).

5.3.3 Matching Problems

The *minimal cost perfect matching problem* on a bipartite graph (or, equivalently, the classical linear assignment problem) can be described as follows. Given two sets $A = \{a_1,..., a_n\}$ and $B = \{b_1, ..., b_n\}$, e.g., sets of machines and locations, the elements from A and B are now to be matched together in pairs, with each pair consisting of one element from A and one from B. Each such pairing arrangement, of which there are $n!$, is called a *perfect matching*. Assume further that there is a given cost d_{ij} for pairing a_i with b_j. The cost of a perfect matching is then defined as the sum of the costs of the pairs (a_i, b_j) included in the matching.

Consider now the same problem in a general complete graph with nodes $A = \{1, ..., n, n+1, ..., 2n\}$ and the $[2n \times 2n]$-dimensional matrix $\mathbf{D} = (d_{ij})$ that expresses the costs incurred by including the pair (a_i, a_j) in the matching.

As an example, let $A = \{1, 2, 3, 4\}$. The perfect matching (1, 2) and (3, 4) corresponds to the permutation $\pi = [3, 4, 1, 2]$, and its total cost is $d_{13} + d_{24}$. Since $\mathbf{D}$ is symmetric, this is the same as $\frac{1}{2}(d_{13} + d_{24} + d_{31} + d_{42}) = \frac{1}{2} \sum_{i=1}^{4} d_{i,\pi(i)}$.

In this example, the perfect matching corresponds to a permutation consisting of two circuits, each of length two. In general, each perfect matching consists of n circuits, each of length two. Consider the perfect matching $(1, n+1)$, $(2, n+2)$, ..., $(n, 2n)$ with the permutation matrix $\mathbf{F} = \begin{bmatrix} \mathbf{0} & \mathbf{I} \\ \mathbf{I} & \mathbf{0} \end{bmatrix}$. Applying the cycle property introduced at the end of the previous section, one can show that any possible perfect matching can be written $\mathbf{XFX}^T$ with some permutation matrix $\mathbf{X}$, in analogy with the previous result. It follows that the minimal cost perfect matching problem can be written as the following quadratic assignment problem:

$$MCPM: \min_{X \in \Pi_{2n}} z = \text{tr}(\mathbf{DXFX}^T)$$

This general matching problem can be solved in polynomial time by Edmonds's algorithm (1965).

5.4 Applications

Although originally describing a facility layout model, the quadratic assignment problem may describe many other situations. We will now describe some applications from the literature with data matrices ranging in size from 4×4 to 36×36.

5.4.1 Relay Team Running

An example from the world of sports was provided by Heffley (1977). To convey the idea, consider the forming of a 4×100m relay team. We may consider a base time of each of the four runners in each of the four legs. This time might be the expected time in the case of a flawless hand-off to the next runner. Define the matrix $\mathbf{C} = (c_{ij})$ where c_{ij} is the base time on the i-th leg for the j-th runner. Let the entries $f_{j\ell}$ of the matrix $\mathbf{F}$ denote the amount of time by which the j-th runner's base

Chapter 5: Layout Models

time is increased in any leg due to the hand-off if he passes the baton to the ℓ-th runner. Defining the upper triangular matrix

$$\mathbf{D} = \begin{bmatrix} 0 & 1 & 0 & 0 \\ 0 & 0 & 1 & 0 \\ 0 & 0 & 0 & 1 \\ 0 & 0 & 0 & 0 \end{bmatrix}$$

results in the following expression for the total relay time $z(\pi)$, where the i-th leg is assigned to runner $\pi(i)$:

$$z(\pi) = \sum_i \sum_k d_{ik} f_{\pi(i)\pi(k)} + \sum_i c_{i\pi(i)}$$
$$= f_{\pi(1),\pi(2)} + f_{\pi(2),\pi(3)} + f_{\pi(3),\pi(4)} + c_{1,\pi(1)} + c_{2,\pi(2)} + c_{3,\pi(3)} + c_{4,\pi(4)}.$$

As a numerical example let us consider the following matrices, with times in seconds:

$$\mathbf{F} = \begin{bmatrix} 0 & .3 & .3 & .2 \\ .5 & 0 & .2 & .5 \\ .3 & .2 & 0 & .3 \\ .5 & .6 & .7 & 0 \end{bmatrix} \quad \mathbf{C} = \begin{bmatrix} 10.5 & 10.1 & 10.2 & 10.4 \\ 10.6 & 10.1 & 10.1 & 10.5 \\ 10.8 & 10.6 & 10.5 & 10.8 \\ 10.6 & 10.2 & 10.1 & 10.6 \end{bmatrix}$$

Forming a team considering only the base times (i.e., solving a linear assignment problem) results in assigning leg 1 to runner 4, leg 2 to runner 2, leg 3 to runner 1, and leg 4 to runner 3. The running time for this assignment is 10.4 + 10.1 + 10.8 + 10.1 = 41.4 seconds, and the total relay time is 41.4 + 0.6 + 0.5 + 0.3 = 42.8 seconds. Note that the best total relay time of 42.3 seconds is obtained for the unique optimal solution determined by the permutation $\pi = [2, 3, 1, 4]$, i.e., with legs 1, 2, 3, and 4 assigned to runners 2, 3, 1, and 4, respectively. Incidentally, the worst of the 4! = 24 possible solutions is obtained for $\pi = [3, 4, 2, 1]$, and its total required time is 43.3 seconds.

5.4.2 Backboard Wiring

One of the earliest applications of the quadratic assignment problem is due to Steinberg (1961). His task was to optimize the layout of a Univac I computer by placing 34 electronic computer components on a circuit board. There are 36 positions for the components on the board which are located in a [4 × 9]-dimensional rectangular array, so that exactly two positions must be left vacant. The problem is therefore unbalanced as described in Section 5.2. The 34 electronic components have to be connected by given numbers of wires. In this model, the entries $f_{j\ell}$ of the flow matrix $\mathbf{F}$ are the number (frequency) of wirings between components j and ℓ. The original frequency matrix $\mathbf{F}$ is symmetric and contains about one-third nonzero elements, which vary in magnitude from 1 to 316, for a total number of 2,620 wires. Alternative distance norms between the positions were considered, reflecting different engineering considerations in tracing the wires: the rectangular, the Euclidean, and squared Euclidean distances. With the distance matrix $\mathbf{D}$ and the frequency matrix $\mathbf{F}$, the corresponding quadratic assignment problem of size $n = 36$ seeks a placement of the electronic components on the board so that the total wire length is minimized. An algorithm for this particular problem was devised, and using a random initial solution, configurations for the Euclidean and the squared Euclidean norm were obtained.

5.4.3 Building Layout Planning

Consider the situation in which people move between rooms in an office building, or between buildings on a university campus. It is then natural to try to arrange the rooms in the building or the buildings on the campus in such a way that the total length of people's movement between rooms and buildings is minimized. This can be modeled as a quadratic assignment problem, in which the items to be located are the rooms in a building (or the buildings on a campus), and the flows correspond to the number of trips of individuals moving between rooms or buildings.

One such application, due to Dickey and Hopkins (1972), is concerned with the optimal location of $n = 16$ new buildings on the campus of Virginia Polytechnic Institute and State University, built during the years 1968 to 1974. In addition to these new buildings, there were already 54 buildings on the campus. Since the campus is mainly reserved for pedestrian travel, an attempt is made to expand the campus in such a way that excessively long walks are avoided for both students and staff to and between classes and other facilities.

The campus layout and expansion problem may be represented in the following mathematical terms. Define

$d_{j\ell}$ as the distance between the new locations j and ℓ,
f_{ik} as the daily number of trips from the new building i to the new building k,
g_{ip} as the distance between the new site i and the existing site (and building) p, and
h_{jp} as the number of daily trips between the new building j and the existing building p.

If the facility $F_{\pi(i)}$ is placed at site i, the total distance between site i, $i = 1, \ldots, 16$, and all existing buildings, $p = 1, \ldots, 54$ is then

$$c_{i,\pi(i)} = \sum_{p=1}^{54} g_{ip} h_{\pi(i),p} \quad , i = 1, \ldots, 16.$$

Therefore, minimizing the total traffic or distance can again be modeled by a quadratic assignment problem with additional linear terms

$$\underset{\pi \in \Pi}{\text{Min}} \ \sum_j \sum_\ell d_{j\ell} f_{\pi(j),\pi(\ell)} + \sum_i c_{i,\pi(i)}$$

Application of a heuristic solution method to the model resulted in a 13% reduction over the proposed arrangement.

A similar application is due to Elshafei (1977). It concerns the Ahmed Maher Hospital in Cairo, Egypt. The planning is done for the outpatient department which occupies a separate building and consists of a reception area and 17 clinics. The flow of patients between these 18 facilities is to be minimized. All facilities require about the same area, except the minor surgery department which needs nearly twice as much floor space. Therefore this department is artificially split into two, which therefore will have to end up in adjacent locations in any acceptable solution.

For this problem, the entries f_{ik} of the matrix **F** correspond to the flow of patients (i.e. the average number of trips from the i-th to the k-th clinic). The flow between the two pseudo-clinics corresponding to the minor surgery one is set equal to a large number in order to force them into adjacent locations. The distances $d_{j\ell}$ from the j-th to the ℓ-th location are measured by taking the length of the paths of the patients. In case of a floor change, the vertical distance is multiplied by a subjective factor of three to approximately model the time delay. The result is a quadratic assignment problem of size $n = 19$. By applying a simple ranking heuristic, the total traffic in the original layout for the outpatient department could be reduced by about 19 percent.

5.4.5 Keyboard Design

Burkard and Offermann (1977) use the quadratic assignment problem in an attempt to optimize the design of keyboards of typewriters and computers. Their goal is to assign the 26 letters $a, b, \ldots, z$ to the keyboard positions, so as to minimize the time to write a text. Recall that the layout of a regular *qwerty* keyboard looks as shown in Figure II.23.

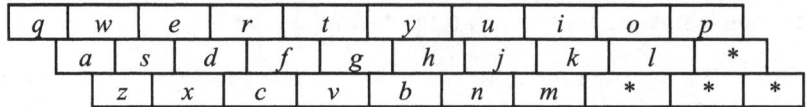

Figure II.23

It has been found that the typing speed is primarily dictated by the frequency of letter pairs in the text and the time to move the hands and press the corresponding keys in succession. Table II.8 adapted from Pollatschek *et al.* (1976), indicates the time (in seconds) required for an average typist's hand movements. For example, consider the letter pair *TT*. It requires no hand movement at all, so that it will take .191 seconds on a manual typewriter and .133 seconds on an electric typewriter (or a computer, for that matter). Similarly, the letter pair *CP* requires two hands and involves movements two rows up, so that the figures in the bottom row of Table II.8 apply.

Table II.8

Movement	Electromechanical		Manual	
	Same hand	Other hand	Same hand	Other hand
No movement	0.133	0.133	0.191	0.191
Same row, with movement	0.180	0.133	0.238	0.191
One row up	0.180	0.133	0.238	0.191
One row down	0.180	0.133	0.263	0.191
Two rows up or down	0.205	0.133	0.306	0.191

A text in English from *The Times* and *The Daily Telegraph* was used and 100,000 letter pairs were counted. Some examples of the observed frequencies are shown in Table II.9.

Table II.9

Pair	he	er	nd	qu	nk	tp
Frequency	3,026	2,123	1,156	107	42	6
Pair	eh	re	dn	uq	kn	pt
Frequency	26	1,835	10	0	24	74

Coding the letters $a, b, ..., z$ by the numbers $1, 2, ..., 26$, and defining f_{ik} as the frequency of the letter pair coded as (i, k) and $d_{j\ell}$ as the time needed to press the ℓ-th key after the j-th, we obtain a quadratic assignment problem of size $n = 26$. Burkard and Offermann (1977) used a heuristic algorithm to solve the problem, and they obtained the solution displayed in Figure II.24.

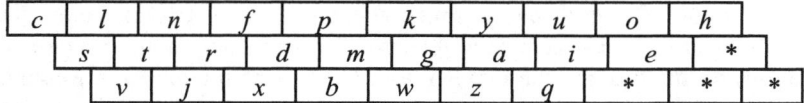

Figure II.24

The value of the objective function for the current keyboard could be reduced from 419,712.2 to 382,343.7 (electromechanical); and from 586,397.9 to 543,005.3 (manual), corresponding to a reduction in the order of 10 percent. Even better solutions have lately become available, see *QAPLIB* at the University of Graz website. However, with the hundreds of millions of existing keyboards all over the world, a changeover would be extremely costly, so that it is unlikely that this keyboard would ever replace the current *qwerty* keyboard. Furthermore, the keyboard in Figure II.24 is only optimized for the English language. Optimal keyboards for other languages will look quite different.

5.5 Solution Methods

This section discusses in some detail solution methods for the quadratic assignment problem. We first describe a classic exact solution, which, due to the complexity of the problem, is limited to fairly small examples. The next section explores a variety of heuristic methods that are designed to solve problems of practical size. Finally, the last section surveys some methods that are intended to solve problems that involve more features than the quadratic assignment problem.

5.5.1 Exact Solution Methods

The quadratic assignment problems described in Section 5.3 have heavily structured matrices. For those cases, special solution methods have been developed. Triangulation and traveling salesman problems are still difficult, but, at least relatively speaking, easier than quadratic assignment problems. As a case in point, instances of traveling salesman problems with up to 1,000 nodes have been solved to optimality. In contrast, the general quadratic assignment problem with arbitrary distance and flow matrices is much more difficult to solve. The solution procedures are enumerative and depend on finding good lower bounds. As mentioned above, lower bounds obtained from relaxation of linear formulations are weak. Therefore, different bounding procedures have been developed for solving quadratic assignment problems. They are of essentially three different kinds. First, there are the so-called Gilmore-Lawler and related bounds, which use minimal scalar products as we discuss below. Secondly, there are the eigenvalue bounds described by Finke *et al.* (1987) and later elaborated upon by Hadley *et al.* (1990, 1992). These bounds are considered the best available; they do, however also require the most effort to compute. Thirdly and finally, there are bounds based on the reformulation of the quadratic assignment problem that require the solution of related linear assignment problems or linear programming relaxations. Contributions of this type are those by Drezner (1995), the references in Resende *et al.* (1995), and Rendl (2002). Below we first discuss some of these bounds, and then show how they can be used in a branch and bound solution procedure.

In order to derive bounds, consider again two formulations of the quadratic assignment problem developed in Section 5.2. above, specifically the permutation formulation

$$QAP': \min_{\pi \in \Pi_n} z(\pi) = \sum_i \sum_k d_{ik} f_{\pi(i),\pi(k)} + \sum_i c_{i,\pi(i)}$$

and the algebraic matrix formulation

$$QAP'': \min_{\mathbf{X}_\pi : \pi \in \Pi_n} z(\mathbf{X}_\pi) = \text{tr}(\mathbf{DX}_\pi \mathbf{F}^T + \mathbf{C})\mathbf{X}_\pi^T.$$

The classical approach, the Gilmore-Lawler bounding procedure, due to Gilmore (1962) and Lawler (1963) (see also Li *et al.*, 1994), is based on the concept of minimal scalar products. To explain this, consider any two *n*-vectors, e.g. $\mathbf{x}$ and $\mathbf{y}$ and their scalar product $\mathbf{x}^T\mathbf{y} = \sum_i x_i y_i$. We want to allow reordering of the components and determine the smallest and largest possible value of the corresponding scalar product. In other words, the objective is to find permutation matrices $\mathbf{X}_\pi^-$ and $\mathbf{X}_\pi^+$ which, respectively, minimize and maximize $\mathbf{x}^T\mathbf{X}_\pi\mathbf{y}$. Calling the minimal value $(\mathbf{x}^T\mathbf{y})_-$ and the maximal value $(\mathbf{x}^T\mathbf{y})_+$ it is easy to prove, by contradiction, the following

Lemma II.5: Let two *n*-vectors $\mathbf{x}$ and $\mathbf{y}$ be given. Rearrange the elements of $\mathbf{x}$ in nondecreasing order and those of $\mathbf{y}$ in nonincreasing order. Then $(\mathbf{x}^T\mathbf{y})_-$ is obtained by forming the scalar product of these two rearranged vectors. Similarly, $(\mathbf{x}^T\mathbf{y})_+$ is found by ordering both vectors in nondecreasing order.

Let now $\mathbf{x}'$ and $\mathbf{y}'$ denote the vectors $\mathbf{x}$ and $\mathbf{y}$ after they have been rearranged in nondecreasing as well as nonincreasing order, respectively. Then we have $\mathbf{x}' = \mathbf{X}^\uparrow \mathbf{x}$, where $\mathbf{X}^\uparrow = (x_{ij}^\uparrow)$, with $x_{ij}^\uparrow = 1$, if the *i*-th smallest element is in the *j*-th position in the original vector **x**, and 0 otherwise. Similarly, we obtain $\mathbf{y}' = \mathbf{Y}^{\downarrow}\mathbf{y}$, where $\mathbf{Y}^{\downarrow} = (y_{ij}^{\downarrow})$ with $y_{ij}^{\downarrow} = 1$, if the i-th largest element is in the j-th position of **y**, and 0 otherwise. Then $(\mathbf{x}^T\mathbf{y})_- = (\mathbf{x}')^T\mathbf{y}' = (\mathbf{X}^{\uparrow}\mathbf{x})^T \mathbf{Y}^{\downarrow}\mathbf{y} = \mathbf{x}^T(\mathbf{X}^{\uparrow})^T \mathbf{Y}^{\downarrow}\mathbf{y}$, so that the permutation matrix $\mathbf{X}_{\pi}^{-}$ is then $\mathbf{X}_{\pi}^{-} = (\mathbf{X}^{\uparrow})^T\mathbf{Y}^{\downarrow}$. Similarly, define $\mathbf{Y}^{\uparrow} = (y_{ij}^{\uparrow})$ with $y_{ij}^{\uparrow} = 1$ if the i-th smallest element is in the j-th position and 0 otherwise. Then $(\mathbf{x}^T\mathbf{y})_+ = (\mathbf{X}^{\uparrow}\mathbf{x})^T \mathbf{Y}^{\uparrow}\mathbf{y} = \mathbf{x}^T(\mathbf{X}^{\uparrow})^T \mathbf{Y}^{\uparrow}\mathbf{y}$, so that the appropriate permutation matrix is then $\mathbf{X}_{\pi}^{+} = (\mathbf{X}^{\uparrow})^T\mathbf{Y}^{\uparrow}$.

Example: Consider the two vectors $\mathbf{x}^T = [2, -1, 4, 0]$ and $\mathbf{y}^T = [3, -2, 4, 5]$. We first order the vectors as $\mathbf{x}' = [-1, 0, 2, 4]$ and $\mathbf{y}' = [5, 4, 3, -2]$. We can then determine

$$\mathbf{X}^{\uparrow} = \begin{bmatrix} 0 & 1 & 0 & 0 \\ 0 & 0 & 0 & 1 \\ 1 & 0 & 0 & 0 \\ 0 & 0 & 1 & 0 \end{bmatrix} \text{ and } \mathbf{Y}^{\downarrow} = \begin{bmatrix} 0 & 0 & 0 & 1 \\ 0 & 0 & 1 & 0 \\ 1 & 0 & 0 & 0 \\ 0 & 1 & 0 & 0 \end{bmatrix},$$

so that the permutation matrix

$$\mathbf{X}_{\pi}^{-} = (\mathbf{X}^{\uparrow})^T\mathbf{Y}^{\downarrow} = \begin{bmatrix} 1 & 0 & 0 & 0 \\ 0 & 0 & 0 & 1 \\ 0 & 1 & 0 & 0 \\ 0 & 0 & 1 & 0 \end{bmatrix},$$

and hence $(\mathbf{x}^T\mathbf{y})_- = (\mathbf{x}')^T\mathbf{y}' = (\mathbf{X}^{\uparrow}\mathbf{x})^T(\mathbf{Y}^{\downarrow}\mathbf{y}) = \mathbf{x}^T(\mathbf{X}^{\uparrow})^T \mathbf{Y}^{\downarrow}\mathbf{y} = \mathbf{x}^T \mathbf{X}_{\pi}^{-} \mathbf{y} = (-1)(5) + 0(4) + 2(3) + 4(-2) = -7$. Similarly, with

$$\mathbf{Y}^{\uparrow} = \begin{bmatrix} 0 & 1 & 0 & 0 \\ 1 & 0 & 0 & 0 \\ 0 & 0 & 1 & 0 \\ 0 & 0 & 0 & 1 \end{bmatrix},$$

we obtain

$$\mathbf{X}_\pi^+ = (\mathbf{X}^\uparrow)^T \mathbf{Y}^\uparrow = \begin{bmatrix} 0 & 0 & 1 & 0 \\ 0 & 1 & 0 & 0 \\ 0 & 0 & 0 & 1 \\ 1 & 0 & 0 & 0 \end{bmatrix}$$

and $(\mathbf{x}^T\mathbf{y})_+ = (\mathbf{X}^\uparrow\mathbf{x})^T \mathbf{Y}^\uparrow\mathbf{y} = \mathbf{x}^T \mathbf{X}_\pi^+ \mathbf{y} = (-1)(-2) + 0(3) + 2(4) + 4(5) = 30$.

Apply now Lemma II.5 to the algebraic formulation QAP''. Any constant matrix $\mathbf{M}$ with the property $\mathbf{M} \leq \mathbf{DX}_\pi\mathbf{F}^T + \mathbf{C}$ for all feasible solutions (permutations) $\mathbf{X}_\pi$ is a valid lower bound on the objective function value $z(\mathbf{X}_\pi)$, i.e.,

$$\min_{\mathbf{X}_\pi \in \Pi} \text{tr} (\mathbf{M} \mathbf{X}_\pi^T) \leq \min_{\mathbf{X} \in \Pi} \text{tr} (\mathbf{DX}_\pi\mathbf{F}^T + \mathbf{C})\mathbf{X}_\pi^T$$

where the left hand side represents, according to our earlier discussion, a linear assignment problem with respect to the matrix $\mathbf{M}$. Such a bound matrix $\mathbf{M}$ may be found as follows. The matrix $\mathbf{DX}_\pi\mathbf{F}^T$ is equal to $(\mathbf{d}_{i\bullet} \mathbf{X}_\pi \mathbf{f}_{j\bullet}^T)$ where $\mathbf{d}_{i\bullet}$ is the i-th row of $\mathbf{D}$ and $\mathbf{f}_{j\bullet}$ is the j-th row of $\mathbf{F}$. Consequently, we have for each of its elements the minimal scalar product $(\mathbf{d}_{i\bullet} \mathbf{f}_{j\bullet}^T)_-$ as a lower bound.

It is, however, possible to sharpen this lower bound. Let π be the permutation corresponding to $\mathbf{X}_\pi$. Then the terms $d_{ii}f_{\pi(i),\pi(i)}$ in the objective function indicate that the diagonal elements of the two matrices are to be multiplied in every feasible solution. This leads to the final form of the Gilmore-Lawler bound GLB. Define the elements m_{ij} by

$$m_{ij} = d_{ii}f_{jj} + (\bar{\mathbf{d}}_{i\bullet} \bar{\mathbf{f}}_{j\bullet}^T)_- + c_{ij}$$

where $\bar{\mathbf{d}}_{i\bullet}$ is the i-th row of $\mathbf{D}$ without the diagonal element d_{ii} and $\bar{\mathbf{f}}_{j\bullet}^T$ is the j-th row of $\mathbf{F}$ without f_{jj}. Then the lower bound GLB is given by the optimal objective function value of the assignment problem with matrix $\mathbf{M}$, i.e.,

$$GLB = \min_{\mathbf{X}_\pi \in \Pi} \text{tr} (\mathbf{M} \mathbf{X}_\pi^T) = \min_{\mathbf{X}_\pi \in \Pi} \sum_i m_{i,\pi(i)} .$$

Example: Consider the quadratic assignment problem with the distance matrix

Chapter 5: Layout Models

$$\mathbf{D} = \begin{bmatrix} 0 & 15 & 5 & 2 \\ 24 & 0 & 3 & 5 \\ 6 & 11 & 0 & 9 \\ 2 & 9 & 5 & 0 \end{bmatrix},$$

the flow matrix

$$\mathbf{F} = \begin{bmatrix} 0 & 7 & 0 & 5 \\ 5 & 0 & 4 & 4 \\ 7 & 5 & 0 & 8 \\ 6 & 9 & 2 & 0 \end{bmatrix}, \text{ and a linear part}$$

$$\mathbf{C} = \begin{bmatrix} 1 & 9 & 0 & 2 \\ 4 & 2 & 2 & 5 \\ 5 & 7 & 7 & 3 \\ 4 & 5 & 2 & 0 \end{bmatrix}.$$

The optimal solution is given by the permutation $\pi = [3, 1, 2, 4]$ with an objective value of $\bar{z} = 421$, while the worst solution has an objective value of $z = 607$. Applying the Gilmore-Lawler bound in its primitive version, the bound matrix is

$$\mathbf{M} = (m_{ij}) = \left(\mathbf{d}_{i\bullet} \mathbf{f}_{j\bullet}^T \right)_{-} + (c_{ij}) = \begin{bmatrix} 10 & 28 & 39 & 22 \\ 15 & 32 & 46 & 28 \\ 30 & 60 & 87 & 54 \\ 10 & 28 & 39 & 22 \end{bmatrix} + \begin{bmatrix} 1 & 9 & 0 & 2 \\ 4 & 2 & 2 & 5 \\ 5 & 7 & 7 & 3 \\ 4 & 5 & 2 & 0 \end{bmatrix} =$$

$$= \begin{bmatrix} 11 & 37 & \mathbf{39} & 24 \\ 19 & \mathbf{34} & 48 & 33 \\ \mathbf{35} & 67 & 94 & 57 \\ 14 & 33 & 41 & \mathbf{22} \end{bmatrix}.$$

For instance, $m_{23} = (\mathbf{d}_{2\bullet} \mathbf{f}_{3\bullet}^T) + c_{23} = [0, 3, 5, 24][8, 7, 5, 0]^T + 2 = 46 + 2 = 48$. The linear assignment problem with matrix $\mathbf{M}$ is then solved, e.g., by the Hungarian method, and the solution indicated by the elements in boldface in the above matrix $\mathbf{M}$. The optimal objective value of the linear assignment problem (and therefore a lower bound on the original quadratic assignment problem) is 130, significantly lower than the optimal objective value of the *QAP*. Computing the Gilmore-Lawler bound in its sharpened version, we obtain

$$\mathbf{M} = (m_{ij}) = (d_{ii}f_{jj}) + (\overline{\mathbf{d}}_{i\bullet}\overline{\mathbf{f}}_{j\bullet}^T)_- + (c_{ij}) =$$

$$\begin{bmatrix} 0 & 0 & 0 & 0 \\ 0 & 0 & 0 & 0 \\ 0 & 0 & 0 & 0 \\ 0 & 0 & 0 & 0 \end{bmatrix} + \begin{bmatrix} 39 & 90 & 126 & 78 \\ 46 & 131 & 179 & 105 \\ 87 & 110 & 166 & 130 \\ 39 & 66 & 96 & 66 \end{bmatrix} + \begin{bmatrix} 1 & 9 & 0 & 2 \\ 4 & 2 & 2 & 5 \\ 5 & 7 & 7 & 3 \\ 4 & 5 & 2 & 0 \end{bmatrix} =$$

$$= \begin{bmatrix} 40 & 99 & 126 & \mathbf{80} \\ 50 & 133 & 181 & 110 \\ 92 & \mathbf{117} & 173 & 133 \\ 43 & 71 & \mathbf{98} & 66 \end{bmatrix}.$$

For instance, $m_{31} = d_{33}f_{11} + (\overline{\mathbf{d}}_{3\bullet}\overline{\mathbf{f}}_{1\bullet}^T)_- + c_{31} = 0 + [6, 9, 11][7, 5, 0]^T + 5 = 87 + 5 = 92$. Again, the optimal solution to the linear assignment problem is indicated by the boldface numbers in the matrix $\mathbf{M}$. The corresponding value of the objective function is $GLB = 345$, a considerably better lower bound.

Note that the solution to the linear assignment problem in the primitive version above corresponds to the permutation [3, 2, 1, 4], which, if inserted in the original problem, has a value of the objective function of $z = 501$. Similarly, the permutation [4, 1, 2, 3] corresponds to the sharpened version; its value of the objective function is $z = 514$. We conclude that for the optimal objective function value $\overline{z}$, we must have $345 = GLB \leq \overline{z} \leq 501$.

Before the bound matrix $\mathbf{M}$ is computed, the matrices $\mathbf{D}$ and $\mathbf{F}$ are usually reduced in a preprocessing phase in order to obtain an even stronger lower bound. This reduction procedure is reminiscent of the procedure used in the Hungarian method for the linear assignment problem and is aimed at reducing the magnitude of the quadratic term in favor of the linear term. For instance, the reduction method proposed by Roucairol (1979) tries to reduce the largest nondiagonal elements as much as possible without making any nondiagonal elements negative. Other reduction strategies have been studied by Burkard (1973) and Edwards (1980). Some of these strategies have proven to be very successful in practice.

We will now discuss eigenvalue bounds and consider the special case where the distance and flow matrices $\mathbf{D}$ and $\mathbf{F}$ are both symmetric, i.e., $\mathbf{D} = \mathbf{D}^T$ and $\mathbf{F} = \mathbf{F}^T$. The corresponding quadratic assignment problem

$$\text{Min } z(X_\pi) = \text{tr}(\mathbf{DX}_\pi \mathbf{FX}_\pi^T)$$

Chapter 5: Layout Models

is then said to be symmetric. Let now $\lambda_1, ..., \lambda_n$ denote the eigenvalues of $\mathbf{D}$ and $\mu_1, ..., \mu_n$ denote the eigenvalues of $\mathbf{F}$. From matrix theory we know that the symmetry of $\mathbf{D}$ and $\mathbf{F}$ implies that all their eigenvalues are real. Let $\lambda = [\lambda_1, \lambda_2, ..., \lambda_n]^T$ and $\mu = [\mu_1, \mu_2, ..., \mu_n]^T$ denote the column vectors of eigenvalues of $\mathbf{D}$ and $\mathbf{F}$, respectively. Finke et al. (1987) have then proved

Lemma II.6: For the symmetric quadratic assignment problem, the following eigenvalue bounds are valid for all permutation matrices $\mathbf{X}_\pi$:

$$(\lambda^T \mu)_- \leq \mathrm{tr}\left(\mathbf{D} \mathbf{X}_\pi \mathbf{F} \mathbf{X}_\pi^T\right) \leq (\lambda^T \mu)_+.$$

Example: With the symmetric matrices

$$\mathbf{D} = \begin{bmatrix} 0 & 19 & 6 & 2 \\ 19 & 0 & 5 & 8 \\ 6 & 5 & 0 & 7 \\ 2 & 8 & 7 & 0 \end{bmatrix} \text{ and } \mathbf{F} = \begin{bmatrix} 0 & 6 & 0 & 5 \\ 6 & 0 & 4 & 7 \\ 0 & 4 & 0 & 8 \\ 5 & 7 & 8 & 0 \end{bmatrix},$$

we obtain the eigenvalue vectors $\lambda^T = [25.1327, -20.1792, 1.1893, -6.1427]$ and $\mu^T = [15.5726, 0.7729, -6.8864, -9.4591]$. We then find $(\lambda^T \mu)_- = -564.913$ and $(\lambda^T \mu)_+ = 625.479$, so that the objective function values z of the quadratic assignment problem with distance matrix $\mathbf{D}$ and flow matrix $\mathbf{F}$ will satisfy $z \in [-564.913; 625.479]$. However, since all elements in $\mathbf{D}$ and $\mathbf{F}$ are nonnegative, the objective function values z will also be nonnegative for all feasible solutions, rendering negative lower bounds useless.

Before establishing $(\lambda^T \mu)_-$ as a lower bound to the symmetric quadratic assignment problem, the authors therefore suggest that a reduction be performed on the $\mathbf{D}$ and $\mathbf{F}$ matrices, in order to improve the bound. This is accomplished by a reduction which reduces the spread of the eigenvalues of each of the $\mathbf{D}$ and $\mathbf{F}$ matrices as much as possible. For details, see Finke et al. (1987), as well as further work by Hadley et al. (1990, 1992). It has been conjectured that eigenvalue bounds improve as the problems get larger.

The remainder of this section describes a so-called *single assignment method* where the items are assigned to slots one at a time. The method was originally introduced by Gilmore (1962) and Lawler (1963). Whereas the formulations in Section 5.2 were based on the notion of assigning items T_i to slots S_j, it is now convenient to modify this and assign items T_j to slots S_i. The procedure can be displayed in a tree structure, where at each node a subset $N_1 \subset N = \{1, 2, ..., n\}$ of the slots S_i have had items $T_{\pi(i)}$ assigned to them, while the slots of the complementary subset $N_2 = N \setminus N_1$ are not yet assigned. Furthermore, for some of the slots S_i in N_2 with no items yet assigned to them, there may be forbidden items T_j. Rather than following the branching procedure by Gilmore and Lawler, who allow more than two branches to lead out of a node, we will describe the binary branching procedure of Gavett and Plyter (1966). Their approach selects a free slot S_i, and assigns to it an allowable item T_j, i.e., we fix the assignment $\pi(i) = j$ on the right branch. The left branch leads to a node where $\pi(i) \neq j$, i.e., where slot S_i has no item assigned to it, but is not allowed to be assigned item T_j further down in this part of the tree.

At each node of the branching tree, a lower and an upper bound on the objective function value are calculated and the best extension of the current partial assignment $\pi(i)$, $i \in N_1$ is sought. At this point, the objective function value can be decomposed as follows:

$$z = \sum_i \sum_k d_{ik} f_{\pi(i),\pi(k)} + \sum_i c_{i,\pi(i)}$$
$$= \sum_{i \in N_1} \sum_{k \in N_1} d_{ik} f_{\pi(i),\pi(k)} + \sum_{i \in N_2} \sum_{k \in N_2} d_{ik} f_{\pi(i),\pi(k)}$$
$$+ \sum_{i \in N_1} \sum_{k \in N_2} (d_{ik} f_{\pi(i),\pi(k)} + d_{ki} f_{\pi(k),\pi(i)}) + \sum_{i \in N_1} c_{i,\pi(i)} + \sum_{i \in N_2} c_{i,\pi(i)}.$$

In this expression, the first double sum and the first single sum are constants that can easily be calculated; the total of these two expressions will be denoted by α. The second double sum constitutes the quadratic part, and the remaining two sums the linear part of a quadratic assignment problem of the reduced size $|N_2| \times |N_2|$.

The lower bounding techniques discussed earlier can now be applied at each node of the tree. Upper bounds are naturally available as the objective function values computed for the assignments obtained by solving the linear assignment problem using the lower bound matrix $\mathbf{M}$. Forbidden assignments $\pi(i) \neq j$ are handled by setting $m_{ij} := \infty$.

We now discuss how at a given node a free slot S_i is chosen and some allowable item T_j is assigned to it. For that purpose, we assume that the current lower bound matrix $\mathbf{M}$ has been reduced as in the Hungarian matrix for the linear assignment problem, resulting in the matrix $\overline{\mathbf{M}} = (\overline{m}_{ij})$. We denote by $L(\mathbf{M})$ the current lower bound on the objective function value (in the reduced problem), and by U the best upper bound that has been found so far in the original problem, so that $\Delta := U - L(\mathbf{M}) - \alpha$ is the gap between the bounds. If now item T_j is assigned to slot S_i, then $L(\mathbf{M})$ must increase by at least $\overline{m}_{ij}$, since the current bound $L(\mathbf{M})$ corresponds to an all-zero cost assignment in the matrix $\overline{\mathbf{M}}$. On the other hand, if the item T_i is *not* assigned to slot S_j, then $L(\mathbf{M})$ must increase by at least the amount r_{ij}, where

Chapter 5: Layout Models

$r_{ij} = \min_{k \neq j} \{\overline{m}_{ik}\} + \min_{k \neq i} \{\overline{m}_{kj}\} \geq 0$ (following an idea from Little *et al.* (1963)). It follows that any $\overline{m}_{ij}$ for which $\overline{m}_{ij} > \Delta$ and/or $r_{ij} > \Delta$ holds, can be replaced by ∞, since such an assignment cannot be optimal. Furthermore, we will select the branch by the criterion $\max_{i,j} \{r_{ij}\}$, noting that all r_{ij} vanish except possibly $r_{i\pi(i)}$.

We are now ready to state the branch and bound procedure in algorithmic form, for solving the quadratic assignment problem with distance matrix **D**, flow matrix **F**, and linear part matrix **C**.

Gavett and Plyter's Branch and Bound Method

Step 1: Find a lower bound matrix **M**, i.e., $\mathbf{M} \leq \mathbf{DX}_\pi \mathbf{F}^T + \mathbf{C}$ for all permutation matrices $\mathbf{X}_\pi$. Let $U := \infty$, $N_1 := \emptyset$, $N_2 := \{1, 2, \ldots, n\}$, and $\alpha := 0$.

Step 2: Solve the linear assignment problem with matrix **M**. Denote by π its optimal solution, by $L(\mathbf{M})$ its optimal objective function value based on the cost matrix **M**, and by $U(\mathbf{M})$ the corresponding objective function value of the (full) quadratic assignment problem. Let $U := \min \{U; U(\mathbf{M})\}$.

Step 3: Let $\overline{\mathbf{M}}$ denote the Hungarian method reduction of **M** and let $\Delta := U - L(\mathbf{M}) - \alpha$ be the bound gap. Compute $r_{ij} = \min_{k \neq j} \{\overline{m}_{ik}\} + \min_{k \neq i} \{\overline{m}_{kj}\}$ $\forall i, j$.

Let $\overline{m}_{ij} := \infty$ $\forall i, j$, for which $\overline{m}_{ij} > \Delta$ and/or $r_{ij} > \Delta$. Select the branch $(\bar{i}, \bar{j})$ by the criterion max $\{r_{ij}\}$. Let $N_1 := N_1 \cup \{\bar{i}\}$, and let $N_2 := N_2 \setminus \{\bar{i}\}$.

Step 4: *Left branch:* $\pi(\bar{i}) \neq \bar{j}$. Let $m_{\bar{i}\bar{j}} := \infty$ and solve the linear assignment problem with the revised matrix **M**. Compute $L(\mathbf{M})$, $U(\mathbf{M})$, and let $U := \min \{U; U(\mathbf{M})\}$. If $\alpha + L(\mathbf{M}) \geq U$, close the branch.

Right branch: $\pi(\bar{i}) = \bar{j}$. Modify the distance matrix **D** by deleting row $\bar{i}$ and column $\bar{i}$, and the flow matrix **F** by deleting row $\bar{j}$ and column $\bar{j}$. Modify the linear assignment matrix **C** by deleting row $\bar{i}$ and column $\bar{j}$. Let then

$$c_{ij} := c_{ij} + \sum_{k \in N_1}(d_{ki}f_{\pi(k),j} + d_{ik}f_{j,\pi(k)}) \quad \forall \ i \in N_2, j \notin \pi(N_1),$$

and let $\alpha = \sum_{i \in N_1} c_{i,\pi(i)} + \sum_{i \in N_1} \sum_{k \in N_1} d_{ik} f_{\pi(i),\pi(k)}$. Find a lower bound matrix **M** for the reduced quadratic assignment problem of size $|N_2| \times |N_2|$, compute π, $L(\mathbf{M})$ and $U(\mathbf{M})$, and let $U := \min \{U, U(\mathbf{M})\}$. If $L(\mathbf{M}) \geq U$, close the branch.

Step 5: Are all branches closed?
    If yes: Stop, the nodes that correspond to the current value U represent optimal solutions.
    If no: Select a branch which is not closed. Go to Step 3.

Gavett and Plyter's branch and bound method will now be illustrated by the following

Example: Consider again the above problem with the matrices

$$\mathbf{D} = \begin{bmatrix} 0 & 15 & 5 & 2 \\ 24 & 0 & 3 & 5 \\ 6 & 11 & 0 & 9 \\ 2 & 9 & 5 & 0 \end{bmatrix}, \mathbf{F} = \begin{bmatrix} 0 & 7 & 0 & 5 \\ 5 & 0 & 4 & 4 \\ 7 & 5 & 0 & 8 \\ 6 & 9 & 2 & 0 \end{bmatrix}, \text{ and } \mathbf{C} = \begin{bmatrix} 1 & 9 & 0 & 2 \\ 4 & 2 & 2 & 5 \\ 5 & 7 & 7 & 3 \\ 4 & 5 & 2 & 0 \end{bmatrix}.$$

The Gilmore-Lawler bound matrix

$$\mathbf{M} = \begin{bmatrix} 40 & 99 & 126 & 80 \\ 50 & 133 & 181 & 110 \\ 92 & 117 & 173 & 133 \\ 43 & 71 & 98 & 66 \end{bmatrix}$$

was already derived above. Solving the linear assignment problem with cost matrix **M** results in $\pi = [4, 1, 2, 3]$, $L(\mathbf{M}) = 345$, and $U = U(\mathbf{M}) = 514$. In Step 3 we use

$$\overline{\mathbf{M}} = \begin{bmatrix} 0 & 17 & 14 & \mathbf{0} \\ \mathbf{0} & 41 & 59 & 20 \\ 17 & \mathbf{0} & 26 & 18 \\ 17 & 3 & \mathbf{0} & \mathbf{0} \end{bmatrix},$$

and $\Delta = 514 - 345 = 169$, to conclude that none of the m_{ij} are modified. We find $r_{21} = 20$, $r_{32} = 20$, $r_{43} = 14$, and all other r_{ij} vanish. We select $(\bar{i}, \bar{j}) = (2, 1)$, i.e., $\pi(2) = 1$ and $N_1 = \{2\}$, $N_2 = \{1, 3, 4\}$. For the left branch (node n_2) we obtain

Chapter 5: Layout Models 285

$$\mathbf{M} = \begin{bmatrix} 40 & 99 & 126 & 80 \\ \infty & 133 & 181 & 110 \\ 92 & 117 & 173 & 133 \\ 43 & 71 & 98 & 66 \end{bmatrix}, \text{ and } \overline{\mathbf{M}} = \begin{bmatrix} 0 & 17 & 14 & 0 \\ \infty & 21 & 39 & 0 \\ 17 & 0 & 26 & 18 \\ 17 & 3 & 0 & 0 \end{bmatrix},$$

so that $L(\mathbf{M}) = 365$, $\pi = [1, 4, 2, 3]$, and $U(\mathbf{M}) = 527$.

For the right branch (node n_3),

$$\mathbf{D} = \begin{bmatrix} 0 & 5 & 2 \\ 6 & 0 & 9 \\ 2 & 5 & 0 \end{bmatrix}, \mathbf{F} = \begin{bmatrix} 0 & 4 & 4 \\ 5 & 0 & 8 \\ 9 & 2 & 0 \end{bmatrix}, \text{ and } \mathbf{C} = \begin{bmatrix} 9 & 0 & 2 \\ 7 & 7 & 3 \\ 5 & 2 & 0 \end{bmatrix},$$

which updates to

$$\begin{bmatrix} 9 & 0 & 2 \\ 7 & 7 & 3 \\ 5 & 2 & 0 \end{bmatrix} + \begin{bmatrix} 243 & 105 & 210 \\ 76 & 77 & 81 \\ 80 & 63 & 79 \end{bmatrix},$$

and adding

$$(\overline{\mathbf{d}}_{i\bullet} \overline{\mathbf{f}}_{j\bullet}^T) = \begin{bmatrix} 28 & 41 & 28 \\ 60 & 93 & 72 \\ 28 & 41 & 28 \end{bmatrix},$$

we obtain the Gilmore-Lawler bound

$$\mathbf{M} = \begin{bmatrix} 280 & 146 & 240 \\ 143 & 177 & 156 \\ 113 & 106 & 107 \end{bmatrix},$$

which reduces to

$$\begin{bmatrix} 134 & 0 & 93 \\ 0 & 34 & 12 \\ 7 & 0 & 0 \end{bmatrix},$$

with row numbers 1, 3, 4, and column numbers 2, 3, 4. With $\alpha = c_{21} = 4$, we therefore obtain $\pi = [3, 1, 2, 4]$, $L(\mathbf{M}) = 396$ and $U = U(\mathbf{M}) = 421$.

We choose the right branch, and return to Step 3. Now $\Delta = 421 - 396 - 4 = 21$ and we update

$$\overline{\mathbf{M}} = \begin{bmatrix} \infty & 0 & \infty \\ 0 & \infty & 12 \\ 7 & 0 & 0 \end{bmatrix}.$$

Furthermore, $r_{13} = \infty$, $r_{32} = 19$, and $r_{44} = 12$, so that we choose the branch $(\bar{i}, \bar{j}) = (3, 2)$, i.e., $\pi(3) = 2$, and now $N_1 = \{2, 3\}$, and $N_2 = \{1, 4\}$. The left branch (node n_4) has

$$\overline{\mathbf{M}} = \begin{bmatrix} \infty & 0 & \infty \\ \infty & \infty & 0 \\ 0 & 0 & 0 \end{bmatrix}$$

with the only feasible solution $\pi = [3, 1, 4, 2]$, which yields $U(\mathbf{M}) = 449$. For the right branch (node n_5) we only have two solutions, viz., $\pi = [3, 1, 2, 4]$ and $[4, 1, 2, 3]$ with objective function values 421 and 514, respectively, so that we choose the first of these two and close this branch.

With the branches closed at the nodes n_4 and n_5, we now explore node n_2 in Step 3. With

$$\overline{\mathbf{M}} = \begin{bmatrix} 0 & 17 & 14 & 0 \\ \infty & 21 & 39 & 0 \\ 17 & 0 & 26 & 18 \\ 17 & 3 & 0 & 0 \end{bmatrix},$$

$\Delta = U - L(\mathbf{M}) = 514 - 365 = 149$, and $r_{11} = 17$, $r_{24} = 21$, $r_{32} = 20$, and $r_{43} = 14$, the matrix $\mathbf{M}$ cannot be further modified, not even if we take into account the tighter upper bound $U = 421$ found at node n_3; this would give an improved $\Delta = 421 - 365 = 56$, which is still not enough for any $\overline{m}_{ij}$ to be set equal to ∞. With r_{24} being the largest r_{ij}, we select the branch $(\bar{i}, \bar{j}) = (2, 4)$, i.e., $\pi(2) = 4$. The left branch from node n_2 is therefore $\pi(2) \neq 4$, leading to node n_6 in Step 4. We now have

Chapter 5: Layout Models

$$\mathbf{M} = \begin{bmatrix} 40 & 99 & 126 & \mathbf{80} \\ \infty & \mathbf{133} & 181 & \infty \\ 92 & 117 & 173 & 133 \\ 43 & 71 & \mathbf{98} & 66 \end{bmatrix},$$

which is reduced to

$$\overline{\mathbf{M}} = \begin{bmatrix} 0 & 34 & 14 & \mathbf{0} \\ \infty & \mathbf{0} & 1 & \infty \\ 0 & 0 & 9 & 1 \\ 17 & 20 & \mathbf{0} & 0 \end{bmatrix}$$

with $L(\mathbf{M}) = 403$, $U(\mathbf{M}) = 514$ for $\pi = [4, 2, 1, 3]$, so that U remains at 421.

The right branch $\pi(2) = 4$ leads to node n_7, for which we obtain

$$\mathbf{D} = \begin{bmatrix} 0 & 5 & 2 \\ 6 & 0 & 9 \\ 2 & 5 & 0 \end{bmatrix}, \mathbf{F} = \begin{bmatrix} 0 & 7 & 0 \\ 5 & 0 & 4 \\ 7 & 5 & 0 \end{bmatrix}, \mathbf{C} = \begin{bmatrix} 1 & 9 & 0 \\ 5 & 7 & 7 \\ 4 & 5 & 2 \end{bmatrix},$$

so that with the linear updates

$$\begin{bmatrix} 1 & 9 & 0 \\ 5 & 7 & 7 \\ 4 & 5 & 2 \end{bmatrix} \text{ and } \begin{bmatrix} 219 & 276 & 168 \\ 73 & 71 & 94 \\ 75 & 81 & 82 \end{bmatrix},$$

as well as

$$(\overline{\mathbf{d}}_{i\bullet} \overline{\mathbf{f}}^T_{j\bullet})_- = \begin{bmatrix} 14 & 30 & 39 \\ 42 & 66 & 87 \\ 14 & 30 & 39 \end{bmatrix},$$

we obtain the Gilmore-Lawler bound matrix

$$\mathbf{M} = \begin{bmatrix} 234 & 315 & \mathbf{207} \\ 120 & 144 & 188 \\ 93 & \mathbf{116} & 123 \end{bmatrix}$$

and its reduced version

$$\mathbf{M} = \begin{bmatrix} 27 & 85 & 0 \\ 0 & 1 & 68 \\ 0 & 0 & 30 \end{bmatrix}$$

with $L(\mathbf{M}) = 443$, $U(\mathbf{M}) = 478$ for $\pi = [3, 4, 1, 2]$ and $\alpha = c_{24} = 5$. With $\alpha + L(\mathbf{M}) = 5 + 443 = 448 > 421 = L$, we close this branch.

At this point, only node n_6 is live. With $\Delta = U - L(\mathbf{M}) = 421 - 403 = 18$, we can modify the matrix $\overline{\mathbf{M}}$ to

$$\overline{\mathbf{M}} = \begin{bmatrix} 0 & \infty & 14 & 0 \\ \infty & 0 & 1 & \infty \\ 0 & 0 & 9 & 1 \\ 17 & \infty & 0 & 0 \end{bmatrix},$$

from which we obtain $r_{14} = r_{22} = r_{31} = 0$ and $r_{43} = 2$. Therefore, $(\bar{i}, \bar{j}) = (4, 3)$, i.e., $\pi(4) = 3$, and the branch $\pi(4) \neq 3$ leads to the node n_8, and $\pi(4) = 3$ to node n_9. In Step 4, at node n_8, we obtain

$$\overline{\mathbf{M}} = \begin{bmatrix} 0 & \infty & 13 & 0 \\ \infty & 0 & 0 & \infty \\ 0 & 0 & 8 & 1 \\ 17 & \infty & \infty & 0 \end{bmatrix},$$

leading to $L(\mathbf{M}) = 405$, so that $\Delta = U - L(\mathbf{M}) = 421 - 405 = 16$, leading to

$$\overline{\mathbf{M}} = \begin{bmatrix} 0 & \infty & 13 & 0 \\ \infty & 0 & 0 & \infty \\ 0 & 0 & 8 & 1 \\ \infty & \infty & \infty & 0 \end{bmatrix},$$

so that $r_{11} = 0$, $r_{23} = 8$, $r_{32} = 0$, and $r_{44} = \infty$, from which we finally obtain

$$\overline{\mathbf{M}} = \begin{bmatrix} 0 & \infty & 13 & 0 \\ \infty & 0 & 0 & \infty \\ 0 & 0 & 8 & 1 \\ \infty & \infty & \infty & \infty \end{bmatrix},$$

and the last row implies that node n_8 can be closed. Turning now to node n_9 in Step 4, we have $\pi(4) = 3$, but also $\pi(2) \neq 1$, and $\pi(2) \neq 4$ from earlier branches. As a result, we find

Chapter 5: Layout Models

$$\mathbf{D} = \begin{bmatrix} 0 & 15 & 5 \\ 24 & 0 & 3 \\ 6 & 11 & 0 \end{bmatrix}, \mathbf{F} = \begin{bmatrix} 0 & 7 & 5 \\ 5 & 0 & 4 \\ 6 & 9 & 0 \end{bmatrix}, \text{ and } \mathbf{C} = \begin{bmatrix} 1 & 9 & 2 \\ 4 & 2 & 5 \\ 5 & 7 & 3 \end{bmatrix}.$$

The Gilmore-Lawler bound then becomes

$$\mathbf{M} = \begin{bmatrix} 1 & 9 & 2 \\ 4 & 2 & 5 \\ 5 & 7 & 3 \end{bmatrix} + \begin{bmatrix} 14 & 18 & 20 \\ 63 & 65 & 82 \\ 35 & 61 & 58 \end{bmatrix} + \begin{bmatrix} 110 & 85 & 135 \\ 141 & 111 & 171 \\ 97 & 74 & 120 \end{bmatrix} = \begin{bmatrix} 125 & 112 & \mathbf{157} \\ 208 & \mathbf{178} & 258 \\ 137 & 142 & 181 \end{bmatrix},$$

which is reduced to

$$\overline{\mathbf{M}} = \begin{bmatrix} 12 & 0 & \mathbf{0} \\ 29 & \mathbf{0} & 35 \\ \mathbf{0} & 6 & 0 \end{bmatrix}.$$

With $\alpha = c_{43} = 2$, $\pi = [4, 2, 1, 3]$, $L(\mathbf{M}) = 472$ and $U(\mathbf{M}) = 514$, we find that node n_9 can also be closed. Now the tree is completely searched, and the algorithm terminates with the unique optimal solution $\pi = [3, 1, 2, 4]$ whose objective function value is $\overline{z} = 421$, which was obtained at node n_3.

The tree is shown in Figure II.25, in which the numbers $(\alpha + L(\mathbf{M}), U)$ are shown next to each node.

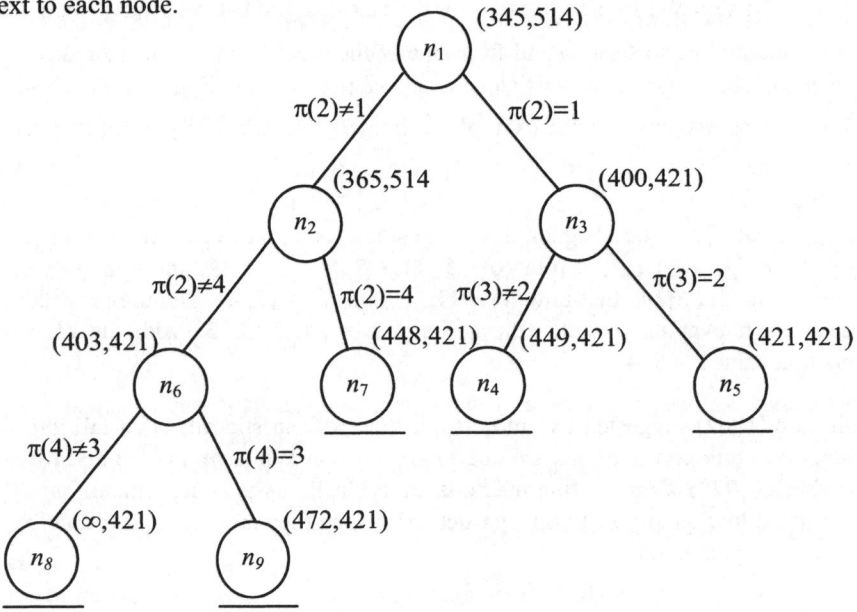

Figure II.25

It must be mentioned that while an optimal solution has been found at node n_5, it is necessary to examine the nodes n_6, n_7, n_8, and n_9 in order to ensure that no alternative optimal solution exists.

5.5.2 Heuristic Solution Methods

Since exact solution techniques, coupled with present-day computer technology, are restricted to quadratic assignment problems with no more than $n = 30$ nodes at the very most, it becomes necessary for most practical problems to consider heuristic procedures for finding near-optimal solutions. Such heuristic methods can be broadly categorized as construction methods, improvement methods, and hybrid methods.

Construction methods use simple rules to choose items to assign to slots, one at a time. One of the most popular classes of construction methods are the *Greedy methods*, which sequentially assign items to slots, so as to minimize the increase in the objective function in each step. A very simple heuristic can be described as follows. Let N_T and N_S denote the sets of items that have already been assigned to slots, and slots to which items have already been assigned, respectively. First, consider only items, ignoring the slots they are going to be assigned to. The next item to be assigned to a slot is then $(\bar{i}, \bar{j}) = \text{argmax } \{f_{ij} + f_{ji} : i \in N_T, j \in N \setminus N_T\}$. Now consider the slots and assume that item $T_{\bar{i}}$ is assigned to slot $S_{\bar{k}}$. The next slot to be chosen is $(\bar{k}, \bar{\ell}) = \text{argmin } \{d_{k\ell} + d_{\ell k} : \ell \in N \setminus N_S\}$. We will then assign item $T_{\bar{j}}$ to slot $S_{\bar{\ell}}$. In other words, for the next assignment we always choose the item with the largest flow to and from one of the other items that has already been assigned. The chosen slot is as close as possible to the existing slot $S_{\bar{k}}$ to which item $T_{\bar{i}}$ was assigned. In our example, if item T_2 had arbitrarily been assigned to slot S_3, i.e., $N_S = \{3\}$ and $N_T = \{2\}$, the next item to be assigned would be determined by max $\{f_{21} + f_{12}, f_{23} + f_{32}, f_{24} + f_{42}\}$ = max $\{5 + 7, 4 + 5, 4 + 9\} = 13$, so that item T_4 will be assigned next. The slot S_ℓ is then determined by min $\{d_{31} + d_{13}, d_{32} + d_{23}, d_{34} + d_{43}\}$ = min $\{6 + 5, 11 + 3, 9 + 5\} = 11$, and item T_4 will be assigned to slot S_1, so that now $N_S = \{1, 3\}$ and $N_T = \{2, 4\}$, and the procedure is repeated. It eventually yields the solution $\pi = [4, 1, 2, 3]$ with the objective function value $z = 514$.

Gilmore (1962) suggested two more sophisticated heuristic rules that fall into this category; other contributions are due to Graves and Whinston (1970) and Müller-Merbach (1970). Construction methods are typically easy to implement, but often only yield low-quality solutions, particularly when they are not followed up by an improvement method.

Limited enumeration methods are another subclass within construction methods. They can be seen as limited branch-and-bound methods which do not search the entire tree. For instance, one may stop the branch-and-bound procedure when the bound gap reaches a sufficiently small, preselected level, or when a preset time limit or tree size is reached. In the branch and bound example in the previous section, a stopping criterion that prescribes termination when the bound gap is no more than 50 would have stopped the process with the tree having only the two nodes n_1 and n_2, and a suboptimal solution π = [4, 1, 2, 3] with the objective function value z = 514. With a limited tree size of four nodes, the search would have stopped with the tree having nodes n_1, n_2, n_3, and n_4, and the (optimal) solution π = [3, 1, 2, 4] with the objective value z = 421. However, at this point the optimality of this solution could not be ascertained. This would have been possible, though, with a five node limit, since at the node n_5 we know that an optimal solution has been reached. More sophisticated stopping rules do, of course, exist. Many such rules employ the empirically derived fact that the lower bound increases more rapidly in the early stages of the tree, so that a "dynamic gap" rule could be devised, involving the bound gap as well as the tree level of the node.

Improvement or *local search methods* start with some tentative solution, obtained by, e.g., a limited enumeration or construction method, and try to improve it through some change in the existing assignment. The simplest such change would be the *pairwise exchange*, for which two slots, say S_i and S_j, exchange the items $T_{\pi(i)}$ and $T_{\pi(j)}$ assigned to them. In other words, the elements i and j in the current permutation π are swapped. There are ½ $n(n–1)$ possible swaps, and one may chose the strategy of accepting the first improving swap found, making the best possible swap, or some other choice. For example, the tentative solution π = [4, 1, 2, 3] obtained by the construction method above would allow ½(4)(4–1) = 6 possible swaps, the best of which is to swap the first and the last element to obtain the permutation π = [3, 1, 2, 4], which happens to be the unique optimal solution with $\bar{z}$ = 421. Exchanges that involve three slots, or in general r, slots could also be used. The basis for such a method is the so-called *r-opt method* that was devised by Lin (1965) for the traveling salesman problem. An appropriate adaptation of this method was applied by Roucairol (1987) to the quadratic assignment problem.

Other improvement methods include those based on the *simulated annealing* search strategy, developed by Kirkpatrick *et al.* (1983), and adapted for the quadratic assignment problem and further developed by Connolly (1990). Finally, Glover's (1988) *tabu search method* has also been successfully applied to the quadratic assignment problem by Skorin-Kapov (1990); the ability of tabu search to avoid being trapped at local optima makes this approach particularly beneficial for "large-scale" problems (which still means n < 50).

Finally, *hybrid methods* combine methods for finding good solutions with improvement methods for getting closer to an optimal solution. A hybrid method was actually used in the numerical example above, where the solution $\pi = [4, 1, 2, 3]$ was first found by a construction method, after which this solution was improved using the pairwise exchange improvement method. Contributions in this area are those by Liggett (1981), Bazaraa *et al.* (1983), Kaku *et al.* (1991), and Drezner (2002); see also Pardalos *et al.* (1994).

A survey of heuristics for facility layout problems, including simulated annealing and genetic algorithms, is provided by Balakrishnan *et al.* (2003).

5.5.3 Solving General Facility Layout Problems

While the previous section has discussed exact and heuristic methods to solve the quadratic assignment problem, this section investigates the macro view of facility layout planning. In particular, we will discuss some of the early decision support systems, some of which use quadratic assignment problems as subalgorithms.

From the discussion in the beginning of this chapter, it is clear that layout planning is a complex procedure, having to consider simultaneously many intricately related aspects of processes, materials, flows, and employees. Since the layout planning procedure often involves several individuals, documentation and communication requirements are essential and substantial. Layout planning is today therefore mostly done in a computer-aided mode, involving database management (*DBM*) and computer-aided design (*CAD*) systems. Several algorithms for generating layouts, using the system layout planning (*SLP*) information discussed above have been developed. Similar to the quadratic assignment problem in the previous section, the main heuristic methods can be classified as construction and improvement algorithms. Construction algorithms start with the basic *SLP* and generate a block layout by successively adding items (machines, departments) to some partial layout, until all items are included. Improvement algorithms start with an initial block layout that is either known or was generated by some construction algorithm. Attempts are then made to improve the initial layout by swapping items in the layout subject to the restrictions specified. Both types of algorithms also use some scoring (objective) function to evaluate the layouts that are generated. The scoring can be *adjacency-based*, using the relationship chart discussed in Section 5.1; *distance-based*, using estimates of the cost of flow between items (machines); or some more sophisticated procedure involving, for instance, equipment utilization.

Historically, a number of heuristic methods for facility layout planning originated in the 1960s, and many variations have appeared since. We will now briefly describe four of these classical programs. The first known and available computerized layout algorithm was *CRAFT* (*computerized relative allocation of facilities technique*), an improvement technique developed by Armour and Buffa (1963) and Buffa *et al.* (1964). Items (departments) are assumed to be of rectangular or piecewise rectangular shape, and rectilinear distances between the departments are computed based on the locations of the centers of gravity of their respective geographical areas. The objective is to minimize the total transportation cost of a layout, similar to the objective of the quadratic assignment problem discussed above. In order to do so, *CRAFT* will start with some initial layout and then consider the effect of all pairwise exchanges between adjacent or equally large departments and pick the exchange which causes the largest overall reduction in the objective function value. *CRAFT* continues selecting such pairwise changes in a greedy fashion, until no further decreases of the objective function are possible. A difficulty is that special procedures have to be used to ensure that contiguous piecewise rectangular shapes will remain contiguous even after pairwise exchanges. A modification of *CRAFT* is *COFAD* (*computerized facilities design*), a program by Tompkins and Reed (1976) that also considers the choice of materials handling alternatives when improved solutions are sought. For a given layout, COFAD will seek the alternative with the best overall equipment utilization before the layout itself is improved.

ALDEP (*automated layout design program*) was developed by Seehof and Evans (1967). It is a construction algorithm that randomly chooses the first item (department) and then successively adds more, selected on the basis of the closeness ratings of the relationship chart. The first department is placed in the upper left-hand corner of the layout. *ALDEP* then picks another department with a high closeness rating (such as *A* or *E* in a relationship chart) to the first department and places it next to the first. The procedure is repeated, successively placing subsequent departments adjacent to the preceding ones, according to a predetermined serpentine-like sweep pattern that covers the entire available area. If there are no departments with high closeness ratings to choose from, one of the remaining departments is chosen at random. When all departments are placed, a total adjacency score for the layout is computed, using the closeness ratings transformed into numerical values according to the transformation $(A, E, I, O, U, X) \to (4^3, 4^2, 4^1, 4^0, -4^5) = (64, 16, 4, 1, 0, -1{,}024)$. The entire process is then repeated as many times as desired, and the layout with the highest score is chosen. *ALDEP* has a tendency to produce layouts with unusual shapes of the departments, necessitating special restrictions for these shapes.

CORELAP (*computerized relationship layout planning*) was developed by Lee and Moore (1967). Like *ALDEP*, it is a construction algorithm, but it uses the *total closeness rating*, *TCR*, for an item. In contrast to *ALDEP*, the numerical values 6, 5, 4, 3, 2, 1 are assigned to the closeness ratings *A, E, I, O, U, X*, respectively. Letting r_{ij} denote the closeness rating between items T_i and T_j, we then define the total closeness rating TCR_i for item T_i as $TCR_i = \sum_{j=1}^{n} r_{ij}$. Items are included first according to TCR, and then successively according to the TCR of those items already included. Moore (1971) extended *CORELAP* to also include the rectilinear distances d_{ij} between items T_i and T_j, once they have been placed, using the total scoring expression $\sum_{i=1}^{n}\sum_{j=1}^{n} r_{ij}d_{ij}$. As for *ALDEP*, there is an arbitrariness in the specification of weights for the closeness ratings, which carries over to the rankings of various layout alternatives, as pointed out by Francis *et al.* (1992).

PLANET (*plant layout analysis and evaluation technique*) by Deisenroth and Apple (1972) is a construction algorithm, but unlike *ALDEP* and *CORELAP* it does not use the relationship chart, but flows between items (departments) in the selection process. Each department must also be assigned a priority rating 1, 2, ..., 9, where "1" represents the highest, and "9" the lowest priority. Departments are selected first based on the priority level and secondly based on the flow cost between departments. The department selected is then placed so as to generate the smallest total flow cost. As with *ALDEP* and *CORELAP*, layouts generated with *PLANET* may result in departments with very irregular shapes. In such cases, the layout may need manual adjustment by the planners to obtain practically acceptable solutions.

For more detailed descriptions of the above and other layout algorithms, the reader is referred to the sources above. Good discussions can also be found in Nahmias (1989) and Francis *et al.* (1992). A useful discussion of adjacency requirements in a graph-theoretical framework is provided by Foulds (1983).

PART III: PROJECT SCHEDULING[1]

This part deals with the organization of activities that consume time and possibly other resources. Activities can be anything from mundane tasks of everyday life to complex tasks in business, engineering or other fields. Clearly, if such activities can only be performed one after the other in some order, then the organization does not present a problem: any feasible sequence of activities will solve the problem. However, when it is possible to perform some of the activities in parallel, the scheduling of activities is no longer trivial, and an efficient tool for planning is needed. The different types of project networks belong to this category of planning tools.

Historically, the types of project networks we will be dealing with date back to the 1950s. With the success of linear programming for solving planning problems involving scarce resources, the stage was set for attempting to manage large-scale projects involving hundreds or even thousands of interrelated activities by scientific methods. The *Critical Path Method*, or *CPM* for short, was developed jointly by the du Pont de Nemours and Remington Rand Univac corporations in an effort to reduce the time required for plant overhaul, maintenance, and construction. Accounts of the early developments can be found in Walker and Sayer (1959) and Moder *et al.* (1983). A similar technique, the *Project Evaluation and Review Technique*, or *PERT*, was developed by the US Navy (PERT, 1958) for the Polaris weapons system. It was the result of a cooperative effort between the consulting firm of Booz, Allen and Hamilton, the Lockheed Aircraft Corporation, and the US Navy Special Projects Office.

This part is organized as follows. In the first chapter, we describe the basic elements included in project networks as well as a method that solves the problem. Chapter 2 allows the consumption of resources other than time.

---

[1] Chapter 2 of this part was coauthored by Professor A. Drexl, Christian-Albrechts-University, Kiel, Germany.

CHAPTER 1 UNCONSTRAINED TIME PROJECT SCHEDULING

This chapter considers the scheduling of projects, in which time is the only parameter of concern. In this simplified model, we then determine ways to outline those activities that decision makers have to regard closely, as their durations are more critical than those of other activities. We then examine the possibility of accelerating the project at the smallest cost possible. Another extension investigates the possibility of uncertain durations of the activities. This allows us to calculate probabilities of finishing the project on time.

1.1 Network Representations

Each project or activity network will be based on three main components: activities, the durations of activities, and the relations between the starting and finishing times of the activities. A brief discussion of these elements follows.

(1) *Activities*. Consider a *project*, which is typically considered to be a term for a fairly large-scale undertaking, usually associated with the macro level. Zooming in to the meso level, we can subdivide the project into separate *tasks*. Finally, on the micro level, each task can be subdivided into *activities*. Activities are considered the atoms of a project. As an example, consider the building of a house. The construction of the house is the project. One of the tasks in that project is the building of the roof, which, in turn, consists of individual activities, such as putting up the rafters, nailing down the shingles, and installing the flashing. While some activities have clear requirements as far as their sequences are concerned (e.g., we cannot nail down the shingles before the rafters have been installed), others can be performed in parallel (e.g., the installation of the doors and the laying of the carpets. As a matter of fact, it is sometimes beneficial to have one macro network, in which each "activity" really represents a partial project, a meso network, in which each of the "activities" represents a task in one of the projects, and finally a micro

network, in which each activity represents what we defined as activity above. Clearly, the boundaries between project, task, and activity are not rigid and are largely determined by the project under consideration.

(2) *Durations of the activities.* Each of the activities referred to above will have a certain duration. In the simplest case, we assume that all of these durations are known with certainty, an assumption that will almost certainly be violated in practice. Considering again the construction example under (1), it is apparent that even the most experienced planner will not know with certainty how long a task, such as obtaining Government permits or the installation of septic tanks, will take, as unforeseen obstacles can occur at any time, e.g., new Government regulations, or a rocky ledge that requires different tools for digging. However, sensitivity analyses can be employed to overcome the difficulty of uncertainty: assuming first that the durations of the activities assume their anticipated values, the decision maker then modifies the durations of individual activities, asking the usual "what if?" questions, thus presenting the decision maker with information regarding possible delays, while retaining the simplicity of a deterministic model. Another possibility is to explicitly include the uncertainty in the model. The difficulty in that is, of course, to specify the probability distribution that governs the duration of an activity. This may be an arduous task in the best case, and plain impossible in the worst. The mathematical difficulties of dealing with a variety of probability distribution in a model and attempting to aggregate them can hardly be overstated. Avoiding this problem and leaving the parameters uncertain is also possible, but will severely affect the quality of the results. After all, in modeling as everywhere else, the quality of the input will determine the quality of the output.

(3) *Precedence relations.* A set of binary relations specified by the decision maker will determine the sequence in which the activities can be performed. As an explanation, let a and b denote two activities. Furthermore, suppose that the decision maker has specified that the activity a must precede the activity b (in symbols: $a \prec b$) in the sense that activity a must be completely finished before b can start. As a natural extension, if a and b must precede c, then a and b must both be completely finished before c can start. Basic logic requires that the precedence relations satisfy the axiom of *transitivity*. In other words, if $a \prec b$ and $b \prec c$, then $a \prec c$ as well. In virtually all cases, a decision maker will specify only a partial list of precedence relations; the remaining relations may then be derived through transitivity. Typically, only immediate predecessor (or successor) activities will be specified.

In addition to the above three main components, other features may be included in project networks. Most prominently, they include resource consumption of the activities. As an example, the pouring of the foundation of a house requires, in

Chapter 1: Unconstrained Time Project Scheduling

addition to time, resources such as manpower, concrete, trucks, money, and others. In the remainder of this chapter we ignore these additional resource requirements; they are introduced in the second chapter of this part.

We are now able to represent the sequence of activities graphically. Basically, there are two distinct ways of going about it, the "activity-on-arc" (*AOA*) and the "activity-on-node" (*AON*) representations. Both use directed graphs as defined and used in the literature; see, e.g., Eiselt and Sandblom (2000). Consider first the traditional activity-on-arc representation. As the name suggests, it represents each activity by a directed arc, while nodes represent events. An event is loosely defined as the beginning or the end of an activity. In addition, there are two special events: the beginning of the project is represented by a source node n_s, and the end of the entire project, represented by a sink node n_t.

Example: Consider the precedence relations in Table III.1.

Table III.1

Activity	Immediate Predecessor
A	–
B	–
C	A
D	A
E	B, C

The corresponding *AOA* network representation is then shown in Figure III.1.

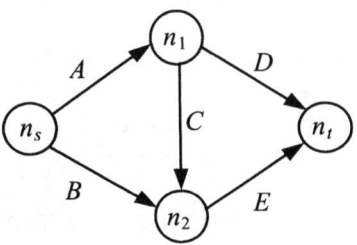

Figure III.1

In this graph, the event n_2 represents the beginning of activity E as well as the end of activities B and C. However, it is not always possible to represent the structure of a given set of precedence relations in an *AOA* network without further tools. Consider, for instance, the precedence relations shown in Table III.2.

Table III.2

Activity	Immediate predecessor
A	–
B	–
C	–
D	A, B
E	B, C
F	A, C

In order to represent these precedence relations in an *AOA* network, so-called *dummy activities* (or dummies for short) must be used. The only purpose of a dummy is to ensure that the precedence relations are properly maintained; consequently, a dummy activity has a duration of zero. The activity-on-arc network that belongs to the precedence relations in Table III.2 is shown in Figure III.2, where dummy activities are shown by broken lines.

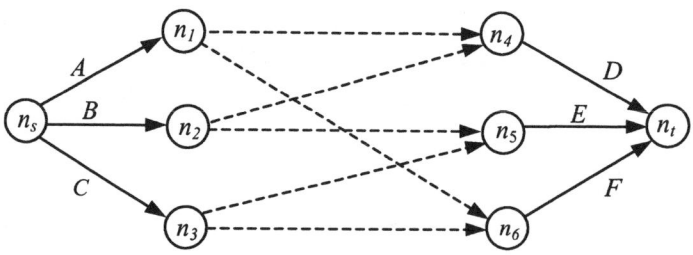

Figure III.2

In addition to logical reasons, dummies are usually also introduced in order to avoid parallel arcs. The reason for this is to enable the machine to uniquely identify an activity. If parallel arcs exist as in Figure III.3a, either activity is broken up into two parts and the structure is replaced by that of Figure III.3b.

Another point worth noting is that logical consistency requires that any project network must be acyclic. Suppose this were not the case, then at least one circuit would exist, e.g., $(n_i, a_{ij}, n_j, a_{jk}, n_k, a_{ki}, n_i)$. Following the usual logic, the activity a_{ki} can only start if its predecessor a_{jk} is completely finished. The activity a_{jk} can, in turn, only begin if its predecessor a_{ij} is completely finished, and a_{ij} can only start if a_{ki} is completely finished, an obvious logical contradiction.

Consider now activity-on-node (*AON*) networks. In such a network, each activity is represented by a node, and the directed arcs identify relations between the activities. Each *AON* network has dummy activities n_s and n_t with zero durations,

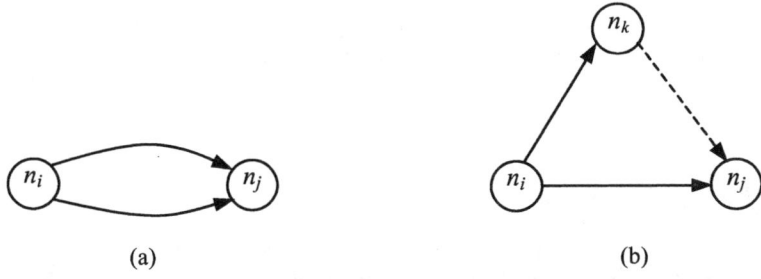

Figure III. 3

representing the project start and finish, respectively, as well as dummy arcs connecting n_s with all nodes that belong to activities without predecessors, and dummy arcs that connect the nodes of all activities without successor to n_t. Other than that, an *AON* network does not require any dummies. As an example, the *AON* networks in Figures III.4 and III.5 represent the precedence relations shown in Tables III.1 and III.2, respectively.

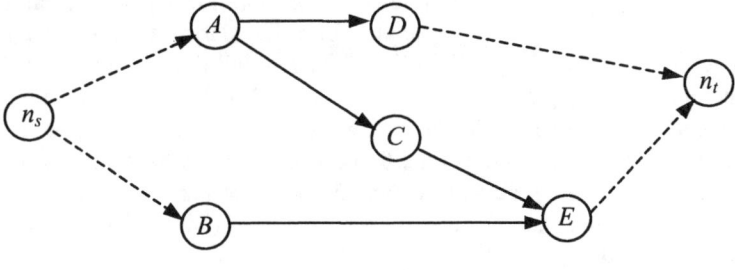

Figure III.4

In general, *AON* networks are easier to construct than their *AOA* counterparts. For that reason, most users today employ *AON* networks. In this chapter, we will discuss both representations, but concentrate on *AON* networks in Chapter 2.

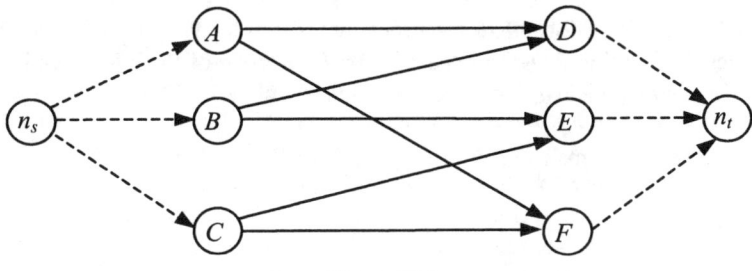

Figure III.5

Both representations have in common that durations are associated with activities. Formally, the duration of an activity a_{ij} in an *AOA* network is denoted by t_{ij}, while the duration of an activity n_i in an *AON* network is t_i. For the remainder of this chapter, we will assume that the durations of all activities are known with certainty.

1.2 The Critical Path Method

The remainder of this chapter describes the *critical path method* (*CPM*). In a project network, a *critical path* from source to sink determines the minimal time required to complete the project, given the durations of the individual activities and their precedence relationships. Furthermore, all activities on the critical paths are referred to as *critical activities*, as any delay in their execution or increase in their duration will necessarily delay the termination of the entire project beyond its minimal completion time. This implies that critical activities must be monitored more closely than other activities.

First, consider *AOA* networks. Let the durations of the activities in the example of Table III.1 and Figure III.1 be $t_A = 3$, $t_B = 5$, $t_C = 4$, $t_D = 8$, and $t_E = 5$ hours, respectively. Consider now the individual events and how early they can possibly occur, starting with the event n_s that arbitrarily starts at time 0. The event symbolized by n_1 can occur only if all preceding activities are completely finished. Here, the only preceding activity is A, which takes 3 hours, so that the event n_1 can occur as early as at time 3. Consider now the event n_2. Its predecessors are the activities B and C; B can start at time 0 and takes 5 hours (so that it is finished at time 5), while activity C can start at time 3 and, as it takes 4 hours to complete, can be done no earlier than at time 7. As the event n_2 (which, incidentally, is also the beginning of the succeeding activity E) can only occur when *all* preceding activities are completely done, the earliest time the event n_2 can take place is at time 7. Similarly, we can determine that the event n_t can occur no earlier than at time $3 + 8 = 11$ (which is the earliest time activity D is completed) and $7 + 5 = 12$ (which is the earliest time activity E is finished). Consequently, n_t, the event symbolizing the end of the entire project, can occur no earlier than at time 12.

We are now able to formalize the procedure. First, let $A = \{a_{ij}\}$ denote the set of activities, and define $EE(n_i)$ as the *earliest event time* of the event symbolized by the node n_i, and let $EE(n_s) := 0$. We can then calculate

$$EE(n_i) = \max_{j : a_{ji} \in A} \{EE(n_j) + t_{ji}\} \ \forall \ n_i \in N$$

It is apparent that the earliest event time $EE(n_i)$ is nothing but the length of the longest path from the source n_s to the node n_i. Furthermore, the earliest event time of the sink $EE(n_t)$ is the earliest time that the entire project can be finished. The

Chapter 1: Unconstrained Time Project Scheduling

process of computing the earliest event times is usually referred to as a *forward sweep* or *forward pass*.

Define now $LE(n_i)$ as the *latest event time*, i.e., the latest possible time that the event n_i can occur without delaying the project beyond $EE(n_t)$. We then, somewhat arbitrarily, define $LE(n_t) := EE(n_t) := T$, the project completion time, as a boundary condition and perform a *backward sweep* (or *backward pass*) by using the recursive relation

$$LE(n_i) = \min_{j:a_{ij} \in A} \{LE(n_j) - t_{ij}\} \ \forall \ n_i \in N,$$

which will result in $LE(n_s) = 0$. This information can be collected in a graph as shown in Figure III.6.

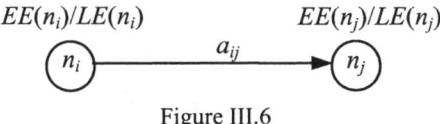

Figure III.6

In the numerical example above, the earliest event times and the latest event times happen to be equal. In general, this will not be the case.

Finally, define the *(total) float* of an activity a_{ij} as

$$TF(a_{ij}) = LE(n_j) - EE(n_i) - t_{ij} \ \forall \ a_{ij} \in A.$$

The total float (sometimes also referred to as *total slack*) denotes the amount of time by which the activity a_{ij} can be delayed (or, equivalently, the duration of a_{ij} can increase) without delaying the entire project, provided that all other activities start at the most beneficial times, i.e., all predecessors of a_{ij} start as early as possible, while all successors of a_{ij} start as late as possible. Loosely speaking, the total float indicates the freedom the planner has when scheduling this activity. Given this definition, the activities a_{ij} with total float $TF(a_{ij}) = 0$ are critical in the sense that they cannot be delayed or increase in duration without delaying the entire project. The activities with zero float make up a path from the source to the sink. This path will be called a *critical path*. It should be noted that the critical path need not be unique, an issue we will return to in Chapter 2 of this part. For an excellent discussion of the concept of critical activities, readers are referred to Elmaghraby (2000). In the numerical example above, the floats of the activities A, B, C, D, and E are 0, 2, 0, 1, and 0, respectively. Hence, the critical path consists of the activities A, C, and E.

Example: Consider a project with precedence relations and durations as shown in Table III.3.

Table III.3

Activity	Immediate Predecessor	Duration
A	-	3
B	-	4
C	A	7
D	B	7
E	A, B	4
F	C	6
G	D, F	8
H	C, E	5
I	D, F	5
J	H	2
K	H, I	4
L	J	6

The project network in this example is shown in Figure III.7, where the necessary dummy activities are shown by broken lines. The appropriate earliest and latest event times are displayed next to the nodes and the activity durations are indicated with the activity names next to the arcs that represent them. All activities on the unique critical path (A, C, F, I, K) are displayed as bold lines.

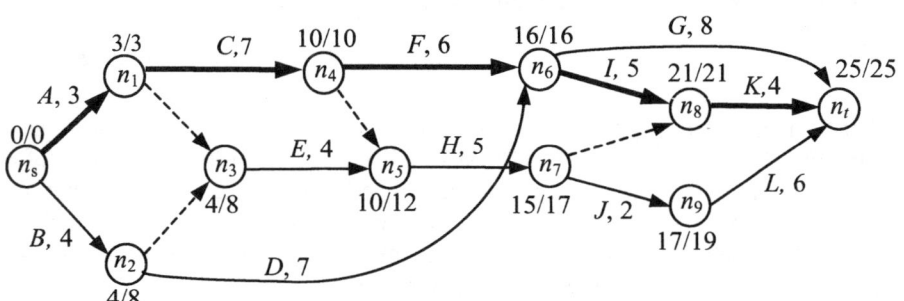

Figure III.7

The project, represented by the network in Figure III.7 can be completed within $T = 25$ hours. None of the critical activities A, C, F, I, and K can be delayed at all without causing the entire project to be late (beyond 25 hours). On the other hand, a noncritical activity such as D may be delayed by up to $LE(n_6) - EE(n_2) - t_{26} = 16 - 4 - 7 = 5$ hours without delaying the project.

Chapter 1: Unconstrained Time Project Scheduling

We will conclude the discussion of *AOA* networks by listing a number of different measures, all of which indicate how loosely (or tightly) the activities fit into a given schedule. One such measure, the total float, was already introduced above. A complete listing is provided below for some activity a_{ij}:

$$\begin{aligned}
\textit{Total float} \quad & TF(a_{ij}) = LE(n_j) - EE(n_i) - t_{ij}, \\
\textit{Free float} \quad & FF(a_{ij}) = EE(n_j) - EE(n_i) - t_{ij}, \\
\textit{Safety float} \quad & SF(a_{ij}) = LE(n_j) - LE(n_i) - t_{ij}, \\
\textit{Interference float} \quad & IF(a_{ij}) = \max\{0; EE(n_j) - LE(n_i) - t_{ij}\}.
\end{aligned}$$

Although many authors use the terms float and slack interchangeably, others such as Elmaghraby (1977) apply the term "float" to activities as defined above, and reserve the term "slack" for events, where the slack of an event n_i is defined as $S(n_i) = LE(n_i) - EE(n_i)$. There is no clear relation between floats and slacks. For example, consider the *AOA* network in Figure 1, this time with activity durations of 2, 7, 6, 8, and 4. The critical path is then $A - C - E$ with all event slacks equal to zero, while (all) floats of activity B equal to 1 and all floats of activity D equal to 2.

Consider now the *AON* representation. Based on the information in Table III.3, the activity network in this representation is shown in Figure III.8.

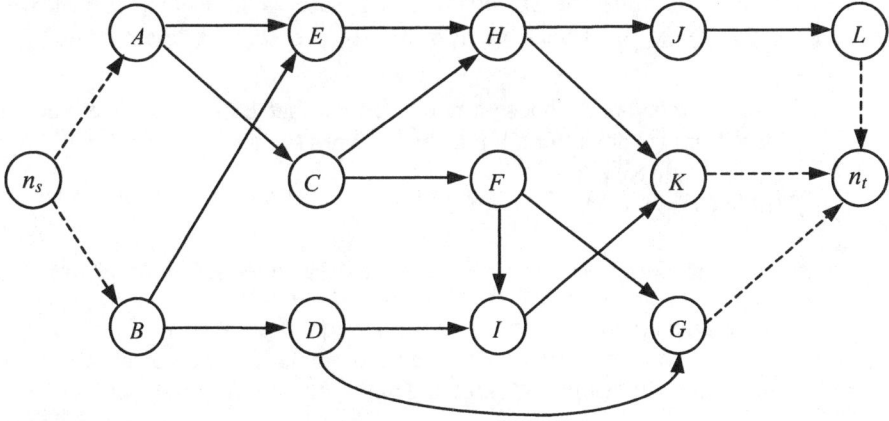

Figure III.8

As *AON* networks are activity-oriented, earliest and latest times are defined for activities rather than events. In particular, for an activity n_k define the earliest starting time $ES(n_k)$, the earliest finishing time $EF(n_k)$, the latest possible starting time $LS(n_k)$, and the latest possible finishing time $LF(n_k)$. Clearly, $EF(n_k) = ES(n_k) + t_k$ and $LF(n_k) = LS(n_k) + t_k$, which is why time is shown twice, at center top and center bottom in the node. The information is included in node n_k as shown in Figure III.9.

ES(n_k)	t_k	EF(n_k)
	n_k	TF(n_k)
LS(n_k)	t_k	LF(n_k)

Figure III.9

Here, the total float is equivalent to the case of *AOA* networks, defined as $TF(n_k) = LF(n_k) - ES(n_k) - t_k$.

Again, the critical path is the longest path in the network. Similar to the *AOA* network representation, it is determined by a forward and a backward labeling phase and the subsequent determination of the total float. In order to formally describe the algorithm, define $\mathcal{P}(n_k)$ as the set of predecessor nodes of n_k and $\mathcal{S}(n_k)$ as the set of successor nodes of n_k. An algorithm that finds the critical path(s) in a given network can then be described as follows.

Critical Paths in *AON* Networks

Step 1: Set $ES(n_s) := EF(n_s) := 0$.

Step 2 (Forward labeling): Choose a node n_j, such that $ES(n_j)$ is not yet known, but $EF(n_i)$ is known for all $n_i \in \mathcal{P}(n_j)$. Does such a node exist?
If yes: Go to Step 3.
If no: Go to Step 4.

Step 3: Compute $ES(n_j) = \max_{n_i \in \mathcal{P}(n_j)} \{EF(n_i)\}$, and calculate $EF(n_j) = ES(n_j) + t_j$.
Go to Step 2.

Step 4: Set $LS(n_t) := LF(n_t) := EF(n_t) := ES(n_t) := T$.

Step 5 (Backward labeling): Choose a node n_i, such that $LF(n_i)$ is not yet known, but $LS(n_j)$ is known for all $n_j \in \mathcal{S}(n_i)$. Does such a node exist?
If yes: Go to Step 6.
If no: Go to Step 7.

Step 6: Compute $LF(n_i) = \min_{n_j \in \mathcal{S}(n_i)} \{LS(n_j)\}$ and calculate $LS(n_i) = LF(n_i) - t_i$.
Go to Step 5.

Chapter 1: Unconstrained Time Project Scheduling

Step 7: For all nodes, calculate the total float $TF(n_k) = LF(n_k) - ES(n_k) - t_k$, $\forall\ n_k \in N$. All activities with zero float are then on the critical path; its length is T.

In order to illustrate this algorithm, consider the following

Example: Determine the critical path in an *AON* network based on the activities, their durations, and the set of precedence relations shown in Table III.3. The resulting network was already developed and shown in Figure III.8. The information concerning earliest and latest starting and finishing times and total floats will be displayed in the format given in Figure III.9. The complete network is presented in Figure III.10, where the critical path is shown by bold lines that connect the critical activities (also shown in boldface).

As in the *AOA* representation, four different types of floats (slacks) can also be determined in the *AON* representation. They are

- the total float $TF(n_k) = LF(n_k) - ES(n_k) - t_k$,

- the free float $FF(n_k) = \min\limits_{n_\ell \in S(n_k)} \{ES(n_\ell)\} - ES(n_k) - t_k$,

- the safety float $SF(n_k) = LF(n_k) - \max\limits_{n_\ell \in \mathcal{P}(n_k)} \{LF(n_\ell)\} - t_k$, and

- the interference float

$$IF(n_k) = \max\left\{0;\ \min\limits_{n_\ell \in \mathcal{S}(n_k)} \{ES(n_\ell)\} - \max\limits_{n_\ell \in \mathcal{P}(n_k)} \{LF(n_\ell)\} - t_k\right\}.$$

Note that these definitions imply that $TF(n_k) \geq FF(n_k)$ and $SF(n_k) \geq IF(n_k)$.

The interpretations of the floats are identical to those discussed above in the *AOA* representation. For instance, consider the activity E in the above example. Given the information in Figure III.10, if all of its predecessors start as early as possible, work on E can commence at time $ES(E) = 4$. Similarly, if all of E's successors start as late as possible, then E may finish as late as $LF(E) = 12$. This allows a time window of $12 - 4 = 8$ hours for an activity whose duration is 4 hours, thus leaving a total float of $TF(E) = 4$ hours. If, on the other hand, all of E's predecessors *and* successors were to start as early as possible, the following scenario presents itself. Activity E's predecessors A and B start at 0 and 0, so that they are completed at time 3 and 4, respectively. As a result, activity E could commence as early as 4. If activity E's successor H also starts as early as possible, it will do so at $ES(H) = 10$, leaving a time window of $10 - 4 = 6$ hours for an activity whose duration is $t_E = 4$ hours, hence E has a free float of $FF(E) = 6 - 4 = 2$ hours. Assume now that all predecessors *and* successors of E start as late as possible (without delaying the

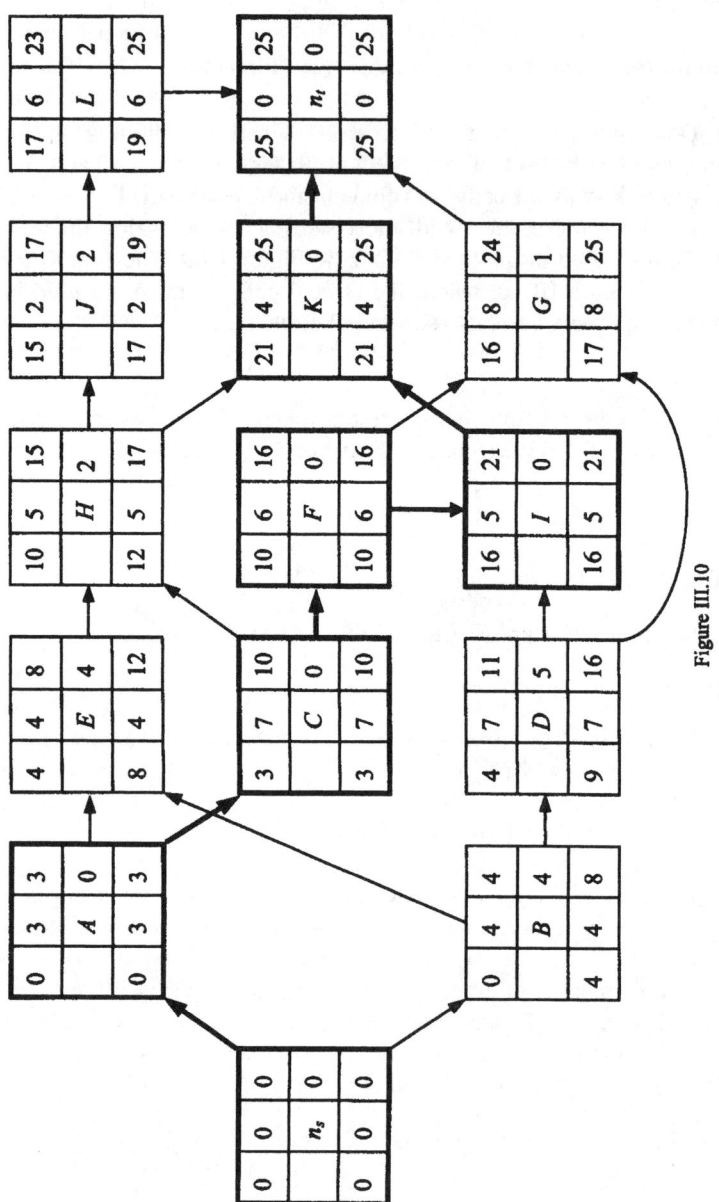

Figure III.10

entire project beyond $T = 25$ hours). Now E's predecessors finish at $LF(A) = 3$ and $LF(B) = 8$, so that E could not commence before 8. Activity E's successor H will also start as late as possible at $LS(H) = 12$, leaving a time window of $12 - 8 = 4$ hours for activity E, hence a safety float of $SF(E) = 4 - 4 = 0$. Finally, if all of E's predecessors start as late as possible, E can start no earlier than at time 8. If its successor H starts as early as possible, it does so at $ES(H) = 10$, leaving a time window of $10 - 8 = 2$ for activity E, whose duration is 4 hours. Consequently, the interference float $IF(E) = \max\{0; 2 - 4\} = 0$.

1.3 Project Acceleration (Crashing)

This section describes a model that starts with an ordinary *CPM* project network, but allows for shortening the durations of one or more activities, so as to finish the project in a prespecified time $\overline{T}$. Clearly, the acceleration of individual activities will incur additional costs, such as labor hired at short notice. It appears reasonable to assume that acceleration of an activity n_k from its current "normal" duration that we now call $\overline{t}_k$, is only possible down to a level $\underline{t}_k$. The costs of the activity at normal time $\overline{t}_k$ are fixed and can be considered overhead costs; here, we are only concerned with the (marginal) cost of acceleration. These costs are assumed to be linear at a rate of Δc_k, i.e., accelerating the activity n_k from its normal duration $\overline{t}_k$ to its "crash time" $\underline{t}_k$ will cost $(\overline{t}_k - \underline{t}_k)\Delta c_k$. The cost function for the acceleration is shown in Figure III.11.

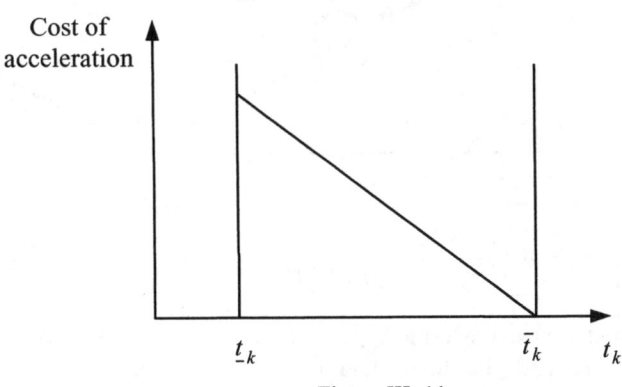

Figure III. 11

The problem is now to determine which activities to accelerate and by how much, so as to reduce the project duration to the desired level of $\overline{T}$ while minimizing the total acceleration costs.

It is readily apparent that the crashing of activities is only meaningful where it leads to the shortening of the critical, i.e., the longest, path in the network. In other words, it would be optimal to accelerate that activity on the critical path whose marginal costs Δc_k are lowest. Doing so will shorten the length of some paths, while others will remain unchanged. If we were to continue to accelerate the same activity, eventually the reduced length of the original critical path will equal that of another path that, while originally shorter, was not affected by the acceleration and now has become critical as well. In the presence of multiple critical paths, we have to choose a set of activities for acceleration, such that each critical path includes at least one of these activities. Put differently, the set of activities (nodes in the activity-on-node and arcs in the activity-on-arc representation) chosen for acceleration must *cover* each critical path. In order to illustrate the main concepts, consider the following

Example: Consider the *AON* project network in Figure III.12. The numbers next to the activity nodes n_k denote the normal time $\bar{t}_k$, the crash time $\underline{t}_k$, and the unit acceleration cost Δc_k. The desired project completion time is $\overline{T} = 11$ hours. Note that $\underline{t}_A = \bar{t}_A = 3$, indicating that activity A cannot be accelerated.

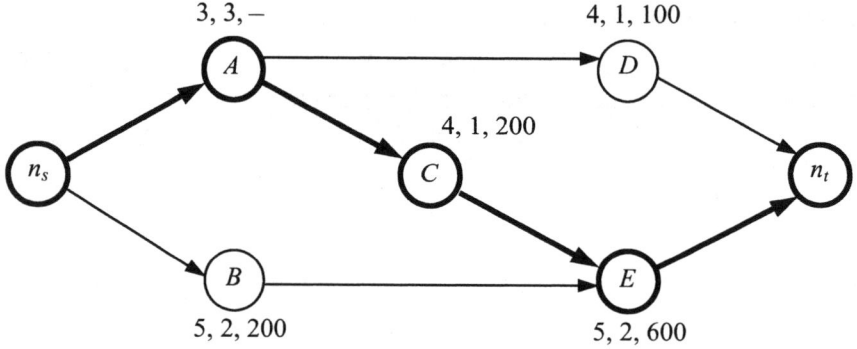

Figure III.12

Given the normal times, the critical path shown by bold lines consists of the activities A, C, E, and the total duration of the project is 12 hours. Any acceleration must be performed on one of the three critical activities. Since A cannot be accelerated and $\Delta c_C = 200$ and $\Delta c_E = 600$, we decide to accelerate activity C by one hour, i.e., from four to three hours. This shortens the length of the critical path to 11 hours, which is the desired completion time. Note that the length of the paths $A - D$ and $B - E$ remain unchanged at 7 and 10 hours, respectively, and that $A - C - E$ is still the unique critical path.

Suppose now that further acceleration is required. Due to its low cost, we could speed up activity C by another hour, i.e., from three to two hours. This leads to the situation shown in Figure III.13 with the two critical paths $A - C - E$ and $B - E$, where the numbers next to the nodes indicate the partly accelerated lengths of the activities.

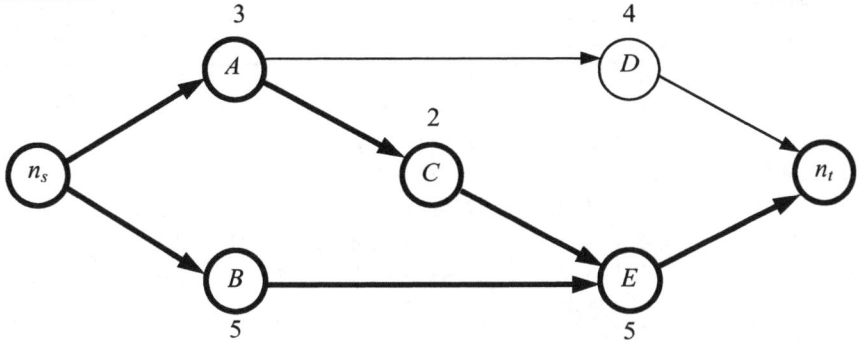

Figure III.13

At this point, the only noncritical path is $A - D$ with length 7; the two other paths $A - C - E$ and $B - E$ are both 10 hours in length and both critical. Any further acceleration requires crashing of activities that are located on both critical paths. In principle, the possible choices are $\{A, B\}$, $\{B, C\}$, and $\{E\}$. However, since activity A cannot be accelerated, the only choices are to simultaneously speed up activities B and C, or to shorten E's duration. As the acceleration of the activities B and C costs $\Delta c_B + \Delta c_C = 200 + 200 = 400$, while the acceleration of E costs $\Delta c_E = 600$, the former option is chosen. This leads to activity durations of 3, 4, 1, 4, and 5 for the activities A, B, C, D, and E, respectively, and a project duration of 9. The critical paths are the same as those shown in Figure III.13. At this point, the options are further limited as the present duration of activity C with 1 hour does not allow any additional acceleration, leaving the crashing of E as the only choice. Table III.4 summarizes the results of this process for all possible project durations.

It is often useful to tentatively crash all activities and thus determine the earliest finishing time that is possible for the project. This will indicate to the decision maker how much acceleration is possible, regardless of how much money is spent on acceleration. Note, however, that in order to reach this shortest possible project duration, it is generally *not* required to crash all activities.

Another fact is worth mentioning. When reducing the total duration of the project as much as possible, invariably many, or even all, activities of a project become critical. That way, there is no slack in the project, which may be considered efficient. However, should any delay occur at all due to some unforeseen circumstances (something not included in a deterministic model such as *CPM*, but typical in real situations), delays will occur immediately.

Table III.4

Project duration $\overline{T}$	Activity durations t_A, t_B, t_C, t_D, t_E	Critical path(s)	Total cost of acceleration
12	3, 5, 4, 4, 5	$A-C-E$	0
11	3, 5, 3, 4, 5	$A-C-E$	200
10	3, 5, 2, 4, 5	$A-C-E, B-E$	400
9	3, 4, 1, 4, 5	$A-C-E, B-E$	800
8	3, 4, 1, 4, 4	$A-C-E, B-E$	1400
7	3, 4, 1, 4, 3	$A-C-E, B-E, A-D$	2000
6	3, 4, 1, 3, 2	$A-C-E, B-E, A-D$	2700
5	not feasible	—	—

In applications, it is not practical to determine all critical paths and determine the least-cost set of activities that cover all critical paths, as was done above. Instead, the problem can be formulated as a linear programming problem. In order to do so, define x_j as the duration of activity n_j, define y_j as the starting time of activity n_j, and define the parameters $\underline{t}_j, \bar{t}_j$ and $\overline{T}$ as before. The cost-minimizing objective function is then

$$\text{Min } z = \sum_j \Delta c_j (\bar{t}_j - x_j) = \sum_j \Delta c_j \bar{t}_j - \sum_j \Delta c_j x_j$$

and as $\sum_j \Delta c_j \bar{t}_j$ is a constant, we can write the problem as

$$\text{P: Max } z' = \sum_j \Delta c_j x_j$$

s.t. $x_j \geq \underline{t}_j \quad \forall\, j$

$\quad\ \ x_j \leq \bar{t}_j \quad \forall\, j$

$\quad\ \ y_j \geq y_i + x_i \ \forall\ n_i \in \mathcal{P}(n_j); \ j = 1, ..., n$

$\quad\ \ y_t \leq \overline{T}$

$\quad\ \ x_j, y_j \geq 0 \ \forall\, j.$

The first two sets of constraints ensure that the time reduction is within the prespecified limits, the third set of constraints guarantees that an activity does not commence before all of its (direct) predecessors are finished, and the last structural constraint ensures that the project is finished when required. Given that

Chapter 1: Unconstrained Time Project Scheduling

in our numerical example $x_A = 3$ (as no reduction is possible), and $y_A = y_B = 0$, the problem can be written as

$$P': \text{Max } z' = 200x_B + 200x_C + 100x_D + 600x_E$$

s.t.
$$\begin{aligned}
x_B &\geq 2 & y_C &\geq 3 \\
x_B &\leq 5 & y_D &\geq 3 \\
x_C &\geq 1 & y_E &\geq x_B \\
x_C &\leq 4 & y_E &\geq y_C + x_C \\
x_D &\geq 1 & y_t &\geq y_E + x_E \\
x_D &\leq 4 & y_t &\geq y_D + x_D \\
x_E &\geq 2 & y_t &\leq \overline{T} \\
x_E &\leq 5
\end{aligned}$$

$$x_B, x_C, x_D, x_E; y_C, y_D, y_E, y_t \geq 0$$

The solutions to the problem are again those shown in Table III.4. Note that the actual reduction costs are calculated by subtracting the fixed amount $\sum_j \Delta c_j \bar{t}_j =$ 5,200 from the optimal values in the problem P'.

1.4 Incorporating Uncertainties (*PERT*)

So far, we have assumed that all components included in the planning process, i.e., the structure of the project network and the durations of its activities, are known with certainty. In this chapter, we will drop this assumption for some of the features of this model. In particular, we retain the assumption that the network structure is known, but do away with the certainty concerning the durations of the activities. If activity durations are random variables, then the results of the planning as well as the process by which they are obtained will depend heavily on the underlying probability distribution. In contrast to most of the existing literature, we will base our arguments not on a particular distribution, but on some general observations.

It is frequently possible in practical situations that managers are able to specify estimates concerning the duration of an activity. More specifically, they may be able to estimate how long an activity is likely to take in the worst case and in the best case, along with an estimate of the normal duration of the activity. We will refer to these three time estimates as the pessimistic, the optimistic and the most likely durations. While the pessimistic and optimistic durations give an indication of the range of the distribution, the normal (or most likely) estimate is the modal value. For any activity n_k, these values will be denoted by t_k^0, t_k^m, and t_k^p, with $t_k^0 \leq t_k^m \leq t_k^p \; \forall \; n_k$.

By definition it is more likely for an activity to assume the modal value than either of the two extremes. How much more likely? That does, of course, depend on the activity. Following the *Empirical Rule* in statistics, we know that any approximately bell-shaped distribution has virtually the entire probability mass within three standard deviations about the mean. Assuming symmetry of the distribution, the mean will coincide with the mode. (Similar results are obtained via Chebyshev's theorem, applicable to any arbitrary distribution, which places 8/9 = 88.89% of the observations within three standard deviations of the mean, or the normal distribution, which has 99.73% of the observations within that range). Suppose now that we were to evenly subdivide the range between -3σ and $+3\sigma$, leading to intervals $[-3\sigma, -1\sigma]$, $[-1\sigma, +1\sigma]$, and $[+1\sigma, +3\sigma]$. If the distribution is approximately normal, then the first and last intervals have close to 16% of the probability mass, with the central interval occupying the remaining 68%. This is very close to a distribution of 1/6, 4/6, and 1/6, which leads to assigning weights of 1, 4 and 1 to the optimistic, most likely, and pessimistic values. This leads to an expected duration of task k of

$$\mu_k = \tfrac{1}{6}[t_k^o + 4t_k^m + t_k^p].$$

Consider now the variance. Invoking the above rules on bell-shaped distributions and placing the optimistic and pessimistic time estimates at three standard deviations below and above the mean of the distribution, respectively, we obtain $t_k^0 = \mu_k - 3\sigma_k$ and $t_k^p = \mu_k + 3\sigma_k$. Solving for σ_k and adding the two equations results in

$$\sigma_k = \frac{t_k^p - t_k^o}{6}.$$

The above expressions for μ_k and σ_k are usually derived assuming that the activity durations follow Euler's beta distribution. This assumption has been criticized by many authors in the literature; a typical example is Elmaghraby (1977).

At this point, it is possible to proceed, albeit with some additional assumptions. Suppose that we were to ignore the variability of the activity durations and apply the deterministic critical path model with the activity durations $t_k = \mu_k$. The project duration is then $EF(n_t) = LF(n_t)$ provided that we satisfy

Assumption 1: The duration of any activity n_k is independent of the durations of all other activities n_ℓ, $\ell \neq k$.

This means that there are no underlying common reasons for activities to have longer (or shorter) than normal durations. This assumption of statistical independence implies that the expected duration of the project equals the sum of the expected durations of all activities on the critical path. In order to proceed with our analysis, we also need

Chapter 1: Unconstrained Time Project Scheduling

Assumption 2: The true critical path is the one whose sum of expected durations is longest.

Due to the nature of the problem, this assumption need not always be satisfied. As an example, consider two (parallel) activities A and B whose durations are bounded by $t_A^o = 1$, $t_A^p = 7$, $t_B^o = 2$, and $t_B^p = 4$, respectively, and whose modal values are $t_A^m = 4$ and $t_B^m = 3$. Even though $\mu_A = 4$ and $\mu_B = 3$, activity B could be truly critical if, e.g., it were to last 3.5 hours while activity A would last only, say, 2 hours. Finally, we also make

Assumption 3: The number of activities on the critical path is large enough so that it is possible to apply the central limit theorem and stipulate normality of the project duration.

With the above three assumptions we are now able to compute probabilities concerning the completion time of the project. Invoking Assumption 1 and defining p^* as the set of subscripts of all critical activities on (one of) the critical paths, we conclude that the variance of the entire project duration is

$$\sigma^2 = \sum_{k \in p^*} \sigma_k^2.$$

As the duration of the project is $X = \sum_{k \in p^*} t_k$, the probability that the project is finished within T hours (a duration specified by the decision maker) is $P(X \le T)$. As per Assumption 3, $X \sim N(\mu, \sigma)$, the random variable $Z = (X - \mu)/\sigma$ is $Z \sim N(0,1)$. We can then write $P(X \le T) = P(Z \le (T-\mu)/\sigma)$ which can now easily be determined. As a numerical illustration, consider the following

Example: Let the precedence relations between the activities as well as their durations be given in Table III.5.

Table III.5

Activity	Immediate predecessors	Activity durations t_k^o, t_k^m, t_k^p
A	–	4, 5, 6
B	A	3, 5, 7
C	–	6, 6, 6
D	B, C	1, 1, 13
E	B, F	1, 2, 3
F	D	3, 4, 5
G	E	1, 4, 13
H	E, F	3, 9, 9

Given the formulas above, the activities $A, B, \ldots, H$ have mean durations of 5, 5, 6, 3, 2, 4, 5, and 8. Their variances are $\frac{4}{36}$, $\frac{16}{36}$, 0, $\frac{144}{36}$, $\frac{4}{36}$, $\frac{4}{36}$, $\frac{144}{36}$, and $\frac{6}{36}$, respectively.

The project network in *AON* representation for the precedence relations in Table III.5 is shown in Figure III.14, where the number next to the nodes indicate the mean duration of the activity.

The critical path is again shown in bold arcs. The mean project duration is $\mu = \mu_A + \mu_B + \mu_D + \mu_F + \mu_E + \mu_H = 27$, and the variance is $\sigma^2 = \frac{1}{36}[4 + 16 + 144 + 4 + 4 + 36] = 5.7\dot{7}$, so that $\sigma \cong 2.4037$. If the decision maker were interested in the probability that the project is finished within, say, 29 hours, then $T = 29$ and $P(X \leq 29) = P(Z \leq .8321) = 79.73\%$. Given all of the (sometimes dubious) assumptions, probability statements such as this should be considered only as a rough estimate. In this example, it appears that there is a reasonably good chance that the project will be finished within 29 hours.

Chapter 1: Unconstrained Time Project Scheduling

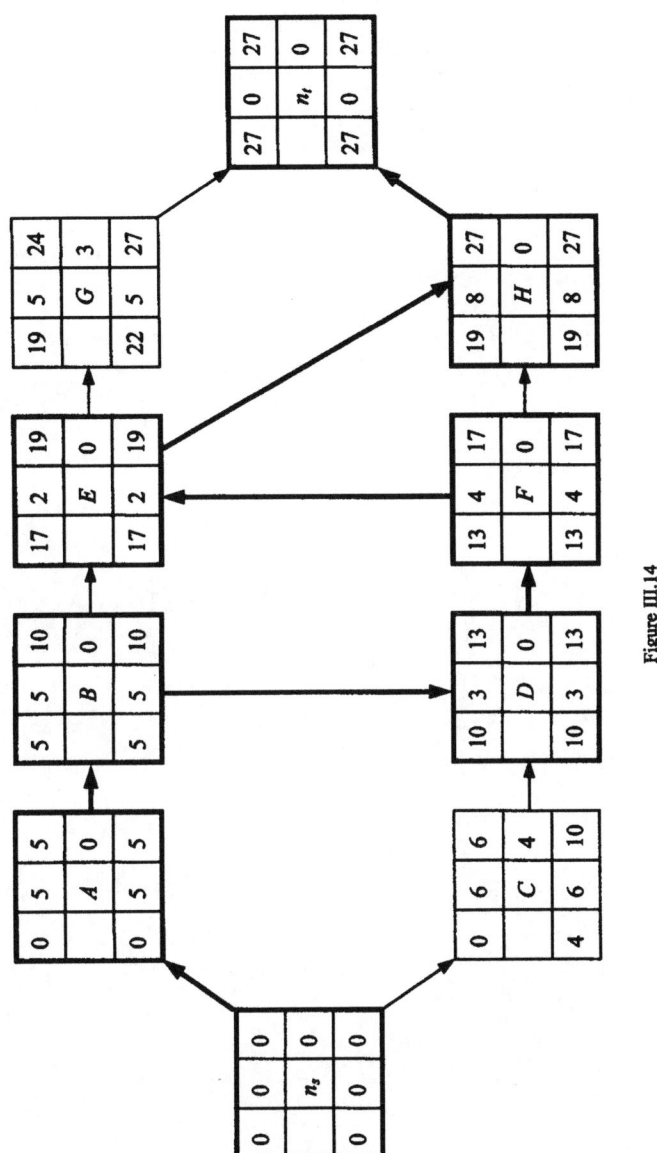

Figure III.14

CHAPTER 2   PROJECT SCHEDULING WITH RESOURCE CONSTRAINTS

This chapter extends the previous results by introducing scarce resources into the model. Not only does this make the problem more difficult, it renders large-scale applications unsolvable by exact methods, especially when multiple resources are considered.

2.1   The Problem and its Formulation

The main purpose of this chapter is to consider extensions of the basic critical path method (*CPM*) of the previous chapter that introduce additional constraints related to resources which may or may not be available. Here, we will consider a single resource. As in the critical path method, the durations of the activities are assumed to be known with certainty, and so is the resource consumption of the activities. The planner's task is then to balance the time and resource requirements of the project. This problem can be approached in two different ways.

The first possibility is to minimize the project completion time T, while making sure that the consumption r_τ of the single resource at any given point in time τ is within the given limit, and the second possibility is to minimize the resource consumption, ensuring that an upper bound $\overline{T}$ on the project duration is not violated. Defining $\overline{R}$ as the number of resource units that are available, the first problem can be written as

$$P_T(\overline{R}): \min_{S \in S(\overline{R})} T$$
$$\text{s.t. } r_\tau \leq \overline{R} \quad \forall \tau = 1, \ldots, T,$$

where $S(\overline{R})$ denotes all feasible schedules that use no more than $\overline{R}$ resource units in any given time period.

The second approach can then be written as

$$P_R(\overline{T}): \min_{S \in S(\overline{R})} \{ \max_{\tau=1,\ldots,\overline{T}} \{r_\tau\} \}$$
s.t. $r_\tau \leq \overline{R} \ \forall \ \tau = 1, \ldots, \overline{T}$.

The two problems are applications of the two versions of the basic economic principle: either maximize the output for a given input, or minimize the input for a given output. While some work has concentrated on the second problem most effort has been devoted to the former of these two problems, $P_T(\overline{R})$, and this is also the focus of this chapter.

In order to motivate the discussion, consider the project network in activity-on-node representation in Figure III.15. The numbers next to a node n_j indicate the activity duration t^j and the resource consumption r^j of activity n_j per time unit, so that $t^j r^j$ is the total resource requirement of activity n_j. For example, activity C has a duration of 3 weeks and a resource consumption of 20 workers per week, i.e., a total of 60 worker-weeks. The duration of the entire project turns out to be 12 weeks and the critical path $n_s - A - D - n_t$ is shown by bold lines.

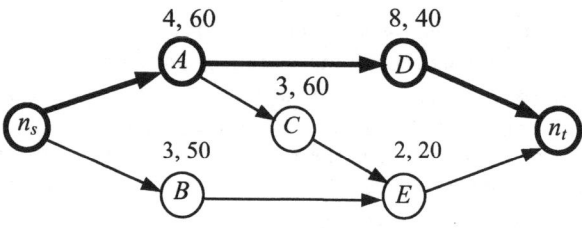

Figure III.15

Below, we consider three different feasible schedules. For the schedule shown in Figure III.16a all activities start as early as possible. Clearly, the critical activities A and D cannot be shifted back and forth; they are shown by solid black bars. On the other hand, the activities B, C, and E have positive float, so that their starting times are not fixed. The resulting resource consumption over time is shown in Figure III.16b, where the abscissa is the time axis, whereas the ordinate is related to the magnitude r of the resource consumption. Figures III.16c and III.16d are similar, except that they relate to a scheduling rule for which all activities start as late as possible. Finally, Figures III.16e and III.16f represent a schedule that has activity B start at time 4, while the activities C and E start as late as possible. It should be noted that the Figures III.16 a, c, and e are usually referred to as *Gantt charts*, named after the American engineer Henry L. Gantt (1861 – 1919), who developed them in 1917.

Chapter 2: Project Scheduling With Resource Constraints

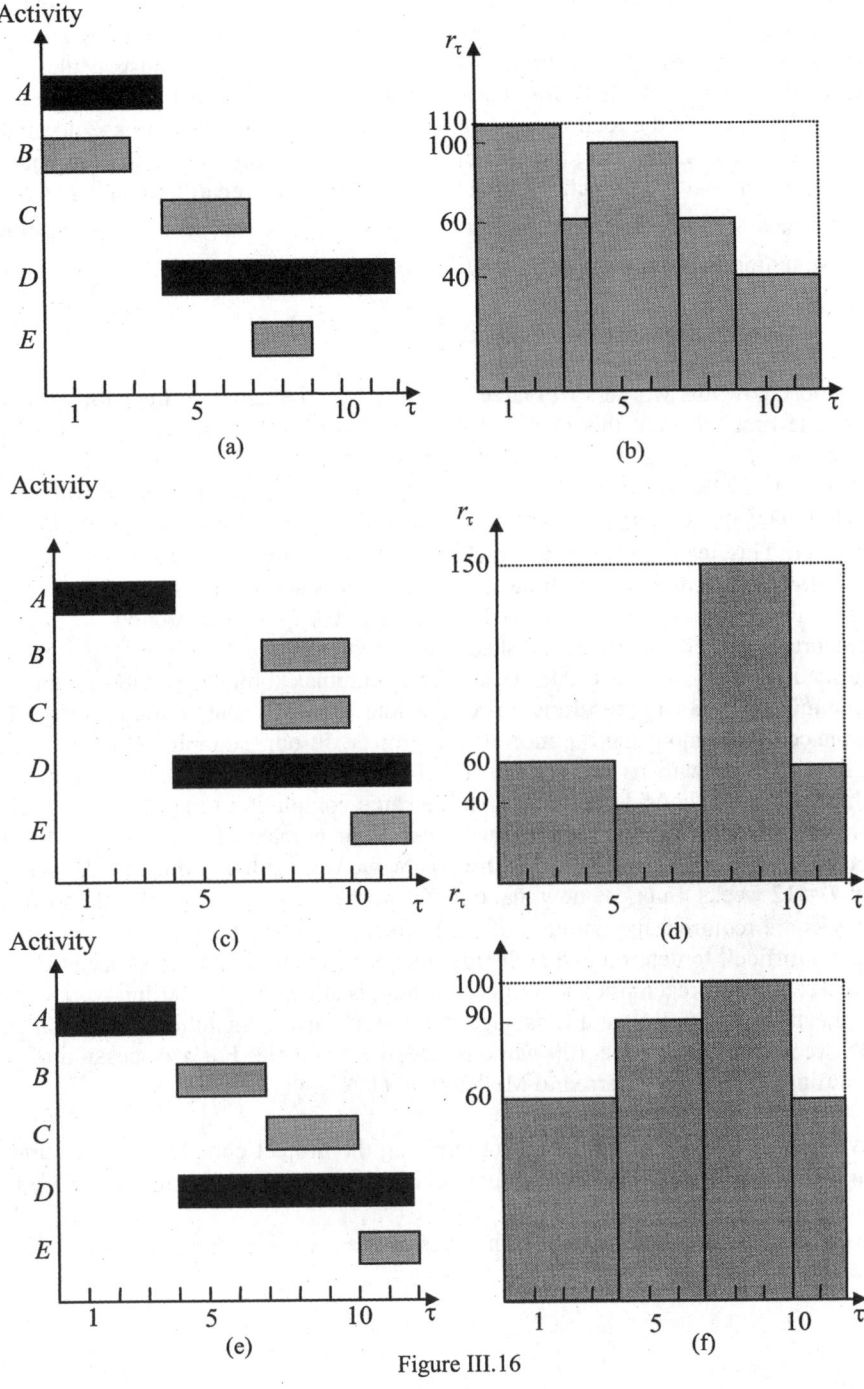

Figure III.16

Which of the three schedules in Figure III.16 is to be preferred depends, of course, on the objective. Suppose that the resource under construction is labor and assume that workers are hired on a weekly basis. In that case, the labor costs of the three schedules are the (shaded) area under the functions shown in Figures III.16b, d, and f. In all cases the total resource consumption is 930 worker-weeks and, under the criterion of minimizing the cost of casual labor, all three schedules are equally good. As a matter of fact, each activity n_j requires—regardless of when it is performed—t^j weeks and r^j workers per week, so that the total resource consumption is always $\sum_j r^j t^j$ = 4(60) + 3(50) + 3(60) + 8(40) + 2(20) = 930, independent of the actual schedule.

Suppose now that workers will have to be hired for the length of the entire project. For the first schedule this means that we will need 110 workers during the first three weeks. During week 4, only 60 of these workers are needed, so that the remaining 50 workers are idle. Formally, in this scenario the number of workers is determined by the largest resource consumption at any point in time during the project. This leads to the problem $P_R(\overline{T})$. Its objective minimizes the maximal resource requirement at any time, a measure that is shown for the three schedules under consideration in Figure III.16 b, d, and f by the box around the highest resource requirement. Here, the shaded area indicates the labor used, whereas the unshaded area indicates the waste. The minimax objective will attempt to minimize this waste, resulting in a schedule whose resource requirements are balanced throughout the duration of the project. In our example, the schedule of Figure III.16e with its requirement of 100 workers performs best. Note that all three schedules considered here have the same completion time of $T = 12$ weeks which cannot be improved upon, regardless of the number of workers hired for the tasks. On the other hand, given 100 workers, we were able to complete the project in $T = 12$ weeks. Suppose now that only 60 workers are available. As 930 worker–weeks are required, the duration of the project is $T \geq 930/60 = 15.5$ weeks, and it is not difficult to determine that the minimal completion time in this case is $T = 18$ weeks. Unless preemption or project splitting is allowed, i.e., starting work on the project, interrupting it, and finishing later, the minimal completion time T remains 18 weeks, unless at least 100 workers become available. For a discussion of job splitting, readers are referred to Moder et al. (1983).

We now consider the problem of minimizing the project completion time subject to given resource constraints. Throughout our discussion, we employ the following

Example: Consider a project with six activities n_1, n_2, ..., n_6 and precedence relations as shown in Table III.6:

Chapter 2: Project Scheduling With Resource Constraints

Table III.6

Activity	Immediate predecessor	t^j	r^j
n_1	–	4	1
n_2	–	1	2
n_3	–	2	1
n_4	n_1, n_2	1	2
n_5	n_3	2	1
n_6	n_5	2	3

The *AON* network for this problem, displaying the earliest and latest starting and finishing times, is shown in Figure III.17b. For convenience, we repeat the legend that applies to *AON* project networks in Figure III.17a.

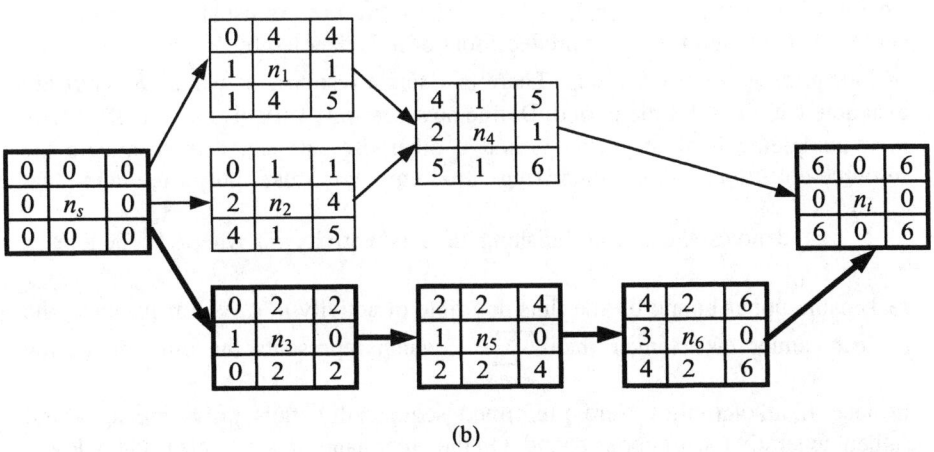

Figure III.17

Scheduling activities at their earliest starting times and applying the critical path algorithm, the resulting Gantt chart is shown in Figure III.18.

Although it is possible to complete the project in $T = 6$ time units, up to five resource units are needed for such a schedule, with activities n_4 and n_6 partially overlapping each other. If no more than $\overline{R} = 5$ resource units were available at any given time, a completion time of $\overline{T} = 6$ is not possible. One approach to deal with problems of this nature is to attempt to minimize the completion time of the project under the constraints that no more resources are used than are available. This is the problem $P_T(\overline{R})$ introduced at the beginning of this chapter.

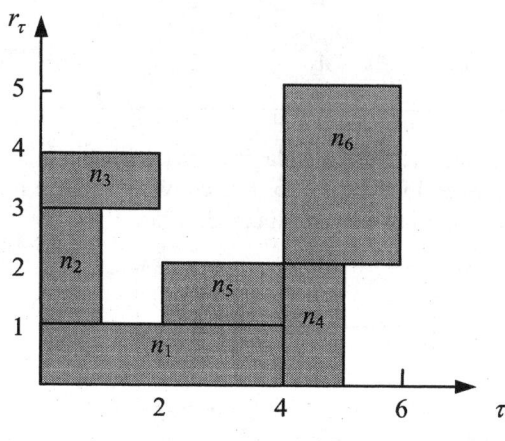

Figure III.18

To formalize, define $N = \{n_j, j = 1, ..., n\} \cup \{n_s, n_t\}$. As usual, the duration of activity n_j is t^j, and the set of predecessors of n_j is denoted by $\mathscr{P}(n_j)$. Preemption or job-splitting is not permitted. There is a single resource of which $\overline{R}$ units are available within each time period. Define now zero-one variables $x_{j\tau} = 1$, if activity n_j is completed in time period τ, and 0 otherwise, we can then formulate the resource-constraint, time-minimizing problem. For that purpose, note that $\sum_{\tau=EF(n_j)}^{M_j} \tau x_{j\tau}$ denotes the actual finishing time of activity n_j, where M_j denotes a reasonable upper bound on the finishing time of activity $n_j \in N$. For instance, the project cannot take longer than $\sum_{n_j \in N} t^j$, which represents the duration of the project, if all activities were preformed sequentially, thus providing a natural (albeit generally loose) upper bound. On the other hand, it is also clear that a lower bound of the completion time of any schedule is $\frac{1}{R} \sum_{n_j \in N} r^j t^j$. The sum is the total number of resource units needed to complete the project (regardless of when), and division by the resource availability per time unit results in the number of days the project would take without any restrictions. Clearly, this number represents a lower bound on the project length. As a result, we know that $\frac{1}{R} \sum_{n_j \in N} r^j t^j \leq M_t \leq$

$\sum_{n_j \in N} t^j$. The resource-constrained scheduling problem can then be formulated as follows.

$$P: \text{Min } z = \sum_{\tau=EF(n_t)}^{M_t} \tau x_{t\tau}$$

s.t.
$$\sum_{\tau=EF(n_j)}^{M_j} x_{j\tau} = 1 \qquad \forall\, n_j \in N \qquad (1)$$

$$\sum_{\tau=EF(n_k)}^{M_k} \tau x_{k\tau} \le \sum_{\tau=EF(n_j)}^{M_j} (\tau - t^j) x_{j\tau} \qquad \forall\, k \in \mathscr{P}(n_j),\, n_j \in N, \qquad (2)$$

$$\sum_{n_j \in N} \sum_{\ell=\tau}^{\tau+t^j-1} r^j x_{j\ell} \le \overline{R} \qquad \forall\, \tau = 1, \dots, T \qquad (3)$$

$$x_{j\tau} = 0 \vee 1 \qquad \forall\, n_j \in N,\, \tau = 1, \dots, T.$$

The objective function in the above problem minimizes the completion time of the project. Constraints (1) ensure that each activity terminates at exactly one point in time. Constraints (2) express that for each predecessor n_k of n_j, the finishing time of activity n_k cannot exceed the starting time of activity n_j. This is nothing but the typical "longest path" constraint of the critical path method. Constraints (3) are the resource constraints, where the left-hand side expresses the total resource use in the time period that ends at τ. The size of the problem depends on the number of activities and the magnitude of the time estimates M_j, the latter requiring good bounds, if the problem is not to become unmanageable.

Blazewicz et al. (1983) have proved that the resource-constrained project scheduling problem is **NP**-hard. Kolisch et al. (1995) have shown that—depending on the characteristics of an instance—some instances can be solved very quickly while others are extremely difficult. Three parameters are known to have a big impact on the computational difficulty of a project instance: network complexity, (i.e., the average number of immediate successors of an activity), resource consumption, (i.e., the average number of resources required per activity) and resource availability.

2.2 Exact Solution Methods

Starting with an early work of Johnson (1967) a variety of branch-and-bound algorithms have been developed for the resource-constrained project scheduling problem. Most of them use partial schedules which are associated with the vertices of the enumeration tree. The branching process consists of extending the partial schedule in different ways. Dominance rules, lower bounds, and immediate selection allows us to decrease the number of alternatives for extending the partial schedule. The methods use different branching schemes and pruning methods. In general, depth-first-search is used in order to keep memory requirements low.

Here, we will informally describe a simple branch and bound method. For simplicity, we will restrict ourselves to problems with a single resource only. The approach described here is based on the concept of *precedence trees*. A precedence tree has a root node n_s, leaves n_t, and each node of the tree is succeeded by the nodes of all those activities that could logically follow it and that have not been scheduled yet. Clearly, each path in the precedence tree has the same length, as each path starts with the source, terminates with the sink, and in between there are the nodes of all n activities in the project. As an example, consider the project network in Figure III.17. One path in the precedence tree starts with n_s, continues with n_1, then n_2, then n_4, n_3, n_5, and n_6, and finally n_t. Another path is $(n_s, n_3, n_5, n_6, n_2, n_1, n_4, n_t)$. In other words, the precedence tree includes all feasible permutations of the activities. It is easy to show that as long as the resource availability per time unit exceeds the largest resource requirement of any individual activity, feasible schedules exist. As a matter of fact, given this condition, each path in the precedence tree represents a feasible solution. To see this, simply arrange all activities sequentially as required in the precedence tree; the resulting schedule requires no more resources than are available and hence it is feasible. However, rather than using a simple sequential schedule, we will schedule each activity as early as possible, while respecting the precedence requirements. As an illustration, consider the feasible path $(n_s, n_1, n_2, n_3, n_4, n_5, n_6, n_t)$. As $r_6 = 3$, we will need at least 3 resource units for any feasible schedule. Scheduling all activities sequentially would give a schedule completion time $\tau = 12$, regardless of the sequence. However, the specified path would yield a schedule with completion time $\tau = 8$, as can be seen in the Gantt chart of Figure III.19.

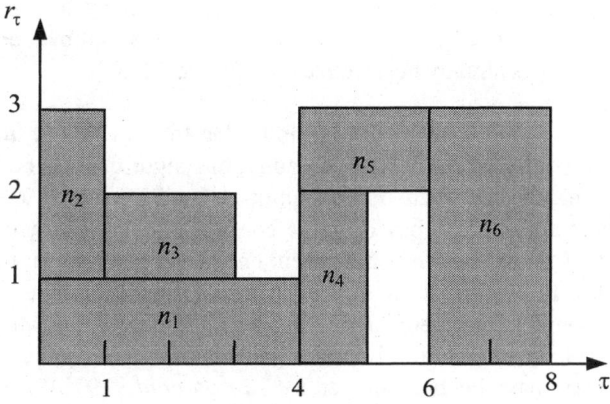

Figure III.19

Note that when n_5 is being scheduled in Figure III.19, one may realize that it could start at $\tau = 3$ rather than $\tau = 4$, yielding a schedule with completion time $\tau = 7$ (which happens to be optimal); see the Gantt chart in Figure III.20.

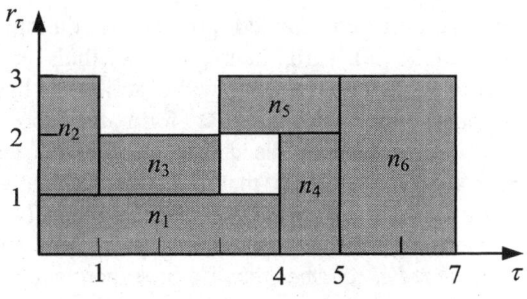

Figure III.20

However, in rescheduling activity n_5, the resulting schedule becomes n_s, n_1, n_2, n_3, n_5, n_4, n_6, n_t.

We can now apply a branch and bound-type procedure to the precedence tree. Starting with a (hopefully) good path in the tree (one that is typically obtained by a heuristic algorithm), we first construct the corresponding schedule and calculate its project duration. We then explore the branches of the tree, building schedules along the paths, pruning the branches (i.e., stopping) whenever the best known completion time is exceeded or to be exceeded before completion. This implies that—as typical in branch and bound procedures—a good-quality starting solution is the key to a successful process.

In our example, the path n_s, n_1, n_2, n_4, n_3, n_5, n_6, n_t will by the completion of activity n_5 already have used $\tau = 8$ time units and cannot be optimal, as a schedule with a duration of 7 has already been found; see Figure III.20.

The above ideas are embodied in an algorithm due to Patterson *et al.* (1989), to which the reader is referred for details. Recently, this algorithm has been enhanced with powerful search tree reduction techniques by Sprecher (1999). Another branch and bound algorithm is based on the concept of delay alternatives used by Christofides *et al.* (1987) which has been enhanced by Demeulemeester and Herroelen (1992). In contrast to the precedence tree algorithm, here each level of the branch-and-bound tree is associated with a fixed time instant at which activities may be started. Other branch and bound algorithms for the resource-constrained scheduling problem have been developed by Stinson *et al.* (1978) Mingozzi *et al.* (1998) and Brucker *et al.* (1998) as well as Igelmund and Radermacher (1983a, 1983b) who have introduced a branching scheme based on so–called minimal forbidden sets. One should remember that with the **NP**-completeness of the underlying problem, the usefulness of this approach is limited in the general instance.

2.3 Heuristic Methods

Since the general resource-constrained project scheduling problem discussed above was shown to be **NP**-hard, heuristic algorithms are required to solve realistic (large-scale) problems. Here, we will describe the basic principles of so-called *schedule generation schemes* that form the core of most heuristic procedures for the problem under discussion. A schedule generating scheme is nothing but a construction heuristic that builds a feasible schedule by adding activities to an already existing partial solution. We usually distinguish between two types of generating schemes, so-called *serial schemes* that use activity incrementation, and *parallel schemes* that use time incrementation.

In serial schedule generation schemes, all activities are first ranked in some order, thus forming a *priority list*. Once such a list exists, it is never updated. Beginning with the starting activity n_s, activities are iteratively scheduled at their earliest precedence and resource feasible time, one after another, following the sequence in the priority list. Here, the priority list is established with the activities ranked in order of nonincreasing activity durations. This makes the scheme a special case of the greedy algorithm. The procedure is initialized by starting with the source node as the initial partial schedule. The method can then be described as follows.

A Serial Scheduling *Greedy* Construction Heuristic

Step 1: Rank the activities in nonincreasing order of their duration. Ties are broken in favor of the activity with a smaller subscript.

Step 2: Select the highest-ranking activity that has not yet been scheduled, say, n_j, given that all of its predecessors are already included in the partial schedule. Activity n_j is scheduled as early as possible and allowable by the resources and added to the current partial schedule.

Step 3: Has the sink n_t been scheduled?
If yes: Stop, a feasible schedule has been found.
If no: Go to Step 2.

Example: Consider again the example from Figure III.17 and suppose that $\overline{R} = 3$ resource units are available. The priority list becomes n_s, n_t, n_2, n_4, n_3, n_5, n_6, n_1 and the activities will be scheduled in the order n_s, n_2, n_3, n_5, n_6, n_1, n_4, n_t with a schedule length of $T = 11$. The Gantt chart without the dummy activities is shown in Figure III.21. Note that the dummy activities n_s and n_t, having vanishing duration times, will not appear in the chart.

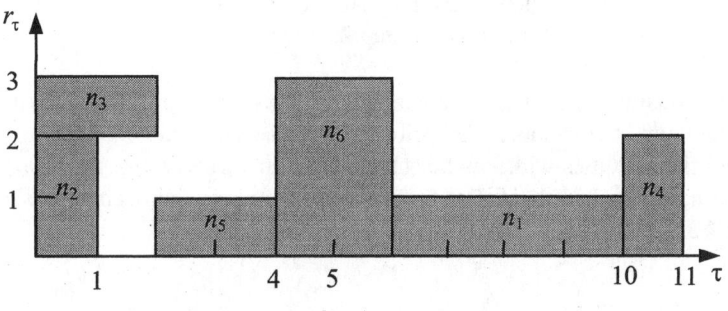

Figure III.21

It is apparent that serial schedule generation schemes depend heavily on the priority lists used. In some of the chapters of the next part, we will discuss list scheduling in a machine scheduling context, where priority lists are also used. We will then use various principles for establishing the priority list. The same is possible here; the lists could be based on the longest duration first or on the highest (lowest) resource requirement, or some other criterion.

Consider now *parallel schedule generation schemes*. Whereas serial schedule generation schemes use activity incrementation, parallel schedule generation schemes are based on time incrementation, where the time point τ is the iteration counter. At each iteration, the priority list is scanned for any precedence and resource feasible activity that could be started at that time. The procedure is repeated until the sink n_t has been scheduled. Again, the procedure is initialized by starting with the source node as the initial partial schedule. The method can then be described as follows.

A Parallel Scheduling *Greedy* Construction Heuristic

Step 1: Rank the activities in nonincreasing order of their duration. Ties are broken arbitrarily. Set the iteration counter (= time) $\tau := 0$.

Step 2: Does there exist at least one activity that has not yet been scheduled, which can start at time τ, with all precedence and resource constraints satisfied?
  If yes: Go to Step 3.
  If no: Set $\tau := \tau + 1$ and go to Step 2.

Step 3: Choose the activity with the highest priority that has not yet been scheduled and that can be started at time τ, so that all precedence and resource constraints are satisfied. Start this activity at time τ.

Step 4: Has n_t been scheduled?
  If yes: Stop, a feasible schedule has been found.
  If no: Set $\tau := \tau + 1$ and go to Step 2.

Example: Consider again the same example as above and suppose again that $\overline{R} = 3$ resource units are available. The priority list is the same, i.e., n_s, n_t, n_2, n_4, n_3, n_5, n_6, n_1 and the activities will now be scheduled in the order n_s, n_2, n_3, n_1, n_5, n_4, n_6, n_t with a schedule length of $T = 8$. The corresponding Gantt chart is shown in Figure III.22.

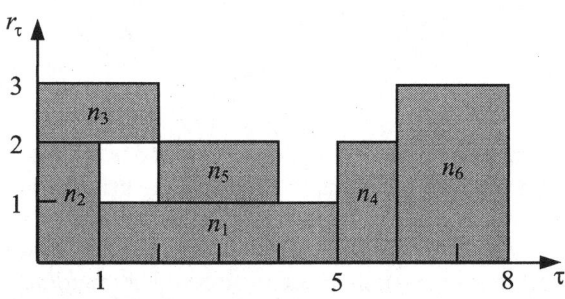

Figure III.22

As for serial schedules, various priority lists could be envisaged. Note that the serial and parallel schemes will always generate feasible schedules, if any exist, which is the case as long as no activity has a larger resource requirement than that maximally available. It is also worth mentioning that in the resource unconstrained case, both types of schemes will generate optimal schedules.

Priority rule-based heuristics combine more or less sophisticated priority lists and schedule generation schemes to construct specific algorithms. They have the advantage of being intuitive, easy to implement, and computationally fast. Their drawback is that the solutions they produce are not as good as those obtained by more elaborate heuristics, such as those developed based on genetic algorithms. For details, see Hartmann (1998).

Finally, we would like to mention that in practical situations of project scheduling, management must be able to deal with tradeoffs between different resources and also between resources and time. For this more general case, branch and bound-type exact algorithms have been developed; see, e.g., Sprecher *et al.* 1997), and for heuristic methods based on genetic algorithms, see Hartmann (1999). Other models, such as those by De Reyck and Herroelen (1999) consider time lags between activity starts, Böttcher *et al.* (1999) take into account partially renewable resources, Salewski *et al.* (1997) include mutually dependent activities, while Möhring *et al.* (1984, 1985) assume that the activity durations are random.

PART IV: MACHINE SCHEDULING MODELS[1]

This part describes deterministic problems of scheduling tasks on machines or processors. In general, these problems may be stated as follows. A given set of jobs or tasks is to be processed on a set of available machines, in such a way that all processing conditions are satisfied and some objective function is optimized. In the following we will use the term *scheduling* in the context of *machine scheduling* problems. Here, we assume that task parameters are deterministic. It is important to note that this assumption does not necessarily mean that we deal only with problems in which all characteristics of a set of resources are known in advance; we also consider problems in which some parameters such as task ready times are unknown in advance, and we need not assume any knowledge about them. From a more pragmatic point of view, deterministic models are, as usual, a simplification of reality that allows the user to actually solve the problems. For a survey of stochastic scheduling problems, the reader is referred to Möhring *et al.* (1984, 1985) and Righter (1994). This simplification is particularly important in machine scheduling problems which, in many industrial applications, have to be solved in real time, thus prohibiting the use of sophisticated stochastic models.

It is probably surprising that stochastic scheduling problems have a longer history than their deterministic counterparts. The first papers on stochastic scheduling were concerned with an analysis of telephone call "traffic" by Erlang (1909); see Brockmeyer *et al.* (1948). These queuing investigations led to the development of the foundations of axiomatic probabilistic theory by Kolmogorov (1933). This type of model is, however, not necessarily suitable for the description of manufacturing scheduling processes where long series of products are to be produced to specific customer orders. It appears that most of the production characteristics, including product processing times and their due dates as defined by the customer, are known in advance. Such processes are therefore amenable to a deterministic analysis of scheduling problems. The first papers in this field appeared in the early 1950s; see Jackson (1955) and Johnson (1954).

---

[1] This part was coauthored by Professor J. Blazewicz, Poznan University of Technology, Poland.

Developments in operations research and combinatorial optimization provided the theoretical basis for the analyses of more complex scenarios.

This part consists of four chapters that describe several issues arising in the context of scheduling. We first present basic concepts and definitions as well as a classification scheme and some examples of typical scheduling models. We then present the most important scheduling algorithms for single machine problems, as well as for parallel and dedicated machine models. Some material resource-constrained scheduling ends the part. Throughout the book a taxonomy proposed by Graham *et al.* (1979) is used to classify scheduling problems. When presenting results, our choice has been motivated first by the importance of a particular result, e.g., in the case of algorithms, their power to solve more than one particular scheduling problem. Secondly, our choice has been influenced by the relevance of the results to practical scheduling problems. This means that not much attention has been paid to enumerative methods. However, throughout this part we refer to existing surveys that cover the issues involved in a detailed fashion.

This part is organized as follows. Chapter 1 introduces basic scheduling problems by way of example, defines problem parameters and classifies the models. Chapter 2 deals with single machine scheduling. The presentation of algorithms and complexity results is divided into three sections, each dealing with one important criterion: makespan (schedule length), mean flow time, and due dates. Chapter 3 considers scheduling on parallel machines. The presentation of the results is again divided into three parts as in the previous chapter, dealing with makespan, mean flow time, and maximal lateness. Finally, Chapter 4 first discusses dedicated machine models, presenting results for open shop, flow shop and the general job shop problem. It then introduces resource-constrained scheduling problems in which, in addition to a machine, each task may also require other resources.

CHAPTER 1 FUNDAMENTALS OF MACHINE SCHEDULING

This introductory chapter is devoted to the discussion of the basic notation and assumptions made in deterministic scheduling theory. We begin by providing some examples of scheduling models that will give insight into the problem. We then give a general description of how scheduling problems can be classified.

1.1 Introductory Examples

In general, *scheduling* is an allocation of resources, such as machines to tasks, so as to optimize a given objective function. This general statement may be interpreted in several ways. Following an approach presented in Baker (1974) we can distinguish two different meanings of the term scheduling. On the macro level, scheduling denotes a *decision making function*, i.e., a process that answers questions such as "What kind of product is to be manufactured (made)?" "On what scale?" "What resources will be used?" The second meaning of the term refers to *scheduling theory*, which provides tools for the efficient solution of scheduling problems. Thus, it can be seen as a collection of principles, models and techniques that provide insight into the scheduling function. A solution to a scheduling problem usually requires two subproblems to be solved. The *allocation problem* decides which resources should be allocated to perform the given tasks. The *sequencing problem* determines when each task will be performed. It is this second meaning of the term scheduling, the *scheduling theory* on the micro level, that we are concerned with here. In order to better illustrate some ideas and concepts used in scheduling theory, we present some examples below.

Example 1: A cabinet maker has orders from four customers. Each customer is willing to pay a certain amount if the ordered piece of furniture is finished before a given due date. Otherwise, the cabinet maker has to reduce his price by a certain penalty for each day exceeding the due date. The time needed to make each individual piece, as well as due dates, prices, and penalties are shown in Table

IV.1. In order to maximize his profit, the cabinet maker has to minimize the total penalty, i.e., the weighted sum of the delays.

Table IV.1

Customer	Pieces of furniture	Price	Duration of task	Due date in	Penalty per day
A	Table	$200	2 Days	4 Days	$20
B	10 Chairs	$300	5 Days	6 Days	$10
C	Wardrobe	$700	7 Days	10 Days	$40
D	Bookcase	$400	4 Days	5 Days	$20

The optimal schedule of the cabinet maker's problems is shown in Figure IV.1 in the form of a *Gantt chart*, i.e., a horizontal bar chart in which the time scale is shown along the horizontal axis. The optimal sequence is A, D, C, B, and the profit associated with this schedule is $(200 + 300 + 700 + 400) - (0 + 120 + 120 + 20) = \$1,340$.

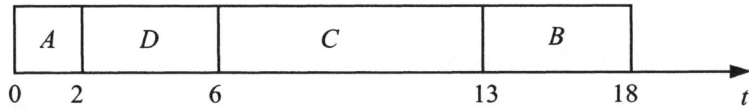

Figure IV.1

Example 2: A set of computer programs is to be processed on a computer system consisting of two parallel, identical processors. Because of certain programming constraints these programs must be processed in a certain order. This situation can be represented by a directed graph in which programs (or tasks) are represented by nodes and precedence relations are shown as directed arcs as in Figure IV.2.

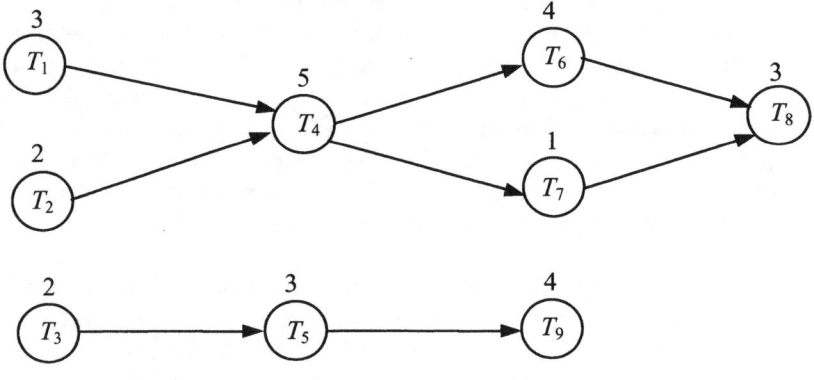

Figure IV.2

Chapter 1: Fundamentals of Machine Scheduling

Each program must wait until all the programs preceding it on any directed path in the graph are completed. Every program needs to run on either one of the two processors during an uninterrupted interval whose length equals the program's processing time in minutes, indicated next to the nodes, and each processor is capable of processing any of the given tasks, but only one at a time. The objective is then to determine an assignment of processors to programs such that the entire project is processed in the shortest possible time.

A schedule of minimal duration is shown in the Gantt chart in Figure IV.3. The entire project takes 15 minutes to complete. Note the idle time from $t = 11$ to 15 on processor 2, shown as a shaded rectangle. Other schedules of the same length exist.

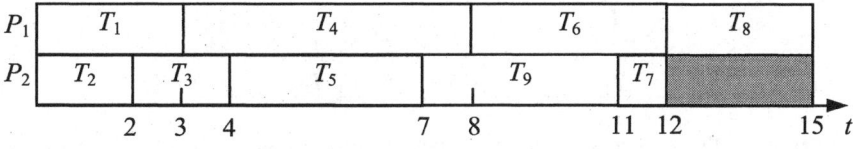

Figure IV. 3

Example 3: Consider a factory producing interlocking windows made to order. Borrowing from Graham (1979), in this hypothetical factory the process of making a window is divided into a number of smaller operations as follows: T_1: Measure and cut the glass panes, T_2: Decorate the glass panes, T_3: Assemble the glass panes in the frame, T_4: Cut and assemble the frame, T_5: Provisionally mount the locking mechanism, T_6: Provisionally attach the hinges, T_7: Align the hinges, T_8: Adjust and tighten the locking mechanism, T_9: Adjust and tighten the hinges, and T_{10}: Align, adjust and permanently seal the glass panes in the frame. A worker will perform each of the above tasks within times (in minutes) shown in Table IV.2.

Table IV.2

Task	T_1	T_2	T_3	T_4	T_5	T_6	T_7	T_8	T_9	T_{10}
Time required	7	7	7	2	3	2	2	8	8	18

Because of space and equipment constraints in the shop, the 20 available workers are paired into 10 teams of two workers each, with each team assembling one window at a time. Since one window requires a total of 64 minutes of assembly time, one may think that a team of two could manage this in 32 minutes. This would mean that in an eight-hour day, each team should be able to assemble 15 windows and with all 10 teams working, 150 windows per day could be produced.

However, when assembling windows, a certain order of operations must be observed. For example, one cannot assemble the glass panes in the frame before the frame is built. Similarly, the locking mechanism cannot be adjusted and tightened before it has been provisionally mounted. The applicable precedence constraints are summarized in Table IV.3.

Table IV.3

Task	Immediate predecessor
T_1, T_4	–
T_2	T_1
T_3	T_1, T_4, T_5
T_5, T_6	T_4
T_7	T_6
T_8, T_9	T_5, T_6, T_7
T_{10}	T_2, T_3, T_5

In addition to the constraints on the work schedule caused by the precedence relations, there are also two rules (known in the factory as "busy" rules) which management requires to be observed during working hours:

Rule 1: No worker can be idle if there is some job he could be doing.

Rule 2: Once a worker starts a job, he must continue working on that job until it is completed.

The current order of assembling windows at the factory for each of the ten identical teams is shown in Figure IV.4a. The schedule shows the activity of each assembler of the team beginning at time zero and progressing to the time of completed assembly, called the *finishing time*, not 32 but 34 minutes later. Although this schedule obeys all the required order-of-assembly constraints given above, it allows each team to complete only slightly over 14 windows per day. Thus the total output of the section is just over 140 windows per day, short of the 150 hoped for.

Surprisingly, this output cannot be improved, even if each task were to require one minute less than shown in Table IV.2. The best schedule with the reduced activity times is shown in Figure IV.4b. As can be seen the length of the optimal schedule *increases* by one minute rather than decreases, as one would have expected. Another scenario would be the following. Suppose that rather than 20 workers processing the tasks in 10 groups as assumed so far, there are 10 additional workers for a total of 30, who work in teams of three each, and assume also that

Chapter 1: Fundamentals of Machine Scheduling   339

the processing times remain the same as those in Table IV.2. The optimal schedule is shown in Figure IV4.c, and the finishing time has now increased to 39 minutes.

These anomalies are caused by the "greedy" rules 1 and 2. Their impact on schedule behavior has been studied by Graham (1966) and we will return to these results when discussing list scheduling in Section 3.1.1.

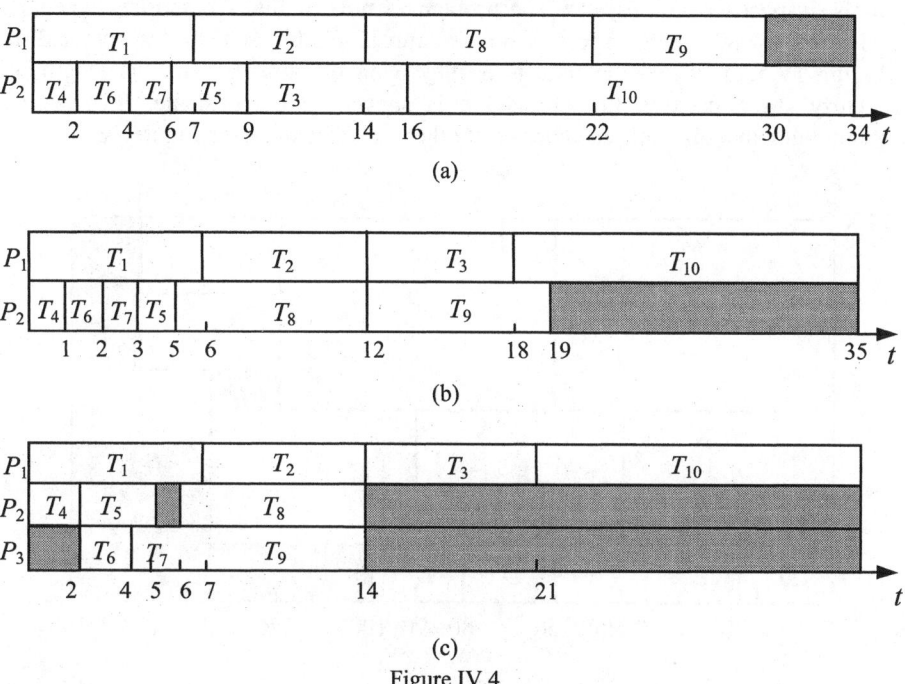

(a)

(b)

(c)

Figure IV.4

Example 4: The use of more than one type of resource in processing a set of tasks may be illustrated by the process of manufacturing ship engines in a factory. Consider a shop of length $L = 100$ ft in which four types of engines are to be produced. The numbers of engines ordered, their sizes (here reduced to a one-dimensional "length") and durations of production periods are displayed in Table IV.4.

Table IV.4

Engine Type	Number of engines ordered	Duration of a production period	Length of engines
A	2	60 Days	50 ft
B	3	45 Days	35 ft
C	3	30 Days	20 ft
D	5	20 Days	15 ft

Suppose now that there are three teams of workers, each capable of producing any of the engines. The objective is to produce all engines in the shortest possible overall time. Because of the shop length, not all engine types can be matched into triples to be produced together. Thus, besides teams of workers, the first resource, it is necessary to consider as a second scarce resource the space that is allocated to engines. More complicated problems could involve additional resources such as tools, inspection stations, etc. The problem is now to find the shortest schedule which is feasible with respect to both resources. Such a schedule is depicted in Figure IV.5, in which the time is displayed on the abscissa and the length of factory space used in the production is shown on the ordinate. It is worth mentioning that optimal schedules exist other than that shown in Figure IV.5.

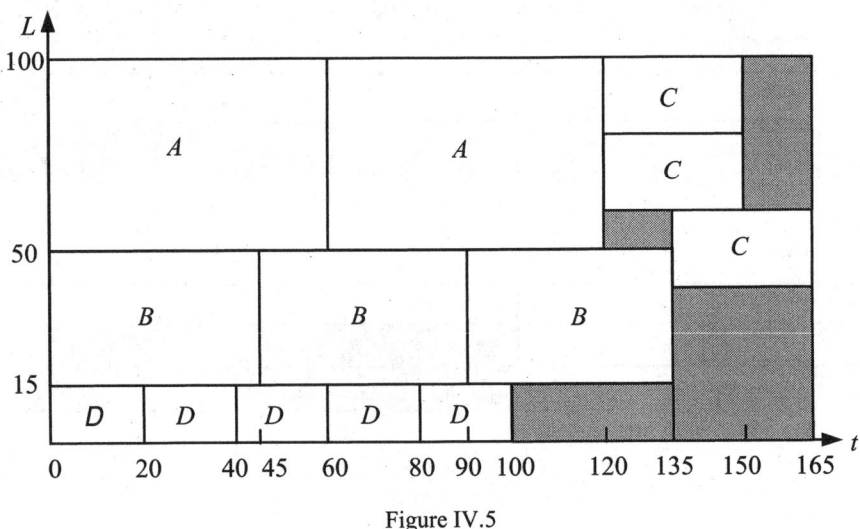

Figure IV.5

1.2 The Models and Their Components

In this section we first introduce and discuss the components of scheduling problems. We then explain their relevance and introduce a taxonomy for identifying and classifying scheduling problems.

1.2.1 Basic Concepts, Notation and Performance Criteria

The general scheduling problem includes a set of *n jobs* or *tasks* $\mathcal{T} = \{T_1, T_2,..., T_n\}$ and a set of *m machines* or *processors* $\mathcal{P} = \{P_1, P_2,..., P_m\}$. There are two general rules to be followed in classical scheduling theory:

Chapter 1: Fundamentals of Machine Scheduling

- each task is to be processed by at most one machine at a time
- each machine is capable of processing at most one task at a time.

First consider the machines. They may be either *parallel*, in which case they can all perform the same functions, or *dedicated*, i.e., specialized for the execution of certain tasks. Three types of parallel machines are distinguished depending on their speeds. If all machines from the set $\mathscr{P}$ have equal task-processing speeds, they are called *identical* machines. If the machines differ in their speeds, but the speed of each machine is constant and does not depend on the task it processes, then they are called *uniform*. Finally, if the processing speeds depend on the specific task and the machine it is processed on, they are called *unrelated*.

In the case of dedicated machines there are three modes of processing: open shop, flow shop, and job shop. In an *open shop* system each task from the set $\mathscr{T}$ must be processed by all machines, but the order of processing is arbitrary. In a *flow shop* system, each task must be processed by all machines and in the same given order. In a *job shop* system the subset of machines which are to process a task and the order of the processing are given and specific for each job.

We now describe the processing of the tasks in detail. In the case of parallel machines each task can be processed by any machine. When tasks are to be processed on dedicated machines, each task $T_j \in \mathscr{T}$ is divided into *operations* O_{1j}, $O_{2j}, \ldots, O_{k_j j}$, each of which may require a different machine. Furthermore, in an open shop system, the number of operations per task is equal to m and their order of processing is such that O_{ij} is processed on machine P_i $\forall$ $i = 1,\ldots, m$, but the order of processing operations is not specified. The case of a flow shop system is similar. Again, each task is divided into m operations with the same allocation of the operations to machines. Moreover, the processing of $O_{i-1,j}$ must always precede the processing of O_{ij} $\forall$ $j = 1, 2,\ldots, n$ and $i = 2,\ldots, m$. In a job shop system, the number of operations per task, their assignment to machines, and the order of their processing, are specified in advance for each job.

Each task $T_j \in \mathscr{T}$ is characterized by the following parameters.

(1) A vector of *processing* times $\overline{\mathbf{p}}_{\bullet j} = [p_{1j}, p_{2j},\ldots, p_{mj}]^T$, where p_{ij} is the time needed by machine P_i to complete task T_j. In the case of identical machines we have $p_{ij} = p_j$, $i = 1, 2,\ldots, m$. If the machines in $\mathscr{P}$ are uniform, then $p_{ij} = p_j/b_i$, $i = 1, 2,\ldots, m$, where p_j is called the *standard processing time* (usually on the slowest machine) and b_i is called the *processing speed factor* of machine P_i (normally one or larger).

(2) An *arrival time* or *ready time* or *release date* r_j. This is the time at which task T_j is ready for processing. If for all tasks from $\mathcal{T}$ the arrival times are equal, then it is often assumed that $r_j = 0, j = 1, 2,..., n$.

(3) A *due date* d_j. This is the time at which task T_j should be completed. A later completion may be allowed, but with a penalty incurred. If the processing of task T_j must be completed at time d_j, then d_j is called the *deadline*.

(4) A *weight* or *priority* w_j. The weight expresses the relative urgency of task T_j.

Below, some definitions concerning task preemptions and precedence constraints among tasks are given.

The mode of processing is called *preemptive* if each task, i.e., each operation in the case of dedicated machines, may be preempted at any time and restarted later at no cost, perhaps on another machine. If the preemption of a task (or an operation, in the case of dedicated machines) is not allowed we call the scheduling problem *nonpreemptive*.

It is possible to include *precedence constraints* among tasks in the set $\mathcal{T}$. For example, $T_k \prec T_j$ means that the processing of T_k must be completed before work on task T_j can commence. In other words, the set $\mathcal{T}$ is ordered by a binary relation $\prec$. The tasks in set $\mathcal{T}$ are called *dependent* if at least two tasks in $\mathcal{T}$ are ordered by such a relation; otherwise, they are called *independent*. A task set $\mathcal{T}$ ordered by a precedence relation is usually represented as a directed *task-on-node* graph, similar to the *activity-on-node* network introduced in Part III, in which nodes represent tasks and arcs specify the precedence relations. An example of a dependent task set is shown in Figure IV.6 where the processing time p_j is shown next to the node T_j. Note that in the case of dedicated machines (except in the open shop system) operations that constitute a task are always dependent, but tasks can be either independent or dependent. Alternatively, again equivalent to the discussion in the previous part of this book, we may use a *task-on-arc* (or *activity-on-arc*) network. Usually, the former approach is applied, but sometimes the latter presentation may be useful and we will encounter such situations below. Task T_j is called *available* at time t if $r_j \leq t$ and all its predecessors with respect to the precedence constraints have been completed at time t. If T_j is the k-th task to be scheduled on processor P_i, we say that T_j is processed in the k-th position on P_i.

We now state some definitions concerning schedules and optimality criteria.

A *schedule* is an assignment of machines in the set $\mathcal{P}$ to tasks in the set $\mathcal{T}$ in time, such that the following conditions are satisfied:

Chapter 1: Fundamentals of Machine Scheduling

- at any moment in time, each machine is assigned to at most one task and each task is being processed by at most one machine,
- task T_j is processed in the time interval $[r_j, \infty]$,
- all tasks are completed;
- for each pair (T_k, T_j), with $T_k \prec T_j$, the processing of T_j starts only after T_k is completed,
- in the case of *nonpreemptive* schedules, no task is preempted; otherwise we assume that the number of preemptions of each task is finite. The resulting schedule is then called *preemptive*.

As an example, consider the set of nine tasks $T_1, \ldots, T_9$ shown in Figure IV.6, whose durations are indicated next to the nodes in the task-on-node network, and whose dependencies are indicated by the arcs. A feasible schedule for processing these tasks on three parallel, identical machines is displayed in Figure IV.7 with the shaded areas indicating idle time of the processors.

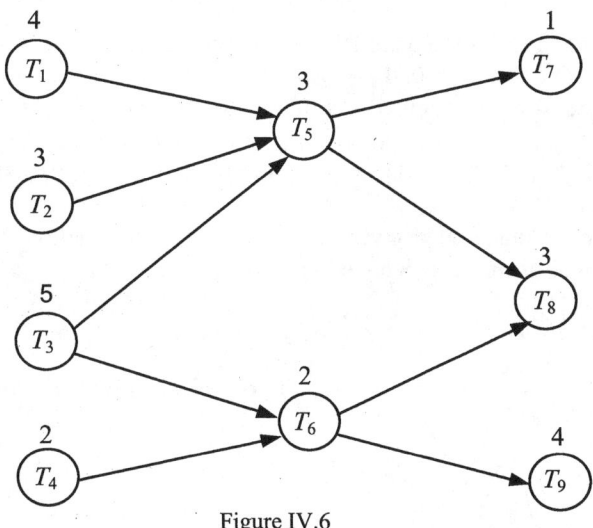

Figure IV.6

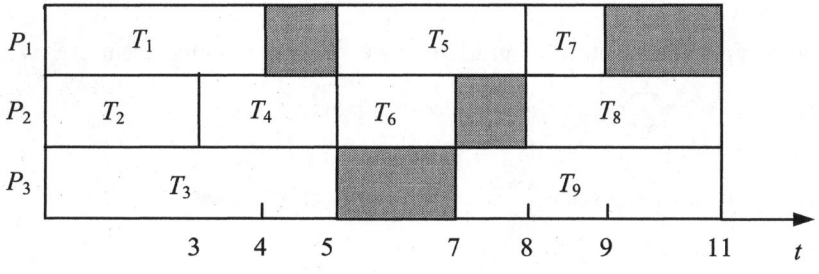

Figure IV.7

In spite of the requirement that each machine be assigned to at most one task, one may sometimes consider a fictitious schedule involving *processor sharing* where m machines process $n > m$ tasks by partitioning the total processing capacity of the m machines equally among the n tasks. A processor-shared schedule can always be converted to a proper schedule by introducing preemptions. Such schedules will be considered in Chapter 3 on parallel machine models.

For any given schedule, the following measures can be calculated for each task T_j, $j = 1, 2,..., n$:

- *completion time* c_j, i.e., the time by which the task is completely processed,
- *flow time* $f_j = c_j - r_j$, i.e., the time elapsed between the arrival of the task and its completion time,
- *lateness* $\ell_j = c_j - d_j$, and
- *tardiness* $t_j = \max\{c_j - d_j, 0\}$.

For the schedule given in Figure IV.7 we find that $\mathbf{c} = (c_j) = [4, 3, 5, 5, 8, 7, 9, 11, 11]$, and, given that $\mathbf{r} = \mathbf{0}$, $\mathbf{f} = (f_j) = \mathbf{c}$. For the computation of lateness and tardiness, due dates are required. Assuming the due dates are $\mathbf{d} = (d_j) = [5, 4, 5, 3, 7, 6, 9, 12, 12]$, then lateness and tardiness for the tasks in the schedule are $\boldsymbol{\ell} = (\ell_j) = [-1, -1, 0, 2, 1, 1, 0, -1, -1]$ and $\mathbf{t} = (t_j) = [0, 0, 0, 2, 1, 1, 0, 0, 0]$, respectively.

In order to evaluate any given schedule, we will employ three alternative performance measures or objective functions.

(1) *Schedule length* or *makespan* $C_{\max} = \max\{c_j\}$;

(2) *Mean flow time* $\overline{F} = 1/n \sum_{j=1}^{n} f_j$ or *mean weighted flow time* $\overline{F}_w = \sum_{j=1}^{n} w_j f_j \bigg/ \sum_{j=1}^{n} w_j$, and

(3) *Maximal lateness* $L_{\max} = \max\{\ell_j\}$.

In some applications, other criteria are more relevant. Among them are *mean tardiness* $\overline{T} = 1/n \sum_{j=1}^{n} t_j$, *mean weighted tardiness* $\overline{T}_w = \sum_{j=1}^{n} w_j t_j \bigg/ \sum_{j=1}^{n} w_j$, and the *number of tardy tasks* $\overline{Y} = \sum_{j=1}^{n} y_j$, where $y_j = 1$, if $c_j > d_j$, and 0 otherwise.

Chapter 1: Fundamentals of Machine Scheduling

As an example, consider again the schedule in Figure IV.7. The values of the performance measures are for the schedule length $C_{\max} = 11$, the mean flow time $\overline{F} = \frac{1}{9}(63) = 7$, and maximal lateness $L_{\max} = \max\{-1, -1, 0, 2, 1, 1, 0, -1, -1\} = 2$. Furthermore, the mean (unweighted) tardiness $\overline{T} = 4/9$, and the number of tardy jobs is $\overline{Y} = 3$. Values of other criteria can be obtained if weights w_j of the tasks are specified.

A *scheduling problem* Π is then defined as a problem with a structure as described above that optimizes some well-defined measure of performance. An *instance I* of problem Π is then obtained by specifying particular values for all the problem parameters.

1.2.2 Interpretation and Discussion of Assumptions

In this section we critically examine the assumptions of machine scheduling problems. First consider the assumptions associated with the set of tasks. As mentioned in Section 1.2.1, deterministic scheduling problems assume that ready times and processing times are known to the modeler. Ready times are typically known in computer systems working in an off-line mode. They are also often known in computer control systems in which measurements are done at predetermined times, i.e., known beforehand.

As far as processing times are concerned, they are usually not known *a priori* in computer systems, unlike in many other applications. Despite this fact, the solution of a deterministic scheduling problem is often important in such systems to give the modeler some information about the average (expected) or the worst case. One possibility to schedule tasks to meet deadlines is to solve the problem with some known upper bounds (in the case of algorithms, one could use their worst-case complexity, even though that may be prohibitively large). Clearly, if all deadlines are met with respect to the worst-case upper bounds, no deadline will be exceeded for the real task processing times. This approach is often used in computer control systems that work in a real-time environment, i.e., in systems in which each task is associated with a deadline that has to be met, where a certain set of control programs must be processed before taking the next sample from the same sensing device.

Alternatively, instead of worst-case values for the unknown processing times, we can take their mean values and, using the procedure described in Coffman and Denning (1973), calculate an estimate of the mean value of schedule length. If it is desired to evaluate a scheduling algorithm, one can first measure the processing times of tasks after processing a task set scheduled according to some algorithm A. Taking these values as an input in the deterministic scheduling problem, one may construct an optimal schedule and compare it with the one produced by algorithm A, which will provide a measure of the quality of the algorithm.

Consider now the different performance measures outlined above. Minimizing schedule length C_{max} is important from the viewpoint of the owner of a set of machines or processors, since it leads to a maximization of the machine utilization factor. This criterion may also be of importance in a computer control system in which a task set arrives periodically and is to be processed in the shortest time possible. The mean flow time criterion $\overline{F}$ is important from the user's point of view, since its optimization results in a schedule that minimizes the amount of time that a job is out of the user's hands, i.e., waiting to be processed or being processed. Note that the mean flow time is $\overline{F} = \frac{1}{n}\sum_{j=1}^{n} f_j = \frac{1}{n}\sum_{j=1}^{n}(c_j - r_j)$, and, with n and r_j given, minimizing $\overline{F}$ is therefore equivalent to minimizing $\sum_{j=1}^{n} c_j$. Similarly, one can show that minimizing mean weighted flow time $\overline{F}_w$ is equivalent to minimizing $\sum_{j=1}^{n} w_j c_j$.

Finally, the maximal lateness criterion L_{max} is important in many computer-controlled systems, in which decisions are to be made in real time. The optimization of this objective leads to the construction of a schedule without late tasks, provided that such a schedule exists. Other criteria involving due-dates are of importance in some economic applications. The three performance measures makespan, mean flow time, and lateness are basic in the sense that they concern three main applications and require three different approaches to the designing of optimization algorithms.

1.2.3 A Classification Scheme

Considering the large number of combinations of parameters and objectives in scheduling models, it is easy to imagine that the number of different scheduling models is huge. In order to facilitate a short and convenient characterization of such problems, it will prove useful to introduce a short and concise classification scheme. Such a scheme was proposed in Graham *et al.* (1979), Lawler *et al.* (1982), and Blazewicz *et al.* (1983). It is presented below and is used throughout this part of the book. The notation comprises three fields $\alpha/\beta/\gamma$.

The first field $\alpha = \alpha_1, \alpha_2$ describes the processor (machine) environment. The parameters have the following meaning:

Chapter 1: Fundamentals of Machine Scheduling

$\alpha_1 \in \{\emptyset, P, Q, R, O, F, J\}$ characterizes the type of machine used. The individual parameters symbolize:

$\alpha_1 = \emptyset$: one machine
$\alpha_1 = P$: identical machines
$\alpha_1 = Q$: uniform machines
$\alpha_1 = R$: unrelated machines
$\alpha_1 = O$: dedicated machines, an open shop system
$\alpha_1 = F$: dedicated machines, a flow shop system
$\alpha_1 = J$: dedicated machines, a job shop system.

The parameter $\alpha_2 \in \{\emptyset, k\}$ denotes the number of machines in the problem as follows:

$\alpha_2 = \emptyset$: the number of machines is assumed to be variable
$\alpha_2 = k$: the number of machines equals some $k \in \mathbb{N}$.

The second field $\beta = \beta_1, \beta_2, \beta_3, \beta_4, \beta_5$ describes task and resource characteristics. Here, the parameter $\beta_1 \in \{pmtn, \emptyset\}$ indicates the possibility of task preemption, so that

$\beta_1 = pmtn$: preemptions are allowed
$\beta_1 = \emptyset$: no preemption is allowed.

The parameter $\beta_2 \in \{\emptyset, res\ \lambda, \sigma, \rho\}$ characterizes additional resources and will be explained in Section 4.4 of this part. For the time being we assume that $\beta_2 = \emptyset$, i.e., no additional resources (or resource constraints) are present.

The parameter $\beta_3 \in \{\emptyset, prec, uan, tree, chain\}$ reflects the precedence constraints, so that

$\beta_3 = \phi$: independent tasks
$\beta_3 = prec$: precedence constraints forming an acyclic graph
$\beta_3 = uan$: uniconnected activity networks, i.e., an acyclic graph in which any two nodes are connected by a directed path
$\beta_3 = tree$: precedence constraints forming a tree
$\beta_3 = chain$: precedence constraints forming a chain.

The parameter $\beta_4 \in \{r_j, \emptyset\}$ describes ready times, where

$\beta_4 = r_j$: ready times are unequal to zero
$\beta_4 = \emptyset$: all ready times are equal to zero.

The parameter $\beta_5 \in \{p_j = p, \underline{p} \le p_j \le \overline{p}, \varnothing\}$ describes task processing times, so that

$\beta_5 = (p_j = p)$: all tasks have processing times equal to p
$\beta_5 = (\underline{p} \le p_j \le \overline{p})$: no processing time p_j is less than $\underline{p}$ and no greater than $\overline{p}$
$\beta_5 = \varnothing$: tasks have arbitrary processing times.

Finally, the third field γ denotes an objective function or performance measure, i.e.,

$\gamma \in \{C_{max}, \Sigma c_j, \Sigma w_j c_j, L_{max}, \Sigma t_j, \Sigma w_j t_j, \Sigma u_j, \Sigma w_j u_j\}$, where we minimize

$C_{max} = \max \{c_j\}$: makespan (schedule length)

Σc_j: total flow time (or, equivalently, $\frac{1}{n}\Sigma c_j = \overline{F}$: mean flow time)

$\Sigma w_j c_j$: total weighted flow time (or, equivalently, the mean weighted flow time

$$\overline{F}_w = \sum_j w_j c_j \Big/ \sum_j w_j)$$

$L_{max} = \max \{\ell_j\}$: maximal lateness

$\frac{1}{n}\Sigma t_j = \overline{T}$: mean tardiness

$\Sigma w_j t_j$: total weighted tardiness (or, equivalently, the mean weighted tardiness $\overline{T}_w = \sum_j w_j t_j \Big/ \sum_j w_j)$

$\Sigma y_j = \overline{Y}$: number of tardy tasks
$\Sigma w_j y_j = \overline{Y}_w$: weighted number of tardy tasks.

In general, if any of the parameters above assumes the symbol $\varnothing$, it is shown in the taxonomy as an empty spot. For example, the problem $P//C_{max}$ denotes a model with identical parallel machines, no preemption, no additional resources, independent tasks, all tasks ready at time zero, arbitrary processing times, and an objective that minimizes makespan. As another example, $O3/pmtn, r_j/\Sigma c_j$ refers to an open shop problem with three machines, preemptions permitted, no additional resource constraints, independent tasks, not all tasks ready at project start, arbitrary processing times, and an objective that minimizes mean flow time. Example 1 in Section 1.1 is of type $1//\Sigma w_j t_j$, Examples 2 and 3 of type $2/prec/C_{max}$, and Example 4 of type $3/res/C_{max}$.

Chapter 1: Fundamentals of Machine Scheduling   349

We can then demonstrate that scheduling problems are closely related in the sense of polynomial transformation; see, e.g., Eiselt and Sandblom (2000) and the next section. Some basic polynomial transformations between scheduling problems are shown in Figure IV.8. For each graph in the figure, the presented problems differ only by one parameter and the arrows indicate the direction of the polynomial transformation, i.e., following the direction of arcs, problems become more difficult. In particular, in Figure IV.8a the problems differ by the type and number of processors. Specifically, Pk indicates a problem with k identical machines, and P a problem with an arbitrary number of identical machines. The definitions for Qk and Rk are similar. In Figure IV.8b, the problems differ by the mode of processing, in Figure IV.8c by the type of precedence constraints, in Figure IV.8d by ready times, in Figure IV.8e by processing times, and in Figure IV.8f by the performance measure. These simple transformations are very useful in many situations when analyzing new scheduling problems. Thus, many of the results presented here can be immediately extended to cover a broader class of scheduling problems.

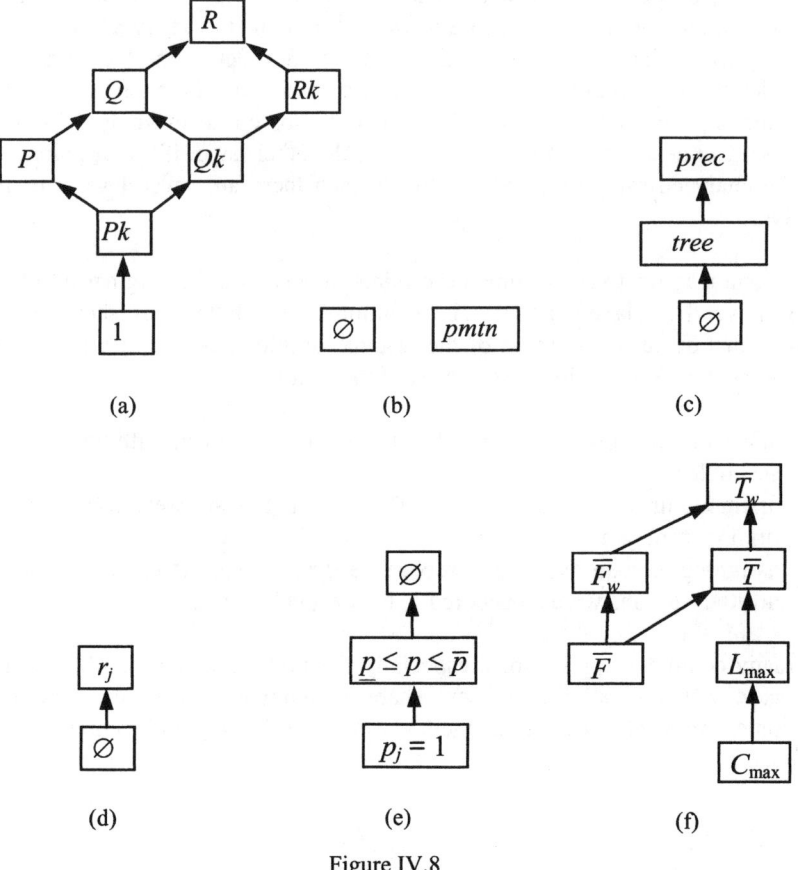

Figure IV.8

1.3 Algorithmic Approaches

Deterministic scheduling problems are a part of the broad class of combinatorial optimization problems. Thus, the general approach to the analysis of these problems can follow similar lines, but one should take into account their peculiarities, which are particularly evident in computer applications. It is obvious that in these applications the time we can devote to solving scheduling problems is seriously limited so that only low order polynomial-time algorithms may be used. Thus, the examination of the complexity of these problems should be a useful area for future study.

In their seminal papers Cook (1971) and Karp (1972) have demonstrated that there exists a large class of combinatorial optimization problems for which most probably there are no *efficient*, i.e., *polynomial* optimization algorithms. Further details can be found in the pertinent literature; see, e.g., Eiselt and Sandblom (2000). It is interesting to note, that as of 1993, out of a collection of 4536 scheduling problems, 3821 had been proved **NP**-hard, 417 were known to be solvable in polynomial time, and 298 were still open (Lawler *et al.*, 1993). When solving scheduling problems, it is customary to first determine the computational complexity of the problem. If the worst-case complexity function is a low-degree polynomial, the problem is "easy" and can be solved accordingly. This issue is discussed in detail in Aho *et al.* (1974). On the other hand, if the decision version of the analyzed problem is **NP**-complete, then there are several ways of further analysis.

First, one may try to relax some constraints imposed on the original problem and then solve the relaxed model. The solution of the latter may then be a good approximation of the solution of the original problem. In the case of scheduling problems such a relaxation may consist of the following:

- allowing preemptions, even if the original problem dealt with non-preemptive schedules
- assuming unit-length tasks, when arbitrary-length tasks were considered in the original problem
- assuming certain types of precedence graphs, e.g. trees or chains, when arbitrary graphs were considered in the original problem.

In many computer applications, especially the first relaxation can be justified in the case where parallel processors share a common primary memory. Such a relaxation may also be advantageous when considering different performance measures.

Chapter 1: Fundamentals of Machine Scheduling

Secondly, when trying to solve hard scheduling problems, users often employ approximation algorithms or heuristics. A necessary condition for such algorithms to be useful in practice is that their worst-case complexity function be bounded from above by a low-order polynomial in the input length. Their usefulness depends on an evaluation of the difference between the solution value they produce and the value of an optimal solution. This evaluation may concern the worst case or mean behavior. Define now $A(I)$ as the value of the objective function found by algorithm A applied to problem instance I of minimization problem Π. Similarly, $OPT(I)$ is the value of the objective function at optimum. We can then state performance ratios of an algorithm; for details see Garey and Johnson (1979).

Definition IV.1: The *performance ratio* of an approximation algorithm A applied to problem instance I is $R_A(I) = A(I)/OPT(I)$. In the case of a maximization problem, the ratio is defined as $R_A(I) = OPT(I)/A(I)$.

As an example, if $A(I) = 13$ and $OPT(I) = 10$, then $R_A(I) = 1.3$, indicating that the approximation solution is 30% worse than the optimal solution in this particular instance of the problem.

Definition IV.2: The *absolute performance ratio* R_A for a approximation algorithm A for a minimization problem Π is defined as $R_A = \inf\{r \geq 1 : R_A(I) \leq r \ \forall I \in \Pi\}$ i.e., the worst-case performance of the algorithm on any instance of the problem.

It is clear that $R_A \geq 1$. As an example, the value of $R_A = 1.2$ for an approximation method indicates that in the worst case, the solution found by this algorithm is 20% worse than the optimal solution.

Definition IV.3: The *asymptotic performance ratio* R_A^∞ for an approximation algorithm A applied to a minimization problem Π is $R_A^\infty = \inf \{r \geq 1: \text{for some } k \in \mathbb{N}, R_A(I) \leq r \ \forall \ I \in \Pi \text{ with } OPT(I) \geq k\}$.

Again, it is clear that $R_A^\infty \geq 1$.

The above definitions express a measure of the quality of approximation algorithms. The closer any of the performance ratios is to 1, the better algorithm A performs. For some combinatorial problems it is highly unlikely that an approximation algorithm of a given accuracy exists; i.e., finding a heuristic whose performance is, say, within 30% of the optimum is as hard as finding a polynomial-time algorithm for the **NP**-complete problem, making the determination of any reasonable solution a daunting task.

Analysis of the worst-case behavior of an approximation algorithm may be complemented by an analysis of its mean behavior. This is particularly meaningful in cases in which algorithms may have a poor worst-case complexity but tend to perform well in practice. Analysis of an algorithm's mean behavior can essentially be done in two ways. The first (theoretical) technique consists of assuming that the parameters of instances of the considered problem Π are drawn from a certain distribution and then analyzing the *mean performance* of algorithm A. Performance criteria are again the *absolute error* of the approximation algorithm, which is the difference between the approximate and optimal values, and the *relative error*, which is the ratio of the two. Asymptotic optimality results in the stronger (absolute) sense are quite rare. On the other hand, asymptotic optimality in the relative sense is often easier to establish; see, e.g., Karp *et al.* (1984), Rinnooy Kan (1987), and Slominski (1982).

The main drawback of the above approach is the difficulty of proofs of the mean performance for realistic distribution functions. Thus, the second (empirical) way of evaluating the mean behavior of approximation algorithms, consists of experimental studies. In this analysis we compare solutions to a set of problems. It is critical that the tests are performed on a large set of instances that should be representative of the type of instance users are interested in. Some remarks on how these tests should be carried out are discussed by Silver *et al.* (1980).

The third and last way of dealing with hard scheduling problems is to use exact enumerative algorithms regardless of their exponential worst-case complexity function. This is a viable option when the problem under consideration is not **NP**-hard in the strong sense and it is possible to solve it by a pseudopolynomial optimization algorithm. Frequently, such an algorithm may behave quite well in practice for reasonably small instances and it can be used even in some computer applications.

Before presenting specific results in scheduling theory, we would like to mention some basic books and survey articles that deal with various aspects of scheduling. As far as books are concerned, the most important are those by: Conway *et al.* (1967), Baker (1974), Coffman (1976), Lenstra (1977), Rinnooy Kan (1976), French (1982), Blazewicz *et al.* (1993, 1996), Brucker (1995), Chrétienne *et* al. (1995), Pinedo (1995), and Pinedo and Chao (1999). Among survey papers one should mention those by Graham *et al.* (1979), Lawler (1982a), Blazewicz (1987), and Lawler *et al.* (1993); see also Brucker and Knust (2003).

CHAPTER 2 SINGLE MACHINE SCHEDULING

This chapter is concerned with single machine scheduling. As mentioned in the previous chapter, three performance criteria will be analyzed in consecutive sections: schedule length (makespan), mean flow time and maximal lateness, respectively. Some other related criteria, such as mean tardiness, will also be discussed.

2.1 Minimizing Makespan

Consider first the simplest scheduling problem $1//C_{max}$, i.e., a problem in which all tasks are assumed to be independent, nonpreemptable, and available at time $t = 0$, and are to be processed by a single machine. It is clear that regardless of the order of the task assignment on the machine, the minimum schedule length (under the assumption that no dummy idle time is created) is $C_{max}^* = \sum_{j=1}^{n} p_j$.

Similar observations are true for some more complicated problems. For example, consider the problem $1/r_j/C_{max}$, i.e., a model with different ready times. The solution to the problem is to assign a task whenever it is available. Any schedule obtained this way will be optimal. Introducing arbitrary precedence constraints does not change the complexity either. It is also worth noting that the possibility of task preemption is not profitable when miminizing makespan. However, this result applies only to the single machine case and does not generalize to multiple machines.

2.2 Minimizing Mean Flow Time

This section deals with the minimization of flow time, i.e., the time that elapses between a task's ready time and its completion time. Problems of this nature are relevant in many industries, wherever costs are incurred during the time that a product waits for processing and while it is processed. Costs for maintenance, cooling, security, and others figure prominently in these scenarios. The firm's objective will then be to minimize these costs, which are nothing but the total weighted flow time of the products.

2.2.1 The Shortest Processing Time Algorithm

We first investigate the problem $1//\Sigma c_j$, i.e., on a single machine, we schedule independent and nonpreemptable tasks of arbitrary lengths and ready times $r_j = 0$ for all j, so as to minimize the mean flow time. For this criterion, it appears reasonable to try to assign first short tasks and then longer ones. This procedure is summarized in the *shortest processing time* (*SPT*) algorithm.

Shortest Processing Time (*SPT*) Algorithm for the $1//\Sigma c_i$ Problem

---

Procedure: Schedule tasks on one machine in nondecreasing order of processing times where ties are broken arbitrarily.

The above approach always leads to an optimal schedule. In order to prove optimality, we will apply an important technique that is used throughout scheduling theory—the so-called *adjacent pairwise interchange*.

Proposition IV.4 (Smith 1956, Baker 1974): For a single machine with zero ready times for all tasks, the mean flow time $\overline{F}$ is minimized by the shortest processing time algorithm.

Proof: Consider some schedule S that is not generated by the *SPT* schedule. That is, somewhere in S there must exist a pair of adjacent tasks T_i and T_j with T_j following T_i, such that $p_i > p_j$. Now construct a new schedule S', in which tasks T_i and T_j are interchanged and all other tasks are completed at the same time as in S. The situation is depicted in Figure IV.9 where B denotes the set of tasks that precede tasks T_i and T_j in both schedules, C_B the time to complete the tasks on B, and A denotes the set of tasks that follow T_i and T_j in both schedules. Let now $f_k(S)$ represent the flowtime of task T_k in schedule S. For schedule S we have

$$\sum_{k=1}^{n} f_k(S) = \sum_{k \in B} f_k(S) + f_i(S) + f_j(S) + \sum_{k \in A} f_k(S) =$$
$$\sum_{k \in B} f_k(S) + (C_B + p_i) + (C_B + p_i + p_j) + \sum_{k \in A} f_k(S).$$

Similarly, for schedule S', we obtain

$$\sum_{k=1}^{n} f_k(S') = \sum_{k \in B} f_k(S') + (C_B + p_j) + (C_B + p_j + p_i) + \sum_{k \in A} f_k(S').$$

Since $\sum_{k \in B} f_k(S) = \sum_{k \in B} f_k(S')$ and $\sum_{k \in A} f_k(S) = \sum_{k \in A} f_k(S')$, we obtain

$$\sum_{k=1}^{n} f_k(S) - \sum_{k=1}^{n} f_k(S') = p_i - p_j > 0.$$

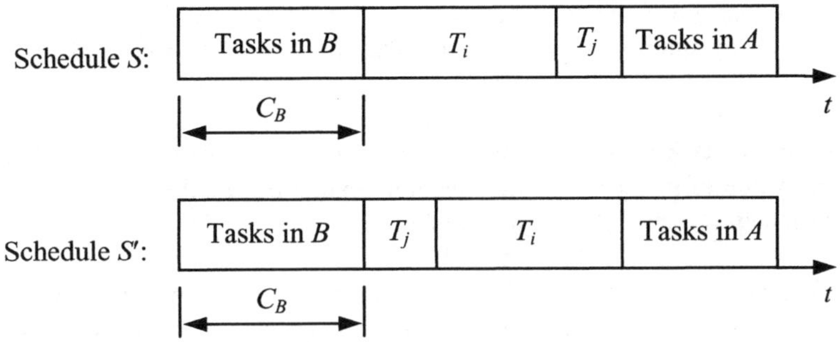

Schedule S: | Tasks in B | T_i | T_j | Tasks in A |

$\overset{C_B}{\longleftrightarrow}$

Schedule S': | Tasks in B | T_j | T_i | Tasks in A |

$\overset{C_B}{\longleftrightarrow}$

Figure IV.9

Thus, the interchange of tasks T_i and T_j transforms schedule S to schedule S', the latter schedule being shorter than S. Hence, no schedule which puts the tasks in any order other than of nondecreasing processing times can be optimal. This proves the lemma. □

Note that Proposition IV.4 is also a corollary of Lemma II.5 of Section 5.5.1 in Part II, dealing with Gilmore-Lawler bounds. If the tasks are first sorted in order of nondecreasing processing times and then renumbered, so that p_j denotes the processing time of the j-th task to be processed, it easily follows that $\sum_j c_j = \sum_{j=1}^{n} (n+1-j) p_j$. This sum is the scalar product of the two vectors $[n, n-1, \ldots, 1]$ and $[p_1, \ldots, p_n]$, which by virtue of Lemma II.5 takes its minimum value when the second of these vectors is sorted in nondecreasing order.

The *SPT* algorithm and the problem $1//\Sigma c_j$ are of complexity $O(n \log n)$ since the most complex activity of the algorithm is sorting the tasks with respect to their processing times, which can be accomplished in $O(n \log n)$ time.

Example: To illustrate the *SPT* algorithm, consider a problem with $n = 10$ tasks whose processing times are **p** = [5, 2, 1, 3, 6, 4, 1, 3, 4, 2]. A sequence generated with the *STP* algorithm, in which ties are broken in favor of the smaller subscript, is shown in Figure IV.10. The mean flow time of the optimal schedule is $\overline{F} = 12.6$.

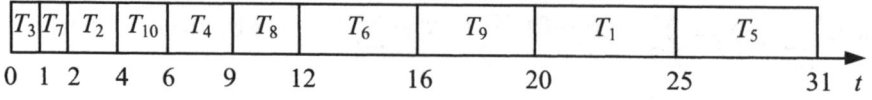

Figure IV.10

The case of equal, but nonzero, ready times $r_j \equiv r$ for all j can be reduced to the case of zero ready times. The start and completion times of all tasks then move ahead by r.

The possibility of task preemption does not change the optimality of the solution, because the value of mean flow time does not depend on whether or not tasks are split in the schedule; see, e.g., McNaughton (1959).

If different ready times are introduced, we obtain the problem $1/r_j/\Sigma c_j$ which is **NP**-hard in the strong sense as Lenstra *et al.* (1977) have proved. The situation can, however, be much improved if the tasks are preemptable. In this case, an optimal schedule can be found by extending the *SPT* rule, by assigning at any given moment an available task with the shortest remaining processing time. Baker (1974) demonstrated that this *SRPT* (*shortest remaining processing time*) algorithm solves the $1/pmtn, r_j/\Sigma c_j$ problem in $O(n \log n)$ time.

2.2.2 The Mean Weighted Flow Time and Other Problems

In this section we first consider minimizing the mean weighted flow time. In this context, weights may indicate the relative importance of a task. Consider the problem $1//\Sigma w_j c_j$ for which another modification of the *SPT* rule can be used. It is called a *weighted SPT* algorithm and it uses a concept of a *weighted processing time* for each task T_j defined as p_j/w_j. An algorithm for this problem was first presented by Smith (1956), and it is also known as *Smith's ratio rule*. It can be described as follows.

Weighted Shortest Processing Time (*WSPT*) Algorithm
for the Problem $1//\Sigma w_j c_j$

Procedure: Schedule tasks in nondecreasing order of their weighted processing times.

Chapter 2: Single Machine Scheduling

An optimality proof by contradiction is easily found by extending the arguments in the proof of Lemma IV.4. The proof can also be found in Pinedo (1995). The analysis of the *WSPT* algorithm is similar to that of the *STP* algorithm, resulting in a complexity of the *WSPT* algorithm of $O(n \log n)$.

Example: Consider a problem with $n = 8$ tasks, processing times $\mathbf{p} = [5, 3, 4, 2, 3, 4, 2, 1]$, and weights $\mathbf{w} = [3, 2, 2, 1, 2, 3, 2, 1]$. The weighted processing times are then $[5/3, 3/2, 2, 2, 3/2, 4/3, 1, 1]$ and an optimal schedule is shown in Figure IV.11. In this example, there are ties between T_7 and T_8, between T_2 and T_5, and between T_3 and T_4. In this example, ties are broken in favor of the task with the smaller subscript. There are eight different optimal solutions in all, each with a mean weighted flow time of $12\frac{1}{4}$.

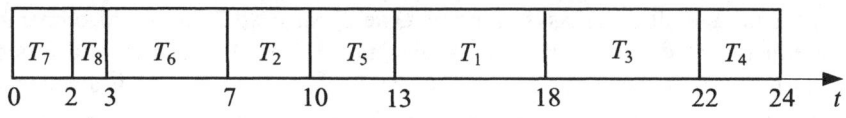

Figure IV.11

Again, introducing different ready times makes the problem **NP**-hard, and this problem remains intractable even for the preemptive case, as Labetoulle *et al.* (1984) have shown.

Consider now an extension involving dependent tasks. The *WSPT* rule can be extended to cover also the case of tree-like precedence constraints as elaborated upon by Horn (1973), Adolphson and Hu (1973), and Sidney (1975). The precedence constraints are said to have an *in-tree* structure if every task except one (the last, called the *root*) has precisely one immediate successor; an *out-tree* structure if every task except one (the first, called the *root*) has precisely one immediate predecessor; and a *chain* structure if it is both an in-tree and an out-tree. In a *parallel chain* structure, tasks form subgroups that have a chain structure. Figure IV.12 illustrates the concepts of in-tree, out-tree, and chain, respectively.

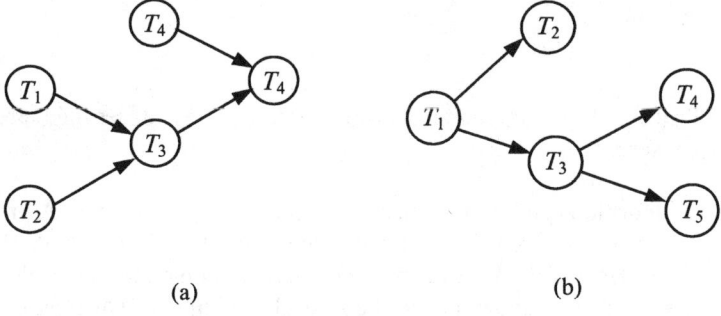

(a)   (b)

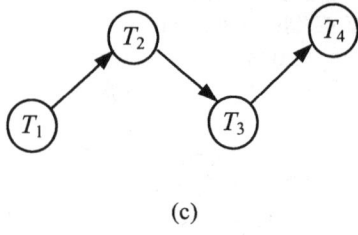

(c)

Figure IV.12

The solution to the problem is based on the observation that two tasks T_i and T_j can be treated as one with processing time $p_i + p_j$ and weight $w_i + w_j$, if $T_i \prec T_j$, $p_i/w_i > p_j/w_j$ and all other tasks either precede T_i, succeed T_j, or are incomparable with either. In order to describe the algorithm we need the concept of a feasible successor set Z_i of each task T_i of an out-tree. Each Z_i, $i = 1, 2,..., n$, is a collection of all sets Z_i^ℓ, $\ell = 1, ...,$ that include T_i and none, some, or all of its successors. Specifically, Z_i^ℓ is defined having the following properties:

- $T_i \in Z_i^\ell$,
- if $T_k \in Z_i^\ell$ and $T_i \neq T_k$, then $T_i \prec T_k$, and
- if $T_k \in Z_i^\ell$ and $T_j \prec T_k$, then either $T_j \in Z_i^\ell$ or $T_j \prec T_i$.

We are now able to state an algorithm for out-trees.

| Algorithm for the 1/out-tree/$\Sigma w_i c_i$ Problem |

Step 1: For each task T_i other than the root of the tree, calculate

$$e_i = \min_{Z_i^\ell \subset Z_i} \left\{ \sum_{T_k \in Z_i^\ell} p_k \bigg/ \sum_{T_k \in Z_i^\ell} w_k \right\}.$$

Step 2: Process tasks in nondecreasing order of e_i, observing precedence constraints.

The above algorithm can be implemented to run in $O(n \log n)$ time. A proof of its optimality is more involved than that for independent tasks. For details the reader is referred to Horn (1973). We will not give a numerical example for this algorithm, since it is quite similar to the next algorithm, for which an example is provided.

Chapter 2: Single Machine Scheduling 359

A slight modification of the above approach allows us to solve a similar problem with precedence structures being in-trees. The approach is based on a concept of so-called antithetical algorithms and schedules; see Baker (1974). The scheduling rule R' is called the *antithetical rule* of scheduling rule R if the task that is assigned position j in the schedule under R is assigned position $(n-j+1)$ under R'. It is possible to prove that any schedule is feasible in the in-tree problem considered here, if and only if its antithetical schedule is feasible in the out-tree problem that results when the precedence constraints of the in-tree are all reversed. Moreover, if a certain measure of performance is minimized by a particular rule R, then this measure is maximized by the antithetical rule R'. Following the above discussion, we can formulate an algorithm for in-trees as follows.

Algorithm for the 1/in-tree/$\Sigma w_i c_i$ Problem

Step 1: For the given in-tree, construct a corresponding out-tree by reversing the direction of all the arcs.

Step 2: For each task T_i other than the root of the tree, calculate

$$g_i = \max \left\{ \left(\sum_{T_k \in Z_i} p_k \right) \Big/ \left(\sum_{T_k \in Z_i} w_k \right) \right\}$$

Step 3: Schedule the root task first and then all other tasks in nonincreasing order of g_i.

Step 4: Construct an antithetical (reversed) schedule to the one obtained in Step 3.

The complexity of the in-tree algorithm is the same as for the out-tree algorithm above.

Example: Consider a problem with precedence structure in the form of an in-tree, where $n = 4$, $\mathbf{p} = [1, 5, 4, 2]$, and $\mathbf{w} = [2, 2, 2, 1]$ and the in-tree structure is given in Figure IV.13a. Reversing the in-tree into the out-tree shown in Figure IV.13b, it follows that $Z_1 = \{\{T_1\}\}$, $Z_2 = \{\{T_2\}\{T_1,T_2\}\}$ $Z_3 = \{T_3\}$, and therefore $g_1 = \max\{p_1/w_1\} = \max\{\frac{1}{2}\} = .5$, $g_2 = \max\{p_2/w_2;(p_1+p_2)/(w_1+w_2)\} = \max\{5/2; 6/4\} = 2.5$, and $g_3 = \max\{p_3/w_3\} = \max\{\frac{4}{2}\} = 2$.

The schedule obtained in Step 3 of the in-tree algorithm is $T_4 - T_2 - T_3 - T_1$, so that the antithetical schedule of Step 4 is $T_1 - T_3 - T_2 - T_4$. The schedule is shown in

Figure IV.13c, and its mean weighted flow time $\overline{F}_w = (2+10+20+12)/(2+2+2+1) = 44/7 = 6\tfrac{2}{7}$.

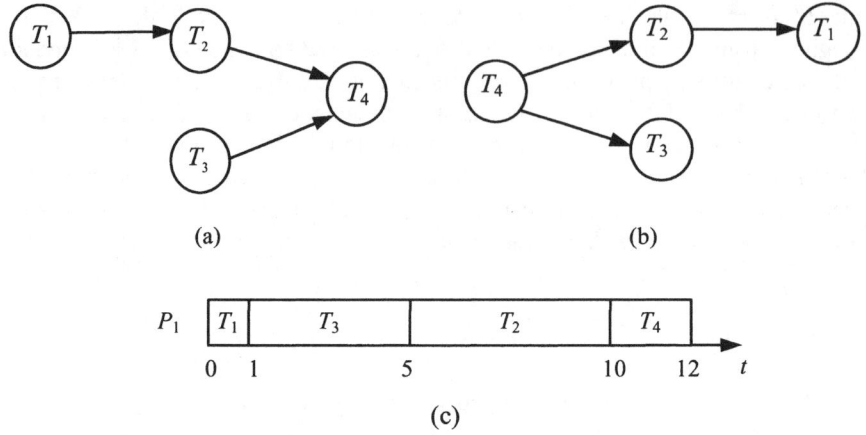

Figure IV.13

This approach was further generalized by Sidney (1975) and Lawler (1978) to the case of series-parallel precedence constraints. Unfortunately, this is as much as one can solve in polynomial time, since more complicated graph structures lead to **NP**-hard problems even if all weights or all processing times are equal; for details the reader is referred to Lawler (1978) and Lenstra and Rinnooy Kan (1978).

2.3 Minimizing Objectives Involving Due Dates

This section discusses problems in which a due date is associated with each task. Due dates may originate from contracts made with customers who demand that tasks are completed at certain times. Failure to do so will incur a penalty. In the case of different penalties for late tasks, an objective could be to minimize the sum of penalties or, using the terminology of machine scheduling, to minimize total weighted tardiness. On the other hand, in some scenarios customers may be willing to accept some tardiness without penalty, as long as it is within reasonable limits. The decision maker could then attempt to minimize the maximal lateness or tardiness.

2.3.1 Earliest Due Date Scheduling

The *Earliest Due Date* scheduling method, or *EDD* for short, is perhaps one of the first algorithms analyzed in the literature. This algorithm was first described by

Jackson (1956) and it can be used to solve problem $1//L_{max}$. The description of the algorithm is as follows.

> **Earliest Due Date (*EDD*) Scheduling Algorithm for the $1//L_{max}$ Problem**

Procedure: Schedule tasks in nondecreasing order of their due dates.

The optimality of schedules generated by the *EDD* scheduling algorithm is proved in the following

Proposition IV.5 (Jackson, 1956): Maximal task lateness L_{max} and maximal task tardiness T_{max} are minimized by the earliest due date scheduling algorithm.

The proof uses the concept of adjacent tasks pairwise interchange and is similar to that of Proposition IV.4. The complexity of the *EDD* scheduling algorithm is O(n log n) because the dominant effort is to sort all tasks with respect to their due dates. Since the algorithm amounts to a simple sorting of due dates, a numerical example appears unnecessary.

The above problem becomes **NP**-hard in the strong sense if the tasks have different ready times, as Lenstra *et al.* (1977) have demonstrated. But the problem $1/r_j/L_{max}$ is important since it appears as a subproblem in procedures for job shop scheduling, which we will cover in Chapter 4. It has therefore received considerable attention, resulting in a variety of branch and bound procedures; see, e.g., Carlier (1982) and Nowicki and Zdrzalka (1986). A simple approach is to schedule the jobs one after another, at each point considering all as yet unscheduled jobs with some enumerative procedure. Obvious pruning rules based on the job release dates r_j are then used; for details, see Pinedo (1995).

However, preemptions allow us again to solve problem $1/pmtn, r_i/L_{max}$ in polynomial time by slightly modifying the *EDD* rule as outlined by Liu and Layland (1973). Such a modification is summarized in the algorithm described below. The procedure is initialized by starting at time $t = 0$ with the task that has the earliest ready date. Ties are broken in favor of the job with the earliest due date.

> **Algorithm for the $1/pmtn, r_i/L_{max}$ Problem**

Procedure: Whenever a task is completed or a new task is ready, preempt the process and continue with the task that has the earliest due date, among all tasks that are ready at this time.

Example: Consider a problem with $n = 8$ tasks. The processing times **p**, ready times **r**, and due dates **d** are summarized in Table IV.5.

Table IV.5

	T_1	T_2	T_3	T_4	T_5	T_6	T_7	T_8
p	5	3	2	4	1	3	6	4
r	2	10	3	0	0	5	15	21
d	15	14	8	16	20	10	28	25

The process is initialized with the task T_4 which is ready at the time $t = 0$ and has an earlier due date than T_5. The process is first preempted at $t = 2$ when job T_1 becomes available. At that time, the task with the earliest due date among the ready, but still unfinished, tasks, i.e., T_1, T_4, and T_5, is T_1. The processing is again preempted at $t = 3$ when T_3 becomes available. As T_3 has the earliest due date among T_1, T_3, T_4, and T_5, it is processed at this time. Continuing in this fashion, we obtain the schedule shown in Figure IV.14 with completion times $\mathbf{c} = [15, 13, 5, 17, 18, 8, 28, 25]$, which, with due dates $\mathbf{d} = [15, 14, 8, 16, 20, 10, 28, 25]$ as specified above, leads to latenesses $\ell = [0, -1, -3, 1, -2, -2, 0, 0]$, so that $L_{\max} = 1$, which occurs for T_4. No other task is late. The results are summarized in Figure IV.14.

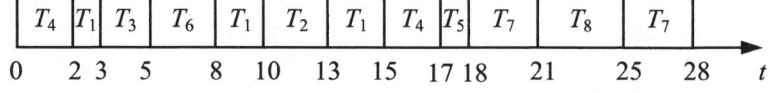

Figure IV. 14

Note that, even though it does not happen in our example, it is possible for idle times to occur.

The computational complexity of this algorithm is $O(n \log n)$. It can be generalized to cover the case of periodically arriving tasks, where each task has a deadline which is defined by the arrival of some subsequent task. Further details can be found in Liu and Layland (1973).

2.3.2 Other Problems

In this section we consider some scheduling problems for which the simple *EDD* rule is not powerful enough to find an optimal schedule. We first examine the $L_{\max}$ criterion and dependent tasks, i.e., the problem $1/prec/L_{\max}$.

Chapter 2: Single Machine Scheduling

In general, when considering dependent tasks and minimization of maximal lateness, one has to assign in some way modified due dates and then schedule tasks in usually nondecreasing order of these new due dates. This approach can be also used in case of problem $1/prec/L_{\max}$ for which the following algorithm has been proposed by Lawler (1973).

Algorithm for the $1/prec/L_{\max}$ Problem

Step 1: Compute modified due dates d_j^* for all tasks such that

$$d_j^* = \min\{d_j, \min_k\{d_k : T_k \succ T_j\}\},$$ i.e., the modified due date equals the earliest due date of the activity itself or any of its successors.

Step 2: Schedule tasks in nondecreasing order of d_j^*, $j=1,\ldots,n$ while observing precedence constraints. Ties are broken in favor of the task with the smaller subscript.

When analyzing the complexity of the above algorithm, we have to take into account that a general precedence graph may include up to $O(n^2)$ arcs. Thus, computing modified due dates will take $O(n^2)$ time. Since the other activities in the algorithm are less complicated, the overall computational complexity of the algorithm is $O(n^2)$. Lawler (1973) has demonstrated that the above algorithm can be generalized to cover the case of an arbitrary nondecreasing cost function. A numerical example for this algorithm will be given below the next algorithm.

There exists a certain similarity of the problem in the case of precedence related tasks and that of scheduling independent tasks. Hence, it is not surprising that introducing different ready times makes the problem $1/r_j, prec/L_{\max}$ **NP**-hard. On the other hand, a possibility of task preemption greatly simplifies the procedure of finding an optimal schedule. Blazewicz (1976) has shown that the problem $1/pmtn, r_j, prec/L_{\max}$ may be solved by combining an approach of Lawler and Moore (1973) for scheduling dependent tasks and an approach of Liu and Layland (1973) for scheduling preemptable tasks. The resulting procedure is summarized below.

Algorithm for the $1/pmtn, r_j, prec/L_{\max}$ Problem

Step 1: Assign modified due dates d_j^* to all tasks according the expression

$$d_j^* = \min\{d_j, \min_k\{d_k : T_k \succ T_j\}\},$$ i.e., the modified due date equals the earliest due date of the activity itself or any of its successors.

Step 2: Schedule available tasks in nondecreasing order of d_j^*, $j=1,...,n$, while observing precedence constraints, and preempting whenever a new task arrives with a smaller modified due date. Ties are broken in favor of the task with the smaller subscript.

Note the close links between the two algorithms above.

Example: Let $n = 10$, **p** = [5, 2, 1, 6, 2, 4, 5, 6, 1, 2], **r** = [0, 0, 10, 5, 12, 10, 3, 8, 9, 10], **d** = [20, 13, 12, 25, 11, 21, 15, 18, 24, 15], and precedence constraints as shown in Figure IV.15.

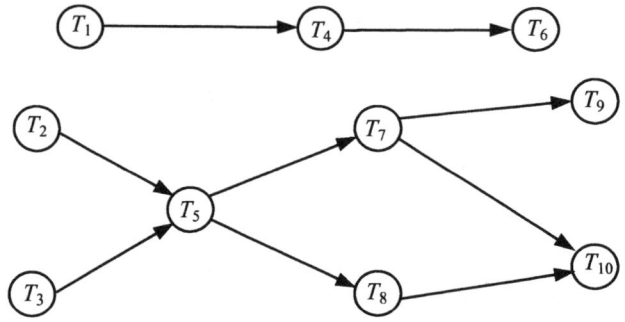

Figure IV.15

We can then compute the modified due dates **d*** = [20, 11, 11, 21, 11, 21, 15, 15, 24, 15]. Applying the algorithm results in the optimal schedule shown in Figure IV.16. Here, **c** = [7, 2, 11, 29, 14, 33, 19, 25, 34, 27], so that ℓ = [−13, −11, −1, 4, 3, 12, 4, 7, 10, 12] and $L_{max} = 12$, where the maximal lateness occurs for tasks T_6 and T_{10}.

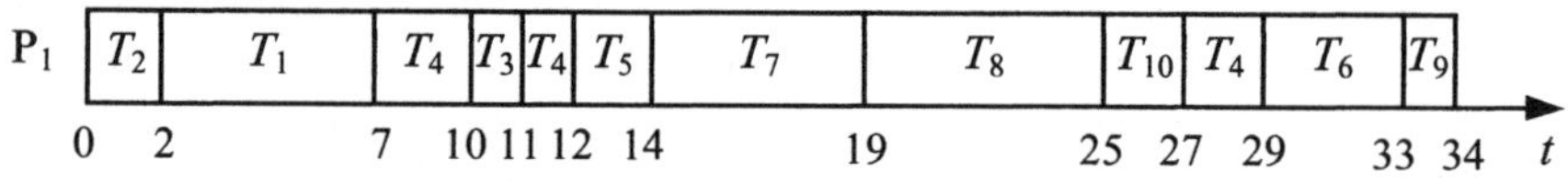

| P_1 | T_2 | T_1 | | T_4 | T_3 | T_4 | T_5 | T_7 | | T_8 | | T_{10} | T_4 | T_6 | T_9 |

0   2   7   10 11 12 14   19   25 27   29   33 34 t

Figure IV. 16

This same example may also serve to illustrate the algorithm stated above for the $1/prec/L_{max}$ problem, if we set the ready date vector **r** = **0** and do not allow preemption. We then obtain the optimal schedule displayed in Figure IV.17. The completion times are then **c** = [23, 2, 3, 29, 5, 33, 10, 16, 34, 18], so that the lateness is ℓ = [3, −11, −9, 4, −6, 12, −5, −2, 10, 3] and $L_{max} = 12$, which occurs for T_6.

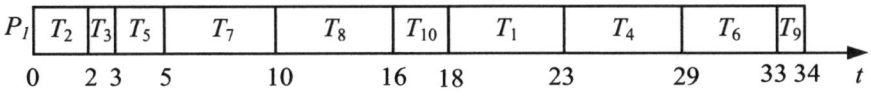

Figure IV. 17

A few results concerning other criteria should also be mentioned. If the objective is to minimize mean tardiness, then the problem $1//\Sigma t_j$ is solvable by a pseudopolynomial algorithm in $O(n^4\Sigma p_j)$ time as shown by Lawler (1977). This problem, however, was proved to be **NP**-hard by Du and Leung (1989b). If chain-like precedence constraints are introduced, then the problem becomes **NP**-hard in the strong sense even for unit processing times as shown by Leung and Young (1989).

The minimization of mean weighted tardiness appears to be a much harder problem. Lenstra et al. (1977) have shown that already $1//\Sigma w_j t_j$ is **NP**-hard in the strong sense. The only solvable problem in that context is $1/r_j, p_j = 1/\Sigma w_j t_j$ which may be reduced to a weighted bipartite matching problem; see Lawler et al. (1992).

As can be expected, there is a large number of enumerative approaches as well as dynamic programming procedures described in the literature. For further details, the reader is referred to Baker (1974). An extensive updated list of enumerative algorithms is given in Lawler et al. (1992).

CHAPTER 3 PARALLEL MACHINE MODELS

This chapter is concerned with the analysis of scheduling problems in a parallel machine environment. As in the previous chapter, the three objectives to be considered are schedule length (makespan), mean flow time, and maximum lateness, respectively. Results for the corresponding problems are presented in the following three sections.

3.1 Minimizing Makespan

In this section we consider several different scheduling models where the objective is to minimize the schedule length. Since the general problem $P//C_{max}$ is **NP**-hard, approximation methods become important. We will describe exact algorithms for special cases that can be solved in polynomial time, and some approximation methods for general problems.

3.1.1 Identical Machines and Tasks of Arbitrary Lengths

This section first analyzes the problem $P//C_{max}$ whose goal is to schedule independent tasks with given lengths on identical machines. This basic scheduling model involving parallel machines is probably one of the most thoroughly analyzed problems in the field. We first analyze the problem's computational complexity. The problem appears to be difficult, since already a problem with only two machines is **NP**-hard, as Karp (1972) demonstrated:

Proposition IV.6: Problem $P2//C_{max}$ is **NP**-hard.

Proof: We prove the proposition by reducing the decision version of the well-known **NP**-complete problem PARTITION (see, e.g., Karp, 1972) to our problem in order to establish the result. PARTITION can be formulated as follows:

Instance: Given a finite set A and a size $s(a_i) \in \mathbb{N}_0$ for each $a_i \in A$.
Question: Is there a subset $A' \subseteq A$ such that $\sum_{a_i \in A'} s(a_i) = \sum_{a_i \in A \setminus A'} s(a_i)$?

Given any instance of PARTITION defined by the set of positive integers $\{s(a_i): a_i \in A\}$, define a corresponding instance of the decision counterpart of $P2//C_{max}$ by setting $n = |A|$, $p_i = s(a_i)$, $j = 1, 2,..., n$, and the threshold value of schedule length $y = \frac{1}{2} \sum_{a_i \in A} s(a_i)$. It follows that there exists a subset A' with the desired property for the given instance of PARTITION if, for the corresponding instance of $P2//C_{max}$, there exists a schedule with $C_{max} \leq y$ and the proposition follows. □

Since there is no hope of finding a polynomial-time algorithm for $P//C_{max}$, one may try to solve the problem along the lines presented in Section 1.3 of this part. First, we may relax some constraints imposed on problem $P//C_{max}$ and allow preemptions of tasks. It appears that problem $P/pmtn/C_{max}$ can be solved very efficiently. McNaughton (1959) has shown that the length of a preemptive schedule cannot be smaller than the maximum of the longest task processing time and the mean processing requirement on a machine and that this bound can actually be achieved, i.e.,

$$C_{max} \geq C^*_{max} = \max\left\{\max_j \{p_j\}; \frac{1}{m}\sum_{j=1}^{n} p_j\right\}$$

The following algorithm is due to McNaughton (1959) and has been referred to as *McNaughton's wrap-around rule*. It constructs a schedule whose length is equal to C^*_{max}.

McNaughton's Algorithm for $P/pmtn/C_{max}$

Step 1: Calculate $C^*_{max} = \max\left\{\max_j \{p_j\}; \frac{1}{m}\sum_{j=1}^{n} p_j\right\}$ and schedule the tasks in any arbitrary order.

Step 2: Break the sequence of tasks at $i C^*_{max}$, $i = 1, ..., m-1$.

Step 3: Schedule all tasks that are processed in the interval $[(i-1) C^*_{max}; i C^*_{max}]$ on processor P_i, $i = 1, ..., m-1$.

Step 4: Any task that has been preempted and is processed on, say, P_i and P_{i+1}, will then logically have to processed first on P_{i+1}, before the processing continues on P_i.

This algorithm always finds a schedule whose length is equal to C^*_{max}. Its complexity is $O(n)$. Since in Step 3 the remaining part of a preempted task will start at time $t = 0$, we will obviously need to sort the preempted parts of a task so that they are processed in chronological order; this is accomplished in Step 4 of the algorithm.

Example 1: As a numerical illustration, consider $m = 2$ processors P_1 and P_2 and $n = 8$ tasks $T_1, ..., T_8$ with processing times $\mathbf{p} = [p_1, ..., p_8] = [3, 4, 2, 4, 4, 2, 13, 2]$ hours. As $\max \{p_j\} = 13$ and $\sum_{j=1}^{8} p_j = 34$, $C^*_{max} = 17$ follows. Scheduling the tasks in the order $T_1,..., T_8$ on the two processors beginning with P_1, we obtain the optimal schedule shown in Figure IV.18, where only T_3 is preempted.

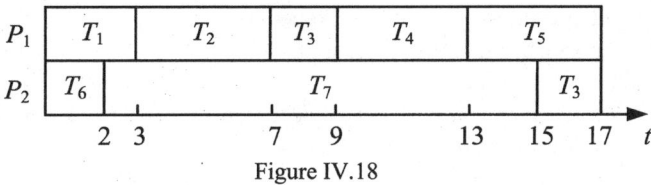

Figure IV.18

Example 2: Let there be $m = 3$ processors, P_1, P_2, P_3, available for the same $n = 8$ tasks, as in Example 1. We then obtain $C^*_{max} = \max\{13, 34/3\} = 13$, and the resulting schedule will preempt task T_7 at time $t = 13$ on processor P_2 and continue with T_7 on P_3 at $t = 0$, as indicated in Figure IV.19.

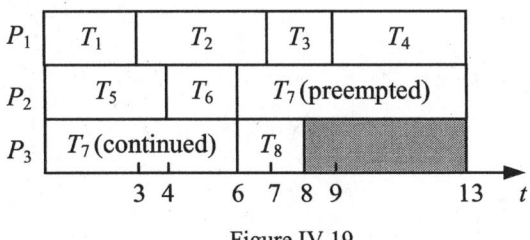

Figure IV.19

From Figure IV.19 it is apparent that the practical implementation of the schedule involves starting the task T_7 at time $t = 0$ on processor P_3, continuing processing until it is finished, while task T_8 follows task T_6 on processor P_2.

The above examples demonstrate that allowing preemptions has made the problem easy. However, there still remains the question of the practical applicability of such a solution. For instance, it appears that in multiprocessor computer systems with a common primary memory, the assumptions of task preemptions are justifiable and the corresponding preemptive schedules can be used in practice. If this is not the case, one may try to find an approximation algorithm for the original problem and evaluate its worst case as well as its mean behavior. We will present such an analysis below.

One of the most frequently used general approximation strategies for solving scheduling problems is known as *list scheduling*. This scheduling tool was first introduced by Graham (1966). It includes a *priority list* of the tasks, and at each step the first available machine is selected to process the first available task on the list. Clearly, the outcome of a given list schedule depends on the order in which tasks appear on the priority list. Unfortunately, list scheduling may result in unexpected behavior of constructed schedules. For instance, the schedule length for problem $P/prec/C_{max}$ with arbitrary precedence constraints may increase if

- the priority list changes,
- task processing times decrease,
- precedence constraints are weakened, or
- the number of machines increases.

These so-called *Graham anomalies* are discussed in Graham (1966), and may be explained in the following numerical

Example: Consider again the problem of the previous example with $n = 8$ tasks, m = 2 processors, and processing times **p** = [3, 4, 2, 4, 4, 2, 13, 2] for the tasks T_1, T_2, ..., T_8. Suppose that the priority list includes the tasks in the ordered sequence $L = (T_1, T_2, ..., T_8)$. Given the precedence relationships shown in Figure IV.20, the resulting list schedule is displayed in Figure IV.21, where $C^*_{max} = 17$. If preemption were allowed, T_7 would have been preempted at $t = 5$ and T_5 scheduled on processor P_2.

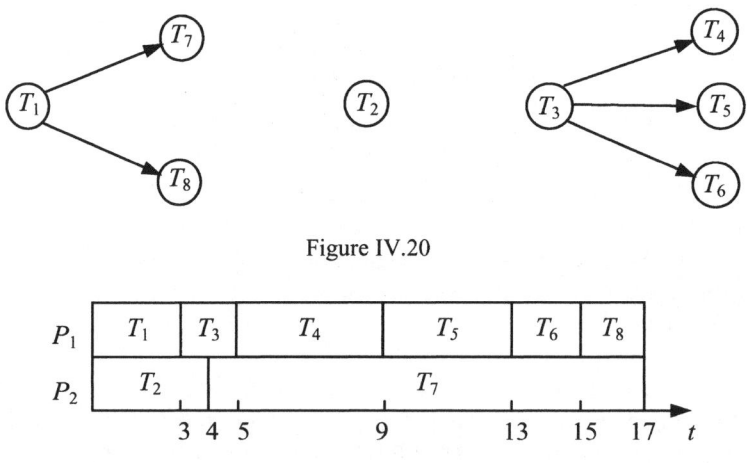

Figure IV.20

Figure IV.21

Change now the priority list to $L' = (T_1, T_2, T_3, T_8, T_4, T_5, T_6, T_7)$. The result is shown in Figure IV.22 with a significant imbalance between the two processors and a sizeable increase in schedule length to $C'_{max} = 23$.

In Figures IV.22 through Figure IV.25, the shaded area indicates idle processors.

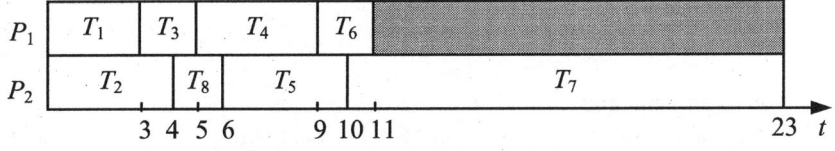

Figure IV.22

Suppose now that the processing times of all tasks in the original problem were to decrease by one, i.e., $p'_j = p_j - 1 \ \forall \ j = 1,...,8$. The resulting list schedule is shown in Figure IV.23. Not only is the schedule quite unbalanced, but despite the reduction of task durations, the schedule length has increased to $C'_{max} = 18$, one hour longer than the original schedule.

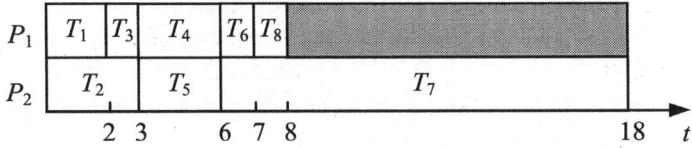

Figure IV.23

Back in the original problem, we now weaken the precedence constraints by deleting the requirement that T_4 must follow T_3. Surprisingly enough, the schedule length increases from the original $C_{max} = 17$ to $C'_{max} = 22$ as shown in Figure IV.24.

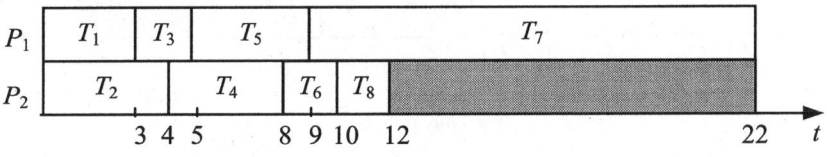

Figure IV.24

Finally, consider again the original problem, but with $m = 3$ rather than the original $m = 2$ processors. The schedule for this problem is shown in Figure IV.25. Note that the schedule length has increased from the original $C_{max} = 17$ to $C'_{max} = 19$, even though a processor was added.

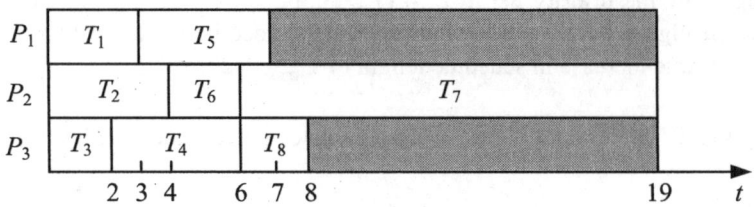

Figure IV.25

The above surprising and counterintuitive results were discovered by Graham (1966). They are caused by the heuristic list scheduling rule which is not guaranteed to find an optimal solution. Graham also determined the absolute performance ratio R_{LS}, defined in Section 1.3, applied to the problem $P//C_{max}$. The result is summarized in

Proposition IV.7: An arbitrary list scheduling algorithm LS applied to problem $P//C_{max}$ with m machines has the absolute performance ratio $R_{LS} = 2 - \frac{1}{m}$.

Recall that $R_{LS} \geq 1$ must always be satisfied, and that, if $R_{LS} = 1$, the list scheduling algorithm will be optimal. For the proof of this theorem, we refer to Graham (1966). To illustrate that this bound is achievable, consider an example with n tasks, in which $n = (m-1)m + 1$, $\mathbf{p} = [1, 1,..., 1, 1, m]$, and $L = (T_1, T_2,..., T_n)$. The optimal and list schedules for $m = 4$ are shown in Figure IV.26a and b, respectively, where $R_{LS} = 1\frac{3}{4}$.

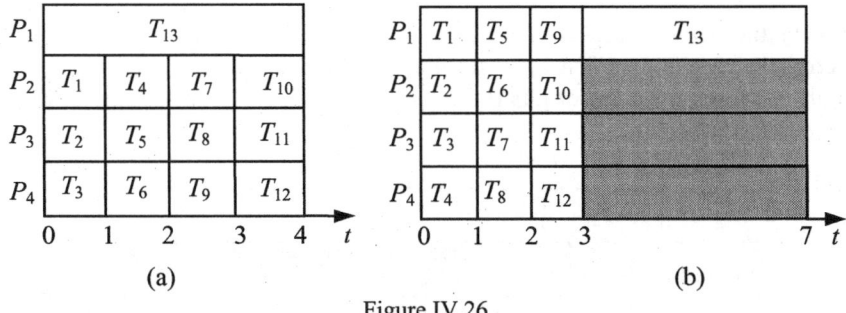

Figure IV.26

It follows from the above discussion that an arbitrary list scheduling algorithm may generate schedules that are quite far from optimality. In the worst case, such a schedule takes almost twice as long as an optimal one. An improvement can be obtained if the task list is suitably ordered. A variety of ordering possibilities exist. The simplest method is the *longest processing time first* (*LPT*) list scheduling algorithm. As its name suggests, it orders tasks on the list in order of nonincreasing p_j. Formally, the method can be stated as follows.

The *LPT* Algorithm for $P//C_{max}$

Step 1: Order tasks in a list in nonincreasing order of their processing times.

Step 2: Whenever a machine becomes free, assign to it the first task on the list that is not yet assigned.

Step 3: Have all tasks been assigned?
If yes: Stop, a schedule has been obtained.
If no : Go to Step 1.

Graham (1969) analyzed the worst-case behavior of the *LPT* algorithm. His result is summarized in

Proposition IV.8: The *LPT* algorithm applied to problem $P//C_{max}$ has the absolute performance ratio

$$R_{LPT} = \frac{4}{3} - \frac{1}{3m}.$$

Example: To show that the lower bound in Proposition IV.8 can be achieved, let $n = 2m + 1$, and $\mathbf{p} = [2m - 1, 2m - 1, 2m - 2, 2m - 2,..., m + 1, m + 1, m, m, m]$. Figure IV.27 shows an optimal and an *LPT* schedule for $m = 3$, achieving the worst-case bound of $R_{LPT} = \frac{4}{3} - \frac{1}{3(3)} = 11/9$.

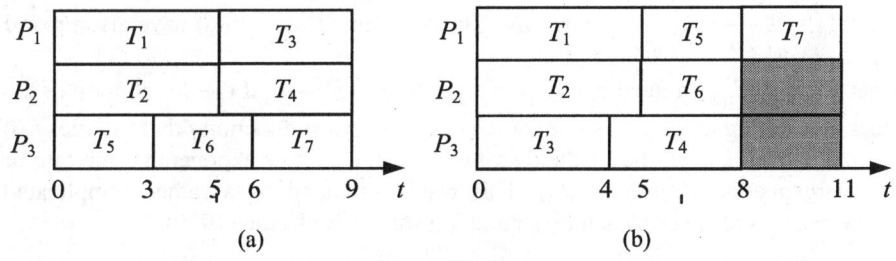

Figure IV. 27

Proposition IV.8 implies that for many machines, i.e., a large value of m, an *LPT* schedule can be, in the worst case, almost one third longer than an optimal one. However, we can expect better performance from the *LPT* algorithm than the theorem indicates, especially when the number of tasks, i.e., the value of n, becomes large. Coffman and Sethi (1976) derived another absolute performance ratio for the *LPT* rule, taking into account the least number k of tasks on any machine.

Proposition IV.9: $R_{LPT}(k) \leq 1 + \dfrac{1}{k} - \dfrac{1}{mk}$.

This result shows that the worst-case performance bound for the *LPT* algorithm for large m approaches $1 + 1/k$.

On the other hand, one may be interested in how good the *LPT* algorithm is on average. Coffman *et al.* (1984) have obtained a pertinent result in which the relative error was found for $P2//C_{max}$, i.e., the problem with two machines, given the assumption that all task processing times p_j are independent samples from the uniform distribution $R\,[0, 1]$.

Proposition IV.10: The expected makespan of the *LPT* algorithm $E(C_{max}^{LPT})$ is bounded by $\dfrac{n}{4} + \dfrac{1}{4(n+1)} \leq E(C_{max}^{LPT}) \leq \dfrac{n}{4} + \dfrac{e}{2(n+1)}$, where $e \approx 2.718$ is the basis of the natural logarithms.

Since $E(p_j) = \frac{1}{2}$ for $j = 1, 2, \ldots, n$ and there are $m = 2$ machines, so that $n/4$ is a lower bound on $E(C_{max}^*)$ we obtain $\dfrac{E(C_{max}^{LPT})}{E(C_{max}^*)} < 1 + O\!\left(\dfrac{1}{n^2}\right)$. Therefore, as the number of tasks, i.e., n, increases, $E(C_{max}^{LPT})$ approaches the optimum no more slowly than $1 + O(1/n^2)$ approaches 1. The above bound can be generalized to cover the case of m machines, for which Coffman *et al.* (1983) have proved that $E(C_{max}^{LPT}) \leq \dfrac{n}{2m} + O\!\left(\dfrac{m}{n}\right)$. Frenk and Rinnooy Kan (1984, 1986) have also proved that $C_{max}^{LPT} - C_{max}^*$ almost surely converges to 0 as $n \to \infty$, if the distribution of the task processing times p_j has a finite mean and a density function f that satisfies $f(0) > 0$. It is also shown that if the distribution is uniform or exponential, the rate of convergence is $O(\log(\log n)/n)$. This result, obtained by a rather complicated analysis, was also confirmed by simulation studies by Kedia (1970).

In conclusion, the *LPT* algorithm behaves quite well, making it useful for practical applications. However, if better performance guarantees are required, other approximation algorithms should be used, e.g., *MULTIFIT* (developed by Coffman *et al.*, 1978), the algorithms proposed by Hochbaum and Shmoys (1987), or the methods by Karmarkar and Karp (1987). A comprehensive treatment of approximation algorithms for this and related problems is provided by Coffman *et* al. (1984); see also Coffman and Lueker (1991), Coffman and Whitt (1995), and Lawler et al. (1993).

Above, we have analyzed the problem $P//C_{max}$ by relaxing it (allowing preemption), then by constructing approximation algorithms and determining their performance ratios. Even though Proposition IV.6 gave a negative answer to the question of the existence of a polynomial-time algorithm for solving $P2//C_{max}$, we have not proved that $P//C_{max}$ is **NP**-hard in the strong sense and could therefore try to find a pseudopolynomial optimization algorithm. Using ideas presented by Rothkopf (1966), a dynamic programming approach was developed, solving $P//C_{max}$ in $O(nC^m)$ time, where C denotes an upper bound on C_{max}^*. For details, readers are referred to Blazewicz (1987). A survey of some other enumerative approaches for the problem in question can be found in Lawler et al. (1993).

3.1.2 Other Algorithms for Identical Machines

Consider now the case of dependent tasks, i.e., problems with precedence constraints. First assume that tasks are to be scheduled nonpreemptively. Since the problem $P//C_{max}$ is already **NP**-hard, there is no hope of finding a polynomial-time optimization algorithm for scheduling tasks of arbitrary length. However, one may try to find a polynomial algorithm for the special case in which the processing times of all tasks equal one. The first algorithm for this type of problem was developed by Hu (1961) for scheduling problems with forests as precedence graphs, which are either in-trees or out-trees. We will first present Hu's algorithm for the case of an in-tree structure, as defined in Section 2.2.2.

The algorithm is based on the concept of a *task level* in an in-tree. A node T_j is said to be at level $k+1$, if there are k arcs on the unique path from T_j to the root of the tree; therefore the root node is at level 1. As it makes extensive use of levels, this method is sometimes called a *level algorithm*. Hu's algorithm is initialized by calculating the levels of all tasks and setting the task starting time at $t := 0$. It can be described as follows.

Hu's Level Algorithm for the P/in-tree, $p_j = 1/C_{max}$ Problem

---

Procedure: Starting with the highest level, assign tasks on that level in arbitrary order without predecessors at time t. Whenever a task is assigned, remove its node and incident arcs. If all tasks on the current level are assigned, move to the next level and repeat; otherwise, set $t := t + 1$ and continue with the present level.

This algorithm can be implemented to run in $O(n)$ time.

Example: Consider a problem with $m = 3$ processors and $n = 12$ tasks, all with processing times $p_j = 1, j = 1, \ldots, 12$, that are connected by precedence relations as shown in Figure IV.28. Task T_{12} is the root and the only task at level 1, tasks T_{10} and T_{11} make up level 2, etc. The levels in the figure are circled by broken lines. The optimal schedule found by the algorithm is shown in Figure IV.29. Note that the optimal schedule length would not change if, for example, task T_1, or tasks T_2 and T_5 were omitted from the problem.

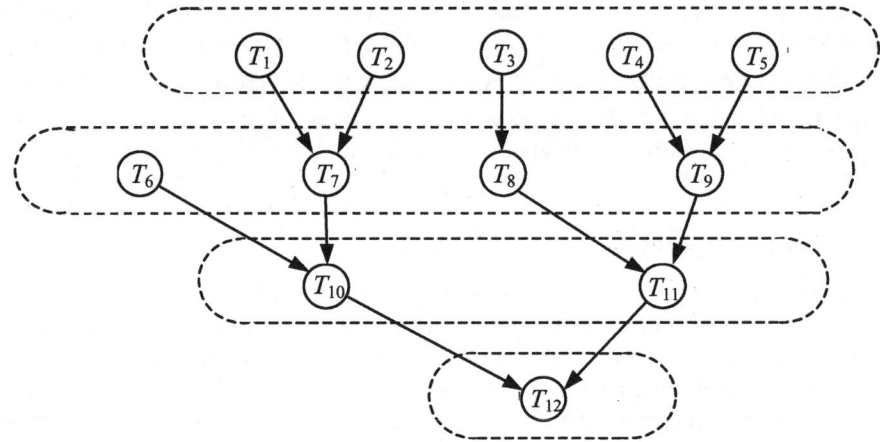

Figure IV.28

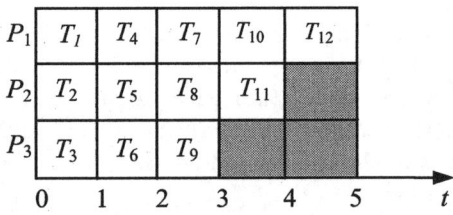

Figure IV.29

A forest consisting of in-trees can be scheduled by adding a dummy task that is an immediate successor of all the roots of the in-trees, and then applying the above algorithm. A schedule for an out-tree (defined in Section 2.2.2) can also be constructed easily by reversing the direction of all arcs, applying the above algorithm to the resulting in-tree and then again reversing the schedule obtained. It is interesting to note that the problem of scheduling opposing forests, i.e., combinations of in-trees and out-trees on an arbitrary number of machines, is **NP-hard**, as Garey *et al.* (1983) have proved.

Chapter 3: Parallel Machine Models   377

However, when the number of machines is limited to two, Coffman and Graham (1972), Fuji *et al.* (1969), and Gabow (1982) have shown that the problem is easily solvable, even for arbitrary precedence graphs. Here, we present the algorithm developed by Coffman and Graham (1972), which can also be extended to cover the preemptive case. The algorithm assigns unique labels to tasks, taking into account the levels of the tasks and the numbers of their immediate successors. The labels are then used to find the shortest schedule where $\mathcal{S}(T_j)$ denotes the set of all immediate successors of task T_j.

Coffman and Graham's Algorithm for $P2/prec, p_j = 1/C_{max}$

Step 1: Assign the label "1" to any individual task T_j with $\mathcal{S}(T_j) = \emptyset$.

Step 2: Suppose that the labels 1, 2,..., $k-1$ have already been assigned. Define S as the set of unlabeled tasks, all of whose successors either do not exist or are labeled. For each task $T_j \in S$, let $\ell(T_j)$ denote a list of labels of tasks that belong to $\mathcal{S}(T_j)$, ordered in nondecreasing order of their values. Let T^* be an element of S, such that $\ell(T^*)$ is lexicographically smaller than $\ell(T_j) \; \forall \; T_j \in S$. Assign label k to T^* and continue assigning labels to all elements in S.

Step 3: Are the nodes of all tasks labeled?
If yes: Go to Step 4.
If no: Go to Step 2.

Step 4: Assign tasks to machines following the procedure in Hu's algorithm, using labels instead of levels.

One can show that the above algorithm can be implemented with a time-complexity which is almost linear in n plus the number of arcs in the precedence graph. This leads to a complexity of $O(n^2)$, if the graph has no transitive arcs; see Sethi (1976). Aho *et al.* (1974) showed that if transitive arcs exist, they can be deleted in $O(n^{2.8})$ time.

Example: Coffman and Graham's algorithm is applied to the precedence graph in Figure IV.30. In Step 1, task T_{13} is arbitrarily assigned the label 1, since $\mathcal{S}(T_{13}) = \emptyset$, and in Step 2, we then have $S = \{T_{10}, T_{11}, T_{12}\}$. With $\mathcal{S}(T_{10}) = \{T_{13}\}$, $\mathcal{S}(T_{11}) = \mathcal{S}(T_{12}) = \emptyset$, we obtain $\ell(T_{10}) = \{1\}$, $\ell(T_{11}) = \ell(T_{12}) = \emptyset$, so that T_{11} and T_{12} arbitrarily receive the labels 3 and 2, respectively, and T_{10} receives the label 4. Now $S = \{T_7, T_8\}$, and with $\mathcal{S}(T_7) = \{T_{11}\}$, $\mathcal{S}(T_8) = \{T_{11}, T_{12}\}$, we find $\ell(T_7) = \{3\}$ and $\ell(T_8) = \{3, 2\}$, so that T_7 receives the label 5, and T_8 is assigned label 6. The procedure continues in this manner. The resulting list of labels for the tasks

$[T_1, T_2, \ldots, T_{13}]$ becomes [11, 12, 13, 8, 9, 10, 5, 6, 7, 4, 3, 2, 1], and the schedule obtained is displayed in Figure IV.31. Note that task T_8 is scheduled after task T_7, even though it has a higher label. This is due to the precedence relationship between T_8 and T_9.

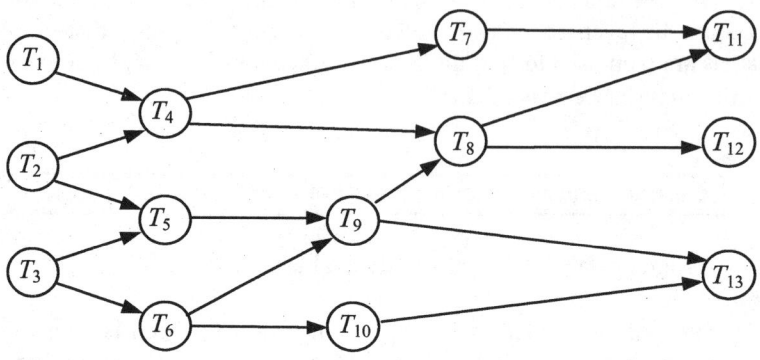

Figure IV.30

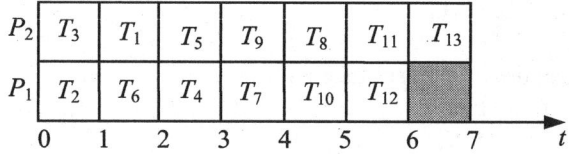

Figure IV.31

It must be stressed that the question concerning the complexity of the problem with a given number of machines, tasks with unit processing times, and arbitrary precedence graphs is still open, despite the fact that much effort has been devoted to solving various special cases; see, e.g., Lenstra and Rinnooy Kan (1984). On the other hand, several authors have dealt with approximation algorithms for these and more complicated problems. Below we quote some of the most interesting results. The application of the level or critical path algorithm applied to problem $P/\text{prec}, p_i = 1/C_{\max}$ has been analyzed by Chen and Liu (1975) and Kunde (1976), who have proved the following performance ratio bound for m machines.

$$R = \begin{cases} 4/3 & \text{for } m = 2 \\ 2 - \dfrac{1}{m-1} & \text{for } m \geq 3 \end{cases}$$

Coffman and Graham's algorithm performs slightly better. Lam and Sethi (1977) have proved that for this algorithm $R = 2 - \frac{2}{m} - \frac{m-3}{mC_{\max}^*}$ for $m \geq 3$. In this context, we refer to the results presented in Section 3.1.1, where Graham's list scheduling anomalies were analyzed. That analysis showed that preemptions can be profitable with respect to two factors. Firstly, they can make problems easier to solve, and secondly, they can shorten the schedule. In fact, Muntz and Coffman (1969, 1970) have shown that these two factors apply in the case of dependent tasks scheduled on machines in order to minimize schedule length. In particular, it is possible to construct an optimal preemptive schedule for tasks of arbitrary length and with other parameters the same as in Hu's algorithm or Coffman and Graham's algorithm. The approach again uses a precedence graph, in which a common sink and dummy arcs have been introduced to the sink from all nodes that have zero outdegree. In contrast to the previous definition, the level of a task T_j is here defined as the sum of all processing times of tasks along a longest path between (and including) T_j and a terminal task, i.e., a task with no successors. Hence, the level of a task indicates the time left until the completion of the schedule.

Note that as time progresses, the level of a task that is currently being executed is decreasing. Furthermore, processor sharing, as discussed in Section 1.2.1 will be used. The procedure can then be summarized as follows.

Muntz and Coffman's Algorithm for $P2/pmtn, prec/C_{\max}$ and $P/pmtn, forest/C_{\max}$

Step 1: Compute the levels of all tasks in the precedence graph, where the level of a task T_j is the length of the longest path from T_j to the root of the tree with respect to the processing times, including p_j.

Step 2: Suppose that h machines and g tasks are available at the highest level that has not yet been considered. Assign tasks at that level to available machines as follows. If $g > h$, assign a fraction $\beta = h/g$ of a machine to each of the g tasks, thus obtaining a fictitious processor-shared schedule. Otherwise assign one task to each machine. If there are any machines left, consider the tasks at the next highest level, observing precedence constraints; otherwise go to Step 3. Repeat as long as there are levels that have not yet been considered.

Step 3: Process the assigned tasks until either a task is finished, or a point is reached at which continuing with the present assignment would mean that a task at a lower level will be executed at a faster rate than a task at a higher level. Preempt all tasks, recalculate the levels, and go to Step 2.

In order to convert the fictitious processor-shared schedule to an optimal schedule, proceed as follows.

> **Converting a Processor-Shared Schedule**

Procedure: For each interval, in which processor sharing occurs, determine the interval length ℓ and calculate the processing time of each task involved in this interval as $\ell\beta$. Schedule the tasks with these processing in the interval with McNaughton's algorithm.

The above algorithm first produces a (possibly) processor-shared schedule, which, by the above procedure, is then converted into a proper optimal schedule. It can be implemented to run in $O(n^2)$ time.

Example: Consider an instance of the problem $P2/pmtn, prec/C_{max}$ with $n = 13$ tasks whose processing times are given by $\mathbf{p} = [3, 2, 4, 1, 5, 6, 2, 4, 3, 5, 4, 3, 3]$, and task precedence constraints as are shown in Figure IV.32.

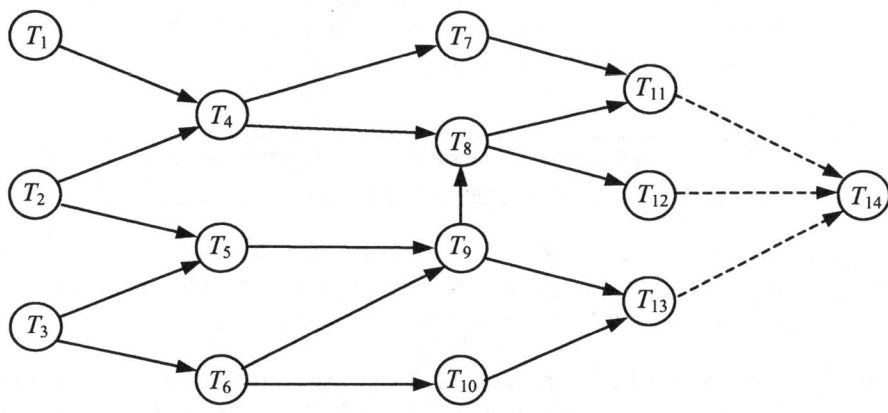

Figure IV.32

First compute the levels of all tasks. By definition they equal the length of the longest paths for a given task to the terminal task. In our example, the longest path from T_1 to the terminal task T_{14} is $(T_1, T_4, T_8, T_{11}, T_{14})$ with length 12. Repeating the procedure for all tasks, we assign the following levels to them: 12, 18, 21, 9, 16, 17, 6, 8, 11, 8, 4, 3, and 3. Continuing with Step 2 of the algorithm, the schedule commences with the task on the highest level, that is, T_3 with a level of 21, on the first machine P_1 at time $t = 0$, which is possible, as T_3 has no predecessors. The task on the next highest level of 18 is T_2, which also has no predecessors, so it is scheduled on the second processor at time $t = 0$. In Step 3, the first event in the processing is that T_2 is finished at time $t = 2$. At this point, we will attempt to schedule the task on the next highest level, which is task T_6 at level 17. However, in order to process T_6 task T_3 must be completed, a precedence condition that is not satisfied at this point. The task on the next level down, i.e., 16, is T_5 which requires T_2 and T_3 to be completed; again, this is not yet satisfied. On the next level further down at 12, is T_1, a task without predecessors. Consequently, T_1 is scheduled right after T_2 is finished. Scheduling continues in this manner. The fictitious processor-shared and resulting optimal schedules are displayed in Figures IV.33 and IV.34, respectively.

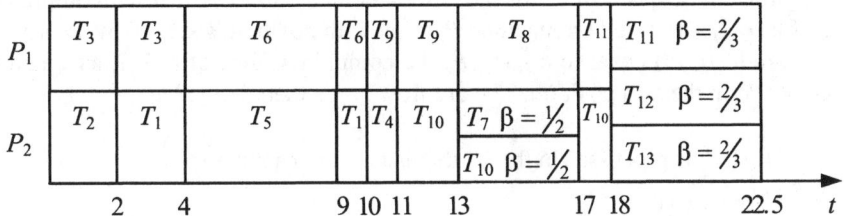

Figure IV.33

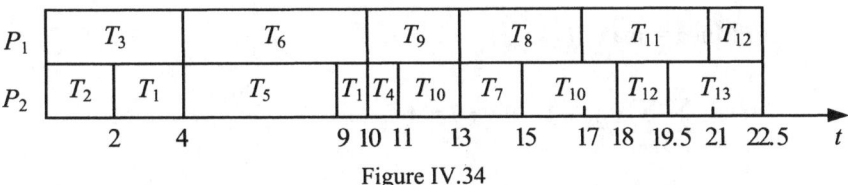

Figure IV.34

At this point we should also mention that there exists another structure of the precedence graph that makes it possible to solve a $Pm/pmtn, uan/C_{max}$ scheduling problem in polynomial time. However, Ullman (1976) has shown that the case of general precedence graphs results in **NP**-hardness of the scheduling problem. Lam and Sethi (1977) have proved that the worst-case behavior of the Muntz and Coffman algorithm for the problem $P/pmtn, prec/C_{max}$ yields a performance ratio of $R = 2 - \frac{2}{m}$ for $m \geq 2$ machines.

3.1.3 Algorithms for Uniform and Unrelated Machines

In this section, we first analyze the case of independent tasks, uniform machines, and nonpreemptive scheduling. In other words, while each task requires its individual processing time and each machine has its individual speed that applies to all tasks, all machines perform the same function. Since the scheduling problem with arbitrary processing times is already **NP**-hard for identical machines, all we can hope to find is a polynomial-time optimization algorithm for tasks with standard processing times of length one. Such an algorithm was described by Graham et al. (1979), where a transportation network approach was employed to solve the problem $Q/p_j = 1/C_{\max}$. The technique is outlined below.

Let there be n sources j, $j = 1, 2,..., n$ and mn sinks (i,k), $i = 1, 2,..., m$; $k = 1, 2,..., n$ where sources correspond to tasks and sinks to machines and positions on them. A task is said to be in the *k-th position* on a machine if it is the k-th task scheduled on this machine. With the standard processing time being $p_j = 1$ for all tasks and with a processing speed of b_i for the i-th machine, it follows that the completion time of task T_j scheduled on machine P_i in the k-th position is k/b_i. Now we set the cost of arc $(j, (i, k))$ equal to $c_{j,ik} = k/b_i$, the completion time of task T_j assigned to processor P_i in the k-th position. The arc flow $x_{j,ik}$ is then defined as

$$x_{j,ik} = \begin{cases} 1, & \text{if } T_j \text{ is processed in the } k\text{-th position on machine } P_i \\ 0 & \text{otherwise} \end{cases}.$$

The scheduling problem can now be modeled as a minimax transportation problem as follows.

$$P: \text{Min } z = \max_{i,j,k} \{c_{j,ik} x_{j,ik}\}$$

$$\text{s.t.} \sum_{i=1}^{m} \sum_{k=1}^{n} x_{j,ik} = 1 \quad \forall j = 1, ..., n$$

$$\sum_{j=1}^{n} x_{j,ik} \leq 1 \quad \forall i = 1, ..., m; k = 1, ..., n$$

$$x_{j,ik} \geq 0 \quad \forall i = 1, ..., m; j = 1, ..., n; k = 1, ..., n.$$

The constraints are of the usual types for the classical transportation problem. The first set ensures that each task is scheduled on exactly one machine in exactly one position, and the second set ensures that no position on any machine is assigned to more than one task. It is well known that this problem can be solved in less than $O(n^3)$ time.

Many other problems in the area of nonpreemptive scheduling of independent tasks are **NP**-hard, so it is not surprising that much work has been done in the search for good heuristics. One such method, a list scheduling algorithm, has been proposed by Liu and Liu (1974a). Here, tasks are ordered on the list in nonincreasing order of their processing times and machines are ordered in nonincreasing order of their processing speeds. Now, whenever a machine becomes free, the first nonassigned task on the list is scheduled on it. In case there is more than one free machine, the fastest one is chosen. The worst-case behavior of the algorithm has been evaluated for the case of $(m + 1)$ machines in the system, m of which have processing speed factors b_i equal to 1 for $i = 1, ..., m$, and the remaining $(m + 1)$st machine has the processing speed factor equal to b. Then it can be shown that the performance ratio $\bar{R}$ will be given by

$$R = \begin{cases} \dfrac{2(m+b)}{b+2} & \text{for } b \leq 2 \\ \dfrac{m+b}{2} & \text{for } b > 2 \end{cases}.$$

The above relation shows that the performance of the algorithm improves as b and m decrease. Other algorithms have been analyzed by Liu and Liu (1974b, c) and by Gonzalez et al. (1977).

By allowing preemptions, it becomes possible to find optimal schedules in polynomial time. More specifically, the problem $Q/pmtn/C_{max}$ can be solved in polynomial time. Here we present the algorithm given by Horvath et al. (1977) even though there is a more efficient technique by Gonzalez and Sahni (1978b). The reason is that the former algorithm allows us to include more general precedence constraints than the latter, and it also generalizes the ideas presented in the Muntz and Coffman algorithm of the previous section. It is based on two concepts:

- the task level, which we define now as the remaining processing requirement of a task, expressed in terms of a standard processing time

- processor sharing.

Suppose that the tasks have been sorted in order of nonincreasing p_j values and processors in order of nonincreasing speed factors b_i. Clearly, no schedule can be shorter than processing the longest job on the fastest machine, which has a duration of p_1/b_1. Similarly, no schedule can be shorter than processing the two longest jobs on the two fastest machines, which requires a time of $(p_1 + p_2)/(b_1 + b_2)$, and so on. This leads to

$$C^*_{max} \geq C = \max \left\{ \frac{p_1}{b_1}, \frac{p_1+p_2}{b_1+b_2}, ..., \frac{\sum_{j=1}^{m-1} p_j}{\sum_{i=1}^{m-1} b_i}, \frac{\sum_{j=1}^{n} p_j}{\sum_{i=1}^{m} b_i} \right\}$$

The following algorithm constructs a schedule with the length equal to the above value of C and may be presented as follows. Note the similarity with Muntz and Coffman's algorithm in the previous section.

Horvath, Lam, and Sethi Algorithm for $Q/pmtn/C_{max}$

Step 1: Let h be the number of available machines and g be the number of tasks at the highest level. If $g \leq h$, assign g tasks to be executed at the same rate on the g fastest machines. Otherwise, assign the g tasks to the h machines, executing at the same rate, thus obtaining a fictitious processor-shared schedule. If there are any machines left, consider the tasks at the next lower level.

Step 2: Preempt the processing and repeat Step 1 whenever either a task is finished or a point is reached at which continuing with the present assignment means that a task at a lower level is being executed at a faster rate than a task at a higher level.

Step 3: Between each pair of successive reassignment points, reschedule the tasks if necessary, i.e., if $g \geq h$, denoting the length of the time interval by y. If $g = h$, assign each task to each processor for y/g time units. Otherwise, go to Step 4.

Step 4: With $g > h$ (more tasks than machines), let p denote the (equal) processing requirement of each task within the current assignment interval, and let b be the processing speed factor of the slowest of the h machines. Is $p/b < y$?

    If yes: Apply the "Converting a Processor-Shared Schedule" procedure, disregarding different machine speeds.

    If no: Divide the current assignment interval into g equal subintervals. Assign the g tasks in such a way that each task occurs in exactly h ($< g$) intervals, each time on a different machine, and go to Step 3.

The complexity of the Horvath, Lam, and Sethi algorithm is $O(mn^2)$.

Example: Consider a problem with $n = 6$ tasks, $m = 2$ processors, processing speed factors $\mathbf{b} = [4, 1]$, and processing times $\mathbf{p} = [20, 24, 10, 12, 5, 4]$. The processor-shared schedule resulting from Steps 1 and 2 in the above algorithm is displayed in Figure IV.35 and the preemptive schedule is shown in Figure IV.36.

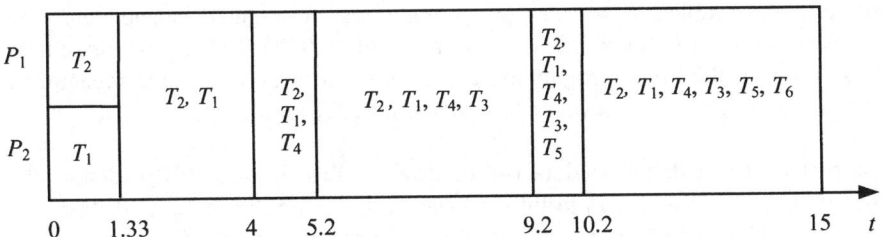

Figure IV.35

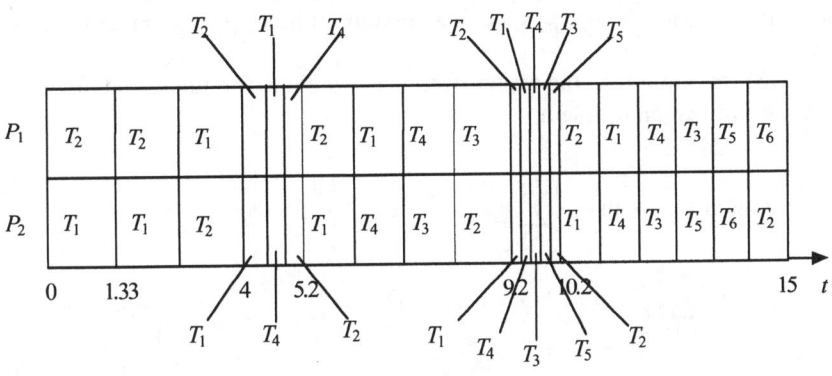

Figure IV.36

When considering dependent tasks, exact optimization algorithms exist only for preemptive scheduling problems. It should be pointed out that the Horvath, Lam, and Sethi algorithm also solves the problem $Q2/pmtn, prec/C_{max}$, i.e., with two uniform machines. Here, the level of a task is defined as in the Muntz and Coffman algorithm, but using standard processing times for all the tasks (for level calculations only). When considering this problem one should also take into account the possibility of solving it for uniconnected activity networks via a slightly modified version of the linear programming approach presented in Section 3.1.2 or another linear programming formulation, which is described below.

Consider now the case of unrelated machines, which turns out to be the most difficult type of problem in this class. Since the machines are unrelated, it is not meaningful to speak about unit-length tasks, and it is not surprising that polynomial-time algorithms are only known for problems allowing preemption. Also, very little is known about approximation algorithms for this case. Some results have been obtained by Ibarra and Kim (1977), but the bounds they derive are not very encouraging. Here, we discuss a preemptive scheduling model.

The problem $R/pmtn/C_{max}$ can be solved in two phases. In the first phase, a linear programming problem is solved. This problem was formulated independently by Blazewicz et al. (1977) and Lawler and Labetoulle (1978). The second phase uses the solution of the linear programming problem and finds an optimal preemptive schedule as described by Lawler and Labetoulle (1978).

The first phase of this procedure can be described as follows. Let p_{ij} denote the time required by machine P_i to fully process task T_j, and define $x_{ij} \in [0, 1]$ as the fraction of the processing time of task T_j that is processed on P_i. Then $p_{ij}x_{ij}$ will be the time spent processing T_j on P_i. Then $\sum_j p_{ij}x_{ij}$ is the actual time that processor P_i is being used, and $\sum_i p_{ij}x_{ij}$ is the total amount of time spent on processing task T_j. In order to minimize the makespan, we can then formulate the following minimax optimization problem

$$P: \text{Min } C_{max} = \max \left\{ \sum_{j=1}^{n} p_{ij}x_{ij}, \sum_{i=1}^{m} p_{ij}x_{ij} \right\}$$

$$\text{s.t. } \sum_{i=1}^{m} x_{ij} = 1 \quad \forall j = 1, 2, \ldots, n$$

$$x_{ij} \geq 0 \quad \forall i = 1, \ldots, m; j = 1, \ldots, n,$$

where the constraints ensure that each task is completely processed. The problem can be rewritten as an equivalent linear programming problem

$$P': \text{Min } C_{max}$$

$$\text{s.t. } C_{max} \geq \sum_{j=1}^{n} p_{ij}x_{ij}$$

$$C_{max} \geq \sum_{i=1}^{m} p_{ij}x_{ij}$$

$$\sum_{i=1}^{m} x_{ij} = 1 \quad \forall j = 1, 2, \ldots, n$$

$$x_{ij} \geq 0 \quad \forall i = 1, \ldots, m; j = 1, \ldots, n,$$

where the first set of constraints ensures that the total amount of processing time on each machine does not exceed the makespan, while the second set guarantees that the makespan is no shorter than the total amount of time required to process each task.

Let now an optimal solution of P consist of values x_{ij}^* and the associated objective function value C_{max}^*. However, at this point we do not yet know the schedule, i.e., the assignment of these task portions to the various machines over time.

The second phase consists of iteratively generating partial schedules by finding feasible flows in a related network. The partial schedules are then combined to form an overall optimal schedule. This process can be crudely estimated to run in $O(m^4n^4)$ time. The approach has been generalized to the case of multiple resources in a variety of settings. For details, see Slowinski (1978), DeWerra (1984), and Blazewicz et al. (1986).

When dependent tasks are considered, linear programming problems similar to those above can again be formulated based on an activity network presentation. If the activity network is uniconnected, an optimal schedule is constructed using the above second phase procedure. It is also worth noting that introducing different ready times into the considered problems is equivalent to minimizing maximal lateness. We will consider such problems in Section 3.3.

3.2 Minimizing Mean Flow Time

This section discusses problems with an objective of minimizing mean flow time. For some special cases there exist low degree polynomial algorithms which solve the problem to optimality. Although for identical machines the general problem is solvable in polynomial time, this is not the case for uniform or unrelated machines, which make the problem **NP**-hard. Allowing preemption in the latter case significantly simplifies the problems.

3.2.1 Identical Machines

McNaughton (1959) has shown that when minimizing mean flow time, preemptions are not profitable in the case of identical machines, at least not as long as equal ready times are assumed. Thus, we limit ourselves to considering only nonpreemptive schedules. When analyzing the nature of the minimal mean flow time criterion, one may expect that by assigning tasks in nondecreasing order of processing times, the mean flow time will be minimized, just as in the case of a single machine for which the shortest processing time (*SPT*) algorithm was described in Section 2.2.1. In fact, Conway et al. (1967) have shown that a proper generalization of this simple rule will also produce an optimization algorithm for

$P//\Sigma c_j$. As usual, we assume that there are m machines P_i and n tasks T_j. Then the algorithm can be described as follows.

An Algorithm for $P//\Sigma c_j$

Step 1: Order all tasks in nondecreasing order of processing times, with ties broken in favor of the task with the smaller subscript and renumber them accordingly.

Step 2: To processor P_i, $i = 1, ..., m$, assign tasks T_j, $j = i, i + m, i + 2m, ...$ in that order.

The complexity of this algorithm is $O(n \log n)$ since task ordering dominates the complexity function.

Example: Consider a problem with $n = 11$ tasks and processing times $\mathbf{p} = [3, 1, 6, 5, 4, 2, 2, 8, 7, 1, 4]$. Sorting the tasks in nondecreasing order results in $[p_2, p_{10}, p_6, p_7, p_1, p_5, p_{11}, p_4, p_3, p_9, p_8] = [1, 1, 2, 2, 3, 4, 4, 5, 6, 7, 8]$. For $m = 2$ machines, we obtain the resulting optimal schedule displayed in Figure IV.37, and for $m = 3$ machines, the schedule will be as shown in Figure IV.38.

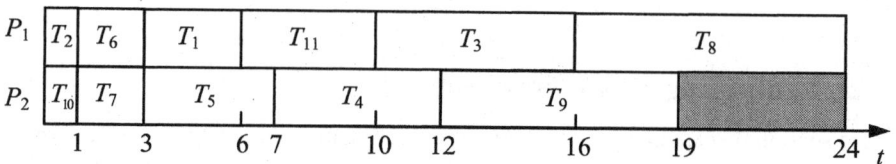

Figure IV.37

Bruno *et al.* (1974) have shown that if different weights are used, the problem $P2//\Sigma c_j$ is already **NP**-hard; introducing different ready times makes it strongly **NP**-hard, even for the case of one single machine.

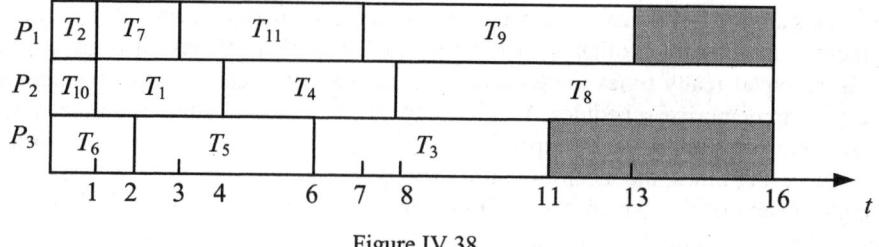

Figure IV.38

Consider now the case of dependent tasks. Here, the problem P/out-tree, $p_j = 1/\Sigma c_j$ is solved by Hu's algorithm for P/in-tree, $p_i = 1/C_{max}$ at the beginning of Section 3.1.2, adapted to the out-tree case. On the other hand, Lenstra and Rinnooy Kan (1978) have proved that problem $P2$/prec, $p_j = 1/\Sigma c_j$ is strongly **NP**-hard, as are almost all problems with arbitrary processing times. The reason is that problems $P2$/in-tree/Σc_j and $P2$/out-tree/Σc_j are already **NP**-hard as demonstrated by Sethi (1977). Unfortunately, no approximation algorithms for these problems have been evaluated with respect to their worst case behavior.

3.2.2 Uniform and Unrelated Machines

The results of Section 3.2.1 indicate that scheduling dependent tasks on uniform or unrelated processors is in general an **NP**-hard problem. It appears that no heuristics have been investigated for this problem.

First consider uniform processors and independent tasks to be scheduled without preemption, i.e., $Q//\Sigma c_j$. Consider now any given schedule on m machines. The tasks processed on machine P_i are $T_{i(k)}$, $k = 1, ..., n_i$, where k is the position of task $T_{i(k)}$ on machine P_i, and n_i is the number of tasks processed on P_i. Then the flow time of $T_{i(k)}$ is $f_{i(k)} = (1/b_i)\sum_{j=1}^{k} p_{i(j)}$ As $n = \sum_{i=1}^{m} n_i$, the mean flow time $\overline{F}$ is given by

$$\overline{F} = \frac{1}{n}\left(\sum_{i=1}^{m} \frac{1}{b_i} \sum_{k=1}^{n_i} (n_i - k + 1) p_{i(k)}\right).$$

The term in the outer brackets is the sum of n terms, each of which is the product of a processing time and one of the following coefficients:

$$\frac{1}{b_1}n_1, \frac{1}{b_1}(n_1 - 1),..., \frac{1}{b_1}; \frac{1}{b_2}n_2, \frac{1}{b_2}(n_2 - 1),..., \frac{1}{b_2}; ...; \frac{1}{b_m}n_m, \frac{1}{b_m}(n_m - 1),..., \frac{1}{b_m}.$$

Conway et al. (1967) have demonstrated that such a sum is minimized by matching these coefficients in nondecreasing order with processing times in nonincreasing order; see also Lemma II.5. An $O(n \log n)$ implementation of this rule has been given by Horowitz and Sahni (1976).

In the case of preemptive scheduling, it is possible to show that there exists an optimal schedule for Q/pmtn/Σc_j, in which $c_j \le c_k$ if $p_j < p_k$. On the basis of this observation, the following algorithm has been proposed by Gonzalez (1977). Here we assume that processors are ordered in nonincreasing order of their processing speed factors b_i. Again, ties are broken in favor of the smaller subscript.

Gonzalez' Algorithm for $Q/pmtn/\Sigma c_j$

Step 1: Place the tasks on the list in shortest processing time order, using standard processing times.

Step 2: Each time the fastest processor completes a task, preemption occurs and all other tasks in progress move to a faster processor, while a new task is started on the slowest processor.

The complexity of the algorithm is $O(n \log n + mn)$. Based on the way the algorithm proceeds, it is also referred to as the *shortest remaining processing time on the fastest machine* rule. For a proof of optimality, see, e.g., Pinedo (1995).

Example: Consider a problem with $m = 3$ machines with processing speed factors $\mathbf{b} = [2, 2, 1]$. Assume that there are $n = 7$ tasks with standard processing times $\mathbf{p} = [12, 4, 24, 20, 16, 8, 8]$. Sorting the tasks in *SPT* order, we obtain $[p_2, p_6, p_7, p_1, p_5, p_4, p_3] = [4, 8, 8, 12, 16, 20, 24]$, and an optimal schedule is shown in Figure IV.39.

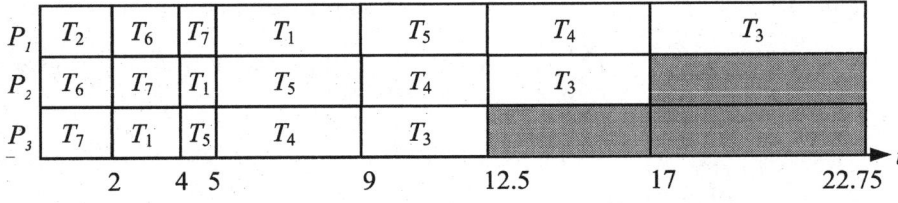

Figure IV.39

Consider now the case of unrelated processors, and more specifically $R//\Sigma c_j$. An approach to the solution of this problem is based on the observation that if task T_j is processed on machine P_i as the last task, it contributes its processing time p_{ij} to the mean flow time $\overline{F}$. The same task processed in the previous position contributes $2p_{ij}$, $3p_{ij}$ in the one before that, and so on. This reasoning, put forward by Bruno et al. (1974), allows the construction of a matrix $\mathbf{Q}$ representing contributions to $\overline{F}$ of particular tasks processed in different positions on different machines. In particular, the $[mn \times n]$-dimensional matrix $\mathbf{Q}$ has in row $i + m(k-1)$ and column j the contribution towards the flow time, if task T_j is processed on machine P_i in the k-th position from the end. Given this definition, we can write

Chapter 3: Parallel Machine Models

$$\mathbf{Q} = \begin{bmatrix} (p_{ij}) \\ 2(p_{ij}) \\ \cdot \\ \cdot \\ \cdot \\ n(p_{ij}) \end{bmatrix}.$$

The problem is now to choose n elements from $\mathbf{Q}$ such that

- exactly one element is chosen from each column (i.e., each task must be processed),
- at most one element is chosen from each row (each position on each machine is used no more than once), and
- the sum of the chosen elements is minimal.

This problem can be formulated as an assignment problem. Defining $x_{ij} = 1$ if element q_{ij} from $\mathbf{Q}$ is chosen, and $x_{ij} = 0$ otherwise, the problem can be written as

$$\text{P: Min } z = \sum_{i=1}^{mn} \sum_{j=1}^{n} q_{ij} x_{ij}$$

$$\text{s.t. } \sum_{i=1}^{mn} x_{ij} = 1 \quad \forall \ j = 1, ..., n$$

$$\sum_{j=1}^{n} x_{ij} \leq 1 \quad \forall \ i = 1, ..., mn$$

$$x_{ij} = 0 \vee 1 \quad \forall \ i = 1, ..., mn; \ j = 1, ..., n.$$

It is well known that problem P can be solved in polynomial time; see, e.g., Bruno et al. (1974).

Example: Consider the following instance of problem $R//\Sigma c_j$, with $m = 3$ processors and $n = 5$ tasks, and where the matrix of processing times is given as

$$(p_{ij}) = \begin{bmatrix} 3 & 2 & 4 & 3 & 1 \\ 4 & 3 & 1 & 2 & 1 \\ 2 & 4 & 5 & 3 & 4 \end{bmatrix}$$

The matrix $\mathbf{Q}$ for this problem is

$$\mathbf{Q} = \begin{bmatrix} 3 & (2) & 4 & 3 & 1 \\ 4 & 3 & 1 & (2) & 1 \\ (2) & 4 & 5 & 3 & 4 \\ 6 & 4 & 8 & 6 & (2) \\ 8 & 6 & (2) & 4 & 2 \\ 4 & 8 & 10 & 6 & 8 \\ 9 & 6 & 12 & 9 & 3 \\ 12 & 9 & 3 & 6 & 3 \\ 6 & 12 & 15 & 9 & 12 \\ 12 & 8 & 16 & 12 & 4 \\ 16 & 12 & 4 & 8 & 4 \\ 8 & 16 & 20 & 12 & 16 \\ 15 & 10 & 20 & 15 & 5 \\ 20 & 15 & 5 & 10 & 5 \\ 10 & 20 & 25 & 15 & 20 \end{bmatrix}$$

The solution of the assignment problem with cost matrix $\mathbf{Q}$ is shown by the bold elements in brackets. They indicate the following solution:

$q_{31} = 1$: T_1 is processed in position 1 on P_3, counted from the end,
$q_{12} = 1$: T_2 is processed in position 1 on P_1, counted from the end,
$q_{53} = 1$: T_3 is processed in position 2 on P_2, counted from the end,
$q_{24} = 1$: T_4 is processed in position 1 on P_2, counted from the end,
$q_{45} = 1$: T_5 is processed in position 2 on P_1, counted from the end.

The corresponding schedule is shown in Figure IV.40, and the total minimal flow time is $F^* = \sum c_j = 10$.

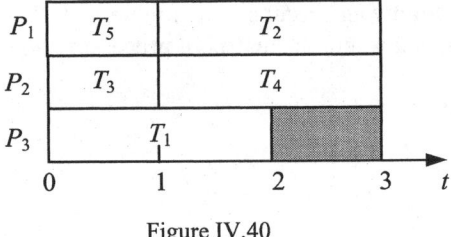

Figure IV.40

It is interesting to note that the problem $R/pmtn/\Sigma C_j$ is strongly **NP**-hard; see Sitters (2001).

3.3 Minimizing Maximal Lateness

Although useful in practice, models for minimizing maximal lateness are generally more difficult to solve than those for minimizing schedule length. A variety of these models are discussed in this section.

3.3.1 Identical Machines

As mentioned in Section 2.3, other than minimizing the maximal lateness L_{max}, problems with objectives involving due date criteria are usually **NP**-hard, even for the case of only a single machine. In this subsection we will discuss minimization of the L_{max} criterion. It seems quite natural that for this type of problem the general rule should be to schedule tasks according to their earliest due dates, using the *EDD* rule. However, this simple rule first introduced by Jackson (1955) produces optimal schedules in only a few cases. Mostly, either more complicated algorithms are necessary or the problems are **NP**-hard. First consider a nonpreemptive scheduling problem with independent tasks. Figure IV.8 in Section 1.2.3 clarifies the relation between problems with C_{max} and L_{max} criteria. In particular, all problems that are **NP**-hard with the C_{max} criterion remain **NP**-hard with the L_{max} criterion. For example $P2//L_{max}$ is **NP**-hard. On the other hand, unit processing times of tasks make the problem easy, and $P/p_j = 1, r_j/L_{max}$ can be solved by an obvious application of the *EDD* rule; see, e.g., Blazewicz (1977). Similarly, the problem $P/p_j = p, r_j/L_{max}$ can also be solved in polynomial time by an extension of the single processor algorithm, see Garey *et al.* (1981). Unfortunately, very little is known about the worst-case behavior of approximation algorithms for the general **NP**-hard problem.

The preemptive mode of processing makes the scheduling problems in this class much easier. The fundamental approach for all such problems is testing for feasibility via the network flow approach, first described by Horn (1974) for the problem $P/pmtn, r_j, d_j/-$. Let the ready times r_j and the deadlines d_j in $P/pmtn, r_j, d_j/-$ be merged and sorted in a list, such that e_k is the k-th ready time or due date and $e_0 < e_1 < ... < e_k$ with $k < 2n$. We can then construct a bipartite network with source and sink nodes added as shown in Figure IV.41.

The first set of nodes corresponds to time intervals in a schedule, i.e., node I_i, $i = 1, 2, ..., k$, corresponds to the time interval $[e_{i-1}, e_i]$. The second set of nodes corresponds to the tasks. The capacity of the arc joining the source n_s of the network with node I_i is equal to $m(e_i - e_{i-1})$ and thus corresponds to the total processing capacity of all m machines in this interval. If task T_j can be processed in interval $[e_{i-1}, e_i]$ obeying its ready time and deadline, then the nodes I_i and T_j are joined by an arc of capacity $e_i - e_{i-1}$. Node T_j is joined with the sink n_t of the network by an arc whose lower and upper bound on the flow equals p_j. Finding a feasible flow pattern is then equivalent to the construction of a feasible schedule

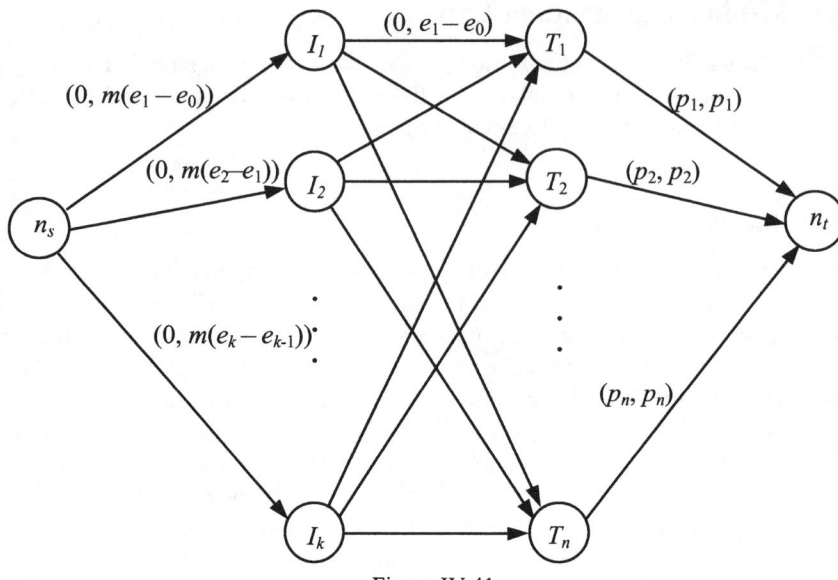

Figure IV.41

and this can be accomplished in $O(n^3)$ time. Then a binary search can be conducted on the optimal value of L_{max}, with each trial value of L_{max} inducing deadlines which are checked for feasibility by means of the network flow pattern already obtained. This procedure can be implemented to solve problem $P/pmtn$, r_j/L_{max} in $O(n^3 \min \{n^2, \log n + \log \max p_j\})$ time, as shown by Labetoulle et al. (1984).

Consider now the case of dependent tasks. A general approach in this case consists of assigning *modified due dates* d_j^* to tasks, depending on the number and due dates of their successors. When scheduling nonpreemptable tasks on a multiple machine system with all processing times $p_j = 1$, polynomial-time scheduling algorithms are available. Consider such a problem with in-tree precedence constraints and assume that the tasks have been sorted and renumbered in some precedence order in such a way that T_1 is the root of the tree. Brucker's (1976) algorithm below minimizes L_{max}. Here, $s(j)$ denotes the subscript of the immediate successor of T_j.

> **Brucker's Algorithm for $P/in\text{-}tree, p_j = 1/L_{max}$**

Step 1: Set $d_1^* := 1 - d_1$.

Step 2: For T_j, $j = 2, 3, ..., n$, compute a modified due date
$d_j^* = \max \{1 + d_{s(j)}^*, 1 - d_j\}$.

Chapter 3: Parallel Machine Models   395

Step 3: Schedule tasks in nonincreasing order of their modified due dates, subject to the precedence constraints. Ties are broken in favor of the task with the smallest subscript.

This algorithm can be implemented to run in $O(n \log n)$ time.

Example: Consider a system with $m = 4$ processors, $n = 32$ tasks, due dates $\mathbf{d} =$ [16, 20, 4, 3, 15, 14, 17, 6, 6, 4, 10, 8, 9, 7, 10, 9, 10, 8, 2, 3, 6, 5, 4, 11, 12, 9, 10, 8, 7, 5, 3, 5], and the precedence structure shown in Figure IV.42. Note that the precedence graph for this problem is actually a forest consisting of three in-trees. In computing the modified due dates, Step 1 was applied three times, once for each of the three roots. The resulting schedule is shown in Figure IV.43.

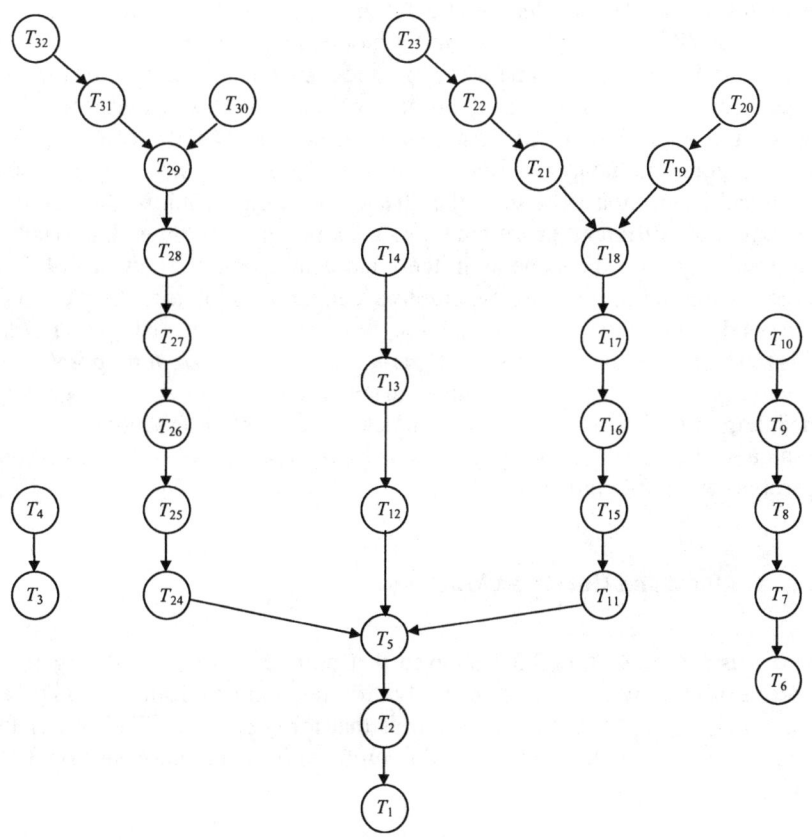

Figure IV.42

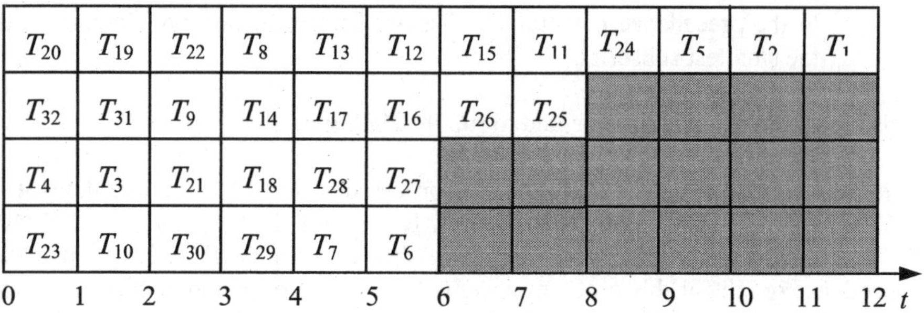

T_{20}	T_{19}	T_{22}	T_8	T_{13}	T_{12}	T_{15}	T_{11}	T_{24}	T_5	T_2	T_1
T_{32}	T_{31}	T_9	T_{14}	T_{17}	T_{16}	T_{26}	T_{25}				
T_4	T_3	T_{21}	T_{18}	T_{28}	T_{27}						
T_{23}	T_{10}	T_{30}	T_{29}	T_7	T_6						

0   1   2   3   4   5   6   7   8   9   10   11   12 t

Figure IV.43

Brucker *et al.* (1977) found that, somewhat surprisingly, out-tree precedence constraints cause the problem to be **NP**-hard. However, for the case of two machines, a different way of computing modified due dates which allows us to solve the problem in $O(n^2)$ time was proposed and then generalized for the case of different ready times (Garey and Johnson, 1976, 1977). This increases the running time of the algorithm to $O(n^3)$ and this is as much as we can get. Further generalizations of nonpreemptive scheduling problems no longer result in polynomial algorithms. On the other hand, preemptions allow the solution of problems with different processing times in polynomial time; but results are otherwise in general the same as in the nonpreemptive case. Lawler (1982b) has presented algorithms that are preemptive counterparts of the Brucker and the Garey and Johnson algorithms and the one presented by Garey and Johnson (1977). He shows that problems *P/pmtn,in-tree/*$L_{\max}$, *P2/pmtn, prec/*$L_{\max}$ and *P2/pmtn, prec, r_j/*$L_{\max}$ are solvable in polynomial time. These preemptive scheduling algorithms are more complex, but employ essentially the same techniques for dealing with precedence constraints as the corresponding algorithms for unit-length tasks.

3.3.2 Uniform and Unrelated Machines

The discussion in Section 3.3.1 showed that nonpreemptive scheduling with the objective of minimizing $L_{\max}$ is generally difficult, and practically the only known polynomial-time optimization algorithm is that for $Q/p_j=1/L_{\max}$. This problem can be solved via the minimax transportation problem formulation of Section 3.1.3 by setting $c_{j,ik} = k/b_i - d_j$.

The following discussion concentrates on preemptive scheduling. First, consider uniform processors. One of the most interesting algorithms in that area has been presented by Federgruen and Groenevelt (1986) for problem $Q/pmtn, r_j/L_{\max}$. It is a generalization of the network flow approach to the feasibility testing of problem

$P/pmtn, r_j, d_j/-$ described in Section 3.3.1. The feasibility testing procedure for problem $Q/pmtn, r_j, d_j/-$ uses a tripartite rather than a bipartite network formulation of the scheduling problem with an additional level of nodes corresponding to machine-period combinations. Finding a feasible flow in such a network corresponds to the construction of a feasible schedule for $Q/pmtn, r_j, d_j/-$. This can be done in $O(mn^3)$ time. In the case of precedence constraints, $Q2/pmtn, prec/L_{max}$ and $Q2/pmtn, prec, r_j/L_{max}$ can be solved in $O(n^2)$ and $O(n^6)$ time, respectively, exploiting Lawler's (1982b) algorithms. As far as unrelated processors are concerned, problem $R/pmtn/L_{max}$ can be solved by a linear programming formulation very similar to that in Section 3.1.3; for details, see Lawler and Labetoulle (1978).It should also be mentioned that the case in which precedence constraints form a uniconnected activity network can also be solved via the same modification of the linear programming problem as described for the C_{max} criterion at the end of Section 3.1.3. For details the reader is referred to Slowinski (1981).

CHAPTER 4   DEDICATED MACHINE AND RESOURCE-CONSTRAINED MODELS

In this chapter we consider scheduling problems in which a task requires processing on more than one machine, i.e., the task can be represented as a set of operations. In contrast to the models presented in the previous chapter, the machines in this section are no longer parallel, i.e., identical or uniform. The dedicated scheduling models in this chapter allow three classes of models, open shops, flow shops, and job shops. To each such problem type we devote a section. Unfortunately, most of the scheduling problems in this class are **NP**-hard, which is especially true for criteria other than $C_{\max}$. In the first two sections of this chapter, we describe polynomial-time optimization algorithms for some open shop and flow shop scheduling problems. In the third section, we first discuss an exact algorithm for a simple job shop problem, followed by a branch and bound procedure and a heuristic for more involved job shop problems. The fourth and last section is devoted to resource-constrained machine scheduling problems.

4.1   Open Shop Scheduling

Recall that in an open shop each of the tasks $T_j, j = 1, \ldots, n$, must be processed on each of the processors $P_i, i = 1, \ldots, m$, but the order of processing a task on the various processors is not specified (hence the expression "open" shop), nor does it have to be the same for different tasks. Recall also that $p_{ij}, i = 1, 2; j = 1, \ldots, n$, denotes the processing time of the j-th job on the i-th machine.

Consider first nonpreemptive scheduling. As each job has to be processed on both machines, $C_{\max} \geq \left\{ \sum_{j=1}^{n} p_{1j}, \sum_{j=1}^{n} p_{2j} \right\}$. This result can be sharpened by

Proposition IV.11: For the $O2//C_{max}$ problem, the optimal schedule length C^*_{max} is given by the expression

$$C^*_{max} = \max\left\{\max_{j}\{p_{1j} + p_{2j}\}, \sum_{j=1}^{n} p_{1j}, \sum_{j=1}^{n} p_{2j}\right\}.$$

A proof of the proposition can be found in Pinedo (1995).

A consequence of Proposition IV.11 is the *Longest Alternate Processing Time* (*LAPT*) rule. It states that whenever one of the machines becomes idle, the remaining job with the longest processing time on the other machine should be scheduled next. The *LAPT* rule is the basis of the following algorithm due to Pinedo (1995), which determines an optimal schedule for the $O2//C_{max}$ problem.

Longest Alternate Processing Time (*LAPT*) Algorithm

Prodecure: Schedule the tasks on the two machines in such a way that whenever a machine becomes idle, the remaining task with the longest processing time on the other machine is scheduled next. Ties are broken arbitrarily.

Example: Consider a problem, in which $n = 5$ jobs are to be scheduled on $m = 2$ machines with the following matrix **P** of processing times:

$$\mathbf{P} = \begin{bmatrix} 3 & 2 & 5 & 1 & 4 \\ 4 & 4 & 1 & 3 & 5 \end{bmatrix}.$$

The optimal schedule found with the *LAPT* algorithm is shown in Figure IV.44. The makespan is $C^*_{max} = 17$.

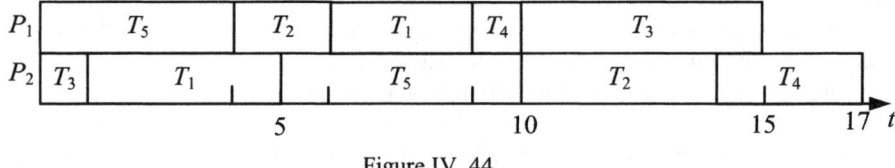

Figure IV. 44

An alternative algorithm for the problem was developed by Gonzalez and Sahni (1976); it runs in $O(n)$ time.

It can be shown that the problem becomes **NP**-hard as soon as the number of machines increases to three. But again, preemptions result in a polynomial time algorithm. Gonzalez and Sahni (1976) solve problem $O/pmtn/C_{max}$ by using the following generalization of the expression in Proposition IV.11:

$$C^*_{max} = \max\left\{\max_j\left\{\sum_{i=1}^{m} p_{ij}\right\}, \max_i\left\{\sum_{j=1}^{n} p_{ij}\right\}\right\}$$

and then applying an algorithm such as that by Lawler and Labetoulle (1978), discussed in Section 3.1.3.

It should also be mentioned that problems $O2//\Sigma c_j$ and $O2//L_{max}$ are **NP**-hard; see Achugbue and Chin (1982), and Lawler *et al.* (1982b), respectively. Furthermore, the problem $O/pmtn$, r_j/L_{max} is solvable via the linear programming approach of Cho and Sahni (1981). As far as heuristics are concerned, Achugbue and Chin (1982) have developed arbitrary list scheduling and shortest processing time algorithms for the $O//\Sigma c_j$ problem. Their asymptotic performance ratios are $R_L^\infty = n$ and $R_{SPT}^\infty = m$, respectively. Since the number n of tasks is usually much larger than the number m of processors, i.e., $n \gg m$, the bounds indicate the advantage of shortest processing time schedules over arbitrary list schedules.

4.2 Flow Shop Scheduling

Recall that in a flowshop, each of the tasks T_j, $j = 1, \ldots, n$, must be processed on each of the processors P_i, $i = 1, \ldots, m$, and all in the same prespecified order. This order is usually defined by the numbering of P_i. One of the classical flow shop scheduling algorithms is due to Johnson (1954). It solves problem $F2//C_{max}$ and can be described as follows.

Johnson's Rule for the $F2//C_{max}$ Problem

Step 1: Choose tasks such that $p_{1j} \leq p_{2j}$ and schedule them in nondecreasing order of their p_{1j}'s. Ties are broken arbitrarily.

Step 2: Schedule all other tasks in nonincreasing order of their p_{2j}'s.

This algorithm requires $O(n \log n)$ time as the sorting effort dominates the complexity function. A simple proof of its optimality, using contradiction, can be found, e.g., in Pinedo (1995).

Example: Consider a case in which $n = 5$, $m = 2$ and the matrix of processing times is

$$\mathbf{P} = \begin{bmatrix} 4 & 4 & 30 & 6 & 2 \\ 5 & 1 & 4 & 30 & 3 \end{bmatrix}.$$

As $p_{1j} < p_{2j}$ for $j = 1, 4$, and 5, we schedule T_5 first, followed by T_1 and T_4 (in that order), and, in Step 2 of the algorithm, T_3 is scheduled next, followed by T_2. The optimal schedule to this problem is shown in Figure IV.45.

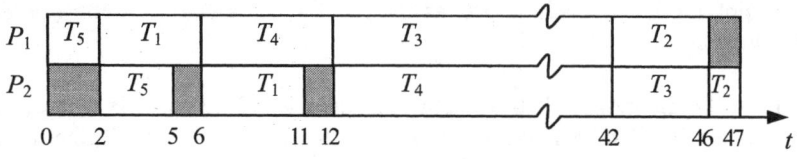

Figure IV. 45

The algorithm can be extended to also cover the special case of three machine scheduling for which either the conditions $\min_j \{p_{1j}\} \geq \max_j \{p_{2j}\}$ or $\min_j \{p_{3j}\} \geq \max_j \{p_{2j}\}$ are satisfied. In this case, an optimal schedule can be obtained by applying Johnson's rule to two machines with processing times ($p_{1j} + p_{2j}, p_{2j} + p_{3j}$).

However, more complicated assumptions concerning problem parameters such as precedence constraints, additional machines or other criteria, make the problem strongly **NP**-hard. For those problems heuristic algorithms are needed, but not much work has been done in this area. One of the exceptions is Gonzalez and Sahni (1976), who evaluate the worst-case behavior of an algorithm. The method is based on Johnson's rule for the problem $F//C_{max}$ and the authors prove its worst-case behavior to have an absolute performance ratio $R = \lceil m/2 \rceil$. Barany (1981) proposed a complicated heuristic whose performance does not depend on n and is proved to have an absolute performance ratio of $R = \frac{1}{2}(m-1)(3m-1)p_{max}$, where p_{max} denotes the longest processing time of any job on any of the given machines. Finally, Röck and Schmidt (1982) develop heuristics that replace m machines by two machines, and define the task processing times as the sums of the appropriate original processing times. The absolute performance ratios are proportional to m.

As far as preemptive scheduling is concerned, the situation is the same as described above, i.e., $F2/pmtn/C_{max}$ is solved by Johnson's rule, and other problems are strongly **NP**-hard. The only exception is $F2/pmtn/\Sigma c_j$, whose computational complexity is an open question.

4.3 Job Shop Scheduling

In the case of job shop scheduling, the number of operations per task T_j, their assignment to machines P_i and order of processing are given. While it is not required that each task is processed on all machines, we will assume *no recirculation*, i.e., no task will visit any machine more than once. Denoting by O_{ij} the operation of task T_j on machine P_i, the sequence of O_{ij} needs to be specified for each task. In this section we first present a polynomially solvable case of the job shop scheduling problem with two machines. This is followed by an exact method and a heuristic for the case of $m \geq 2$ machines.

4.3.1 Basic Ideas

In order to motivate the discussion consider the following

Example 1: Assume that there are two tasks T_1 and T_2, and three machines P_1, P_2 and P_3. Each task consists of three operations, and the task processing times p_{ij} are given by the matrix $\mathbf{P} = \begin{bmatrix} 3 & 2 \\ 1 & 1 \\ 3 & 2 \end{bmatrix}$, and the processing sequences are O_{11}, O_{21}, O_{31}, for the first task, and O_{12}, O_{32}, and O_{22} for the second task. A feasible but not optimal schedule is shown in Figure IV.46.

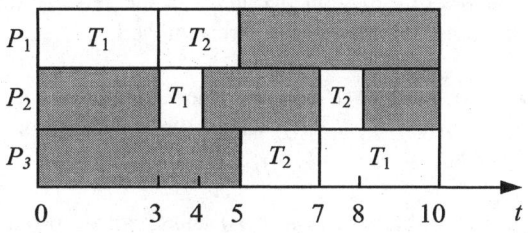

Figure IV. 46

It is well known that job shop scheduling is not only a model with a broad application area but also one of the computationally hardest problems to be investigated so far in this part; the job shop scheduling problem is already **NP**-hard for only two machines. No heuristic with performance guarantee has yet been developed. There are, however, some special cases for which polynomial time algorithms exist:

- two-machine job shops in which all operations have unit processing time. They are solvable in a time linear in the total number of tasks; see, e.g., Hefetz and Adivi (1982) and Brucker (1981),

- job shops that involve two tasks. These problems are solvable in $O(n_1 n_2 \log n_1 n_2)$ time; see Akers (1956), and Brucker (1988), where n_1 and n_2 are the number of operations of the first and the second task, respectively, and
- job shops with two machines, in which tasks do not have more than two operations; Jackson (1956) showed these problems to be solvable in $O(n \log n)$ time.

Below we present an algorithm by Jackson for solving the last of these three problems $J2/n_j \leq 2/C_{\max}$. The algorithm divides tasks into four *classes* J_1, J_2, J_{12} and J_{21}, where J_1 and J_2 denote the sets of tasks that require processing only on machine P_1 or P_2, respectively. The set J_{12} includes all jobs that are to be processed first on P_1 and then on P_2, and the set J_{21} is defined similarly. Notice that once a job from J_{12} has been completed on machine P_1, its scheduling on machine P_2 is irrelevant (for the purpose of schedule lengths), as long as P_2 remains busy. A corresponding argument applies to the set J_{21}. Determining the sequence of jobs in the set J_{12} can be done by Johnson's rule. Similarly, Johnson's rule is applied to schedule the jobs in J_{21}. Note, however, that the roles of P_1 and P_2 are then exchanged, since tasks in J_{21} are processed first on P_2 and later on P_1. Jackson's algorithm can then be described as follows.

Jackson's Algorithm for the $J2/n_j \leq 2/C_{\max}$ Problem

Step 1: Apply Johnson's rule to sequence tasks in the sets J_{12} and J_{21}, where for J_{21} the processing times p_{1j} and p_{2j} are switched.

Step 2: Assign tasks to machine P_1 in the order J_{12}, J_1, J_{21} and assign tasks to machine P_2 in order J_{21}, J_2, J_{12}, where the tasks in sets J_1 and J_2 appear in arbitrary order.

Example 2: Consider a problem with $n = 10$ tasks, whose processing times and machine sequencing are as shown in Table IV.6.

Table IV.6

Tasks T_i	T_1	T_2	T_3	T_4	T_5	T_6	T_7	T_8	T_9	T_{10}
p_{1j}	3	2	2	3	4	–	1	1	1	3
p_{2j}	1	–	4	1	4	3	5	–	2	4
Sequence P_1, P_2	1, 2	1, –	2, 1	2, 1	1, 2	–, 2	2, 1	1, –	2, 1	1, 2

Given the information in Table IV.6, we find that $J_1 = \{T_2, T_8\}$, $J_2 = \{T_6\}$, $J_{12} = \{T_1, T_5, T_{10}\}$, and $J_{21} = \{T_3, T_4, T_7, T_9\}$. Applying Johnson's rule to the sets of tasks

Chapter 4: Dedicated Machine and Resource-Constrained Models   405

J_{12} and J_{21} results in the ordered sets $J_{12} = \{T_{10}, T_5, T_1\}$ and $J_{21} = \{T_4, T_3, T_7, T_9\}$, where the order of T_7 and T_9 could be reversed. The orders of the sets J_1 and J_2 remain as shown above. The schedule that is arrived at with Jackson's algorithm is shown in Figure IV.47. Note that the order of T_2 and T_8 could be reversed.

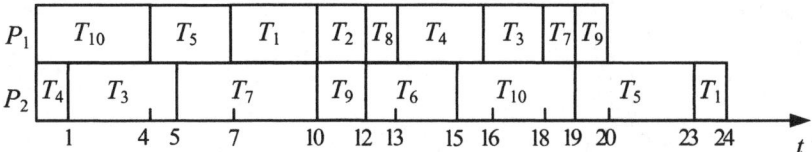

Figure IV.47

The complexity of Jackson's algorithm is $O(n \log n)$, since its most complex function is ordering the tasks. Unfortunately, this is it as far as polynominal time algorithms go; other problems of scheduling in job shops are strongly **NP**-hard.

Consider now the general job shop problem $J//C_{max}$ with $m \geq 2$ processors. Again, we assume no recirculation. A useful representation of the problem is by means of a directed so-called *disjunctive graph* $G = (N, C, D)$, which consists of a set of nodes N, each of which corresponds to an operation O_{ij}, as well as an artificial source n_s and an artificial sink n_t. The set C is a set of so-called *conjunctive arcs*, and D is a set of so-called *disjunctive arcs*. The conjunctive arcs of C indicate the processing order of the operations O_{ij} of a task T_j. The set C also includes artificial arcs from the source to all nodes that correspond to the first operation in the processing sequence of each machine, and artificial arcs from the last operations in the processing sequence of each machine to the artificial sink. The disjunctive arcs of D indicate which machines the operations have in common, i.e., for every pair of operations requiring the same machine, two directed arcs with opposite directions are introduced. The disjunctive arcs in D form a set of cliques, i.e., complete subgraphs, each induced by all nodes that have one machine in common. Each arc leading out of a node $O_{ij} \in N$ has a weight equal to the corresponding operation processing time; the weights of the arcs leading out of n_s are zero.

A (possibly partial) *schedule* corresponds to an acyclic subgraph of G that includes all arcs in C and some of D. The set of disjunctive arcs in this subgraph is called a *selection* and will be denoted by S. If S is incident to all nodes O_{ij} and contains exactly one disjunctive arc from each pair we call S a *complete selection*. Each complete selection corresponds to a complete schedule, and the longest path in G with only the disjunctive arcs used in the complete selection is the length of the schedule. The disjunctive graph for the numerical problem introduced at the beginning of this section is shown in Figure IV.48, in which the solid arcs belong to the set C, while the broken arcs belong to D.

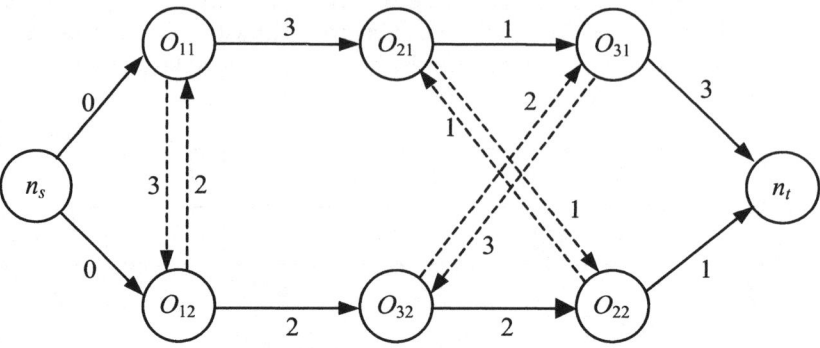

Figure IV.48

A complete selection that corresponds to the solution in Figure IV.46 is displayed by the graph in Figure IV. 49, and includes the disjunctive arcs (O_{11}, O_{12}), (O_{21}, O_{22}), and (O_{32}, O_{31}), indicating that on machine P_1, task T_1 is processed before T_2; the same holds for machine P_2, while on machine P_3, task T_2 is processed before T_1. The length of the corresponding schedule is the length of the longest path in this graph, which includes all arcs in C and those disjunctive arcs that belong to the chosen schedule. More specifically, the length of the longest path to some node O_{ij} denotes the point in time at which task T_j starts being processed on machine P_i. In our example, the longest path from n_s to n_t is $(n_s, O_{11}, O_{12}, O_{32}, O_{31}, n_t)$, indicated by bold lines in Figure IV. 49, and has a length of 10, which is also the schedule length or makespan.

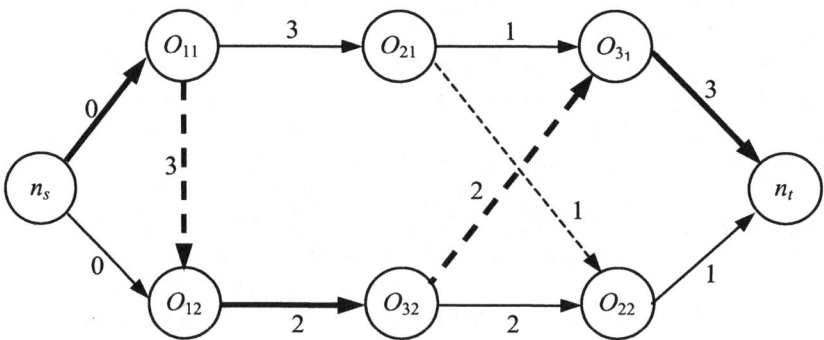

Figure IV.49

Chapter 4: Dedicated Machine and Resource-Constrained Models   407

4.3.2 A Branch and Bound Algorithm

This section describes a branch and bound method to solve $J//C_{max}$ problems. It is based on so-called active schedules, which are defined as follows.

Definition IV.12: A *schedule* is said to be *active* if no operation can be finished earlier by some resequencing, without delaying some other operation.

In an active schedule, jobs arriving at a machine will be processed as soon as possible, according to the sequence prescribed by the schedule. It is easy to verify that the schedule in Figure IV.46 is active. At least one of the active schedules will be optimal. There exists a simple algorithm for generating all active schedules, and this method will serve as a subalgorithm in a branch and bound method. In this method, Ω denotes the set of schedulable operations, i.e., all operations O_{ij} that have not been scheduled yet, while their predecessors have already been scheduled. Furthermore, r_{ij} denotes the earliest possible starting time of such an operation O_{ij}. The algorithm is initialized with Ω containing the first operation O_{ij} of each job T_j, $\forall\ j = 1, \ldots, n$, and with $r_{ij} := 0$ for all $O_{ij} \in \Omega$. It can then be formally stated as follows.

> **A Generator of Active Schedules**

Step 1: Find $t^* = \min\ \{r_{ij} + p_{ij}\colon O_{ij} \in \Omega\}$ and let $O_{i^*j^*} := \underset{O_{ij} \in \Omega}{\arg\min}\{t^*\}$.

Step 2: Find $\Omega' := \{O_{i^*j} \in \Omega\colon r_{i^*j} < t^*\}$. For each $O_{i^*j} \in \Omega'$, schedule O_{i^*j} next on machine i^*, replace in Ω each operation O_{i^*j} by its immediate successors, update $r_{ij} := r_{ij} + p_{ij}$ for each newly scheduled job, and go to Step 1.

The algorithm terminates when all active schedules have been generated, which is done in a tree-like fashion with nodes corresponding to the operations $O_{i^*j^*}$ in Step 1 and one, two, or more branches leading to the operations in Ω'. The leaves of this tree are the active schedules.

Example 3: When applied to the job-shop scheduling problem in Example 1, the algorithm is initialized with $\Omega = \{O_{11}, O_{12}\}$ and $r_{11} = r_{12} = 0$. In Step 1, $t^* = \min\ \{r_{11} + p_{11}; r_{12} + p_{12}\} = \min\{0 + 3; 0 + 2\} = 2$, so that $O_{i^*j^*} = O_{12}$. In Step 2, $\Omega' = \{O_{11}, O_{12}\}$. Scheduling now O_{11}, we find $\Omega = \{O_{12}, O_{21}\}$, $r_{12} = 0$, $r_{21} = 3$, so that $t^* = 2$ and $O_{i^*j^*} = O_{12}$, etc. Starting the schedule with O_{12} instead, we find $\Omega = \{O_{11}, O_{32}\}$, $r_{11} = 0$, $r_{32} = 2$, so that $t^* = 3$ and $O_{i^*j^*} = O_{11}$, etc. Continuing in this fashion, the algorithm generates four active schedules, as displayed by the tree in Figure IV.50, where the nodes are indicated by the operations to be scheduled at that point.

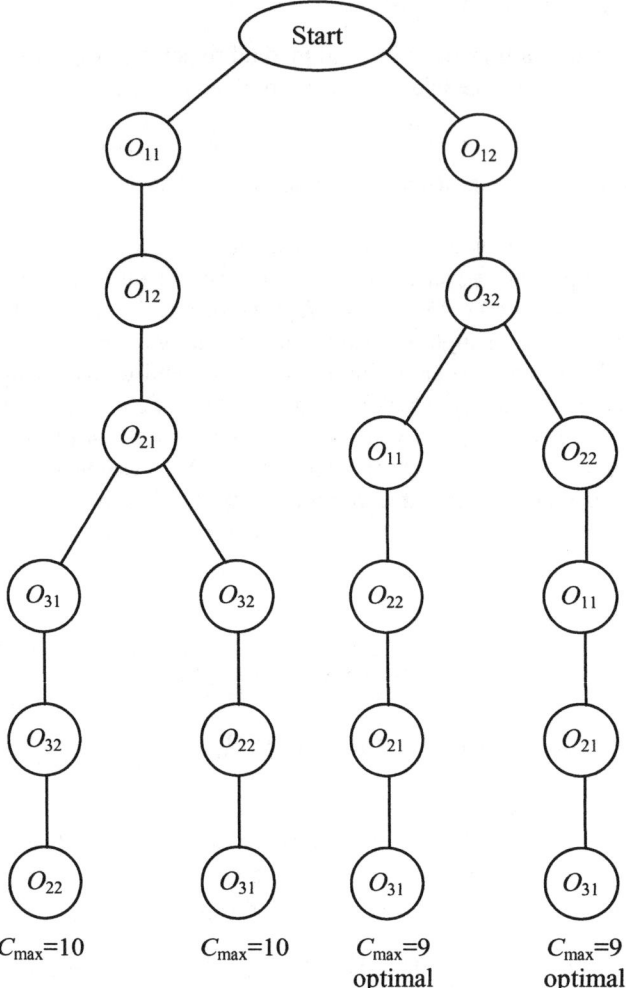

Figure IV.50

Note that the two optimal leaves of the tree, with $C_{max} = 9$, actually represent identical schedules. This is due to the tie occurring at the branching of the node marked O_{32}, where $t^* = \min \{r_{11} + p_{11}; r_{22} + p_{22}\} = \min \{2 + 3; 4 + 1\} = 5$. The optimal schedule is displayed in Figure IV.51. The nonoptimal schedule in Figure IV.46 corresponds to the leaf marked O_{31} with $C_{max} = 10$, and to see how this schedule can be constructed from the corresponding path in the tree, we proceed as follows. Operation O_{11}, the first operation along the path, is first scheduled on P_1 and completed at time $t = 3$. The next on the path is O_{12}, which is therefore processed on P_1 between $t = 3$ and $t = 5$. Next, O_{21} is processed on P_2, between $t = 3$ and $t = 4$; then O_{32} is processed on P_3, from $t = 5$ to $t = 7$, etc. Every active schedule can be constructed in analogous fashion, by tracing the corresponding path in the tree genereated by the algorithm.

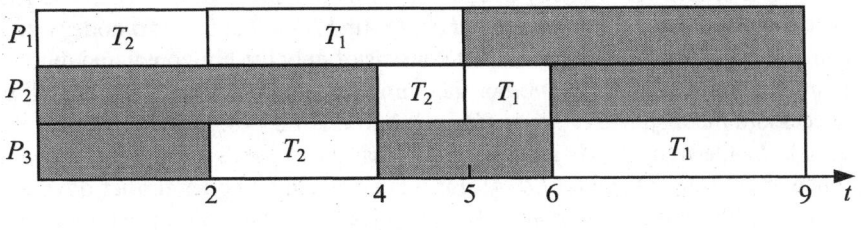

Figure IV.51

Returning to the general discussion, we can now use the active schedule generation as a basis for a branch and bound procedure in the tree that has been generated. As upper bounds we use C_{max} for the best complete schedule found so far, which is obtained when at least one leaf of the tree has been generated. It follows that a depth-first search strategy should be employed. Lower bounds can be computed at each node by determining the length of the longest path from source to sink in the disjunctive graph associated with that particular node in the tree. This disjunctive graph consists of all nodes and conjunctive arcs, plus disjunctive arcs emanating from the node under consideration, as well as from all its predecessor nodes at that point in the tree.

To summarize, consider an instance of the job shop scheduling problem $J//C_{max}$ without recirculation. We can then formally write

---

A Branch and Bound Algorithm for the Problem $J//C_{max}$

Step 1: Construct an active schedule tree using the Generator of Active Schedules algorithm. Using a depth-first strategy, develop the tree until at least one complete schedule has been found. Use C_{max} for this schedule as the best upper bound *UB*.

Step 2: Compute a lower bound *LB* for each node as the length of the longest path from source to sink in the disjunctive graph associated with that node.

Step 3: Prune the tree at each node for which *LB* > *UB*.

Step 4: Continue to grow the active schedule tree, updating *UB* whenever a complete schedule is found with a better (i.e., lower) value of C_{max} than the best known such value so far. Go to Step 2 and repeat the process until all branches of the tree are closed.

Example 4: Applied to Example 1, this algorithm first develops one of the chains in the tree in Figure IV.49 (which chain depends on the branching strategy chosen). If a branch-left strategy is chosen, we obtain $UB = C_{max} = 10$ at the leaf indicated by O_{22}. Computing lower bounds, we find $LB = 7$ at the start node, which has an associated disjunctive graph that consists of only the horizontal and dummy arcs in the graph in Figure IV.48. For the immediate successor node O_{11} and all of its successors the disjunctive arc is (O_{11}, O_{12}) so that $LB = 8$. Developing next the right side of the tree, the disjunctive graph that corresponds to O_{12} includes the disjunctive arc (O_{12}, O_{11}), and $LB = 9$ for this node and all nodes further down the tree. Now, with $UB=C_{max}=9$ at any of the two leaves on this side of the tree, $LB = UB = 9$, and we have found an optimal solution. We can then construct the optimal schedule, which is displayed in Figure IV.51.

A note is in order regarding the condition $LB>UB$ in Step 2 of the algorithm, which may appear contradictory. The upper bound UB refers to the schedule length C_{max} achieved at a leaf and will serve as an upper bound of what the optimal (minimal) C_{max}^* might be. The lower bound LB, on the other hand, refers to the best possible (smallest) schedule length that could be achieved at some node or any of its successors in the achive schedule tree. If $LB>UB$ at some node, that or any of its successors will then surely not correspond to an optimal schedule; the tree can safety be pruned at such a node.

There are more sophisticated methods for obtaining sharper lower bounds than those obtained in Step 2 above, involving formulating and solving related single-machine scheduling subproblems of the type $1/r_j/L_{max}$; see Section 2.3.1 and also Section 4.3.3 below. For details, readers are referred to Pinedo (1995) or Pinedo and Chao (1999). A survey of branch and bound techniques for the job shop scheduling problem is found in Pinson (1995).

There exist many enumerative exact algorithms for the general job shop scheduling problem. The notorious 10-machine, 10-task problem introduced by Fisher and Thompson (1963) still serves as a benchmark for these approaches. No optimal solution to this problem was found until 1987, when Carlier and Pinson developed an algorithm, which took five hours of computing time. One of the most efficient optimization algorithms developed since then is described in Brucker *et al* (1992); using this algorithm the Fisher and Thompson problem was solved on a SUN workstation within 16 minutes with the branch and bound algorithm described by Roy and Sussman (1964). These long computing times are not altogether surprising if we keep in mind that the number of feasible schedules for the m-machine, n-job problem could in principle be as many as $(n!)^m$, even without counting obviously absurd schedules, such as those in which all processors are simultaneously idle.

4.3.3 The Shifting Bottleneck Heuristic

Since the general job shop scheduling problem $J//C_{\max}$ is not only strongly **NP**-hard in theory, but also notoriously difficult to solve in practice, there is a great need for heuristic methods to find reasonably good solutions. Adams *et al.* (1988) have developed a successful method, called the *shifting bottleneck heuristic*. It is described below.

Using the disjunctive graph of the problem under consideration, the method iteratively schedules the machines one after another. In each iteration, the next machine to be included is the one causing the largest maximal lateness and is therefore in a sense the bottleneck; thus the name of the algorithm. Furthermore, in each iteration the algorithm also attempts to locally improve the present solution. As such, it can be considered a greedy construction heuristic with a nested improvement phase.

Formally, let $\mathcal{P}$ denote the set of all machines, and assume that in the current iteration a subset $\mathcal{P}'$ of the machines have been scheduled, in the sense that disjunctive arcs have been selected for the machines in $\mathcal{P}'$. The purpose of the current iteration is to choose from among the machines in $\mathcal{P}\setminus\mathcal{P}'$ the next one to be scheduled and also to determine the sequencing of the operations on this machine. Denote the current disjunctive graph by G', which has disjunctive arcs for all machines in $\mathcal{P}'$, but no disjunctive arc for any machine in $\mathcal{P}\setminus\mathcal{P}'$. This is as if every operation on every machine in $\mathcal{P}\setminus\mathcal{P}'$ could be scheduled in parallel with any other operation on any other machine in $\mathcal{P}\setminus\mathcal{P}'$ without being restricted by any precedence constraints. The schedule length of the current (fictitious) schedule is then the length of a longest (critical) path in G'; we denote it by $C_{\max}(\mathcal{P}')$.

Consider now an operation O_{ij} on a machine $P_i \in \mathcal{P}\setminus\mathcal{P}'$. In order to schedule this operation without increasing the total schedule length of the current partial schedule, one could consider its release date to be the length of the longest path in G' from the source n_s to O_{ij} and its due date could be no later than $C_{\max}(\mathcal{P}')$ – (length of the longest path from O_{ij} to the sink n_t) + p_{ij}. Given these release and due dates for all operations O_{ij} on machine P_i, we can then solve the single-machine problem $1/r_j/L_{\max}$ on this machine. Note that even this fairly simple subproblem that has to be solved $O(m)$ times during the algorithm is strongly **NP**-hard. The resulting minimal lateness $L_{\max}(i)$ is the minimal amount by which the total schedule length increases if P_i were to be scheduled as the next machine. The heuristic is now to select and schedule the machine P_k from $\mathcal{P}\setminus\mathcal{P}'$ whose $L_{\max}(k)$ is the largest, i.e., the "bottleneck" machine. Inserting the appropriate disjunctive arcs in G', the current partial schedule length will increase by at least the amount $L_{\max}(k)$, i.e., we have

$$C_{\max}(\mathcal{P}' \cup \{P_k\}) \geq C_{\max}(\mathcal{P}') + L_{\max}(k).$$

Inequality could occur due to precedence interactions with machines in $\mathscr{P}'$. Although we are now in principle ready to continue with the next iteration, we first try to improve on the tentative solution obtained by resequencing all the machines already scheduled in $\mathscr{P}'$ when the new machine P_k is added. Specifically, all machines $P_i \in \mathscr{P}'$ are resequenced, one by one, by solving the single machine problem $1/r_j/L_{max}$ with release and due dates computed as above, based on the disjunctive graph G', but with disjunctive arcs for P_i excluded and disjunctive arcs for P_k included. After resequencing all machines in $\mathscr{P}'$ in this way, and thereby possibly improving the schedule, we go to the next iteration of the entire procedure.

We now summarize the discussion in algorithmic form. Recall that the problem to be solved in an instance of $J//C_{max}$ with no recirculation, with machine set $\mathscr{P}$ and disjunctive graph G. The algorithm is initialized with the graph G' being the graph G with all disjunctive arcs removed. Furthermore, set $\mathscr{P}' := \emptyset$ and let $C_{max}(\mathscr{P}')$ be the length of the longest path from source to sink in G'. The algorithm can then be described as follows.

The Shifting Bottleneck Heuristic

Step 1: For each machine $P_i \in \mathscr{P} \setminus \mathscr{P}'$ find $L_{max}(i)$ by solving the problem $1/r_j/L_{max}$ where the operations O_{ij} have release dates r_{ij} determined by the longest path from the source to O_{ij} in the graph G', and the due dates d_{ij} are determined as $C_{max}(\mathscr{P}')$ − (length of the longest path from O_{ij} to the sink n_t) + P_{ij}.

Step 2: Find a bottleneck machine P_k by computing $L_{max}(k) = \max \{L_{max}(i): P_i \in \mathscr{P} \setminus \mathscr{P}'\}$. Schedule the operations O_{kj} on machine P_k according to the solution obtained in Step 1. Insert the appropriate disjunctive arcs in G' and let $\mathscr{P}' := \mathscr{P}' \cup \{P_k\}$.

Step 3: Resequence one by one all machines in $\mathscr{P}' \setminus \{P_k\}$ by deleting the corresponding disjunctive arcs from G', solving the appropriate problem $1/r_j/L_{max}$ with release and due dates determined in the graph G' as indicated in Step 1. The machine operations are then resequenced accordingly and the corresponding disjunctive arcs in G' are inserted.

Step 4: Is $\mathscr{P}' = \mathscr{P}$?
    If yes: Stop, a complete schedule has been found.
    If no: Go to Step 1.

Chapter 4: Dedicated Machine and Resource-Constrained Models

Example 5: Applying the shifting bottleneck heuristic to the three-machine, two-job problem of Example 1, we initialize the algorithm with $C_{max}(\mathcal{P}') = C_{max}(\emptyset) = 7$. In Step 1 we solve, by enumeration, the problem $1/r_j/L_{max}$ for P_1 (with $r_{11} = r_{12} = 0$, $d_{11} = 3$, $d_{12} = 4$, from which we obtain $L_{max}(1) = 1$ for the optimal sequence O_{11}, O_{12}). Similarly, for P_2 (with $r_{21} = 3$, $r_{22} = 4$, $d_{21} = 4$, $d_{22} = 6$) the optimal sequence is O_{21}, O_{22} with $L_{max}(2) = 0$, and for P_3 (with $r_{31} = 4$, $r_{32} = 2$, $d_{31} = 7$, $d_{32} = 6$) the optimal sequence is O_{32}, O_{31} with $L_{max}(3) = 0$. In Step 2 we therefore find that P_1 is the bottleneck machine which is scheduled. With no machine to resequence in Step 3, the current disjunctive graph G' with $C_{max} = 8$ is displayed in Figure IV.52.

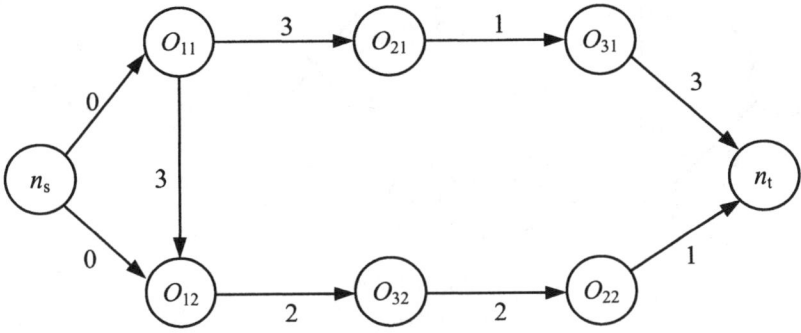

Figure IV.52

In the second iteration, we solve, again by enumeration, the problem $1/r_j/L_{max}$ for machine P_2 ($L_{max} = 0$ for the optimal schedule O_{21}, O_{22}), and for machine P_3 ($L_{max} = 2$ for the two optimal schedules O_{31}, O_{32} and O_{32}, O_{31}; we break the tie by choosing the one with the earliest due date first, i.e., O_{32}, O_{31} since $d_{31} = 8$ and $d_{32} = 7$, according to the earliest due date rule; see Section 2.3.1). In Step 2 we therefore pick P_3 as the bottleneck machine and schedule it accordingly for a $C_{max} = 10$. In Step 3 we resequence P_1 (the schedule O_{11}, O_{12} is replaced by O_{12}, O_{11}) and have now obtained a schedule with $C_{max} = 9$ with a disjunctive graph G' as displayed in Figure IV.53.

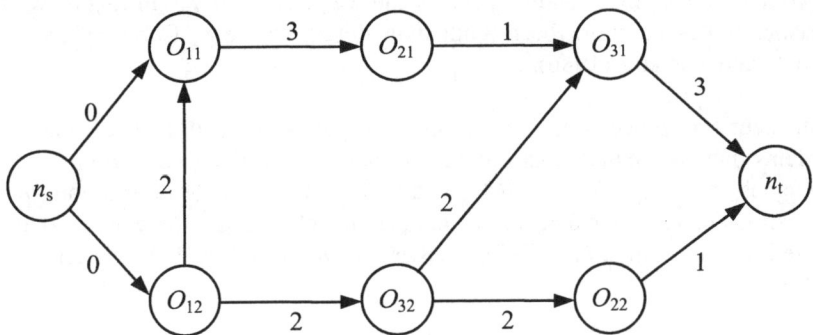

Figure IV.53

Finally in the third iteration, P_2 is scheduled. Both O_{21}, O_{22} and O_{22}, O_{11} are optimal with $L_{max} = 0$; we choose the second one (the first is actually a nonactive schedule). Resequencing P_1 and P_3 will not change any schedule. We end the algorithm with the optimal schedule already displayed in Figure IV.51; the corresponding optimal disjunctive graph G' with $C_{max} = 9$ is displayed in Figure IV.54.

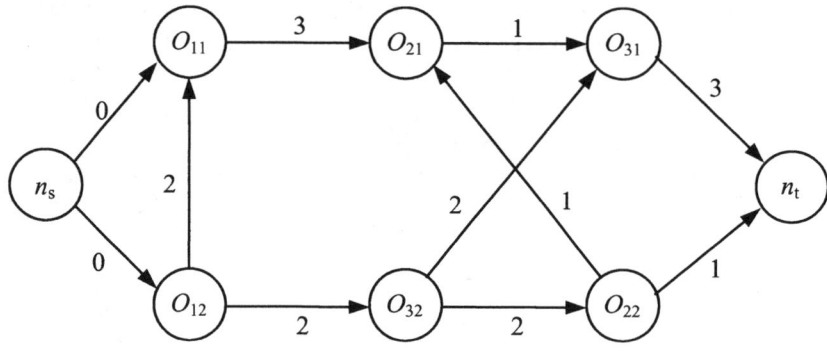

Figure IV.54

We would like to point out that the $1/r_j/L_{max}$ subproblems that need to be solved repeatedly actually may have additional restrictions due to precedence constraints between operations on machines already scheduled. Such precedence constraints could also have the effect that, when processing two consecutive operations on a machine, the second operation could not start immediately after the first one is completed, but only at some later time, causing a delay. We therefore refer to such constraints as *delayed precedence constraints*. For a discussion, see Pinedo (1995). The shifting bottleneck heuristic has turned out to perform very well in practice, as shown by extensive numerical testing. For instance, the 10-machine, 10-job problem of Fisher and Thompson, referred to earlier, was rapidly solved to optimality. Note that the NP-hard single-machine $1/r_j/L_{max}$ problem appears as a subproblem in the branch and bound procedure as well as in the shifting bottleneck heuristic. It has therefore been studied in detail; see, e.g., Carlier (1982) and Nowicki and Zdrzalka (1986).

Other heuristic procedures for solving the job shop and related scheduling problems include simulated annealing and tabu search. For a description of these techniques, see, e.g., Eiselt and Sandblom (2000). For a description of the applications of these local search techniques to scheduling problems, readers are referred to Anderson *et al.* (1997), Vaessens *et al.* (1996), and Ovacik and Uzsoy (1997).

4.4 Resource - Constrained Machine Scheduling

This section discusses resource-constrained machine scheduling problems. The model considered here is very similar to the one defined in Section 1.2. The only difference is the existence of s types of resources available in quantities $m_1, m_2,..., m_s$. For its processing, each task $T_j, j = 1, ..., n$, requires one processor as well as certain amounts of the resources, specified by a resource requirement vector $[R_1(T_j), R_2(T_j),..., R_s(T_j)]$, where $R_k(T_j) \in [0; m_k]$ with $k = 1, 2,..., s$, denotes the number of units of the k-th resource required for the processing of task T_j. In the case of job shop scheduling each operation is characterized by a resource requirement vector. We assume here that all required resources are allocated to a task before its processing begins, or resumes in the case of preemptive scheduling. Moreover, we assume that all resources are returned by a task after its completion, or in case of its preemption. For that purpose, these assumptions are often used in practice, despite the fact that they may lead to inefficient use of the resources.

Before discussing basic results in this area, we extend the classification scheme $\alpha/\beta/\gamma$ of Section 1.2.3 to include limited resources. In particular, this aspect is characterized by the parameter β_2 in the β field, where $\beta_2 \in \{\varnothing, res\ \lambda, \sigma, \rho\}$, which indicates

$\beta_2 = \varnothing$: there are no resource constraints, and
$\beta_2 = res\ \lambda, \sigma, \rho$: there are specified resource constraints, where $\lambda, \sigma, \rho \in \{\cdot, k\}$ denote the number of *resource types*, *resource limits* and *resource requirements*.

If $\lambda, \sigma, \rho = \cdot$, then there is an arbitrary number of resource types, resource availability, and resource requirements, respectively, provided by the user as problem input. On the other hand, if λ, σ, and ρ assume some fixed and given number, this number is stated in its space in the taxonomy.

There is a large body of work on resource constrained scheduling. One of the more important problems in resource-constrained scheduling deals with the nonpreemptive scheduling of independent tasks. Garey and Johnson (1975) suggest solving the problem of scheduling unit-length tasks on two machines with arbitrary resource constraints and requirements by the following algorithm. Note that a *maximal matching* in a graph is a maximal cardinality set of edges with no vertex in common.

> **Garey and Johnson's Algorithm for Problem $P2/res\cdots, p_j = 1/C_{max}$**

Step 1: Construct an n-node undirected graph G with each node symbolizing a distinct task and with an edge joining T_i and T_j if and only if $R_k(T_i) + R_k(T_j) \le m_k, k = 1, 2,..., s$.

Step 2: Find a maximal matching F in the graph G. An optimal schedule is of length $C^*_{max} = n - |F|$.

Step 3: Process in parallel all pairs of tasks joined by edges belonging to the set F. Other tasks are processed individually.

The main idea in this algorithm is the correspondence between a maximal matching in a graph displaying resource constraints, and a minimal-length schedule. Its complexity therefore depends on the complexity of the algorithm determining the maximal matching.

Example: Consider a problem with $m = 2$ machines, $n = 5$ tasks, $s = 2$ types of resources, of which $m_1 = 2$ and $m_2 = 3$ units are available. The resource requirements of the five tasks are 1, 0, 2, 1, 2 for the first resource and 2, 2, 0, 1, 1 for the second resource. The resulting maximal matching $F = \{(T_1, T_4), (T_2, T_5)\}$ is indicated by bold arcs in the graph in Figure IV.55, and a corresponding schedule with $C^*_{max} = n - |F| = 5 - 2 = 3$ is shown in Figure IV.56. Note that the sequence of tasks is arbitrary, e.g., another optimal schedule processes T_3 on P_1 first, then T_2 and T_5, and finally T_1 and T_4 in parallel.

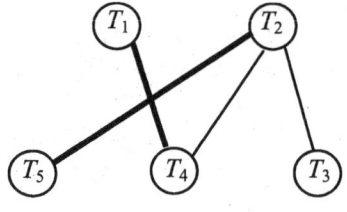

Figure IV.55

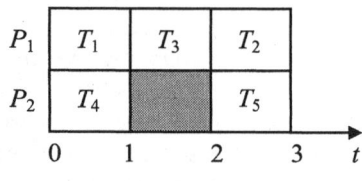

Figure IV.56

The case of only one single resource can be solved even more efficiently. Here, an optimal schedule is produced by ordering tasks in nonincreasing order of their resource requirements and assigning tasks in that order to the first free machine on which a given task can be processed, considering its resource constraint. Thus, problem $P2/res\ 1\cdot\cdot,\ p_j = 1/C_{max}$ can be solved in $O(n \log n)$ time. As a special case, consider zero-one resource requirements, i.e., $R_1(T_j) = 0 \vee 1$. The resulting problem can be solved in $O(n)$ time even for arbitrary ready times and an arbitrary number of machines, i.e., for problem $P/res\ 1\cdot 1,\ r_j,\ p_j = 1/C_{max}$. This is done by first assigning tasks with unit resource requirements up to m_1 in each slot, and then filling these slots with tasks with zero resource requirements. For further details, the reader is referred to Blazewicz (1978).

So far we have outlined two polynomial-time algorithms that solve nonpreemptive resource-constrained scheduling problems with tasks of unit length. Two more such algorithms are described in Blazewicz et al. (1986). It appears that these are all the polynomial results that we can obtain, since other problems of nonpreemptive scheduling under resource constraints have been proved **NP**-hard. Exploring the sensitivity of the parameters that have an influence on the hardness of the problem leads to the following observations. First, different ready times cause strong **NP**-hardness of the problem even for two processors and very simple resource requirements, i.e., problem *P2/res*1··, r_j, $p_j = 1/C_{max}$ is already strongly **NP**-hard, as shown by Blazewicz (1986). Secondly, an increase in the number of processors from two to three results in strong **NP**-hardness of the problem. That is, problem *P3/res* 1··, $p_j = 1/C_{max}$ is strongly **NP**-hard, as proved by Garey and Johnson (1975). Note that this is the famous 3-partition problem, the first problem to be proved strongly **NP**-hard. Finally, even the simplest precedence constraints result in the **NP**-hardness; that is, the problem *P2/res* 111, *chain*, $p_j = 1/C_{max}$ is **NP**-hard in the strong sense as pointed out by Blazewicz et al. (1983). A summary of **NP**-hardness results of different resource-constrained scheduling problems and their interrelationships can be found in Lawler et al. (1993).

Given that many resource-constrained scheduling problems are **NP**-hard and that they often have to be solved in real time, heuristics are called for, preferably those with guaranteed worst-case behavior. Below, we present three simple list scheduling heuristics, which differ from each other by the ordering of tasks on the list. It is also worth pointing out that resource-constrained scheduling with unit processing times of the tasks is equivalent to a variant of a bin packing problem (see, e.g., Eiselt and Sandblom, 2000) in which the number of items per bin cannot exceed *m*.

The two heuristic methods described below are very flexible: they can solve any scheduling problem with identical machines, arbitrary processing times, an arbitrary number of resources, and, if required, even precedence constraints.

Heuristic 1: *First fit (FF)*. Each task T_j is assigned sequentially to the earliest time slot for which no resource or machine limit is violated.

Heuristic 2: *First fit decreasing (FFD)*. A variant of the first algorithm applied to a list ordered in nonincreasing order of the *maximal relative resource requirement* $R_{max}(T_j)$, where $R_{max}(T_j) = \max \{R_\ell(T_j) / m_\ell : 1 \leq \ell \leq s \}$.

Example: Consider a scheduling problem $Pm/res\cdots/C_{max}$ with $m = 3$ processors, $n = 7$ tasks, and $s = 2$ resources, of which $m_1 = 4$ and $m_2 = 3$ units are available. The processing times are **p** = [3, 1, 5, 4, 6, 2, 4] and resource requirements [0, 2, 3, 2, 1, 0, 3] and [2, 1, 1, 0, 0, 1, 2]. The first fit heuristic then finds a schedule as shown in Figure IV.57; its makespan is $C_{max} = 14$.

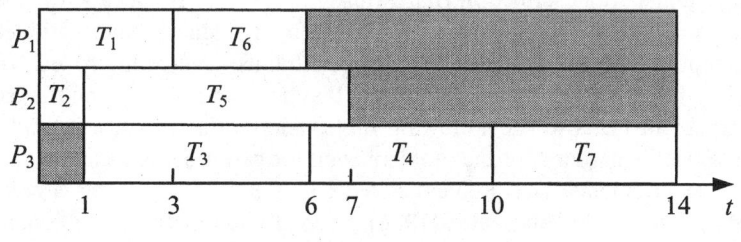

Figure IV.57

In order to apply the first fit decreasing heuristic, it is required to first calculate the maximal relative resource requirements. In this example, they are $R_{max}(T_j) = $ ⅔, ½, ¾, ½, ¼, ⅓, and ¾, respectively. The resulting schedule is then shown in Figure IV.58; its makespan is $C_{max} = 14$.

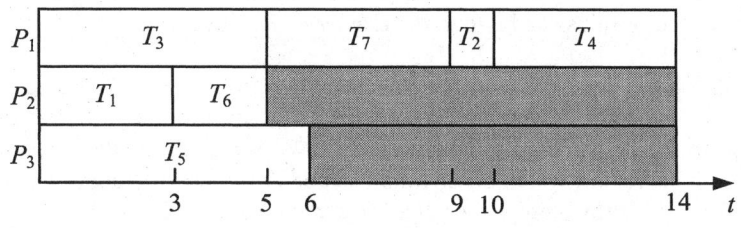

Figure IV.58

Krause *et al.* (1975) have derived asymptotic performance ratios for the two heuristic methods above, as applied to the problem $P/\text{res } 1\cdots, p_j=1/C_{max}$:

$$\frac{27}{10} - \left\lceil \frac{37}{10m} \right\rceil < R_{FF}^{\infty} < \frac{27}{10} - \frac{24}{10m}, \text{ and}$$

$$R_{FFD}^{\infty} = 2 - \frac{2}{m}.$$

Notice that for very large values of m, the performance bound improves from 2.7 to 2, as we apply the *FFD* heuristic as opposed to the *FF* heuristic. Garey and Graham (1975) have shown that for the problem $P/\text{res } \cdots/C_{max}$, the asymptotic performance bound of the *FF* algorithm is

$$R_{FF}^{\infty} = \min\left\{\frac{m+1}{2}, s+2-\frac{2s+1}{m}\right\},$$

while for the problem $P/\text{res } \cdots, prec/C_{max}$, the performance ratio is $R_{FF}^{\infty} = m$. Unfortunately, no results are reported on the probabilistic analysis of approximation algorithms for resource constrained scheduling.

Chapter 4: Dedicated Machine and Resource-Constrained Models

Finally, consider the problem $Pm/pmtn$, $res \cdots /C_{max}$, i.e., the same problem as above but with m machines instead of two. It is still possible to solve the problem in polynomial time via linear programming. Similarly, the related problem $P/pmtn$, $res \cdot\cdot 1/C_{max}$ can be solved by the generalization of the linear programming approach described at the end of Section 3.1.3. The latter approach is able to handle different ready times as well as the L_{max} criterion. Both approaches can also be adapted to cover the case of a uniconnected activity network in the same way as that described in Section 3.1.2. Finally, when analyzing preemptive scheduling, Krause *et al.* (1975) have analyzed the worst-case behavior of certain approximation algorithms for the problem $P/pmtn$, $res1\cdot\cdot/C_{max}$, finding the performance bounds $R_{FF}^{\infty} = R_{FFD}^{\infty} = 3 - \dfrac{3}{m}$. Surprisingly, using an ordered list does not improve these bounds.

REFERENCES

Achugbue, J.O., Chin, F.Y. (1982): Scheduling the Open Shop to Minimize Mean Flow Time. SIAM Journal on Computing 11, 709-720

Adams, J., Balas, E., Zawack, D. (1988): The Shifting Bottleneck Procedure for Job Shop Scheduling. Management Science 34, 391-401

Adolphson, D., Hu, T.C. (1973): Optimal Linear Ordering. SIAM Journal of Applied Mathematics 25, 403-423

Aho, A.V., Hopcroft, J.E., Ullman, J.D. (1974): The Design and Analysis of Computer Algorithms. Addison-Wesley, Reading

Ahuja, R.K., Magnanti, T.L., Orlin, J.B. (1993): Network Flows. Prentice-Hall, Englewood Cliffs, NJ

Akers, S.H. (1956): A Graphical Approach to Production Scheduling Problems. Operations Research 4, 244-245

Alonso, W. (1964): Location and Land Use. Harvard University Press, Princeton, NJ

Anderson, E.J., Glass, C.A., Potts, C.N. (1997): Local Search in Combinatorial Optimization: Applications in Machine Scheduling. In: Aarts, E.H.L., Lenstra, J.K. (eds.): Local Search in Combinatorial Optimization. J. Wiley, Chichester, UK

Armour, G.C., Buffa, E.S. (1963): A Heuristic Algorithm and Simulation Approach to Relative Location of Facilities. Management Science 9, 294-309

Baird, D.G., Gertner, R.H., Picker, R.C., (1994): Game Theory and the Law. Harvard University Press, Cambridge, MA

Baker, K.R. (1974): Introduction to Sequencing and Scheduling. J. Wiley & Sons, New York

Balakrishnan, J., Cheng, C.-H., Wong, K.-F. (2003): FACOPT: a User Friendly FACility Layout OPTimization System. Computers & Operations Research 30, 1625-1641

Balinski, M.L. (1965): Integer Programming: Methods, Uses, Computations. Management Science 12, 253-313

Ball, M.O., Lin, F.L. (1993): A Reliability Model Applied to Emergency Service Vehicle Location. Operations Research 41, 18-36

Banzhaf, J.F. III (1965): Weighted Voting doesn't Work: A Mathematical Analysis. Rutgers Law Review 19, 317-343

Barany, L. (1981): A Vector-Sum Theorem and its Application to Improving Flow Shop Guarantees. Mathematics of Operations Research 6, 445-452

Barda, O.H., Dupuis, J., Lencioni, P. (1990): Multicriteria Location of Thermal Power Plants. European Journal of Operational Research 45, 332-346

Barzilai, J., Cook, W., Golany B. (1987): Consistent Weights for Judgments Matrices of the Relative Importance for Alternatives. Operations Research Letters 6, 131-134

Barzilai, J., Golany, B. (1994): AHP Rank Reversal, Normalization and Aggregation Rules. INFOR 32, 57-64

Bazaraa, M.S., Kirca, O. (1983): A Branch-and-Bound Based Heuristic for Solving the Quadratic Assignment Problem. Naval Research Logistics Quarterly 30, 287-304

Beckmann, H.J. (1972): Spatial Cournot Oligopoly. Papers of the Regional Science Association 28, 37-47

Belton, V., Gear, A.E. (1983): On a Shortcoming of Saaty's Method of Analytic Hierarchies. Omega 11, 228-230

Belton, V., Stewart, T.J. (2002): Multiple Criteria Decision Analysis: An Integrated Approach. Kluwer Academic Publishers, Boston

Belton, V., Vickers, S.P. (1993): Demystifying DEA–A Visual Interactive Approach Based on Multiple Criteria Analysis. Journal of the Operational Research Society 44, 883-896

Benati, S., Laporte, G. (1994): Tabu Search for the $(r|X_p)$ Medianoid and $(r|p)$ Centroid Problems in the Plane. Location Science 2, 193-204

Benayoun, R., Roy, B., Sussman, B. (1966): ELECTRE: Une méthode pour guider le choix en presence de points de vue multiples. SEMA Note 49

Bhadury, J., Eiselt, H.A. (1998): Reachability of Locational Nash Equilibria. Operations Research Spektrum 20, 101-107

Bhadury, J., Eiselt, H.A., Jaramillo, J.H. (2003): An 'Alternate' Heuristic for Medianoid and Centroid Problems in the Plane. Computers & Operations Research 30, 553-565

Bielza, C., Shenoy, P.P. (1999): A Comparison of Graphical Techniques for Asymmetric Decision Problems. Management Science 45, 1552-1569

Blazewicz, J. (1976): Scheduling Dependent Tasks with Different Arrival Times to Meet Deadlines. In: Gelenbe, E., Beilner, H. (eds.): Modelling and Performance Evaluation of Computer Systems. North Holland, Amsterdam, 57-65

Blazewicz, J. (1976): Scheduling Dependent Tasks with Different Arrival Times to Meet Deadlines. Modeling and Performance Evaluation of Computer Systems, North-Holland, Amsterdam, 57-65

Blazewicz, J. (1977): Simple Algorithms for Multiprocessor Scheduling to Meet Deadlines. Information Processing Letters 6, 162-164

Blazewicz, J. (1978): Complexity of Computer Scheduling Algorithms Under Resource Constraints. Proceedings of the 1. Meeting AFCET-SMF on Applied Mathematics, Palaiseau, 169-178

Blazewicz, J. (1987): Selected Topics in Scheduling Theory. Annals of Discrete Mathematics 31, 1-60

Blazewicz, J., Barcelo, J., Kubiak, W., Röck, H. (1986): Scheduling Tasks on Two Processors with Deadlines and Additional Resources. European Journal of Operational Research 26, 364-370

Blazewicz, J. Cellary, W., Słowinski, R., Weglarz, J. (1986): Scheduling Under Resource Constraints: Deterministic Models. Annals of Operations Research 7, J.C. Baltzer, Basel

Blazewicz, J., Cellary, W., Weglarz, J. (1977): A Strategy for Scheduling Splittable Tasks to Reduce Schedule Length. Acta Cybernetica 3, 99-106

Blazewicz, J., Ecker, K. (1983): A Linear Time Algorithm for Restricted Bin Packing and Scheduling Problems. Operations Research Letters 2, 80-83

Blazewicz, J., Ecker, K., Pesch, E., Schmidt, G., Weglarz, J. (1996): Scheduling Computer and Manufacturing Processes. Springer-Verlag, Berlin

Blazewicz, J., Lenstra, J.K., Rinnooy Kan, A.H.G. (1983): Scheduling Subject to Resource Constraints: Classification and Complexity. Discrete Applied Mathematics 5, 11-24

Blazewicz, J., Ecker, K., Schmidt, G., Weglarz, J. (1993): Scheduling in Computer and Manufacturing Systems. Springer-Verlag, Berlin

Böttcher, J., Drexl, A., Kolisch, R., Salewski, F. (1999): Project Scheduling Under Partially Renewable Resource Constraints. Management Science 45, 543-559

Brandeau, M.L., Chiu, S.S. (1989): An Overview of Representative Problems in Location Research. Management Science 35, 645-674

Brans, J.P., Vincke, Ph. (1985): A Preference Ranking Organization method. Management Science 31, 647-656

Brockmeyer, E., Halstrom, H.L., Jensen, A. (1948): The life and works of A K Erlang. Copenhagen.

Brucker, P.J. (1976): Sequencing Unit-Time Jobs with Treelike Precedence on m Machines to Minimize Maximum Lateness. Proceedings of the IX International Symposium on Mathematical Programming, Budapest

Brucker, P. (1981): Minimizing Maximum Lateness in a Two-Machine Unit-Time Job Shop. Computing 27, 367-370

Brucker, P. (1988): An Efficient Algorithm for the Job-Shop Problem with Two Jobs. Computing 40, 353-359

Brucker, P.J. (1995) Scheduling Algorithms. Springer-Verlag, Berlin

Brucker, P.J., Garey, M.R., Johnson, D.S. (1977): Scheduling Equal-Length Tasks Under Treelike Precedence Constraints to Minimize Maximum Lateness. Mathematics of Operations Research 2, 275-284

Brucker, P., Jurisch, B., Sievers, B. (1992): A Branch and Bound Algorithm for the Job-Shop Scheduling Problem. Osnabrücker Schriften zur Mathematik, Universität Osnabrück

Brucker, P., Knust, S. (eds.) (2003): Sequencing and Scheduling. Special issue, European Journal of Operational Research 149/2

Brucker, P., Knust, S., Schoo, A., Thiele, O. (1998): A Branch & Bound Algorithm for the Resource-Constrained Project Scheduling Problem. European Journal of Operational Research 107, 272-288

Bruno, J., Coffman, Jr., E.G., Sethi, R. (1974): Scheduling Independent Tasks to Reduce Mean Finishing Time. Communications of the ACM 17, 382-387

Buffa, E.S., Armour, G.C., Vollman, T.E. (1964): Allocating Facilities with CRAFT. Harvard Business Review 42, 136-159

Burkard, R. E. (1973): Die Störungsmethode zur Lösung quadratischer Zuordnungsprobleme. Operations Research Verfahren 16, 84-108

Burkard, R.E., Offerman, J. (1977): Entwurf von Schreibmaschinentastaturen mittels quadratischer Zuordnungsprobleme. Zeitschrift für Operations Research 21, B121-B132

Camerini, P.M., Galbiati, G., Maffioli, F. (1988): Algorithms for Finding Optimum Trees: Description, Use and Evaluation. Annals of Operations Research 13, 265-397

Campbell, J.F. (ed.) (1996): Hub Location. Location Science 4/3, special issue

Campbell, J.F., Ernst, A.T., Krishnamoorthy, M. (2002): Hub Location Problems. In: Drezner, Z., Hamacher, H.W. (eds.): Facility Location: Applications and Theory, 373-403. Springer-Verlag, Berlin

Carlier, J. (1982): The One Machine Sequencing Problem. European Journal of Operational Research 11, 42-47

Cela, E. (1998): The Quadratic Assignment Problem: Theory and Applications. Kluwer Academic Publishers, Dordrecht, The Netherlands

Charnes, A., Cooper, W.W., Rhodes, E. (1978): Measuring Efficiency of Decision Making Units. European Journal of Operational Research 2, 429-444

Chen, N.-F., Lui, C.L. (1975): On a Class of Scheduling Algorithms for Multiprocessor Computing Systems. In: Feng, T.-Y. (ed.): Parallel Processing, Lecture Notes in Computer Science, 24, Springer-Verlag, Berlin, 1-16

Cho, Y., Sahni, S. (1981): Preemptive Scheduling of Independent Jobs with Release and Due Times on Open, Flow and Job Shops. Operations Research 29, 511-522

Chrétienne, Ph., Coffman, E.G. Jr., Lenstra, J.K., Liu, Z. (eds.) (1995): Scheduling Theory and its Applications. J. Wiley & Sons, Chichester, UK

Christofides, N., Alvarez-Valdés, R., Tamarit, J.M. (1987): Project Scheduling with Resource Constraints: A Branch and Bound Approach. European Journal of Operational Research 29, 262-273

Christofides, N., Beasley, J.E. (1982): A Tree Search Algorithm for the p-Median problem. European Journal of Operational Research 10, 196-204

Christofides, N., Mingozzi, A., Toth, P. (1980): Contributions to the Quadratic Assignment Problem. European Journal of Operational Research 4, 243-247

Church, R.L. (1984): The Planar Maximal Covering Location Problem. Journal of Regional Science 24, 185-201

Church, R.L., Garfinkel, R.S. (1978): Locating an Obnoxious Facility on a Network. Transportation Science 12, 107-118

Church, R.L., ReVelle, C.S. (1974): The Maximal Covering Location Problem. Papers of the Regional Science Association 32, 101-118

Church, R.L., ReVelle, C.S. (1976): Theoretical and Computational Links Between the p-Median, Location Set Covering, and the Maximal Covering Location Problems. Geographical Analysis 8, 406-415

Church, R.L., Roberts, K.L. (1983): Generalized Coverage Models and Public Facility Location. Papers of the Regional Science Association 53, 117-135

Church, R.L., Sorenson, P. (1994): Integrating Normative Location Models into GIS: Problems and Prospects with the p-Median Model. Technical Report 94-5, National Center for Geographic Information and Analysis, University of California at Santa Barbara

Coffman, E.G., Jr. (ed.) (1976): Computer and Job/Shop Scheduling. J. Wiley, New York

Coffman, E.G., Jr., Denning, P.J. (1973): Operating Systems Theory. Prentice-Hall, Englewood Cliffs, NJ

Coffman, E.G. Jr., Frederickson, G.N., Leuker, G.S. (1983): Probabilistic Analysis of the LPT Processor Scheduling Heuristic. (Unpublished paper)

Coffman, E.G., Jr., Frederickson, G.N., Leuker, G.S. (1984): A Note on Expected Makespans for Largest-First Sequences of Independent Tasks on Two Processors. Mathematics of Operations Research 9, 260-266

Coffman, E.G., Jr., Garey, M.R., Johnson, D.S. (1978): An Application of Bin-Packing to Multiprocessor Scheduling. SIAM Journal on Computing 7, 1-17

Coffman, E.G., Jr., Garey, M.R., Johnson, D.S. (1984): Approximation Algorithms for Bin-Packing: an Updated Survey. In: Ausiello, G., Luccertini, M., Serafini, P. (eds.): Algorithm Design for Computer System Design. Springer Verlag, Wien, 49-106

Coffman, E.G., Jr., Graham, R.L. (1972): Optimal Scheduling for Two Processor Systems. Acta Informatica 1, 200-213

Coffman, E.G., Jr., Sethi, R. (1976): A Generalized Bound on LPT Sequencing. RAIRO-Informatique 10, 17-25

Coffman, E.G., Lueker, G.S. (1991): Probabilistic Analysis of Packing and Related Partitioning Problems. Wiley, New York, NY

Coffman, E.G., Whitt, W. (1995): Recent Asymptotic Results in the Probabilistic Analysis of Schedule Makespans. In: Chrétienne, Ph., Coffman, E.G., Lenstra, J.K., Liu, Z. (eds.): Scheduling Theory and its Applications. Wiley, Chichester, 15-31

Cohon, J.L. (1978): Multiobjective Programming and Planning. Academic Press, New York

Connolly, D.T. (1990): An Improved Annealing Scheme for the QAP. European Journal of Operational Research 46, 93-100

Conway, R.W., Maxwell, W.L., Miller, L.W. (1967): Theory of Scheduling. Addison-Wesley, Reading, MA

Cook, S.A. (1971): The Complexity of Theorem-Proving Procedures. Proceedings of the 3rd Annual ACM Symposium on the Theory of Computing, 151-158

Cooper, L. (1963): Location-Allocation Problems. Operations Research 11, 331-343

Cooper, L. (1964): Heuristic Methods for Location-Allocation Problems. SIAM Review 6, 37-53

Cooper, L. (1973): *N*-Dimensional Location Models: an Application to Cluster Analysis. Journal of Regional Science 13, 41-54

Current, J.C., O'Kelly, M. (1992): Locating Emergency Warning Sirens. Decision Sciences 23, 221-234

Current, J.R., Schilling, D. (1987): Elimination of Source A and B Errors in *p*-Median Location Problems. Geographical Analysis 19, 95-110

Current, J.R., Schilling, D. (1990): Analysis of Errors Due to Demand Aggregation in the Set Covering and Maximal Covering Location Problems. Geographical Analysis 22, 116-126

Current, J.R., Storbeck, J.E. (1988): Capacitated Covering Models. Environment and Planning 15, 153-163

D'Aspremont, C., Gabszewicz, J.J., Thisse, J.-F. (1979): On Hotelling's 'Stability in Competition.' Econometrica 47, 1145-1150

Dalton, H. (1920): The Measurement of Inequity of Incomes. The Economic Journal 30, 348-361

Daskin, M.S. (1983): Maximum Expected Covering Location Model: Formulation, Properties, and Heuristic Solution. Transportation Science 17, 48-70

Daskin, M.S. (1995): Network and Discrete Location. Wiley-Interscience, New York, NY

Daskin, M.S., Stern, E.H. (1981): A Hierarchical Set Covering Model for Emergency Medical Service System Design. Transportation Science 15, 137-152.

De Reyck, B., Herroelen, W., (1999): The Multi-Mode Resource-Constrained Project Scheduling Problem with Generalized Precedence Constraints. European Journal of Operations Research 119, 538-556

De Werra, D. (1984): Preemptive Scheduling, Linear Programming and Network Flows. SIAM Journal on Algebraic and Discrete Mathematics 5, 11-20

Deeney, M. (1999): Optimal Location of Mesometric Facilities. PhD Dissertation. The Johns Hopkins University, Baltimore, MD

Deisenroth, M.P., Apple, M.A. (1972): Computerized Plant Layout Analysis and Evaluation Technique (PLANET). Technical Papers, AIIE 25 Anniversary Conference & Commission, Norcross, GA, 75-87

Demeulemeester, E., Herroelen, W., (1992): A Branch and Bound Procedure for the Multiple Resource-Constrained Project Scheduling Problem. Management Science 38, 1803-1818

Dickey, J.W., Hopkins, J.W. (1972): Campus Building Arrangement Using TOPAZ. Transportation Research 6, 59-68

Domschke, W., Drexl, A. (1985): Location and Layout Planning: An International Bibliography. Volume 238 in Lectures Notes in Economics and Mathematical Systems, Springer-Verlag, Berlin-Heidelberg-New York

Drezner, Z. (1982): Competitive Location Strategies for two Facilities. Regional Science and Urban Economics 12, 485-493

Drezner, Z. (1992): A Note on the Weber Location Problem. Annals of Operations Research 40, 153-161

Drezner, Z. (1995): Lower Bounds Based on Linear Programming for the Quadratic Assignment Problem. Computational Optimization and Applications 4, 159-165

Drezner, Z. (2002): A New Genetic Algorithm for the Quadratic Assignment Problem. California State University, Fullerton, CA

Drezner, Z., Klamroth, K., Schöbel, A., Wesolowsky, G.O. (2002): The Weber Problem. In: Drezner, Z., Hamacher, H. (eds.): Facility Location: Applications and Theory. Springer-Verlag, Berlin, 1-36

Du, J., Leung, J.Y.-T. (1989): Scheduling Tree-Structured Tasks on Two Processors to Minimize Schedule Length. SIAM Journal of Discrete Mathematics 2, 176-196

Eaton, D.J., Church, R.L., Bennet, V.L., Hamon, B.L., Lopez, L.G.V. (1981): On Deployment of Health Resources in Rural Valle de Cauca, Columbia. TIMS Studies in Mangement Sciences 17, 331-359

Eaton, D.J., Daskin, M.S., Simmons, D., Bulloch, B., Jansma, G. (1985): Determining Emergency Medical Services Vehicle Deployment in Austin, Texas. Interfaces 15, 96-108

Edmonds, J. (1965): Maximum Matching and a Polyhedron with 0,1 Vertices. Journal of Research of the National Bureau of Standards (B) 69, 125-130

Edwards, C.S. (1980): A Branch and Bound Algorithm for the Koopmans-Beckmann Quadratic Assignment Problem. Mathematical Programming Study 13, 35-52

Edwards, W. (1971): Social Utilities. The Engineering Economist, New York Summer Symposium Series 6, 119-129

Edwards, W. (1977): How to Use Multiattribute Utility Measurement for Social Decision Making. IEEE Transactions on Systems, Man, and Cybernetics SMC-7, 326-340

Efroymson, M.A., Ray, T.L. (1966): "A Branch and Bound Algorithm for Plant Location. Operations Research 14, 361-368

Efroymson, M.A., Ray, T.L. (1966): Branch and Bound Algorithm for Plant Location. Naval Research Logistic Quarterly 16, 331-344

Ehrgott, M. (2001): Stationing of Rescue Helicopters in South Tyrol. Department of Engineering Science, University of Auckland, Auckland, New Zealand

Ehrgott, M., Gandibleux, X. (2002): Multiple Criteria Optimization: State of the Art Annotated Bibliographic Surveys. Kluwer Academic Publishers, Boston, MA.

Eiselt, H.A. (1989): Modeling Business Problems with Voronoi Diagrams. Canadian Journal of Administrative Sciences 6, 43-53

Eiselt, H.A. (1992): Location Modeling in Practice. American Journal of Mathematical and Management Sciences 12, 3-18

Eiselt, H.A. (2001): Consistency in Estimating Weights. INFOR 39, 64-70

Eiselt, H.A., Bhadury, J. (1995): Stability of Nash Equilibria in Locational Games. Rechereche opérationnelle/Operations Research 29, 19-33

Eiselt, H.A., Langley, A. (1990): Some Extensions of Domain Criteria in Decision Making Under Uncertainty. Decision Sciences 21, 138-153

Eiselt, H.A., Laporte, G. (1995): Objectives in Location Problems. In: Drezner, Z. (ed.): Facility Location: A Survey of Applications and Methods. Springer-Verlag, New York, 151-180

Eiselt, H.A., Laporte, G. (1996): Sequential Location Problems. European Journal of Operational Research 96, 217-231

Eiselt, H.A., Laporte, G. (1998): Demand Allocation Functions. Location Science 6, 175-187

Eiselt, H.A., Laporte, G., Thisse, J.-F. (1993): Competitive Location Models: A Framework and Bibliography. Transportation Science 27, 44-53

Eiselt, H.A., Pederzoli, G., Sandblom, C.-L. (1987): Continuous Optimization Models. W. de Gruyter, Berlin

Eiselt, H.A., Sandblom, C.-L. (2000): Integer Programming and Network Models. Springer-Verlag, Berlin

Eiselt, H.A., Sandblom, C.-L., Barnsley, C. (1997): A Computational Investigation into Shapes of Polytopes. Computers & Industrial Engineering 32, 19-28

Elmaghraby, S.E. (1977): Activity Networks: Project Planning and Control by Network Models. Wiley, New York

Elmaghraby, S.E. (2000): On Criticality and Sensitivity in Activity Networks. European Journal of Operational Research 127, 220-238

Elshafei, A.N. (1977): Hospital Lay-Out as a Quadratic Assignment Problem. Operational Research Quarterly 28, 167-179

Elzinga, J., Hearn, D. (1972): Geometrical Solutions for Some Minimax Location Problems. Transportation Science 6, 379-394

Erkut, E., Baptie, T., von Hohenbalken, B. (1990): The Discrete p-Maxian Location Problem. Computers & Operations Research 17, 51-61

Erkut, E., Neuman, S. (1989): Analytical Models for Locating Undesirable Facilities. European Journal of Operational Research 40, 275-291

Erlang, A.K. (1909): The Theory of Probability and Telephone Conversations.

Erlenkotter, D. (1978): A Dual-Based Procedure for Uncapacitated Facility Location. Operations Research 26, 992-1009

Fechner, G.T. (1860): *Elemente der Psychophysik*. Breitkopf & Härtel, Leipzig, Germany

Federgruen, A., Groenevelt, H. (1986): Preemptive Scheduling of Uniform Machines by Ordinary Network Flow Techniques. Management Science 32, 341-349

Finke, G., Burkard, R.E., Rendl, F. (1987): Quadratic Assignment Problems. Annals of Discrete Mathematics 31, 61-82

Fisher, H., Thompson, G.L. (1963): Probabilistic Learning Combinations of Local Job-Shop Scheduling Rules. In: Muth, J.F., Thompson, G.L. (eds.): Industrial Scheduling. Prentice Hall, Englewood Cliffs, NJ, 225-251

Fisher, M.L. (1981): The Lagrangean Relaxation Method for Solving Integer Programming Problems. Management Science 27, 1-18

Forman, E.H., Gass, S.I. (2001): The Analytic Hierarchy Process—an Exposition. Operations Research 49, 469-486

Foulds, L.R. (1983): Techniques for Facilities Layout: Deciding Which Pairs of Activities Should be Adjacent. Management Science 29, 1414-1426

Francis, R.L., Lowe, T.J., Tamir, A. (2002): Demand Point Aggregation for Location Models. In: Drezner, Z., Hamacher, H. (eds.): Facility Location: Applications and Theory. Springer-Verlag, Berlin, 207-232

Francis, R.L., McGinnis, L.F. White, J.A. (1992): Facility Layout and Location: An Analytical Approach. 2nd ed., Prentice-Hall, Englewood Cliffs, NJ

Francis, R.L., White, J.A. (1974): Facility Layout and Location, Prentice-Hall, Englewood Cliffs, NJ

French, S. (1982): Sequencing and Scheduling: An Introduction to the Mathematics of the Job-Shop. Horwood, Chichester

Frenk, J.B.G., Rinnooy Kan, A.H.G. (1984): The Asymptotic Optimality of the LPT Scheduling Heuristic. Report Erasmus University, Rotterdam

Frenk, J.B.G., Rinnooy Kan, A.H.G. (1986): The Rate of Convergence to Optimality of the LPT Rule. Discrete Applied Mathematics 14, 187-197

Fudenberg, D., Tirole, J. (1993): Game Theory. The MIT Press, Cambridge, MA

Fujii, M., Kasami, T., Ninomiya, K. (1969, 1971): Optimal Sequencing of Two Equivalent Processors. SIAM Journal on Applied Mathematics 17, 784-789, and 20, 141

Fujiwara, O., Makjamroen, T., Gupta, K.K. (1987): Ambulance Deployment Analysis: a Case Study of Bangkok. European Journal of Operational Research 31, 9-18

Gabow, H.N. (1982): An Almost-Linear Algorithm for Two Processor Scheduling. Journal of the ACM 29, 766-789

Garey, M.R., Graham, R.L. (1975): Bounds for Multiprocessor Scheduling with Resource Constraints. SIAM Journal on Computing 4, 187-200

Garey, M.R., Johnson, D.S. (1975): Complexity Results for Multiprocessor Scheduling Under Resource Constraints. SIAM Journal on Computing4, 397-411

Garey, M.R., Johnson, D.S. (1976): Scheduling Tasks with Nonuniform Deadlines on Two Processors. Journal of the ACM 23, 461-467

Garey, M.R., Johnson, D.S. (1977): Two-Processor Scheduling with Start-Times and Deadlines. SIAM Journal on Computing 6, 416-426

Garey, M.R., Johnson, D.S. (1979): Computers and Intractability: A Guide to the Theory of NP-Completeness. Freeman, New York

Garey, M.R., Johnson, D.S., Tarjan, R.E., Yannakakis, M. (1983): Scheduling Opposing Forests. SIAM Journal of Algebraic and Discrete Methods 4, 72-93

Gavett, J.W., Plyter, N.V. (1966): The Optimal Assignment of Facilities to Locations by Branch and Bound. Operations Research 14, 210-232

Geoffrion, A., McBride, R. (1978): Lagrangian Relaxation Applied to Capacitated Facility Location Problems. AIIE Transactions 10, 40-47

Gilmore, P.C. (1962): Optimal and Suboptimal Algorithms for the Quadratic Assignment Problem. SIAM Journal of Applied Mathematics 10, 305-313

Glover, F. (1988): Tabu Search. Report 88-3. Center for Applied Artificial Intelligence (CAAI). Graduate School of Business, University of Colorado

Golden, B.L., Wasil, E.A. (2003): Analytic Hierarchy Process. Focused issue of Computers & Operations Research 30/10

Goldman, A. (1971): Optimum Center Location on Simple Networks. Transportation Science 5, 212-221

Goldman, A.J., Dearing, P.M. (1975): Concepts of Optimal Location for Partially Noxious Facilities. Bulletin of the Operations Research Society of America 23, Supplement 1, 331

Gonzalez, T., Ibarra, O.H., Sahni, S. (1977): Bounds for LPT Schedules on Uniform Processors. SIAM Journal on Computing 6, 155-166

Gonzalez, T., Sahni, S. (1976): Open Shop Scheduling to Minimize Finish Time. Journal of the ACM 23, 665-679

Gonzalez, T., Sahni, S. (1978): Preemptive Scheduling of Uniform Processor Systems. Journal of the Association of Computing Machinery 25, 92-101

Goodchild, M.F. (1979): The Aggregation Problem in Location-Allocation. Geographical Analysis 11, 240-255

Graham, R.L. (1966): Bounds for Certain Multiprocessing Anomalies. Bell System Technical Journal 25, 1563-1581

Graham, R.L. (1969): Bounds for Multiprocessing Timing Anomalies. SIAM Journal on Applied Mathematics 17, 263-269

Graham, R.L. (1979): Combinatorial Scheduling Theory. In: Steen, L.A. (ed.): Mathematics Today: Twelve Informal Essays. Springer-Verlag, New York

Graham, R.L., Lawler, E.L., Lenstra, J.K., Rinnooy Kan, A.H.G. (1979): Optimization and Approximation in Deterministic Sequencing and Scheduling Theory: a Survey. Annals of Discrete Mathematics 5, 287-326

Graves, G.W., Whinston, A.B. (1970): An Algorithm for the Quadratic Assignment Problem. Management Science 17, 453-471

Hadley, S., Rendl, F., Wolkowicz, H. (1990): Bounds for the Quadratic Assignment Problem Using Continuous Optimization Techniques. In: Integer Programming and Combinatorial Optimization. University of Waterloo Press, Waterloo, 237-248

Hadley, S., Rendl, F., Wolkowicz, H. (1992): A New Lower Bound via Projection for the Quadratic Assignment Problem. Mathematics of Operations Research 17, 727-739

Hakimi, S.L. (1964): Optimum Locations of Switching Centers and the Absolute Centers and Medians of a Graph. Operations Research 12, 450-459

References

Hakimi, S.L. (1965): Optimum Distribution of Switching Centers in a Communication Network, and Some Related Graph Theoretic Problems. Operations Research 12, 462-475

Hakimi, S.L. (1983): On Locating New Facilities in a Competitive Environment. European Journal of Operational Research 12, 29-35

Hakimi, S.L. (1990): Locations with Spatial Interactions: Competitive Locations and Games. In: Mirchandani, P.B., Francis, R.L. (eds.): Discrete Location Theory, Wiley-Interscience, New York, 439-478

Hall, M. (1959): The Theory of Groups. Macmillan, New York

Halpern, J. (1976): The Location of a Center-Median Convex Combination on an Undirected Tree. Journal of Regional Science 16, 237-245

Hamacher, H.W., Nickel, S. (1996): Multicriteria Planar Location Problems. European Journal of Operational Research 94, 66-86

Handler, G.Y. (1974): Minimax Network Location Theory and Algorithms. Ph.D. dissertation, Massachusetts Institute of Technology, Cambridge, MA

Hanjoul, P., Peeters, D. (1985): A Comparison of Two Dual-Based Procedures for Solving the p-Median Problem. European Journal of Operations Research 20, 387-396

Hanne, T (2001): Intelligent Strategies for Meta Multiple Criteria Decision making. Kluwer, Boston, MA

Harsanyi, J.C. (1964): A Solution for Non-Cooperative Games. In: Dresher, M., Shapley, L.S., Tucker, A.W. (eds.): Advances in Game Theory. Annals of Mathematics Studies 52, Princeton University Press, Princeton, NJ

Harsanyi, J.C. (1975): Can the Maximin Principle Serve as a Basis for Morality? A Critique of John Rawls's Theory. American Political Science Review 69, 594-606

Hartmann, S. (1998): A Competitive Genetic Algorithm for Resource-Constrained Project Scheduling. Naval Research Logistics 45, 733-750

Harwitz, M., Lentnek, B., Rogerson, P., Smith, T.E. (1998): Optimal Search on Spatial Paths with Recall, Part I: Theoretical Foundations. Papers in Regional Science 77, 301-327

Harwitz, M., Lentnek, B., Rogerson, P., Smith, T.E. (2000): Optimal Search on Spatial Paths with Recall, Part II: Computational Procedures and Examples. Papers in Regional Science 79, 293-305

Haywood, O. G. Jr. (1954): Military Decision and Game Theory. Journal of the Operations Research Society of America 2, 365-385

Hefetz, H, Adiri, I. (1982): An Efficient Optimal Algorithm for the Two Machine Unit-Time Jobshop Schedule-Length Problem. Mathematics of Operations Research 7, 354-360

Heffley, D.R. (1977): Assigning Runners to a Relay Team. In: Ladany, S.P., Machol, R.E. (eds.): Optimal Strategies in Sports. North-Holland, Amsterdam, 169-171

Helms, B.P., Clark, R.M. (1971): Location Models for Solid Waste Management. Journal of Urban Planning and Development Division (ASCE) UP1, 1-13

Hillsman, E.L. (1984): The p-Median Structure as a Unified Linear Model for Location-Allocation Analysis. Environment and Planning B 16, 305-318

Hillsman, E.L., Rhoda, R. (1978): Errors in Measuring Distances From Populations to Service Centers. Annals of Regional Science 12, 74-88

Hochbaum, D.S., Shmoys, D.B. (1987): Using Dual Approximation Algorithms for Scheduling Problems: Theoretical and Practical Results. Journal of the Association for Computing Machinery 34, 144-162

Hogan, K., ReVelle, C.S. (1986): Concepts and Applications of Backup Coverage. Management Science 32, 1434-1444

Hopmans, A.C.M. (1986): A Spatial Interaction Model for Branch Bank Accounts. European Journal of Operational Research 27, 242-250

Horn, W.A. (1973): Minimizing Average Flow Time with Parallel Machines. Operations Research 21, 846-847

Horn, W.A. (1974): Some Simple Scheduling Algorithms. Naval Research Logistics Quarterly 21, 177-185

Horowitz, E., Sahni, S. (1976): Exact and Approximate Algorithms for Scheduling Non-Identical Processors. Journal of the ACM 23, 317-327

Horvath, E.G., Lam, S., Sethi, R. (1977): A Level Algorithm for Preemptive Scheduling. Journal of the ACM 24, 32-43

Hotelling, H. (1929): Stability in Competition. Economic Journal 39, 41-57

Hu, T.C. (1961): Parallel Sequencing and Assembly Line Problems. Operations Research 9, 841-848

Hua Lo-Keng and Others (1962): Application of Mathematical Methods to Wheat Harvesting. Chinese Mathematics 2, 77-91

Huber, D., Church, R. (1985): Transmission Corridor Location Modelling. The Transportation Journal, ASCE, 3, 114-129

Huriot, J.M., Thisse, J.-F. (eds.) (2000): Economics of Cities. Cambridge University Press, Cambridge, UK

Hurter, A.P., Martinich, J.S. (1989): Facility Location and the Theory of Production. Kluwer Academic Publishers, Boston, MA

Huxley, S.J. (1982): Finding the Right Spot for a Church Camp in Spain. Interfaces 12, 108-114

Hwang, C.L., Yoon, K. (1981): Multiple Attribute Decision Making: Methods and Applications. Springer-Verlag, New York

Ibarra, O.H., Kim, C.E. (1977): Heuristic Algorithms for Scheduling Independent Tasks on Nonidentical Processors. Journal of the ACM 24, 280-289

Igelmund, G., Radermacher, F.J. (1983): Algorithmic Approaches to Preselective Strategies for Stochastic Scheduling Problems. Networks 13, 29-48

Ignizio, J.P. (1976) Goal Programming and Extensions. Heath, Lexington, MA

Ignizio, J.P. (1982): Linear Programming in Single- and Multiple-Objective Systems. Prentice-Hall, Englewood Cliffs, NJ

Jackson, J.R. (1955): Scheduling a Production Line to Minimize Maximum Tardiness. Research Report 43, Management Research Project. University of California, Los Angeles

Jackson, J.R. (1956): An Extension of Johnson's Result on Job Lot Scheduling. Naval Research Logistics Quarterly 3, 201-203

Jacobs, D.A., Silan, M.N., Clemson, B.A. (1996): An Analysis of Alternative Locations and Service Areas of American Red Cross Blood Facilities. Interfaces 26/3, 40-50

Jacobsen, S.K., Madsen, O.B.G. (1980): A Comparative Study of Heuristics for a Two-Level Routing-Location Problem. European Journal of Operational Research 5, 278-287

Johnson, S.M. (1954): Optimal Two- and Three-Stage Production Schedules with Setup Times Included. Naval Research Logistics Quarterly 1, 61-68

Johnson, T.J.R. (1967): An Algorithm for the Resource-Constrained Project Scheduling Problem. PhD Dissertation, M.I.T., Boston, MA

Kahneman, D., Tversky, A. (1979): Prospect Theory: An Analysis of Decisions Under Risk. Econometrica 47, 263-291

Kaku, B.K., Thompson, G.L., Morton, T.E. (1991): A Hybrid Heuristic for the Facilities Layout Problem. Computers & Operations Research 18, 241-253

Kariv, O., Hakimi, S.L. (1979a): An Algorithmic Approach to Network Location Problems. Part I: The p-Centers. SIAM Journal of Applied Mathematics 37, 513-538

Kariv, O., Hakimi, S.L. (1979b): An Algorithmic Approach to Network Location Problems. Part II: The p-Median. SIAM Journal of Applied Mathematics 37, 539-560

Karmarkar, N., Karp, R.M. (1982): The Differencing Method of Set Partitioning. Report UCB/CSD 82/113, Computer Science Division, University of California, Berkeley, CA

Karp, R.M. (1972): Reducibility Among Combinatorial Problems. In: Miller, R.E., Thatcher, J.W. (eds.): Complexity of Computer Computations. Plenum Press, New York, 85-103

Karp, R.M., Lenstra, J.K., McDiarmid, C.J.H., Rinnooy Kan, A.H.G. (1984): Probabilistic Analysis of Combinatorial Algorithms: An Annotated Bibliography. In: O'hEigeartaigh, M.D., Lenstra, J.K., Rinnooy Kan, A.H.G. (eds.): Combinatorial Optimization: Annotated Bibliographies. J. Wiley, Chichester, UK

Keeney, R.L. (1992): Value-Focused Thinking. Harvard University Press, Cambridge, MA

Keeney, R.L. (2002): Common Mistakes in Making Value Trade-Offs. Operations Research 50, 935-945

Keeney, R.L., Raiffa, H. (1976, 1993): Decisions with Multiple Objectives, Preferences and Value Tradeoffs. Cambridge University Press, Cambridge, MA

Kimes, S.W., Fitzsimmons, J.A. (1990): Selecting Profitable Hotel Sites at *La Quinta* Inns. Interfaces 20, 12-20

Kirca, O., Erkip, N. (1988): Selecting Transfer Station Locations for Large Solid Waste Systems. European Journal of Operational Research 35, 339-349

Kirkpatrick, S., Gelatti, C.D., Vecchi, M.P. (1983): Optimization by Simulated Annealing. Science 220, 671-680

Klamroth, K. (2002): Single-Facility Location Problems with Barriers. Springer-Verlag, New York

Köksalan, M., Süral, H., Kirca, Ö. (1995): A Location-Distribution Application for a Beer Company. European Journal of Operational Research 80, 16-24

Kolisch, R., Sprecher, A., Drexl, A., (1995): Characterization and Generation of a General Class of Resource-Constrained Project Scheduling Problems. Management Science 41, 1693-1703

Kolmogorov, A.N. (1933): Grundbegriffe der Wahrscheinlichkeitsrechnung. Berlin, Germany

Koopmans, T.C., Beckmann, M.J. (1957): Assignment Problems and the Location of Economic Activities. Econometrica 25, 53-76

Korpela, J., Tuominen, M. (1996): Benchmarking Logistics Performance with an Application of the Analytic Hierarchy Process. IEEE Transactions on Engineering Management 43, 323-332

Krause, K.L., Shen, V.Y., Schwetman, H.D. (1975): Analysis of Several Task-Scheduling Algorithms for a Model of Multiprocessing Computer Systems. Journal of the Association for Computing Machinery 22, 522-550

Kuby, M. (1987): Programming Models for Facility Dispersion: The p-Dispersion and Maxisum Dispersion Problems. Geographical Analysis 19, 315-329

Kuby, M. (1989): A Location-Allocation Model of Lösch's Central Place Theory: Testing a Uniform Lattice Network. Geographical Analysis 21, 316-37

Kuhn, H.W., Kuenne, R.E. (1962): An Efficient Algorithm for the Numerical Solution of the Generalized Weber Problem of Spatial Economics. Journal of Regional Science 4, 21-34

Kunde, M. (1976): Beste Schranken beim LP Scheduling. Bericht 7603, Institut für Informatik und Praktische Mathematik, Universität Kiel

Labetoulle, J., Lawler, E.L., Lenstra, J.K, Rinnooy Kan, A.H.G. (1984): Preemptive Scheduling of Uniform Machines Subject to Release Dates. In: Pulleyblank, W.R. (ed.): Progress in Combinatorial Optimization. Academic Press, New York

Lam, S., Sethi, R. (1977): Worst Case Analysis of Two Scheduling Algorithms. SIAM Journal on Computing 6, 518-536

Larichev, O.I., Olson, D.L. (2001): Multiple Criteria Analysis in Strategic Siting Problems. Kluwer Academic Publishers, Boston, MA

Lawler, E. L. (1963): The Quadratic Assignment Problem. Management Science 9, 586-599

Lawler, E.L. (1973): Optimal Sequencing of a Single-Machine Subject to Precedence Constraints. Management Science 19, 544-546

Lawler, E.L. (1977): A Pseudopolynomial Algorithm for Sequencing Jobs to Minimize Total Tardiness. Annals of Discrete Mathematics 1, 331-342

Lawler, E.L. (1982a): Recent Results in the Theory of Machine Scheduling. In: Bachem, A. Grötschel, M., Korte, B. (eds.): Mathematical Programming: The State of Art-Bonn 1982. Springer Verlag, Berlin, 202-234

Lawler, E.L. (1982b): Preemptive Scheduling of Precedence-Constrained Jobs on Parallel Machines. In: Dempster, M.A.H., Lenstra, J.K., Rinnooy Kan, A.H.G. (eds.): Deterministic and Stochastic Scheduling. Reidel, Dordrecht, 101-123
Lawler, E.L., Labetoulle, J. (1978): Preemptive Scheduling of Unrelated Parallel Processors by Linear Programming. Journal of the ACM 25, 612-619

Lawler, E.L., Lenstra, J.K., Rinnooy Kan, A.H.G. (1981,1982b): Minimizing Maximum Lateness in Two Machine Open Shop. Mathematics of Operations Research 6, 153-158. Erratum in Mathematics of Operations Research 7, 635

Lawler, E.L., Lenstra, J.K., Rinnooy Kan, A.H.G. (1982): Recent Developments in Deterministic Sequencing and Scheduling: A Survey. In: Dempster, M.A.H.,

Lenstra, J.K., Rinnooy Kan, A.H.G. (eds.): Deterministic and Stochastic Scheduling. Reidel, Dordrecht, 35-73

Lawler, E.L., Lenstra, J.K., Rinnooy Kan, A.H.G., Shmoys, D.B. (1993): Sequencing and Scheduling: Algorithms and Complexity. In: Graves, S.C., Rinnooy Kan, A.H.G., Zipkin, P.H. (eds.): Logistics of Production and Inventory. North-Holland, Amsterdam, 445-522

Lee, J. (1991): Analysis of Visibility Sites on Topographic Surfaces. International Journal of Geographical Information Systems 5, 413-429

Lee, R.C., Moore, J.M. (1967): CORELAP-Computerized Relationship Layout Planning. Journal of Industrial Engineering 18, 194-200

Lee, S. (1972): Goal Programming for Decision Analysis. Auerbach, Philadelphia, PA

Lenstra, J.K. (1977): Sequencing by Enumerative Methods. Mathematisch Centrum, Amsterdam

Lenstra, J.K., Rinnooy Kan, A.H.G. (1978): Complexity of Scheduling under Precedence Constraints. Operations Research 26, 22-35

Lenstra, J.K., Rinnooy Kan, A.H.G., Brucker, P. (1977): Complexity of Machine Scheduling Problems. Annals of Discrete Mathematics 1, 343-362

Lenstra, J.K., Rinnooy Kan, A.H.G., Brucker, P. (1984): Complexity of Machine Scheduling Problems. Annals of Discrete Mathematics 1, 343-362

Leung, J.Y.-T., Young, G.H. (1989): Minimizing Total Tardiness on a Single Machine with Precedence Constraints. Technical Report, Computer Science Program, University of Texas, Dallas

Li, Y., Pardalos, P., Ramakrishnan, K., Resende, M.G.C. (1994): Lower Bounds for the Quadratic Assignment Problem. Annals of Operations Research 50, 387-410

Ligget, R.S. (1981): The Quadratic Assignment Problem: An Experimental Evaluation of Solution Strategies. Management Science 27, 442-458

Lin, S. (1965): Computer Solution of the Traveling Salesman Problem. Bell System Technical Journal 44, 2245-2269

Little, J.D.C., Murty, K.G., Sweeney, D.W., Karel, C. (1963): An Algorithm for the Traveling Salesman Problem. Operations Research 11, 972-989

Liu, C.L., Layland, J.W. (1973): Scheduling Algorithms for Multiprogramming in a Hard Real-Time Environment. Journal of the ACM 20/1, 46-61

Liu, J.W.S., Liu, C.L. (1974a): Bounds on Scheduling Algorithms for Heterogeneous Computing Systems. In: Rosenfeld, J.L. (ed.): Information Processing 74, North-Holland, Amsterdam, 349-353

Liu, J.W.S., Liu, C.L. (1974b): Bounds on Scheduling Algorithms for Heterogeneous Computing Systems. Technical Report UIUCDCS-R-74-632, Department of Computer Science, University of Illinois at Urbana-Champaign, IL

Liu, J.W.S., Liu, C.L. (1974c): Performance Analysis of Heterogeneous Multi-Processor Computing Systems. In: Gelenbe, E., Mahl, R. (eds.): Computer Architecture and Networks, North-Holland, Amsterdam, 331-343

Lootsma, F.A. (1988): Numerical Scaling of Human Judgment in Pairwise-Comparison Methods for Fuzzy Multi-Criteria Decision Analysis. Mathematical Methods for Decision Support, NATO ASI Series F, Computer and System Sciences, Springer-Verlag, Berlin, Germany, vol. 48, 57-88

Lootsma, F.A. (1990): The French and the American School in Multi-Criteria Decision Analysis. Recherche Opérationnelle/Operations Research 24, 263-285

Love, R.F., Morris, J.G. (1979): Mathematical Distances of Road Travel Distances. Management Science 25, 130-139

Love, R.F., Morris, J.G., Wesolowsky, G.O. (1988): Facilities Location: Models and Methods. North-Holland Publishing Company, New York-Amsterdam-London

Love, R.F., Truscott, J.H., Walker, J.H. (1985): Terminal Location Problem: a Case Study Supporting the Status Quo. Journal of the Operational Research Society 36, 131-136

Love, R.F., Yerex, L. (1976): Application of a Facilities Location Model in the Prestressed Concrete Industry. Interfaces 6, 45-49

Luce, R.D., Raiffa, H. (1957): Games and Decisions, Wiley, New York

Manser. M.H. (1988): Dictionary of Eponyms. Penguin London, UK

Maranzana, F (1964): On the Location of Supply Points to Minimize Transport Costs. Operational Research Quarterly 15, 261-270

Mareschal, B., Brans, J.-P. (1988): Geometrical Representations for MCDA. European Journal of Operational Research 34, 69-77

Marsh, M.T., Schilling, D.A. (1991): A Comparison of Equity Measures in Facility Siting Decisions. In: Proceedings of the First International Decision Sciences Institute conference, 244-247

Marsh, M.T., Schilling, D.A. (1994): Equity Measurement in Facility Location Analysis: a Review and Framework. European Journal of Operational Research 74, 1-17

Mc Naughton, R. (1959): Scheduling with Deadlines and Loss Functions. Management Science 6, 1-12

McHarg, I. (1969) Design with Nature. The Natural History Press, Garden City, New York

Mesa, J.A., Boffey, T.B. (1996): A Review of Extensive Facility Location in Networks. European Journal of Operational Research 95, 592-603

Miehle, W. (1958): Link-Length Minimization in Networks. Operations Research 6, 232-243

Mingozzi, A., Maniezzo, V., Ricciardelli, S., Bianco, L., (1998): An Exact Algorithm for the Resource-Constrained Project Scheduling Based on a New Mathematical Formulation. Management Science 44, 714-729

Minieka, E. (1977): The Centers and Medians of a Graph. Operations Research 25, 641-650

Minieka, E. (1981): A Polynomial Time Algorithm for Finding the Absolute Center of a Network. Networks 11, 351-355

Moder, J.J., Phillips, C.R., Davis, E.W. (1983): Project Management with CPM, PERT, and Precedence Diagramming (3rd ed.) Van Nostrand Reinhold, New York

Möhring, R.H., Radermacher, F.J., Weiss, G. (1984): Stochastic Scheduling Problems 1: General Strategies. Zeitschrift für Operations Research 28, 193-260

Möhring, R.H., Radermacher, F.J., Weiss, G. (1985): Stochastic Scheduling Problems II: Set Strategies. Zeitschrift für Operations Research 29, 65-104

Morris, J.G. (1978): On the Extent to Which Certain Fixed Charges Depot Location Can Be Solved by Linear Programming. Journal of the Operational Research Society 29, 71-76

Morris, P. (1994): Introduction to Game Theory. Springer-Verlag, New York

Müller-Merbach, H. (1970): Optimale Reihenfolgen. Springer-Verlag, Berlin

Muntz, R.R., Coffman, E.G., Jr. (1970): Preemptive Scheduling of Real Time Tasks on Multiprocessor Systems. Journal of the ACM 17, 324-338

Muntz, R.R., Coffman, E.G., Jr., (1969): Optimal Preemptive Scheduling on Two-Processor Systems. IEEE Transactions on Computers, C-18, 1014-1029

Murray, A.T., Gottsegen, J.M. (1997): The Influence of Data Aggregation on the Stability of p-Median Model Solutions. Geographical Analysis 29, 200-213

Muther, R. (1961): Systematic Layout Planning. Industrial Education Institute, Boston, MA

Nahmias, S. (1989): Production and Operations Analysis. Irwin, Boston, MA

Nambiar, J.M., Gelders, L.F., van Wassenhove, L.N. (1981): A Large-Scale Location-Allocation Problem in the Natural Rubber Industry. European Journal of Operational Research 6, 183-189

Nambiar, J.M., Gelders, L.F., van Wassenhove, L.N. (1989): Plant Location and Vehicle Routing in the Malaysian Rubber Smallholder Sector: a Case Study. European Journal of Operational Research 38 14-26

Narula, S.C. (1984): Hierarchical Location-Allocation Problems: A Classification Scheme. European Journal of Operational Research 15, 93-99

Nash, J. (1950): Equilibrium Points in N-Person Games. Proceedings of the National Academy of Sciences 36, 48-49

Neebe, A. (1978): A Branch and Bound Algorithm for the p-Median Transportation Problem. Journal of the Operational Research Society 29, 989-995

Nemhauser, G.L., Wolsey, L.A. (1988): Integer and Combinatorial Optimization. J. Wiley & Sons, New York

Neumann, J. von, Morgenstern, O. (1944): Theory of Games and Economic Behavior. Princeton University Press, Princeton, NJ

Niemi, R.G., Weisberg, H.F. (eds.) (1976): Controversies in American Voting Behavior. W.H. Freeman & Son, San Francisco, CA

Nowicki, E., Zdrzalka, S (1986): A Note on Minimizing Maximum Lateness in a One-Machine Sequencing Problem with Release Dates. European Journal of Operational Research 23, 266-267

O'Kelly, M.E. (1986): The Location of Interacting Hub Facilities. Transportation Science 20, 92-106

Okabe, A., Boots, B., Sugihara, K. (1992): Spatial Tessellations Concepts and Applications of Voronoi Diagrams. John Wiley & Sons, New York, N.Y

Okabe, A., Suzuki, A. (1987): Stability of Spatial Competition for a Large Number of Firms on a Bounded Two-Dimensional Space. Environment and Planning A 16, 107-114

Okabe, A., Suzuki, A. (1997): Locational Optimization Problems Solved Through Voronoi Diagrams. European Journal of Operational Research 98, 445-456

Olson, D.L. (1996): Decision Aids for Selection Problems. Springer-Verlag, New York

Opricovic, S. (1998) Multicriteria Optimization in Civil Engineering. Faculty of Civil Engineering, Belgrade

Ostresh, L.M. Jr. (1978): On the Convergence of a Class of Iterative Methods for the Weber Problem Using Generalized Distance Functions. Operations Research 26, 597-609

Ovacik, I., Uzsoy, R. (1997): Decomposition Methods in Large Scale Job Shops. Kluwer, Boston
Owen, G. (1982): Game Theory, 2nd ed. Academic Press, New York

Pardalos, P.M., Wolkowicz, H. (eds.) (1994): Quadratic Assignment and Related Problems. Volume 16 in the DIMACS Series in Discrete Mathematics and Theoretical Computer Science. American Mathematical Society, Providence RI

Patterson, J.H., Slowinski, R., Talbot, F.B., Weglarz, J. (1989): An Algorithm for a General Class of Precedence and Resource Constrained Scheduling Problems. In: Slowinski, R. Weglarz, J. (eds.): Advances in Project Scheduling. Elsevier Science, Amsterdam, 3-28

PERT: Program Evaluation Research Task (1958): Phase I Summary Report. Special Projects Office, Bureau of Ordnance, Department of the US Navy, Washington, DC

Pigou, A.C. (1912): Wealth and Welfare. MacMillan, London

Pinedo, M. (1995): Scheduling. Theory, Algorithms and Systems. Prentice Hall, Englewood Cliffs, NJ

Pinedo, M., Chao, X. (1999): Operations Scheduling with Applications in Manufacturing and Services. McGraw-Hill, New York

Pinson, E. (1995): The Job Shop Scheduling Problem: A Concise Survey and Some Recent Developments. In: Chrétienne, Ph., Coffman, E.G. Jr., Lenstra, J.K., Liu, Z. (eds.): Scheduling Theory and its Applications. J. Wiley, New York, NY, 177-293

Plastria, F. (1995): Continuous Location problems. In: Drezner, Z (ed.): Facility Location: A Survey of Applications and Methods. Springer-Verlag, New York-Berlin-Heidelberg, 225-262

Pollatschek, M.A., Gershoni, N., Radday, Y.T. (1976): Optimization of the Typewriter Keyboard by Computer Simulation. Angewandte Informatik 10, 438-439

Prescott, E.C., Visscher, M. (1977): Sequential Location Among Firms with Foresight. Bell Journal of Economics 8, 378-393

Psaraftis, H.N., Tharakau, C.G., Ceder, A. (1986): Optimal Response to Oil Spills: the Strategic Decision Case. Operations Research 34, 203-217

Rawls, J. (1971): A Theory of Justice. Harvard University Press, Cambridge, MA

Rendl, F. (2002): The Quadratic Assignment Problem. In: Drezner, Z., Hamacher, H.W. (eds.): Facility Location: Applications and Theory. Springer-Verlag, Berlin, 439-457

Resende, M.G.C., Ramakrishnan, K.G., Drezner, Z. (1995): Computing Lower Bounds for the Quadratic Assignment Problem with an Interior Point Algorithm for Linear Programming. Operations Research 43, 786-791

ReVelle, C.S. (1986): The Maximum Capture or Sphere of Influence Problem: Hotelling Revisited on a Network. Journal of Regional Science 26, 343-357

ReVelle, C.S., Bigman, D., Schilling, D., Cohon, J., Church, R.L. (1977): Facility Location: A Review of Context Free and EMS Models. Health Services Research 12, 129-146

ReVelle, C.S., Hogan, K. (1989): The Maximum Availability Location Problem. Transportation Science 23, 192-200

ReVelle, C.S., Marianov, V. (1991): A Probabilistic FLEET Model with Individual Vehicle Reliability Requirements. European Journal of Operational Research 53, 93-105

ReVelle, C.S., Swain, R. (1970): Central Facilities Location. Geographical Analysis 2, 30-42

Righter, R. (1994): Stochastic Scheduling. In: Shaked, M., Shantikumar, G. (eds.): Stochastic Orders. Academic Press, San Diego, CA

Rinnooy Kan, A.H.G. (1976): Machine Scheduling Problems: Classification, Complexity and Computations. Nijhoff, The Hague

Rinnooy Kan, A.H.G. (1987): Probabilistic Analysis of Algorithms. Annals of Discrete Mathematics 31, 365-384

Röck, H., Schmidt, G. (1982): Machine Aggregation Heuristics in Shop Scheduling. Bericht 82-11, Fachbereich 20 Informatik, Technische Universität Berlin

Rosing, K.E. (1992): An Optimal Method for Solving the (Generalized) Multi-Weber Problem. European Journal of Operational Research 58, 414-426

Rosing, K.E., Harris, B. (1990): Algorithmic and Technical Improvements: Optimal Solutions to the (Generalized) Multi-Weber Problem. E.G.I. Discussiestukken Nr. 90-10, Economisch Geografisch Instituut, Erasmus Universiteit, Rotterdam, Nederland

Rosing, K.E., ReVelle, C.S. (1997): Heuristic Concentration: Two Stage Solution Construction. European Journal of Operational Research 97, 75-86

Rosing, K.E., ReVelle, C.S., Rosing-Vogelaar, H. (1979): The p-Median and its Linear Programming Relaxation: An Approach to Large Problems. Journal of the Operational Research Society 30, 815-823

Rothkopf, M.H. (1966): Scheduling Independent Tasks on Parallel Processors. Management Science 12, 347-447

Roucairol, C. (1987): A Parallel Branch and Bound Algorithm for the Quadratic Assignment Problem. Discrete Applied Mathematics 18, 211-225

Roy, B. (1971): Problems and Methods with Multiple Objective Functions. Mathematical Programming 1, 280-283

Roy, B. (1978): ELECTRE III: Un algorithme de classement fonde sur une representation floue des preferences en presence de critères multiples. Cahiers des Centre d'Etudes Recherche Opérationnelle 20, 3-24

Roy, B., Sussman, B. (1964): Les problèmes d'ordonnancement avec contraintes disjunctives. Note DS no 9 bis, SEMA, Montrouge

Saaty, T.L. (1980): The Analytic Hierarchy Process. Mc Graw-Hill, New York

Saaty, T.L. (1994): How to Make a Decision. Interfaces 24/6, 19-43

Salewski, F., Schirmer, A., Drexl, A. (1997): Project Scheduling Under Resource and Mode Identity Constraints: Model, Complexity, Methods, and Applications. European Journal of Operational Research 102, 88-110

Sankaran, J.K., Srinivasa Raghavan, N.R. (1997): Locating and Sizing Plants for bottling Propane in South India. Interfaces 27/6, 1-15

Schilling, D., ReVelle, C., Cohon, J., Elzinga, J. (1980): Some Models for Fire Protection Locational Decisions. European Journal of Operational Research 5, 1-7

Schmenner, R.W. (1982): Making Business Location Decisions. Prentice Hall, Englewood Cliffs, NJ

Schniederjans, M.J., Kwak, N.K., Helmer, M.C. (1982): An Application of Goal Programming to Resolve a Site Location Problem. Interfaces 12, 65-72

Seehof, J.M., Evans, W.O. (1967): Automated Layout Design Program. Journal of Industrial Engineering 18, 690-695

Selten, R. (1965): Spieltheoretische Behandlung eines Oligopolmodels mit Nachfrageträgheit. Zeitschrift für die gesamte Staatswissenschaft 12, 301-324

Sethi, R. (1976): Scheduling Graphs on Two Processors. SIAM Journal for Computing 5, 73-82

Sethi, R. (1977): On the Complexity of Mean Flow Time Scheduling. Mathematics of Operations Research 2, 320-330

Shamos, M.I., Hoey, D. (1975): Closest Point Problems. Proceedings of the 16th Annual Symposium on Foundations of Computer Science 151-162

Shapley, L.S. (1953): A Value for n-Person Games. In: Contributions to the Theory of Games II, Annals of Mathematic Studies 28, Princeton University Press, 307-317

Sidney, J.B. (1975): Decomposition Algorithms for Single-Machine Sequencing with Precedence Relations and Deferral Costs. Operations Research 23, 283-298

Silver, E.A., Vidal, R.V., DeWerra, D. (1980): A Tutorial on Heuristic Methods. European Journal of Operational Research 5, 153-162

Simon, H.A. (1979): Rational Decision Making in Business Organizations. The American Economic Review 69, 493-513

Simon, H.A. (1982): Models of Bounded Rationality. MIT Press, Cambridge, MA

Sitters, R.A. (2001): Two NP-hardness Results for Preemptive Minsum Scheduling of Unrelated Parallel Machines. Proceedings of the 8th International IPCO Conference, Lecture Notes in Computer Science, vol. 2081, 396-405

Skorin-Kapov, J. (1990): Tabu Search Applied to the Quadratic Assignment Problem. ORSA Journal of Computing 2, 33-45

Slominski, L. (1982): Probabilistic Analysis of Combinatorial Algorithms: A Bibliography with selected Annotation. Computing 28, 257-267

Slowinski, R. (1978): Scheduling Preemptible Tasks on Unrelated Processors with Additional Resources to Minimize Schedule Length. In: Bracchi, G., Lockemann, P.C. (eds.): Lecture Notes in Computer Science 65. Springer-Verlag, Berlin, 536-547

Slowinski, R. (1981): Multiobjective Network Scheduling with Efficient Use of Renewable and Nonrenewable Resources. European Journal of Operational Research 7, 265-273

Smith, W.E. (1956): Various Optimizers for Single Stage Production. Naval Research Logistics Quarterly 3, 59-66

Sprecher, A. (1999): Solving the RCPSP Efficiently at Modest Memory Requirements. Management Science, (to appear).

Sprecher, A., Hartmann, S., Drexl, A. (1997): An Exact Algorithm for Project Scheduling With Multiple Modes. OR Spektrum 19, 195-203

Stackelberg, H. von (1943): Grundlagen der theoretischen Volkswirtschaftslehre (translated as The Theory of the Market Economy). W. Hodge, London

Starr, M.K. (1962): Product Design and Decision Theory. Prentice-Hall, Englewood Cliffs, NJ

Starr, M.K., Zeleny, M. (1977): State and Future of the Arts. In: Starr, M.K., Zeleny, M. (eds.): Multiple Criteria Decision Making. Volume 6 in TIMS Studies in the Management Sciences, North-Holland, Amsterdam, 5-30

Steinberg, L. (1961): The Backboard Wiring Problem: A Placement Algorithm. SIAM Review 3, 37-50

Steiner, J. (1835): Aufgaben und Lehrsätze. Journal für die reine und angewandte Mathematik 13, 361-364

Stinson, J.P., Davis, E.W., Khumawala, B.M. (1978): Multiple Resource-Constrained Scheduling Using Branch and Bound. AIIE Transactions 10, 252-259

Teitz, M.B., Bart, P. (1968): Heuristic Methods for Estimating the Generalized Vertex Median of a Weighted Graph. Operations Research 16, 955-961

Thünen, J.H. von (1826): Der isolirte Staat in Beziehung auf Landwirthschaft und Nationalökonomie. Hamburg. Translated by Wartenberg, C.M. and edited with an introduction by Peter Hall. [1st. ed.] Pergamon Press, Oxford, New York

Tompkins, J.A., Reed, R. (1976): An Applied Model for the Facilities Design Problem. International Journal of Production Research 14, 583-595

Toregas, C., ReVelle, C.S. (1973): Binary Logic Solutions to a Class of Location Problem. Geographical Analysis 5, 145-155

Toregas, C.; Swain, C.; ReVelle, C.; Bergmann, L. (1971): The Location of Emergency Service Facilities. Operations Research 19, 1363-1373

Tzeng, Gwo-Hshiung, Teng, Mei-Hwa, Chen, June-Jye, Opricovic, S. (2002) Multicriteria Selection for a Restaurant Location in Teipei. International Journal of Hospitality Management 21, 171-187

Ullman, J.D. (1976): Complexity of Sequencing Problems. Chapter 4 in E.G. Coffman, Jr., ed.: Computer and Job/Shop Scheduling Theory. Wiley, New York

Vaessens, R.J.M., Aarts, E.H.L., Lenstra, J.K. (1996): Job Shop Scheduling by Local Search. INFORMS Journal on Computing 8, 302-317

Van der Veen, J.A.A., Venugopal, V., (2000): Win-Win Situations in Supply-Chain Partnerships: a Tutorial. OR Insight 13, 22-28

Vazsonyi, A. (2002): Which Door has the Cadillac. Writers Club Press, Lincoln, Nebraska.

Vergin, R.C., Rodgers, J.D. (1967): An Algorithm and Computational Procedure for Locating Economic Facilities. Management Science 13, B-240-254

Walker, M.R., Sayer, J.S., (1959): Project Planning and Scheduling. Report 6959, E.I. du Pont Nemours & Co., Wilmington, Delaware

Weaver, J.R., Church, R.L. (1983): A Comparison of Solution Procedures for Covering Location Problems. Proceedings of the Fourteenth Annual Pittsburgh Conference on Modeling and Simulation, 14 (part 4), 1417-1422

Weaver, J.R., Church, R.L. (1985): A Median Location Model with Nonclosest Facility Service. Transportation Science 19, 58-74

Weiszfeld, E.V. (1937): Sur le point pour lequel la somme des distances de *n* points donnés est minimum. The Tohuku Mathematical Journal 43, 836-843

Weber, A. (1909): Über den Standort der Industrien. 1. Teil: Reine Theorie des Standortes. Tübingen. Translated as: On the Location of Industries. University of Chicago Press, Chicago, IL, 1929

Wersan, S.J., Quon, J.E., Charnes, A. (1962): Systems Analysis of Refuse Collection and Disposable Practices. American Public Works Association Yearbook, 195-211

Wesolowsky, G.O. (1973): Location in Continuous Space. Geographical Analysis 5, 95-112

Wesolowsky, G.O. (1993): The Weber Problem: History and Procedures. Location Science 1, 5-24

Wild, R. (1979): Production and Operations Management. Holt, Rinehart and Winston, London, UK

Wirasinghe, S.C., Waters, N.M. (1983): An Approximate Procedure for Determining the Number, Capacities and Locations of Solid Waste Transfer Stations in an Urban Region. European Journal of Operational Research 12, 105-111

Zeleny, M. (1974): Linear Multiobjective Programming. Volume 95 in: Lecture Notes in Economics and Mathematical Systems. Springer-Verlag, Berlin-Heidelberg-New York

Zeleny, M. (1981): A Case Study in Multiobjective Design: De Novo Programming. In: Nijkamp, P., Spronk, J. (eds.): Multiple Criteria Analysis: Operational Methods. Gower Publishing, Hampshire

Zeleny, M. (1986): Optimal System Design with Multiple Criteria: De Novo Programming approach. Engineering Costs and Production Economics 10, 89-94

Zeleny, M. (1995): Trade-Offs-Free Management via De Novo Programming. International Journal of Operations & Quantitative Management 1, 3-13

Zelinka, B. (1968): Medians and Peripherians of Trees. Archives of Mathematics, Brno 87-95

SUBJECT INDEX

A

activity-on-arc (*AOA*) representation 299
activity-on-node (*AON*) representation 299
agglomeration, central 251
ALDEP 293
algorithm 3
alpha-reliability coverage 177
"alternate" method 235
analytic hierarchy process (*AHP*) 68-70
anti-median 238-242
antithetical schedule 359-360
arc, conjunctive 405
arc, disjunctive 405
arms race 138
attraction functions 161

B

backward sweep 303, 306
Banzhaf power index 149-150
barriers (to transportation) 254
battle of the sexes 136-137
Bayes (expected value) criterion 90-93
Bayes's theorem 17
bimatrix games 133-140
block plan 259-260
brand positioning problem 166
Brucker's algorithm 394-396

C

capacitated plant location problem (*CPLP*) 207
capture problem 250-252
cellular layout 257
centdian problems 189
center problems 178-188
center, absolute 179, 180-185
center, general 178
center, general absolute 179
center, node 178, 186
centroid 251-252
certainty equivalent (*CE*) 76
chance nodes 81-84
characteristic function 143
circle intersect point set (*CIPS*) 212-213
circuit 10
clustering heuristic 233-234
coalition 143
coefficient of pessimism 88
COFAD 293
Coffman and Graham's algorithm 377-379
Colonel Blotto game 132-133
column, essential 173
communication design problem 214
competitive region 249
completion time 344
complexity 4-5
compromise ranking algorithm (*VIKOR*) 54
composite objective function 28
concordance matrix 50-53
consequence nodes 81-82
consistency 62-63, 66-70
constant-sum game 112

constraint method 28
construction heuristics 290-291
core (of a game) 144, 147
CORELAP 293-294
corridor location problems 252
counter coalition 143
covering problems 170-178, 212-214
CRAFT 292-293
crashing 309-313
critical path method (*CPM*) 302-309
customer choice models 161
cycle (of a graph) 10
cycle property 269
cyclic permutations 268

D

data envelopment analysis (*DEA*) 40-43
decision nodes 81-84
decision trees 83-84
demand allocation 161
demand, elastic 160
demand, inelastic 160
determinant of a matrix 6-9
discordance matrix 51-53
dispersion problems 243
distance functions 37, 158
dominance 75, 122, 173
due date 342
dummy activities 300-301

E

earliest due date (*EDD*) scheduling method 360-361
Edwards procedure 31
efficient frontier 26
eigenvalue 7
eigenvalue bounds 280-281
eigenvector 7
ELECTRE 50-54
Elzinga-Hearn algorithm 219-220
equity objectives 244-246
evaluation matrix 30
event time, earliest 303
event time, latest 303

expected value of perfect information (*EVPI*) 102
expected value of sample information (*EVSI*) 109

F

facilities, extensive 252-254
facilities, undesirable 237-242
feature space 166
first fit decreasing heuristic (*FFD*) 417-419
first fit heuristic (*FF*) 417-419
fixed layout problem 257
float, free (*FF*) 305, 307
float, interference (*IF*) 305, 307
float, safety (*SF*) 305, 307
float, total (*TF*) 303-305, 307
flow shop scheduling 341, 401-402
flow time 344
Floyd-Warshall algorithm 10
folding algorithm 193
forbidden areas 253
forest 10
forward sweep 303, 306
from-to charts 259

G

GAIA plane 60-61
game of chicken 135-136
game, fair 132
games in extensive form 114-115
games in normal form 113-114
game, mixed extension 127-133
game, sequential 113
game, simultaneous 113
game of chicken 135-136
Gantt charts 320-321, 336
Gary and Johnson's algorithm 415-417
Gavett and Plyter's branch and bound method 283-290
geometric mean method 70-71
Gilmore-Lawler bound (*GLB*) 278-290
Gini index 244-246
goal programming 40
Gonzalez algorithm 390-392

Graham anomalies 370-372
grand coalition 143
graph 9-10
graph, disjunctive 405
Greedy algorithm 3-4
Greedy algorithm for p-median problems 201-202
group layout 257

H

Hakimi property 189
heuristic concentration 205
hinterland 249
Hodges-Lehmann rule 93-94
Horvath, Lam, and Sethi algorithm 384-385
hub-and-spoke system 164
hub location problems 164, 247-248
Hurwicz criterion 88
Hu's level algorithm 375-376
hyperboloid approximation 227

I

ideal point 37
imperfect information 101-107
improvement cone 25
improvement heuristics 291
imputation 144
indicator probabilities
influence diagrams 81-82
information, imperfect 113
information, incomplete 113
information set 115
insurance premium 81
integer programming problem (IP) 12-13
in-tree 357-360, 375-376, 394-396
iterated dominance 116, 122-124, 173

J

Jackson's algorithm 404-405
job shop 341
Johnson's rule 401-402

K

kernel (of a graph) 52

L

Lagrangean relaxation 13
Lagrangean relaxation for p-median problems 196-201
Laplace rule 93
largest empty circle 242
lateness 344
layout models 255-294
leaf 10
Likert scale 30
linear programming problem (LP) 11
list scheduling 370
location-allocation heuristic 203-204
location, hierarchical 162
location models, conditional 163-164
location models, continuous 157, 211-236
location models, discrete 157
location models on networks 157
location problem, competitive 248-251
location set covering problem 171-175, 212-213
longest alternate processing time algorithm ($LAPT$) 400-401
longest processing time first (LPT) algorithm 372-374
Lorenz curve 244-246
$LSCP$ reduction algorithm 173-175

M

machines, dedicated 341
machines, identical 341
machines, parallel 341
machines, uniform 341
machines, unrelated 341
makespan 344
market capture 167
matching problems 269-270, 415-416
matrix 5-9
maxian 238-242

maximal covering location problem 175-178, 212
maximax rule 88
maximin rule 87
McNaughton's algorithm 368-369
median problems 188-205
medianoid 251-252
multiattribute value functions 46-50
multipurpose shopping 164
multi-stage games 140-143
Multi-(facility) Weber problem 230-235
Muntz and Coffman's algorithm 379-381

N

Nash equilibrium 118-121, 250-251
network 9-10
network, uniconnected 347
newspaper boy problem 90
node property 189
normalization technique 64-66
n-person games 143-150

O

open shop scheduling 341, 399-401
opportunity loss 91, 102
optimum, competitive 251
optimum, social 251
outranking relation 50-61
out-tree 357-360, 376

P

parallel scheduling schemes 329-331
Pareto-optimality 26
partial information 76
PARTITION 367-368
path 10
payoff, anticipated 76
perfect information 101-103
performance ratio, absolute 351
performance ratio, asymptotic 351
permutation matrix 265

PLANET 294
precedence trees 357-360
preference cones 43-46
price search 164
principle of insufficient reason 93
principle of minimum differentiation 251
principle of transfers 246
priority list 328
Prisoner's dilemma 134-135
probability 13-17
probability, indicator 104-109
probability, posterior 104-109
probability, prior 104-109
problem instance 3
process layout 257
processor shared schedule 380-381
processor sharing 383
production-line layout 257
project acceleration 309-313
project evaluation and review technique (*PERT*) 313-317
PROMETHEE 54-61

Q

quadratic assignment problem (*QAP*) 260-292
query problem 154
quota 148

R

random index (*RI*) 67-68
rank reversal 63
reaction function 116-118
reaction graph 119-120
ready time 342
recirculation 403
reference point methods 37-40
regret criterion 89-90
relationship diagram 256-260
resource limits 415
resource types 415
resource requirements 415
risk aversion 76-80
risk premium (*RP*) 76

S

saddle point 125
sample information 101-109
satisfice 19
Savage-Niehans rule 89-90
schedule, active 407-409
scheduling, nonpreemptive 342-343
scheduling, preemptive 342-343
schedule length 344
selection, complete 405
selection problem 19
serial scheduling schemes 328-328
Shapley value 144,148
shifting bottleneck heuristic 411-414
shortest processing time algorithm (SPT) 354-356
shortest remaining processing time algorithm ($SRPT$) 356, 390-392
simple plant location problem 205-207
simplex 9, 89-99
single assignment method 281-290
Smith's ratio rule 356-357
social optimum 251
solution method, exact 3
solution method, heuristic 3
solution, nondominated 26
space relationship diagrams 257-260
St. Petersburg paradox 80
Stackelberg game 116-118
Starr's domain criterion 96-99
states of nature 74
Steiner-Weber problem 220-228
strategies, mixed 127-133
strategies, pure 132
subgame perfect equilibrium 120-121
supply chain optimization 139-140
systematic layout planning (SLP) 256

T

tardiness 344
task level 375
tax game 137-138
TOPSIS method 38-39
tornado diagrams 83, 85-87, 91-93
transportation problem 11
traveling salesman problem 268-269
tree 10

triangulation problems 267
two-person zero-sum games 122-133

U

unlocation problems 164

V

value tree 34-35
value of the game 125
Varignon frame 227-228
vector 5
vector optimization problem 24-29
vertex substitution method 204-205, 235-236
Voronoi diagram 167, 218-219
voting games 148-150

W

Wald criterion 87
Weber problems 220-235
weighted majority game 148-150
weighting method 28
weighted shortest processing time algorithm (*WSPT*) 356-358
Weiszfeld method 225-227

Z

zero-sum game 112

Druck: Strauss Offsetdruck, Mörlenbach
Verarbeitung: Schäffer, Grünstadt